READINGS IN CANADIAN GEOGRAPHY

Revised Edition

Edited by Robert M. Irving, Ph.D.

Holt, Rinehart and Winston of Canada, Limited
Toronto   Montreal

Distributed in the United States of America by Mine Publications, Inc.

Readings in Canadian Geography, Revised Edition

SENIOR EDITOR:
ROBERT M. IRVING, PH.D.
Department of Geography
University of Waterloo

CONSULTING EDITOR:
RALPH R. KRUEGER, PH.D.
Department of Geography
University of Waterloo

ISBN 0-03-925493-3

Distributed in the United States of America by Mine Publications, Inc.,
25 Groveland Terrace, Minneapolis, Minnesota 55403.

Printed in Canada

1 2 3 4 5   76 75 74 73 72

Preface

Teachers of Canadian geography are finding it increasingly difficult to make adequate reading material available to the rapidly growing number of students. It becomes a lesson in frustration to both the university teacher and student to assign readings that are available in limited quantity only. As enrolment increases and geography classes become larger, the problem will become more acute. Teachers in secondary schools are more seriously hindered, since often they do not even have access to the basic geographical journals. Although the impetus for this volume came primarily from the problems of making geographical literature available to students at the University of Waterloo, similar sentiments have been expressed by university colleagues across Canada and by many secondary school teachers who are teaching a course on the geography of Canada.

The goal, therefore, has been to assemble a book of readings that will serve to enrich the teaching of geography courses on Canada. The articles have been selected on the basis of content, rather than on the nature and quality of the research methodology. It will also be noted that the level of presentation ranges from articles that will be readily understood by secondary school students to those that will challenge university undergraduates.

As is indicated by the table of contents, I have focussed not on one theme, but rather, have covered a number of topics on Canadian geography. The introductory paper stands alone as an interpretation of Canada by Andrew H. Clark, who has contributed much to the geographical literature of Canada and has been a longtime observer of the Canadian scene. The core of the volume is contained in four sections: Population and Settlement; Canadian Cities; Agriculture; and Resource Development. Here the focus is on analyzing patterns, changes, and problems. The concluding section is a summary statement by the Economic Council of Canada of the regional disparities in the economic growth of Canada.

The articles have been selected from a wide range of publications reflecting the research of not only geographers, but also economists, demographers, anthropologists, resource specialists, and urban and regional planners. Most of the articles have been published since the late 1950's. Even though some of the statistical data are now out-of-date, the papers are still valid in terms of the concepts and principles they illustrate. Many excellent articles have been excluded owing to space limitations. In many cases that final choice was a difficult one, and I recognize that other geographers would quite possibly have made different selections. Several other articles have not been included because they have already been published in other anthologies. My intent has been to make the volume complementary to other books of readings on Canadian themes. It should also be noted that articles dealing with the physical geography of Canada have been excluded. Indeed, there are many papers that would serve as the basis for an additional book dealing with this theme alone.

Most of the articles appear here as they did in the original publication. A few have been moderately edited and have been checked by their authors prior to publication. Some of the authors have taken this opportunity to expand or revise their texts from the original and to make minor corrections.

Except where noted, the translation of French-language articles has been done by the editor. I am grateful to my French-speaking colleagues who read and re-read the translations. Unfortunately, in the process of translation, it was not possible to capture their style of expression. For the initial translation of Professor Ehler's article I am indebted to Miss Gail Morley of the University of Waterloo.

In the preparation of this volume I sought and received suggestions from several colleagues in different Canadian universities. Chief among these is Professor Ralph R. Krueger, University of Waterloo, who has been associated with the preparation of this volume from the time the initial idea was proposed by him in 1965. He has assisted in the evaluation and selection of many of the articles, has offered advice as to the organization and format of the volume, and has commented critically on the editorial introductions.

Professor Robert A. Murdie, University of Waterloo, has also read and criticized parts of the manuscript, and has sharpened my perspective at various stages in the preparation of the volume. I am indebted to Professor Peter B. Clibbon, Institut de Géographie, Université Laval, whose comments on the French-language literature were particularly appreciated. I also extend recognition to Professor Hans Stolle and Mr. Alan Hildebrand of the Department of Geography and Planning, University of Waterloo, for redrafting almost all of the maps and graphs for use in this volume.

I extend my appreciation to all of the authors, editors and publishers who granted me permission to reproduce their articles in this volume.

Although I have received valuable assistance and constructive criticisms from numerous colleagues, particularly on organization and content, during the preparation of this book, I accept full responsibility for the final product.

University of Waterloo   ROBERT M. IRVING

May, 1968.

Preface To The Second Edition

Since 1967, when the manuscript for the first edition of *Readings in Canadian Geography* was completed, there have been many changes in Canada and there has been a considerable amount of new literature published on themes relevant to geographers and others interested in Canadian development. In this, the second edition, I have attempted to reflect some of these factors and, simultaneously, to achieve a better balance among the major sections. Of special note are the sections dealing with resources development and regional disparities, both of which have been reorganized and expanded. In total, ten new papers have been added and they deal with a broad range of themes including the urbanization process in Canada, foreign investment, ecology of resource development, water export, federal government involvement in water quality control, and the changing federal concept of regional development. In addition, five papers have been rewritten and updated and several other authors have added brief postcripts. To make space for the new papers, several of the original papers that were out-of-date or had been found to be less useful in the classroom have been deleted from this edition.

Mr. Alan Hildebrand, chief cartographer in the Division of Environmental Studies, has prepared the graphic material, and to Alan I express my appreciation for a job well done.

University of Waterloo   ROBERT M. IRVING

January 1972.

To The Young Canadians

                    Jane
                    Stephen
                    Jennifer
                    Lisa

When nothing's doing,
and nobody is around,
Think . . .
When the young ones in the world grow up,
Will they have clean water to swim in?
Fresh water to drink?
Clean air to breathe?
Love to live with?
Silence to rest in?
Think . . . will they?

Jenny

Contents

PART VI

DIVERSITY AND PERSONALITY

An Introduction

Canadians have long been digging up their ancestral roots to inspect the nature and direction of their country's growth. Size, physical diversity, sparseness and unevenness of population distribution, the presence of two founding cultures, and proximity to the United States, have combined to create many problems. Over one hundred years after Confederation, Canadians are still asking the question, "What are we?" and, "Does Canada have a national identity?"

Andrew H. Clark, the author of the introductory article in this volume, is Canadian-born but has pursued his professional career as a geographer in the United States. Nevertheless, a great deal of his research has been on Canadian topics, and he has been a keen observer of developments and trends in Canada. This continuing interest in Canada and in the interpretation of Canada to others is reflected in the following article written, not for Canadians, but as a contribution to a New Zealand geographical publication.

Since 1959, when this interpretation was written, many changes have taken place in Canada. It has one anthem and one flag and the political climate has changed rather drastically. The events in Quebec illustrate the degree of political and social ferment—*Front de Liberation du Québec,* the La Porte-Cross tragedy, bilingualism, and the Montreal riots of 1969. But the focus does not rest upon Quebec alone; it encompasses also the problem of regional economic disparities and disenchantment and concern, as manifested in the exploratory talks regarding Maritime Union (1970) or the question of "One Prairie Province—A Question for Canada" conference (1970). Westerners and Easterners still have a distant and suspicious feeling about Central Canada—the seat of political and economic power dominated by the two urban giants—Montreal and Toronto. For other Canadians, the amount of foreign investment and control of the economy is of concern. But for the general public, perhaps the greatest concern in the early 1970's is whether Canada can prevent the urban malaise and environmental degradation seen in the United States from spreading across the border. In retrospect, Clark has touched on many of these issues and it is for this reason his analysis warrants careful reading and interpretation.

Geographical Diversity and the Personality of Canada

ANDREW H. CLARK

Perhaps the simplest definition of the task of the geographer is that he has the responsibility to describe and explain "what other places are like." Many places, and by extension, many areas, contain great varieties within our usual broad categories of the realms of nature and culture. As part of the character of place and area any such variety is an important element of description, and often of great significance in explanation. Of more specific geographical interest, however, is the unevenness of distribution within area, creating what may be best described as the geographical diversity of the area. It is to the study of such geographical diversity, indeed, that most of our professional efforts are devoted. It is, therefore, of great importance that we should not, by implication, neglect it in the employment of such portmanteau adjectives as "British," "Russian," "American" or—to the point of this essay—"Canadian."

It is well known that the Kingdom of Canada is a very large country. Despite its size, of course, it does have certain of the homogeneities of any national state: one national government,[1] one flag and one anthem,[2] one code of criminal law,[3] one national system of defence. And there are other characteristics that are very widespread. Almost all Canadians have seen snow (even if tens of thousands on the West Coast have seen it only rarely or distantly) and most of them live with it for many months each year. Almost all Canadians are very close neighbours of one of the economically and militarily most powerful countries in the world, and again, almost all of them are more or less unhappy about the implications of that juxtaposition. Most Canadians, too, are uneasily aware that they lie between this power and its increasingly near, and often threatening, rival for world leadership; they share the fear of being crushed between the upper millstone of an amoral and irresponsible arrogance and the nether of a heavy-handed avuncularity.

But, even if the long list of fairly uniform attributes might seem to give sharp definition to the adjective "Canadian," the stubborn facts of the country's diversity, particularly in its geographical aspects, blur it again, until the image of Canada's personality often becomes a vague and formless thing in the eyes of the rest of the world. The true shape of that personality is to be seen only by close examination of the warp and woof of its fabric, and particularly perhaps, the variegated individual threads of the very patterns of diversity that, in total, make the shape of the whole fabric so hard to discern. Such an examination would imply no less than a complete geography of Canada, but a few selective observations may at least indicate some of the more important threads that must be isolated and followed.

Traditionally, Canada is divided for geographical description into six rather vague and somewhat overlapping areas: Maritime Canada, Quebec, Ontario, "The Prairies," "The Coast" and "The North." The remarks that follow will be directed to each of these in turn, considering elements of its individuality that contribute to the geographical diversity of the nation, and occasionally probing into aspects of such diversity within

● ANDREW H. CLARK *is V. C. Finch Research Professor of Geography, Department of Geography, University of Wisconsin, Madison, Wisconsin.*

This article has been reprinted from: H. L. McCaskill (ed.), *Land and Livelihood: Geographical Essays in Honour of George Jobberns* (Auckland, New Zealand Geographical Society, 1962), pp. 23-47, and is reproduced with permission of the author and the New Zealand Geographical Society.

The article was written in 1959 prior to Canada's adoption of the maple leaf flag and when conditions were different from today.

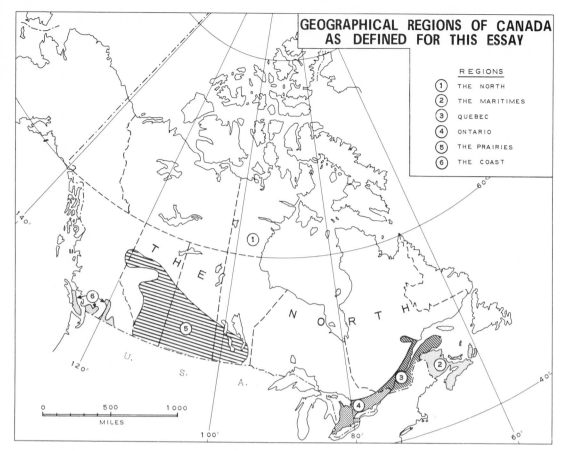

Figure 1

the regions themselves. The selection of "threads" has been made on the basis of highly personal judgements as to what will contribute, in such a brief review, to a better understanding of the whole.

THE NORTH

"The North" overlaps all the other "regions" except, perhaps, that it may be said to have a definable border against "The Prairies" of Manitoba, Saskatchewan and Alberta. Certainly almost all of Newfoundland-Labrador, most of Quebec and Ontario, much of British Columbia, and the northern reaches of the "Prairie Provinces" themselves are as much a part of it as are the Yukon, the Northwest Territories and the vast archipelago reaching to within a few degrees of the pole.

Perhaps it is the pervasiveness of The North (in western Ontario it abuts on the American border and reaches south of any part of the western provinces) that has created an image of Canadian homogeneity: "Our Lady of the Snows," a vastness of coniferous forests, rock-knobbed, lake-strewn and swamp-bound. Certainly Canadians have always taken something of a pride in its size and physical ruggedness, have dreamed of Eldorados of minerals, pulpwood and hydroelectric power, and have realized some part of the dream. But The North is itself as varied in character as many very large countries.

A map of the vegetation cover of Canada's north on a world, continental, or even national, scale is deceptively simple. We have not devised any really satisfactory way of mapping vegetation associations at such scales that conveys the reality to the reader. One can go for miles or days overland in the northern "coniferous" forest seeing little or nothing but aspen, birch, bare rock, muskeg, tundra-like shrub, and only occasionally a conifer. Vast areas are dominated by spruce, fir and larch, of course, but the distribution is extraordinarily spotty. And the concept of The North as a forested land not only suffers from the hiatuses everywhere found in the forest itself, but from the existence of the equal or greater areas of bare rock, sand, gravel, and scattered lines and patches of stunted trees, shrubs, grass or herbs that constitute the Arctic barrens of The North's poleward flank.

To many who know Canada less well, The North and the Canadian or Laurentian Shield are more or less synonymous. Most of us would be sophisticated enough to real-ize that The Shield is far from coincident with The North, which includes vast areas of bedrock of Paleozoic or even later[4] ages, thus greatly diluting the characteristic of "pre-Cambrianism" that is so strongly at-tached to The North in our minds. But it may not be as widely recognized that in its surface features The Shield is another gen-eric concept that has ceased to have much meaning.[5]

The early centuries of the vaguely com-bined concepts of The North and The Shield came from a rather limited acquaintance with its southern borders, especially along the Great Lakes—St. Lawrence waterway. The impression was deepened by the obser-vations of millions of passengers who passed through and over the same areas along Can-ada's transcontinental railways. Only re-cently has extensive air reconnaissance and

Figure 2

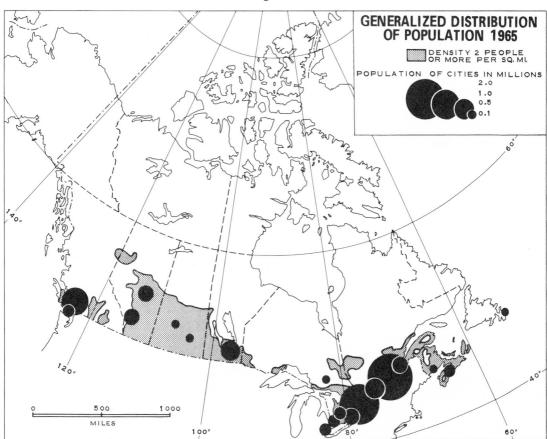

GENERALIZED DISTRIBUTION OF POPULATION 1965

DENSITY 2 PEOPLE OR MORE PER SQ. MI.

POPULATION OF CITIES IN MILLIONS
2.0
1.0
0.5
0.1

MILES

5

photography revealed the degree to which a substantially distorted view of uniformity in The Shield, and by transference, in The North, had become prevalent. From the stark, mile-high, ice-scalloped Torngats of Labrador to the ill-drained flats of Hudson Bay's south coast, from the glacial peaks of Ellesmere Island to the vast swampy areas of the district of Keewatin, or from the rugged western cordilleras to the intricate maze of the Mackenzie delta, The North contains almost as great a variety of landforms as it does rocks and vegetation. Its known mineral resources are great but most spottily and irregularly distributed; most of its area is less rich in useful mineral resources than is Western Europe or the United States. To think of The North as mineral-rich is, properly, to think of very minute parts of it, haphazardly dotted about its vastness in a patternless array, as having mineral resources rather than any other economic resources of significance. It is rich in minerals chiefly in comparison with its poverty in so many other things.

It is clear, therefore, that the immense variety of Canada's North makes any broadly painted stereotype rather more confusing than helpful in our understanding of it. The coldness of winters, the lack of system in its drainage, the sparsity of people, the difficulty of surface travel—these rather obvious attributes are, indeed, generally applicable. But it is also true that many of the more popular adjectives used to describe it are shredded to meaninglessness on the jagged teeth of its diversity. And what is true of The North is equally true of the other parts of Canada.

THE MARITIMES

The writer's principal regional interest in Canada has been in the maritime provinces. To complicate his problem, the accession of Newfoundland (including Labrador) to Canada (1949) has created a new entity, Atlantic Canada, and in appropriate time that may become a regional concept with much value to it. But regions of popular consciousness[6] cannot be created by fiat of government or public-relations men. The maritime provinces of Canada are by tra-

dition the western-most three of the four maritime provinces of British North America as it was for nearly a century between the American Revolution and the establishment of the Dominion of Canada. The concept of the three that joined in confederation between 1867 and 1873—New Brunswick, Nova Scotia and Prince Edward Island—as a unit, "The Maritimes," is likely to remain strong in Canadian popular consciousness for many years to come.

Nevertheless, any easy concept of a corner of Canada with a substantial measure of uniformity fails to stand up under close inspection.[7] All three provinces have French-speaking populations (approaching a balance with those of English speech in New Brunswick). Prince Edward Island and Nova Scotia in addition, received large contingents of Gaelic speakers in the late 18th and early 19th centuries from the Highlands and Hebrides of Scotland; 60 years ago they were reputed to hold more speakers of the sacred tongue than Scotland itself.[8] Also to the countryside of Prince Edward Island and New Brunswick (and especially to Halifax in Nova Scotia) came large contingents of similarly poverty-stricken Irishmen from the southern counties of the Emerald Isle, directly or via Newfoundland. One of Nova Scotia's counties was the home of the famous Lunenburgers, largely German - speaking, who came to Halifax shortly after its founding in 1749 (the hope was to dilute the Acadians with "foreign" Protestants), only soon to be relocated rather unceremoniously on a more westerly sector of the uninviting southern coast of the province. Once quadrilingual, The Maritimes are reduced in variety to two languages today, but the traditions of the non-English groups[9] remain strong in these seagirt lands.

It is hard to give a capsule description of The Maritimes that comprehends their physical variety. Most of Prince Edward Island is potential farming land; most of New Brunswick and Nova Scotia is a rocky or boggy, forested, cut-over or burned-over wilderness.[10] The farms once extended much more widely than they do today in the "soft rock" areas, particularly adjacent to the Bay

6

of Fundy and the Gulf of St. Lawrence, but the migration of settlers from this once extensive farmland has been accelerating throughout the past 75 years and in a selective way that has left smaller discrete areas of relatively intensive farming irregularly scattered over the three provinces.

There are minerals in The Maritimes and there is a wealth of fishery resources off their shores, but most Maritimers, even outside the largest cities and towns, are neither miners nor fishermen. It is hard to characterize most rural Maritimers—perhaps most of them divide their interests between a little forestry, a little farming, and dozens of other ways of picking up an often rather scanty living from selling their skills and labour where they can.

Perhaps the most pervasive attitude of the people of The Maritimes in the past century has been one of disillusionment with their lot. Rather reluctant joiners of Confederation, there is a widespread feeling among them that the persistence of straitened economic circumstances through subsequent decades has not been unconnected with the loss of the substantial degree of independence they enjoyed as British colonies with representative and responsible local government. But their attitude toward being Canadians is extremely varied. Among those of Loyalist and Highland Scottish traditions particularly, the pride in being British has been generally much stronger than consciousness of being Canadians. With the changing times, as Britain's star slowly sets and that of Canada rises in the peculiar collection of republics, kingdoms, colonies and whatnot that forms the Commonwealth, they are struggling to don the rather artificial mantle of Canadianism.[11] Perhaps those in whom the British tradition stirred less strong feelings—the Southern Irish, the New England Yankees, and especially *les Acadiens*—the concept of a truly Canadian loyalty and nationality has been easier to grasp. But each of these groups also has other interests and traditions that hamper the development of a feeling of national identity.

Especially strong as a secondary attachment of all Maritimers is that to New England. To New England's cities flowed much of the great turn-of-the-century haemorrhage of its youth and we may hazard a guess that more families than not in The Maritimes have intimate ties of not-too-remote cousinship with the area that mothered the region's first large post-Acadian settlement. Until very recently, it was Boston, from the Back Bay to the farthest fringes of its industrial satellites, that was the cultural mecca of Maritimers, not either of Canada's two "millionaire" cities, Montreal or Toronto. Perhaps an exception could be made of New Brunswick's northern counties into which *les Canadiens* have been overflowing,[12] but they in turn have been exporting sons and daughters, as Quebeckers have been doing for a century by the tens of thousands, to the mills of Maine, New Hampshire, Vermont and Massachusetts.

QUEBEC

Bismarck is reported to have said that the greatest political fact of modern times was the inherited and permanent one that North America speaks English.[13] Nothing in the course of events of mid-twentieth century has reduced the strength of his observation. But he might well have added that, *for Canada*, the most important political fact was that a third of its citizens, heavily concentrated in the province of Quebec, did not. It is, indeed, proper for any prospect of Canada as a whole to single out the province of Quebec as a very individualistic area.

The geographer at once enters the *caveat* that there are islands of English speakers in the large cities and (although a fast diminishing one) in the Eastern Townships bordering New England, and that there are large islands of French speakers west of Quebec, especially in Ontario and Manitoba, as well as to the east of it in The Maritimes. But Quebec is the home of the French Canadians, the political and cultural areal unit of their individuality, their *pays* in the truest sense of the word. It is, and

7

will long remain, a region of cultural separateness. That Quebec is French-speaking, and much more uniformly of one particular religion, may actually contribute less to what a Canadian means by "Quebec" than the acute consciousness, no less vivid to other Canadians than to Quebeckers, of their sense of perpetual injury and frustration.

For *les canadiens* the conquest[14] of just two centuries ago is still the most critical event in all of history. Although early British actions, such as the Proclamation of October 17, 1763, seemed designed to subjugate them, succeeding actions of the conquerors (for example that great charter of *canadien* rights and privileges, the Quebec Act of 1774; the Constitutional Act of 1791; the Union Act of 1840; and, finally, the British North America Act of 1867) all were meticulously drafted to protect language, religion, French civil law, and in all, the full cultural independence of the French Canadians. If the goal, the dream, the chief motivation of *les canadiens* has been *la survivance,* surely they have it. Comprising nearly one-third of Canada's people and maintaining that proportion despite periods of heavy non-French immigration by their abundant fertility, and with such stout defences of law and custom, their continued apprehensions would seem ill-founded even though few immigrants have come from France since 1763, and almost all immigrants to Canada wind up speaking English rather than French in the second generation.

The difficulty is that the world, and that large sector of its land surface that is Canada, changes with the passing years, and *les canadiens* have a deep suspicion, often fanned by ill-conceived remarks of English-speaking Canadians, that the national goals imply the imposition of more and more pressures for national uniformity. Never very far below the surface, a mood of rebellious suspicion in Quebec can seem to slowly die away only to flame up again, as it has many times since Confederation in 1867, from the episode of the hanging of Louis Riel, through the fight for separate tax-supported schools in other provinces and the conscription issues of two world wars, to the latest federal-provincial dispute on taxation.

Most educated and intelligent Canadians, English- or French-speaking, realize that the basis of Confederation, the British North America Act, was designed for conditions of a century ago and needs a major overhauling today. But no really satisfactory or workable machinery for its amendment exists except by a sort of treaty between individual provinces and the federal government. Needless to say, the British government, the authority responsible for the act that embodies that constitution, cannot and will not interfere in the affairs of its sister, and in every legal respect equal, kingdom.[15] Stripped of the ponderous language of the constitutional lawyers and historians, it can be fairly said that Canada has a constitution, which, in the most important respects of the necessary reshuffling of federal-provincial powers and responsibilities, is unamendable as long as those speaking for the people of any one province refuse to agree to changes. The deepest instinct of *les canadiens,* perhaps not without good justification, is that any change that surrenders any power to the federal government is likely to threaten *la survivance.* And the actions of the Provincial Government of Quebec reflect those suspicions sharply and directly in all federal-provincial negotiations.

Thus that stout, rather large island of cultural individuality stubbornly maintains itself in the lands along the St. Lawrence. To preserve language and faith, the French Canadian continues to receive an education that has been described as much better designed for life in the Kingdom of Heaven than in *La Province de Québec.*[16] But if many of its leaders live in a world of backward-looking introspection, its young folk face a scientifically-oriented, technologically-complex present. Most *canadiens,* like other Canadians, now live in the city, not the country: they are factory hands, shop clerks, bus drivers, not *habitants.* And they are disturbed and confused, in that they have been taught to be suspicious of the very basis of the society and economy into which they must integrate themselves.

Yet, if Quebec has a widespread tendency to homogeneity in such deeply imbedded cultural characteristics, emphasized by contrast with the rest of the country, it would be a disservice to the purpose of this essay if no mention were made of the vast differences that do exist even within the relatively small parts of Quebec that are at all closely settled. The pioneering "colony" settlements of the Clay Belt and the Laurentians[17] are often sharply distinct from the old seigneurial lands along the St. Lawrence, some of which have been farmed for more than three centuries. The huge industrial complexes of Montreal, Three Rivers and the Saguenay valley are in equally vivid contrast to the quiet, pastoral, now slowly emptying countryside.[18] But in Canada's geographical variety this is a ubiquitous theme, in no way restricted to Quebec; it is the great individuality of Quebec within Canada that is to be remembered here.

ONTARIO

Upper Canada, the Ontario of this century, has its own well-defined personality and one that has stood out the more sharply because of rivalry or conflict with, and physical location beside, Lower Canada, or Quebec. It is only a little more than a century since Ontario's population topped that of Quebec, but it has maintained or increased that demographic lead as it has also grown to be the industrial, commercial and financial heart of Canada. Nearly one Canadian in three lives in Ontario, and on the average, he lives there with a very much higher standard of living than the other two to east or west. Loyalist-founded, a focus of Ulster migration, empire-minded to a degree rivalling that of the most home-sick New Zealander, Ontario yet received one of its largest early bodies of settlers directly from New York (and often ultimately from New England), an accretion that was much more a part of the westward movement of the late eighteenth and early nineteenth centuries than that of a predominantly Loyalist emigré group.[19] A great focus of British immigration in the nineteenth century, Ontario often served more as a funnel for the peopling of Indiana, Illinois, Michigan, Wisconsin, Minnesota and Iowa, than as a permanent home of the new settlers. To its cost, thus, it often cared for the settlers for a generation only to see sons and daughters move west and south to another flag. Partly out of anxiety to check this flow, it fostered a manufacturing industry, an encouragement that somewhat ironically brought in its train that large and fundamental element of the Canadian industrial structure, the American branch factory. The national policy of tariff protection may have saved the new nation, while drawing strength from the peripheries to be concentrated at the centre.[20] It certainly created jobs for Canadian factory workers, chiefly in Central Canada and more in Ontario than in Quebec, but the goods they produced, whatever the nationality of factory ownership, have been patterned, to an overwhelming degree, on the lasts of American mass culture.

The strongly marked Americanism of Ontario people is not limited to the traditions of the earliest settlers, to the clothes they wear, the food they eat, the cars they drive, the books they read, the crops they grow, the games they play, or the crimes they commit. It penetrates to the heart of their lives and aspirations in many other ways. Canadian magazines and theatres, art galleries and orchestras, struggle manfully toward individuality and independence, but—with a few notable exceptions like the Stratford Theatre—if the artists are successful it is often more by emulation or imitation of the American exports with which they must compete than by contrast.[21] The very Ontario accent, once distinctive in the shortness of its vowels, is now much nearer the norm of standard North American English, spoken by most Americans, than is that of New England or the American South. The World's Series of baseball has never had a Canadian team participant, but it draws as large, or larger, an audience on Canadian radio and television than does Canada's traditional annual gladiatorial extravaganza, ice-hockey's Stanley Cup play-offs. By popular

9

pressure, the Kennedy-Nixon presidential campaign debates were carried on national Canadian networks and commanded audiences comparable to those of a national Canadian political campaign. There is no need to repeat examples. Ontario, which some American demagogues of the past pretended to see as an arrowhead plunged in the heart of the United States, seems in most obvious external ways to be assimilating rapidly to its enveloping neighbour.

Many observers have seen the resulting spirit and personality of Ontario to have marked schizophrenic tendencies. Here is a land, long noted for an almost blind, often nearly chauvinistic, loyalty to things British, becoming, in spite of itself, more and more a part of the culture of the very people whose bordering location, actions and attitudes have been a major cause of the strength and persistence of British attachment. And, of course, these Ontario British-Americans aren't nearly as much at ease with flesh-and-blood Britishers, excepting Royalty and a handful of public figures who symbolize the ancient tradition, as they are with Americans. It would be amusing, if it were not for the overtones of sadness, to listen to the same Ontarian, perhaps on successive nights, ripping first into Englishmen for one set of faults, and then into Americans for another set (some of them duplicated) thus negatively emphasizing, in turn, the American and British roots of his joint inheritance. One of the least attractive characteristics of Ontario's spirit is an almost perpetual, if alternating, petulance toward the two nations from which its culture, tradition and blood derive, and with whom, in the nature of things, it must live closely for all of its days.

To be sure, Ontarians do have political and judicial systems markedly different from those of their American neighbours, which, rightly or wrongly, but passionately and vocally, they believe to be greatly superior. It is true that, as in the United States, "centre" in politics is defined as somewhere near the middle of the contemporary spectrum of political philosophy, and since neither major party can advocate programmes too far away from the centre and hope to gain power (or, certainly, to implement such programmes and stay in power), the net result is a philosophy and practice of government that is remarkably similar to that obtaining south of the border. The independence of the federal judiciary is probably comparable in the two countries, so the real difference lies in the relationships between legislative and executive power. Divorced from the labels "British" or "American," it is doubtful if, in theory or fact, Canadians would plump heavily for one or the other side of the very debatable question of the separation of the two. But Ontarians have seized on the distinction as a symbol of difference and make rather more of it than do other Canadians. It is true that anti-Americanism is becoming something of a national Canadian sport and is believed to have strongly influenced the last two Canadian national elections, but it is sufficiently more self-conscious, if not strident, in Canada's "keystone province," to be an element of Ontario's character that distinguishes it from the other areas.

Considering closely settled Ontario only, as distinct from the province's own vast share of The North, Ontario probably exhibits more intra-regional uniformity than most of the other Canadian areas as we have divided them; more uniformity that is, if we ignore the contrasts of urban and rural, industrial and agricultural. The controlling reason may be that Ontario's traditions and relative wealth have developed one of the better systems of public education in the world, and it is a pervasive system that reaches every corner of the province. Its public university at Toronto has maintained standards and leadership that might be the envy of any state-supported institution, and these have been followed closely by the private universities. For these, and a variety of other, cogent reasons, Ontarians have developed remarkably strong and uniform senses of achievement and provincial identity.

THE PRAIRIES

Divided from settled Ontario by a thousand miles of lightly peopled or empty wilderness, where The North bends south, and from The Coast by some of the world's most rugged mountains, is the Canadian segment of the great northward loop of the North American Great Plains and prairie grasslands (with some bordering desert and parkland) between the 49th parallel and the western expression of The North. It embraces a relatively small part of Manitoba's extended boundaries, half of Saskatchewan, almost that much of Alberta and even a corner of British Columbia. The rest of the three provinces is "north," or mountain slopes, but it is in the former grasslands and their parkland borders that almost all of the three million "Prairieites" live. Most of them are now economically connected to something other than farming, but the aura of grain, cattle, sheep, pigs and poultry is pervasive in the work, dreams, politics and literature of all the people. And although manufacturing has made great strides in the regional metropolis, Winnipeg, and in other fast-growing urban centres like Regina, Saskatoon, Calgary and Edmonton, it is new staples of oil, other minerals, and forest products that have joined the farmers' yields as great regional surplus products that must be sold abroad. The resultant push for freer channels of foreign trade, often in sharp opposition to the protectionist attitudes of Ontario and Quebec, has affected the economic and political viewpoints of prairie people more than, perhaps, any one other factor.

Landbound, isolated from fellow Canadians east and west, most Prairie folk, as most of their fellow countrymen, live relatively close to the American boundary, and constantly compare their lot with that of the Minnesotans, Dakotans, and Montanans whom they can, and do, visit so easily. Polyglot, and as geographically and culturally diverse in origin, they have been forced into a good deal of unity of feeling in regional consciousness by the same pressures. Saskatchewan, with more people of ultimate German than ultimate English origin, is the one most representative of its region. That an adopted son (Diefenbaker) became Canada's first prime minister with a name neither English, Irish, Scottish nor French, in the same era that the United States acquired its first president with a name neither English, Irish, Scottish nor Dutch, is relevant here; less than half (much less in Saskatchewan) of prairie people stem ultimately from the British Isles, and of the latter a large, if unknown, proportion has a descent that leads, geographically and for at least a generation or two, through the American midwest of the 19th century.

The Prairies got a firm groundwork of settlers from Ontario and The Maritimes (and some from Quebec) in their earliest decades (as Wisconsin, Iowa and Minnesota had their New Englanders, perhaps once or more removed), but the mid-20th century character of The Prairies reflects the kaleidoscopic patterns of the cultural mosaic created by the blocks of Germans, Ukrainians, Poles, Scandinavians, Dutch, Belgians and dozens of other groups who flooded in with the building of its huge railroad net.[22] And because of selective migration of those of English-speaking origin towards the cities, especially east and south of The Prairies, the countryside, still the basic source of prairie wealth one way or another, is becoming denuded of British stock. Hard-to-spell-and-pronounce names of mayors, aldermen, members of the provincial legislative assembly, or members of the federal parliament, are appearing in increasing numbers in the pages of Prairie newspapers.

It is true that the post-war immigration to Canada has brought more immigrants from continental Europe to Ontario, Quebec and British Columbia by far than to The Prairies. There are few farming opportunities anywhere in Canada in a period of a rapid nation-wide decrease of farming people, and the new immigrants have been attracted much more largely by the magnets of the manufacturing, mining, construction and forest industries. The result is that Western Canada's melting pot has simmered now for forty years without the addition of many new ingredients. But the result is still a highly distinctive and only partly blended Prairie mixture. Cultural origin and religion still are important distinguishing elements; the hold of an Anglo-Saxon, Protestant tradition is much weaker than in Ontario; and, perhaps more than anywhere else in Canada, these people think of themselves as *Canadians* "new" or "old," but *Canadians*, yet with interests often sharply at odds with those of The Maritimes, Quebec, Ontario, or The Coast.[23]

A large element of lineal descent from Americans (citizens or residents) and constant intercommunication across the border has submerged most anti-American feeling, in the sense of "American" versus "British" traditions. Nevertheless, the erratic actions of the American government in matters of quotas and tariffs on Canadian primary products have not engendered any strong affection for the southern neighbours. Except for Ontarians, no Canadians are more thoroughly "American" in speech and culture; in no part of Canada would most individual Americans feel more at home; but the casually friendly attitude toward Americans is not matched by an equal regard for their government or its actions. If the cultural anti-Americanism of The Prairies is so subdued that the army of invading oilfield roustabouts from Texas and neighbouring parts scarcely notices it, political anti-Americanism remains at a rather high pitch. In provincial politics, local issues often maintain local parties in power (sometimes deriving from the agrarian radicalism of another era), but there has been a marked tendency toward a more strongly nationalistic bias in federal politics.

In the days of the wheat and railroad booms, and recently with oil and natural gas expansion, The Prairies have dreamed of playing a much greater role in Canada, demographically, economically, and ultimately, politically. But a glance at their resource characteristics suggests that their role in the near future of the Canadian nation is likely to remain small and as likely to decline as to increase. Yet they occupy one of the largest settled (if generally lightly so)

parts of Canada and in their strong individuality can never be ignored by the other parts of the country.

THE COAST

If Prairie settlement, outside of the major cities, is rather evenly distributed, any relief map of British Columbia, as any representation of its population distribution or economic production, makes vivid its extreme spottiness of opportunities for settlement or economic activity. Indeed, the vast majority of British Columbians live but a few miles from the American border in Victoria or Nanaimo on Vancouver Island, in Greater Vancouver and the lower Fraser Valley, or at Trail. It is no accident that as far east as Kingston, Ontario, British Columbia is "The Coast" in popular consciousness. The dispersed minority are found along the narrow valleys of the tributaries of the Columbia and Fraser drainage systems, sparsely along the railroads, in widely separated lumbering, fishing, mining or refinery camps along the rugged, deeply fiorded coast, or in even more widely scattered ranches and farms in the basins of the central interior.

The isolation of The Coast's people from the rest of British North America was relieved more in theory than in fact by their Confederation into the Dominion of Canada some ninety years ago. Even the belated completion of the promised transcontinental railway (the CPR) a decade and a half later, gave the local population little sense of national identity or security. That rested on Britain, and especially the British navy, well into the 20th century, and that dependence, together with the long history of conflict with Americans for the area,[24] created an unusually strong attachment to things British in the minds and hearts of its people. The outward Britishness of Victoria on Vancouver Island is not matched by the rather more typically North American bounce of the bustling metropolis of Vancouver, nor has Vancouver escaped a large immigration of strains of population deriving from neither the British Isles nor France. But, as elsewhere in Canada, the newcomers to Vancouver have adopted the regional spirit and are hardly less rigorously attached to the concept, "British," than their fellows on the island. As with the Maritimers, the Coasters wear their "Canadianism" lightly and in the face of American cultural encroachments, have tended to wave the Union Jack rather than the beaver-bedecked red ensign once widely used by other Canadians. But they, too, feel a sense of nearly irreparable loss as the political and economic star of their sentimental "homeland," Britain, wanes.

In some ways the Coasters may be thought to have even less reason for strong British sentiment than their Prairie cousins across the mountains. In the past they have expressed themselves strongly on such matters as the "surrender" of country north and west of the Columbia River and the "sell-out" by the British commissioner on the Alaskan boundary award,[25] but much of their export trade in lumber, fruit, salmon, metals and ores has always been with Britain, and the emphasis on British connections has seemed to make the nearby boundary with Washington State somewhat more substantial and protective.[26] The realities of the world situation do not escape the very practical men of affairs who have built a thriving outpost on the edge of civilization in a short century, and there is a strong effort to direct feeling more strongly to Canada and to link themselves more closely to their nearest potential markets and suppliers of raw material in The Prairies. Of special concern have been pipelines for oil and gas, improved highways, and rapid expansion of the tourist and "retirement" attractions of Canada's "California."

The Coast, and the whole province of British Columbia, represent the most dynamic area of Canadian development as the decade of the 1960's opens. There is a sense of optimism, a ferment, a fluctuating but at times often almost explosive pace of development, that very clearly distinguishes the region from its nearest competitors in The North, The Prairies and the central provinces. To all Canadians, but especially to the Coasters themselves, this is the land of greatest promise for Canada's immediate future. And, again paradoxically, a region in which old-world traditions have hitherto maintained great strength, is showing, externally at least, more and more characteristics of new-world frontier brashness.

CONCLUSION

These are the regions of Canada, diverse in natural character, spirit, economic resources, history of development, and peoples. They share remarkably few things to help bind them into nationhood and they show a number of marked centrifugal tendencies. Any such survey must sooner or later recognize that one of the outstanding problems of Canada is its almost symbiotic cultural and economic ties with the United States, and the preoccupation of Canadians with that circumstance. Most of Canada's trade, import and export, is with her only land neighbour. But the ratio of the roughly five-billion-dollar trade total is about three to two in favour of the United States, leaving Canada a very large deficit not made up by her substantial world export trade. The periods of higher value of the Canadian than the American dollar are a source of satisfaction only to the unthinking, for it reflects the continued flood of American capital into the Canadian economy where it represents a good fifth of the total investment. And the flow is most strongly directed, not to secondary or tertiary industries, despite all those branch factories, but to the supply of raw materials needed in the American economy—iron, nickel, pulp and paper, petroleum and natural gas. Steadily the American colossus, however benevolent its intent, gains an ever firmer foothold in Canada and without most Americans being really aware of what is happening. After all, American exports to Canada represent only about one percent of the national income of the United States. But Canadian exports to the United States, if smaller in value, represent 10 to 15 percent of her national income, and no Canadian can ignore them.

Canadian-American relations, never a major U.S. political issue, are always near the surface in Canada, and almost once a decade become a major election issue that may topple a government (as it did the Liberals in 1911 and in 1957). It is probably not on the cards that Canada can, or will, do very much to change things, but it is nearly certain that Canadian attitudes toward the United States are one of the most significant, apparent aspects of the Canadian character. To live in the same house with a giant, however kindly the latter's intentions, is to face the constant danger of being trampled. Canadians cannot forget a fact of life of which most Americans are blissfully unaware.[27] The pre-eminence of the position of the United States in Canadian life and thought means that one of the most interesting elements of Canadian regional diversity is the varying attitude toward the American "menace." But however much it varies, it is everywhere a matter of concern, and paradoxically perhaps, the stronger is the cultural and economic pressure of the United States, the more are Canadians drawn together in a feeling of nationhood. The question remains whether the probable blossoming of Canadianism can ever wholly include *canadienisme*, and whether, by the time of flowering, Canada will have retained much of its cultural and economic individuality.[28] One must certainly hope so, just as one may hope that Canada's present marked regional cultural diversity will not be ironed out under the steamroller of modern mass-communication media and the pervasiveness of American pressures.

REFERENCES

[1] But ten provincial governments with their own regional interests, and their own peculiarities of statute and administration.

[2] Or lack of either, according to the point of view. There has never been any official decision to adopt an anthem other than "God Save the Queen" (the lugubrious "O Canada" has had heavy going) or a flag other than the Union Jack (although the British Red Ensign with a maple leaf, the Canadian shield, or other emblem in the lower right-hand quarter, has had widespread use in private, semi-official, or even official ways).

[3] There is only one kind of first-degree murder in Canada compared with half-a-hundred kinds in the United States; but there are wide varieties in civil law.

[4] Especially southwest of Hudson Bay, in the Archipelago, and west of Canada's own, internal line of "Great Lakes." (We may forget that Lake Winnipeg is larger than Lake Ontario, or that both Great Bear and Great Slave lakes exceed Lake Erie in area, and rank among the first ten interior water bodies of the world in size.) And, of course, The North includes the Mackenzie Lowland and vast extents of the western cordilleras.

[5] I am indebted for a well documented and argued statement of this fact to Alan Grey, a former New Zealander and now a research assistant at the University of Wisconsin in Madison.

[6] The writer made some attempt to examine this problem for the South Island of New Zealand in *The Invasion of New Zealand, by People, Plants and Animals: The South Island*, New Brunswick, 1949.

[7] A point stressed in two of the writer's recent publications: *Three Centuries and the Island*, Toronto, 1959, about Prince Edward Island, and "Old World Origins and Religious Adherence in Nova Scotia," *Geographical Review*, Vol. 50, 1960, pp. 317-44.

[8] One of the writer's grandmothers, born in Prince Edward Island about 1850, spoke no English until she went to school and was still as fluent in Gaelic as in English in her eighties.

[9] The English-origin groups themselves were sharply divided into descendants of pre-Revolutionary New England immigrants, of Loyalist emigrés during and after the American Revolution, of Methodist-tinged Yorkshiremen, and many other smaller groups, a good deal of whose individualities have persisted into the 20th century.

[10] Little of the forest is not in its second—or "n'th"—growth since the arrival of Europeans. Starting with a substantial variety of combinations of hardwoods and softwoods in variegated patterns, the additional factor of various stages of succession after cutting or burning has made the pattern of forest or woodland character in The Maritimes chaotic. These woody, swampy, or rocky areas, comprising so much of The Maritimes, were long referred to familiarly as

"The moose farm"; alas the moose are also sadly declining, from disease or inadequate nutrition.

[11] The figure is borrowed from H. F. Angus, writing more than two decades ago, in *Canada and Her Great Neighbour*, Toronto, 1938. To quote him directly (p. 7) . . . "To a dispassionate observer, there may be something pathetic in this *attitude* of Canadians, for with sufficient detachment we can think of Canadians creating a nationality as analogous to primitive people learning to disfigure themselves with clothes, because they mistake the clothes for civilization."

[12] Most French-speaking New Brunswickers today are of *Canadien* rather than *Acadien* tradition. But the latter still has great vigour and to assume that the distinction is without significance is somewhat akin to identifying Highlanders with southern Irish because of a common Gaelic background.

[13] Ironically, this was quoted by J. W. Dafoe in *Canada, an American Nation*, New York, 1935, p. 3, to make a point of the common ground of Americans with English-speaking Canadians.

[14] French-Canadians prefer the expression "cession."

[15] The complex history of the attempts to make it legally and theoretically possible for Canadians to amend their own constitution (as it now is) is told in many places. See especially P. Gerin-Lajoie: *Constitutional Amendment in Canada*, Toronto, 1950.

[16] The figure is used by Mason Wade: *The French Canadian Outlook*, New York, 1946, p. 137.

[17] "Les Laurentides" in the terminology of the distinguished school of *Canadien géographes*.

[18] Any picture of a teeming countryside with a shortage of farmland is purely fictitious now, whatever validity it may have had in the past.

[19] This point is made by many historians, most recently by the late J. B. Brebner: *Canada, a Modern History*, Ann Arbor, 1960.

[20] With notably weakening effects on the peripheries, especially The Maritimes, but probably on The Prairies and The Coast as well.

[21] On this theme note the varied views of Bruce Hutchison: *The Unknown Country*, Toronto, 1948; Vincent Massey: *On Being Canadian*, Toronto, 1948; and E. K. Brown: *On Canadian Poetry*, Toronto, 1944. J. B. Brebner, *Canada* . . . (1960), p. 53, has this to say: "The serious periodicals, *Dalhousie Review* (Halifax), *Revue de l'Universite Laval* (Quebec), *Culture* (Quebec), *Relations* (Montreal), *Revue de l'Universite d'Ottawa*, *Queen's Quarterly* (Kingston), and *University of Toronto Quarterly* publish little that is experimental in arts and letters, but *Canadian Forum* (Toronto), *Northern Review* (Montreal), *Canadian Review of Music and Art* (Toronto), and *Here and Now* (Toronto) are all alert to the contemporary."

[22] A magnificent series of studies, heavily focussed on the Prairies, is W. A. Mackintosh and W. L. G. Joerg (edit.): *Canadian Frontiers of Settlement*, 8 vols.; Toronto, 1934-8. See esp. Vols. IV-VIII.

[23] The writer lived on The Prairies (in Manitoba) for the first 21 years of his life; the next six were spent in Toronto. He became steeped in a regional point of view (a curious mixture of parochialism and internationalism) so strongly a characteristic of the late John W. Dafoe's (then) great newspaper, the *Manitoba Free Press* of Winnipeg. It was our family's daily bible in matters of politics and economics. At the very least it was well written and edited; its editorial pages were as well read as its news columns—in both respects rare among North American newspapers of the 20th Century.

[24] That left the British Columbians hemmed in by the "crosscut" of the 49th parallel and the result of the "unfair" Alaskan boundary award.

[25] One of several occasions when Theodore Roosevelt forbore to speak softly while carrying a big stick.

[26] Without such "protection" (in terms of tariffs, rail routes, and expenditure of Canadian federal funds) Vancouver might well have declined to become a regional satellite of Seattle, like Tacoma.

[27] Vincent Massey (Canada's first native born and untitled Governor-General) in *On Being Canadian*, Toronto, 1948, p. 117, quotes John Fiske (the eloquent, if not always scholarly, 19th century American historian) as saying that the United States was bounded on the south by the precession of the equinoxes, on the east by the dawn of history, on the west by the day of judgment, and on the north by the Aurora Borealis. (Fiske's *Bounding the United States*, in one edition makes the eastern boundary "primeval chaos"!) It has often seemed to

observers of American attitudes toward Canada that if there is anything more than the northern lights at higher latitudes Americans in general are happily ignorant of it. Massey was charitable enough to call it benevolent ignorance and to couple it with neighbourly respect and only minor forgetfulness.

[28] Again a figure is borrowed, this time from J. M. S. Careless, present occupier of the chair in history at the University of Toronto. In a discussion of Canadian nationalism in the Canadian Historical Association Report for 1954, he said that he felt that continued preoccupation of Canadians with the question of "Canadianism" was either immature or obsolete and amounted to a "sort of ritualistic contemplation of the Canadian navel." In a change of metaphor, he added, "this particular plant might survive better if Canadians were not always anxiously pulling it up by the roots to see if it is growing."

POPULATION AND SETTLEMENT

An Introduction

Sparse numbers and uneven distribution are two significant characteristics of Canada's population. The settlement patterns that have emerged in this vast area have been molded by the interaction of physical, cultural, historical and economic factors. To examine the characteristics of and changes in settlement patterns, the articles in this section are divided into four groups: (i) the Canadian ecumene and the frontiers of settlement, (ii) the origins of the two "classic" settlement types in Canada, (iii) the changing pattern of population distribution and the impact of economic change in selected rural areas, and (iv) the Canadian Eskimo and Indian—their changing economy and problems.

After discussing the various definitions of ecumene, Louis-Edmond Hamelin establishes the types of Canadian ecumene on the basis of the form and function of settlement. He divides the ecumene into four zones—Southern Canada, the Pioneer Fringe, the Sporadic Ecumene and the High Arctic—and discusses the problems and characteristics of each zone. Both Eckart Ehlers and George McDermott are concerned with the Canadian Pioneer Fringe. Ehlers contrasts the Canadian frontiers of settlement with the classical interpretation of the United States' frontier as developed by Frederick Jackson Turner. McDermott, on the other hand, analyzes the changing frontiers of settlement in the Great Clay Belt of Ontario and Quebec. Although both parts of the Clay Belt have similar physical conditions, the settlement frontier on the Ontario side is retreating, whereas on the Quebec side the frontier is advancing. McDermott attributes this difference to the contrasting philosophies that underlie the frontier development policies in the two provinces.

The second group contains two articles that analyze the development of two contrasting types of settlement patterns in Canada. John Warkentin examines the origin of settlement patterns in Manitoba, and in so doing points to the major variations in survey types throughout the Prairies. The origins of the rectangular survey and its impact on the rural settlement pattern take precedence in this study. However, attention is drawn also to the Mennonite farm-village settlements, the French long-lot system, and the Hutterite settlement patterns. An entirely different settlement pattern developed in the St. Lawrence valley of Quebec where the *rang* (range or long-lot) pattern of survey was introduced by the French. Pierre Deffontaines discusses the origins, characteristics, advantages and disadvantages, and the spread of the rang pattern of settlement.

In the third group, the changing population distribution patterns are described. Isabel Anderson summarizes the changing population growth rates and distribution patterns from 1921 to 1961. The late W. B. Baker's commentary highlights the kinds of changes wrought in the Prairies (indeed, these changes are occurring with varying magnitude throughout most of rural Canada) by mechanization, increasing size of farm unit, rural depopulation, an increase of sidewalk farming, and the impact of these changes on rural trading patterns, local government, and rural institutions and values.

In the fourth group of papers, M. R. Hargrave and James Van Stone examine some of the economic changes being experienced by northern Eskimos and Indians. Hargrave describes the changing settlement patterns of the Mackenzie Eskimos in the western part of the Arctic in relation to economic change. Van Stone, on the other hand, focusses on the factors effecting change in the Indian trapping economy of the Canadian subarctic. In the first of the concluding papers to this section, the late Diamond Jenness, who spent the bulk of his professional lifetime working and living among the indigenous people of Canada, reflects upon the Eskimo's future. In contrast, Abe Okpik, a member of the Northwest Territorial Council, expresses the Eskimo point of view on the origins and implications of these changes for the indigenous population.

Types of Canadian Ecumene

Louis-Edmond Hamelin

In Western Europe, where almost all national space is effectively occupied by man, the average population density of a country is a relatively true reflection of human concentration. However, in countries which have large unoccupied areas, the population density, calculated on the basis of total area, does not express a true index of human concentration.

In Canada, if one calculates population density on the basis of provincial land area only (excluding the two Territories and the area occupied by fresh water bodies), the population density "increases" from 5.2 to 9.5 persons per square mile. The establishment of a geographic density is even more demanding. Using 1941 census data, Benoit Brouillette showed that the crude population density in Québec was 6.36 inhabitants per square mile, but that of the settled municipalities exceeded 82 inhabitants. The first problem therefore, is to establish the area in which one wishes to study the population. In other words, where and what is the ecumene?

From the time that the German geographers, Ritter and Humboldt, introduced the word ecumene into western literature, the term has been subject to varied interpretation and even spelling, e.g., ecoumene, oecumene, ecumene, oikoumene, oekoumene, oekumene, etc. Similarly there is a wide range of definitions. According to the "ancients," ecumene described the habitable world; J. M. Houston would further stipulate that it is habitability with respect to settlement. In the same vein, Vidal de la Blache speaks of "inhabited land," a definition completed by G. T. Trewartha who distinguishes inhabited from uninhabited regions, the latter being the "non-ecumene." A slightly different idea is expressed by some authors who define the ecumene as being the "utilized" land, which is a much broader concept than restricting it to inhabited areas only. To this latter concept Max Sorre (1952) would add the navigable parts of the ocean. Likewise, R. Gajda (1960) speaks of "work areas." Finally, the ecumene provides a method of expressing differences such as those between autonomous regions (France) and their dependencies, or those between the most and least viable sections of nations. The latter concept, expressed by Whittlesey, has been adopted by the Pakistani geographer, Zaidi, who defines ecumene as "the effective state area to most of its inhabitants," the rest of the country being declared "extra-ecumenical." Certain authors have even tried to quantify the minimum limits of the ecumene by establishing a demographic threshold of 2 inhabitants per square mile living at a distance of less than 10 miles from a railway system or five miles from a road. Thus, in the evolving notion of the ecumene there is reference to habitation, utilization, organization and the productivity of a territory, but these characteristics have never been clearly defined.

TYPES OF ECUMENE

Two themes will be considered in establishing the types of Canadian ecumene: function and form.

● Louis-Edmond Hamelin, f.r.s.c., *is Directeur, Centre d'Etudes Nordiques, Université Laval.*

This article has been translated by Robert M. Irving, Department of Geography, University of Waterloo, from "Typologie de l'ecoumène Canadien," *Memoires de la Societé Royal du Canada*, Section 1, Quatrième série, Tome IV, juin 1966, and reprinted here with permission of the author and of the Royal Society of Canada.

Functional Types

In developing and systematizing the traditional ideas of the ecumene, one should distinguish four interrelated categories. First is the *habitation ecumene:* this comprises land which is effectively occupied by city and country houses, stores, shopping centres and their parking areas, industries and warehouses, bridges, harbours and airports, railways and bus stations, services such as motels and trailer parks, cemeteries, and open space in the cities. In area these strongly "humanized" lands, which are not restricted only to southern Canada, do not even represent one percent of the total area of Canada. Population density is high; for example, it reaches 150 persons per acre in certain parts of Montreal.

The second type is the *exploitation ecumene,* which is often adjacent to the habitation ecumene. It includes agricultural land, areas of mineral and petroleum extraction, fishing banks ("the true home of the fishermen," according to R. Perret), areas of periodic foresty, hunting territory frequented by indigenous natives or tourists, ski centres, and even semi-urban land where land speculation competes with agriculture. Detailed regional studies in the exploitation ecumene would establish sub-categories on the basis of the number of people that these activities support, the productivity, the duration of operations, and income. From these studies one could establish various classes of intensity and "extensity." An oil well, a polder or a horticultural enterprise would be in the first category, whereas the raising of reindeer in the Mackenzie River area or low-bush blueberries at Lac-Saint-Jean would be extensive forms of exploitation.

The third type is the *linking ecumene,* a major element in a discontinuously settled country. This is often a discreet ecumene but it is vital in what E. Juillard calls the *désenclavement* of Canada. One of the best developed linking ecumenes in Canada is that which forms the bridge between Georgian Bay (Eastern Ontario) and Manitoba and includes the following elements: former portages, abandoned waterways, two transcontinental railways (which are often parallel), telegraph, power, and telephone lines, radio communications, television relay towers, radar links, and fire towers; and in more recent years the airline routes, the Trans-Canada Highway, and pipelines. Even though all of these elements are not present throughout this ecumene, they nevertheless clearly express human intervention. Moreover, the notion of the linking ecumene must be expanded to include water navigation as, for example, on the Mackenzie River and the Great Lakes-St. Lawrence waterway system. Clearly, while the linking ecumene compensates for distance, it deters, but at the same time enables scattered habitation and exploitation ecumenes to develop. Distance today is no longer a matter of miles, but of time, and even more important, equipment, that is to say, cost.

Between the readily utilized sections of the ecumene is the fourth type, the *sub-ecumene,* of which there are two sub-types: (a) small, unattractive areas which interrupt the continuity of southern Canada, but which are considered parts of the latter; and, (b) immense repugnant empty areas in the interior of the Arctic and subarctic.

There are several ways of applying these functional types of the ecumene to Canada. The *ecumene of propriety* is based on the belief that man, in one way or another, has an impact on the entire country. While the ecumene of propriety is most commonly associated with the habitation, exploitation and linking ecumenes, it is by no means restricted to it. Some scholars argue that because Canada has been completely photographed and is militarily inspected, the whole country is the ecumene. Indeed, it is true that there are few areas that have been or are completely avoided by either Indians and Eskimos, prospectors, trappers, or airlines (commercial and military). However, these practically abandoned sections of the country are more closely related to the concept of a *passive or negative ecumene* than to the concept of an active ecumene. If, on the other hand, one believes that in order for an ecumene to exist man must have a certain amount of effective control over land, then it is necessary to exclude the almost empty and infrequently visited areas. On this basis, about 1,500,000 square miles (including fresh water bodies) must be removed, or about 40 percent of Canada.

The *active ecumene* comprising the habitation, exploitation, and linking ecumenes, has two levels of intensity — the intensive and secondary. First, let us consider the intensive ecumene. Interpreted in this manner, the habitation, exploitation (excluding hunting areas of the Eskimo and Indian) and linking ecumenes occupy about 10 percent of Canada; the population density reaches 52 per square mile or 10 times greater than the crude density of Canada. Hence, when one considers only the effectively occupied area, Canada loses some of its giganticism and takes on the scale of a moderately sized country. On this basis Canada is only twice the size of France, which is a more realistic comparison than 18 times if one employs absolute size. The principal or active ecumene is complemented by a secondary or sporadic ecumene which is spread over half of Canada. It is characterized by dispersed and slight human penetration and is found in the Middle North (Mid-North, J. W. Watson, 1963), in the lower Arctic, and even southern Canada.

These two ecumenes, the active and passive, comprise the *total ecumene* of Canada.

In 1952, I wrote in *La Géographie Difficile* that, in determining the ecumene one must consider more than just the lived-in regions. This observation introduced the concept of *subterranean* (mines), *aerial* (passengers and freight) and *hydrographic ecumenes*. In addition to land and fresh water bodies, the hydrographic ecumene includes the Gulf of St. Lawrence, the inlets of British Columbia, the waters of the Arctic archipelago, the Bay of Fundy, and Hudsonia (Foxe-Hudson-James Bay water bodies). On a functional basis these water bodies are linking and exploitation ecumenes. The hydrographic ecumene includes the channels used by submarines,

canals, fishing banks, beaches, the bottom of the continental shelf, pack ice (seal hunting), and ice islands (observation posts). In addition, all fresh water bodies and marshes (for game) would be a part of this ecumene.

The dynamic nature of the pioneer fringe has given birth to the expressions *potential ecumene* and *potential sub-ecumene*. The notion of future development is one of the most risky because one ignores the structural and speculative changes which can be brought to bear by technology, by needs, by many pressures and by man's capacity for innovation; even the ecological relationships can be modified. Because of this, the amount of land available for expansion of the ecumene is largely unforeseeable. For example, in 1934 Mark Jefferson stated that he saw "no future" with respect to the enlargement of the Québec ecumene; however, less than 20 years later, one could speak of "iron ore galore" (T. A. Retty) in Québec-Labrador. Moreover, 15 years ago who would have thought that the section of the Great North to receive the greatest investment would have been the DEW line? Predictions can also fall in the realm of the fantastic, so much so that J. C. Langelier, in 1887, predicted a population of 31 million along the southern edge of Hudson-James Bay. Clearly, the establishment of the potential ecumene in new lands, requires a great deal of prudence—and luck.

Forms of Inhabited and Exploited Areas

In certain cases, a broad correlation exists between the functional types of ecumene and their form. For example, an elongated form of settlement may be found along railway lines, whereas settlements associated with minierial exploitation sites tend to be of a dispersed oasis type. Beyond these, the ecumene takes four general forms.

1. The *bloc* form corresponds to an older type of settlement in which the physical environment enables expansion. Southern "Alsama" (Alberta-Saskatchewan-Manitoba), with its vast area of cultivated land, furnishes the best Canadian example. On

Figure 1

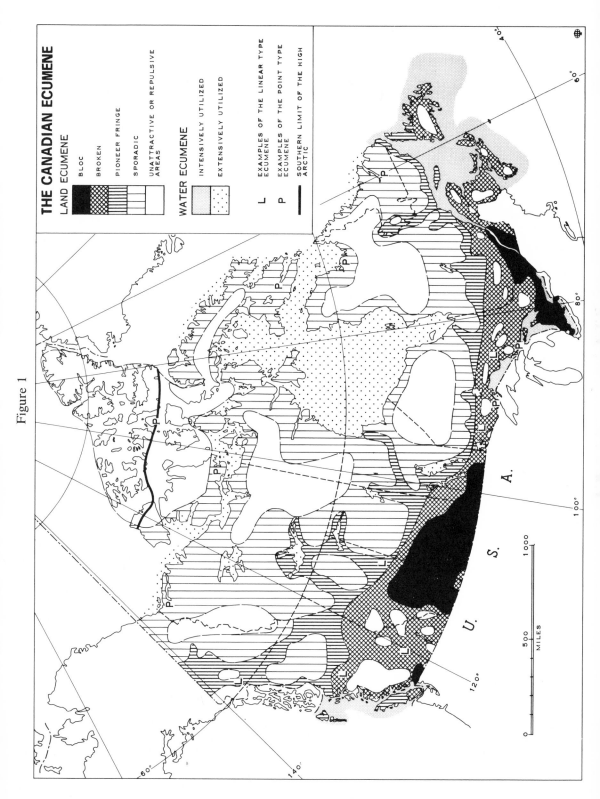

THE CANADIAN ECUMENE

LAND ECUMENE

- BLOC
- BROKEN
- PIONEER FRINGE
- SPORADIC
- UNATTRACTIVE OR REPULSIVE AREAS

WATER ECUMENE

- INTENSIVELY UTILIZED
- EXTENSIVELY UTILIZED

L — EXAMPLES OF THE LINEAR TYPE ECUMENE

P — EXAMPLES OF THE POINT TYPE ECUMENE

—— SOUTHERN LIMIT OF THE HIGH ARCTIC

U. S. A.

MILES

0   500   1000

the regional level, the bloc type should have a diameter of approximately 50 miles, and if there are sub-ecumene areas, they should occupy less than 10 percent of the total area. The expression, "bloc," does not imply that the population must be distributed uniformly within the ecumene, or that it should decrease evenly from the centre to the periphery; the term refers only to the form of spatial extension and it states nothing about the intensity of occupation.

2. The *linear* form is found in each of the chief functional types of the ecumene. Examples are found along the railway and highway systems of central Ontario, in the valleys of British Columbia, along the North Coast of the lower Saint Lawrence River, in the spread of suburban growth along major roads, and in some rural settlements in the Maritimes. By definition, the linear form must extend for about 50 miles along which interruptions are short. A special case of the linear ecumene is brought about by ring occupation around lakes, for example in the area north of Montreal; tourism is responsible for this sub-type.

3. The *point* form is found throughout two-thirds of northern Canada. There are three sub-types.

(a) *Isolated point.* Here, residential or exploitation centres are located hundreds of miles from one another and are not linked by road or railway. Eskimo settlements, polar radar bases, abandoned and old extraction sites (Rankin in the N.W.T.), and administrative centres (Inuvik) fall in this category. Isolation is an important characteristic because between these oases the land is almost completely devoid of man or at best it is a secondary ecumene.

(b) *Linked point.* This sub-type consists of exploitation or service towns situated along development roads; for example, Wabush and Schefferville, Churchill, and Whitehorse, where the arrangement of the settlements resembles knots in a rope. These northern settlements, like the isolated point settlements, can form over-organized villages, so much so that H. B. Hawthorn

spoke of "the instant creation of complete communities."

(c) *Temporary point.* This sub-type is characteristic of a weakly developed ecumene and reflects either systematic nomadism or the partial disappearance of many marginal settlements.

4. The *dispersed* form of ecumene reflects, as does the preceding one, the discontinuity of the habitat. Whereas the point type corresponds to villages and small towns, the dispersed type refers to regions in which people are scattered over a few dozen square miles. In the case of southern Canada, the dispersed type is characterized at the local scale by frequent, multiform interruptions of the residential, exploitation, linking ecumenes. As Pierre Biays and K. H. Stone have noted, these regions can have two pioneer zones—an external margin and an internal margin. The sub-categories here are numerous, and many sectors of southeast Canada and of the near North exhibit this type of bizarre-shaped and broken-up ecumene.

SCHEMATIC ZONATION OF THE ECUMENE

The functional and morphological types furnish a systematic picture of the Canadian ecumene. In general, however, they do not correspond to any specific geographic locations. The characteristics of the habitation, exploitation and linking ecumenes are not associated only with the southern part of the country, nor are the interruptions of the habitat unique to the northern zones. Thus, these general types of ecumene should not be used as a basis for strict regionalization. On the contrary, the zonal arrangement of the Canadian ecumene has its own typology, and can be divided into four zones (modified from R. Gajda):

1. Southern Ecumene

The southern ecumene (certain authors call this "useful Canada" or southern Canada; G. Taylor speaks of the "populous zone") is a discontinuous and uneven ribbon of settlement adjacent to the United States border, and includes the peripheral waters of the St. Lawrence (lakes, estuaries and Gulf) and the Pacific Coast. In terms of land area, it encompasses about 700,000 square miles, or 20 percent of Canada's total area, and contains about 99 percent of Canada's population. The average density is 30 inhabitants per square mile. The problems associated with the southern ecumene are many. It occupies a narrow area which faces the rocky Ontario Shield, scarcely 100 miles away. Between Maine and the St. Lawrence estuary, the belt of habitation is only about ten miles deep. The fact that seven Canadian capital cities are less than 60 miles from the United States border serves to emphasize the location of this ecumene. In this context Canada appears to be only the second floor of a North American building, of which the United States is the foundation. This state of proximity to a much more powerful neighbour is unfavourable for the independent development of Canada. The southern ecumene is further handicapped because of its fragmentation; the inhabited area is open and discontinuous. This is reflected in two ways. First, there is a series of large, discontinuous hearths of settlement. A massive intrusion—the Ontario Shield—has created an almost complete break in the ribbon of settlement, and forms the basis for the division of southern Canada east of the Cordillera into two parts, the Great Lakes-St. Lawrence region (East), and the Alsamian bloc (West). In its turn, the Cordillera cuts western Canada into two more very unequal parts (in terms of area). The historian A. R. M. Lower has pointed out that these breaks in the ecumene are poor conductors of trans-Canadian affairs. Southern Canada, then, is a collection of settlement hearths separated by sparsely peopled or entirely uninhabited areas. From east to west these nuclei are: the shores of the Gulf of St. Lawrence (average density of 45 people per square mile), the Lévis-Windsor corridor (80), southern Alsama (15), and the Lower Fraser (30). Isolation tends to give rise to as many "sociological"

Canadas" as there are hearths of population. Second, spatial discontinuity is further emphasized by differences in economic development and psychological attitudes; in fact, inter-regional communication is so poor that W. K. Lamb envisaged *The Third Solitude*. Between these living cells, emptiness prevails; southern Canada is only an assemblage of distinct parts, and the juxtaposition of these parts appears to be closely related to the anatomical structure of the country.

Geological structure and the inherent quality of the land are not the only factors responsible for the discontinuity of the habitation ecumene in southern Canada. Other factors are important too. For example, the built-up zone is disrupted by great local variation in relief due to the Quaternary processes. Even in the interior of the St. Lawrence Lowlands, the agricultural ecumene is broken up by sandy deltas, rock ridges and marshy depressions. Indeed, after three centuries of occupation, the railway and highway connections between

THE "PRINCIPAL ECUMENE" SHOWN AS A PERCENTAGE OF THE TOTAL AREA

(by Province)	
Province	Percentage
Newfoundland	6.5
Prince Edward Island	100.0
Nova Scotia	49.7
New Brunswick	61.3
Québec	11.6
Ontario	19.6
Manitoba	17.3
Saskatchewan	47.5
Alberta	30.0
British Columbia	14.4
Yukon	0.9
Northwest Territories	0.3
Canada (average)	12.4

Source: R. Gajda, 1960, p. 8. The principal ecumene corresponds closely to the ecumene used by Gajda. Do not confuse principal ecumene and southern Canada. These data have not changed much since 1960.

Québec and Drummondville still cross an empty and desolate landscape. To the north of Montreal, only 19 percent of Joliette County is politically organized. Disastrous economic fluctuations and an absence of planning have also contributed to the discontinuity of the habitat. The closing of mine sites, the abandonment of marginal land and homesteads, and the plunder of forest areas have created neglected, and for a long time, almost useless areas.

Fortunately, the linking ecumene joins the dispersed nuclei of the habitation ecumene.

2. *Pioneer Fringe*

The zone of pioneer "speculators" extends north from the nodes of southern Canada. The northern limit of this area is marked by a line linking the northern-most points reached by year-round roads and railways; this boundary is extremely dynamic, changing from year to year in response to the opening of new "access roads to resources." These roads connect Schefferville in Québec-Labrador, Moosonee on James Bay, Churchill in Manitoba, Yellowknife in the Mackenzie basin and Dawson in the Yukon, the latter town being 5,000 miles from Labrador. Between each of these pioneer posts, the edge of exploitation is extremely uneven. In Ontario it dips as far south as 50° N. lat., while in the Yukon it extends to the Arctic Circle (67° N. lat.). This zone, the Middle North (Hamelin, 1962), includes the majority of Canadians who reside outside of southern Canada. The habitat is either the point or linear form type. Large areas of seemingly unused land —the empty spaces—or areas which are tenuously occupied, extend between the residential and exploitation centres. Because these intermediate areas are widely scattered, the crude population density of the Middle North is only about 2 people per square mile. Economically, these pioneer zones are either regions of subsistence economy, or colonies of mineral or forestry exploitation. While these linear penetrations

leave parts of the Middle North untouched, (and in the Arctic they do not quite reach "Eskimoland" or Inukland), they enable Canada to stretch northwards. Due to "northing" (I. Bowman) the ecumene has the form of a rake, the teeth of straight roads pointing towards the pole. The northward extension of the ecumene poses new problems of transcontinental liaison in Canada, this time situated in subarctic latitudes and farther removed from the American border.

3. *The Sporadic Ecumene*

In the North, the principal ecumene undergoes an almost complete rupture. The last foothold of the continent—the northern half of the Québec-Labrador Peninsula, northern Ontario and the interior of Keewatin—is a collection of areas almost devoid of habitation. Excluding certain points of contact along the Mackenzie River and the Foxe-James-Hudson areas, the Middle North is cut off from the habitation ecumene of the Far North. Here, the habitat is mostly "Eskimoland," a discontinuous littoral between the northern edge of the continent and the arctic archipelago. Whites outnumber the Eskimos and the total population is only some ten thousand. Before the recent concentration of Eskimos into larger settlements such as Frobisher, it was more realistic to establish population density on the basis of the numbers along the coast rather than per square mile. In addition to "Eskimo country" the sporadic ecumene includes other slightly used areas separated by empty spaces and occasional pioneer penetrations. In many cases the population is nomadic, and the density throughout is very low.

4. *The High Arctic*

The High Arctic, of which Parry Straits near 75° N. lat. is the approximate southern boundary, is practically non-ecumene; it is almost as "absolute void" (Max Sorre). The residential oases are small in number, contain few people and are totally artificial. The contrast between the presence of small, well organized urban hearths and the general absence of man cannot be more marked. On the basis of function these empty spaces comprise a passive ecumene, a territory to defend; hence, they become travelled lands. Petroleum prospecting still does not permit us to speak of exploitation. Excluding Resolute, where Eskimo settlement is the result of a recent government decision, Eskimos no longer inhabit the Canadian Queen Elizabeth archipelago.

OTHER GENERAL CHARACTERISTICS

From this systematic analysis of the types and zonation of the Canadian ecumene other characteristics of the inhabited areas of Canada emerge.

Paradoxically, the country remains poorly adapted to vast areas, and K. Hare has stated that the second largest country in the world has had to create an island to host EXPO! Moreover, as Canada grows northward the residential ecumene becomes smaller for more and more Canadians. Between 1951 and 1961 the majority of new citizens (by birth and immigration) settled in southern Ontario or in the Montreal Plain. In the words of C. A. Doxiadis, Canada has become an "ecumenopolis." Since urbanization emphasizes areal concentration, it appears that Canada is growing as it shrinks. Indeed, it would be astonishing if the progressive disproportion between the head and body did not cause difficult problems of balance and interregional communication. Clearly an over-concentration in the habitation ecumene will require a better linking ecumene between the cities.

A lengthy and fruitful dissertation could be undertaken to contrast the individuality of the ecumene and the diverse structure of the country. From this, one could see how land survey types affect the different forms of residential and exploitation ecumenes. For example, the distribution of population, which is more linear in the case of the *rang*, is dispersed in the township survey system. But we do not know enough about the changes in farming and land use that have resulted in the replacement of

English speaking farmers by French Canadian "habitants" in the Eastern Townships of Québec and in Eastern Ontario. On the basis of "racial" origin, each ethnic group in Canada seems to have its own ecumene: the Eskimos in the Arctic; French Canadians in Laurentia; British in Ontario and British Columbia; and New Canadians in Alsama. Nowadays, urbanization is threatening the polyethnicity of these formerly quasi-exclusive ecumenes. The large Ontario cities are a case in point. One might well ask whether this situation will hasten assimilation of the demographic stocks. However, the bi-ethnic nature of the country makes it difficult, if not impossible, to propose a simple definition of "Canadianization" which would serve as a reference point. From region to region, Canadianism remains differential. Moreover, history has not clearly defined the frontiers of the individual ethnic realms. Has French Canada a meaningful dimension for all of Canada or only for Québec? Does English Canada exist in the same way in Alsama as in Ontario? It is apparent that the intimate relationships between man and land still lack definition in Canada. Is this lack of a sense of identity simply a reflection of the country's youth or is it a bad omen?

If one applies the concept of the ecumene, which is the most realistic barometer to evaluate the relationship between administrative and political frontiers, it is obvious that agreement is practically non-existent. For example, 60° N. latitude is not the only good boundary provincialists may hope for as the northern boundary of the western provinces. Realignment of the regional limits must be one of the first steps in the reorganization of Canada and these adjustments are desirable at all levels—counties, regions, provinces and even the country.

For the most part, Canada is no longer characterized by a frontierism of land clearing. Traditionally, pioneer life was agricultural and quite self-sufficient; now, in the Middle North, pioneerism is related to mineral and oil exploitation, forestry and even fishing; agriculture is no longer enlarging the country. Even though the pioneer zone is far removed from the southern centres of activity, it reflects big city initiative. J. M. S. Careless has demonstrated how metropolitanism has replaced 19th century frontierism. Indeed, the reputation of agriculture in Canada has been overrated to the point that in Québec one speaks of the "myth" of agriculture (P. Dagenais). Today, Canada is no longer a new country, at least not in the same sense that it was around 1910.

All pioneer zones are extremely variable both in space and time. There are two basic reasons for this dynamism. If the zone is an agricultural one, the conditions are so marginal that it has scarcely a chance to become established when other seemingly better work opportunities loom on the horizon. If development is based on forestry or mining, which is the case along the penetration roads in the Subarctic, extraction is a function of capital and even more of outside markets. Hence, outside decisions determine the vitality of the pioneer zone; a major administrative decision can signify the opening of new centres at a cost of hundreds of millions of dollars, or it signifies the abandonment of towns (Uranium City), or stagnation (Abitibi). Change is so rapid that each year it is necessary to redraw the map of the pioneer zone, especially the non-agricultural zone. Man's presence along the margins of the exploitation, linking, and residential ecumenes is often temporary, fragile, and non-consolidated; P. Burton has written "impermanence." In such an ecumene, the place-name studies are very complex.

No part of Canada has a temperate climate; it is a climate of extremes. The migration of tourists in winter and summer to visit the holiday areas, the annual migration of Labradoreans, occupations that connote seasonal displacement, like prospecting, and exclusive winter roads in the Middle North furnish many examples of the temporal modification of the residential, exploitation, and linking ecumenes. In a large part of Canada there is a "winter" ecumene and a "summer" ecumene. Moreover, climatic variations and the generally harsh conditions, result in a high index of discomfort for a large proportion of the people across most of the country; in the course of a year most days are disagreeable; the Canadian ecumene is not a climatic paradise.

The attractiveness of the lowlands on the one hand, and the topographic and climatic repulsiveness of the mountains (as well as certain modes of life) on the other, have encouraged people to live in the lower altitudes. After calculations made by Staszewski, it can be shown that in 1951 the distribution of Canadians on a vertical basis differed from the world average. Seventy percent of the population (in contrast to a world average of 56 percent) lived in regions less than 200 meters above sea level, and 2 percent (compared with 8 percent) lived in regions over 1,000 meters above sea level. Data on a provincial basis also serve to illustrate that Canadians avoid mountainous and difficult localities.

According to the principles of a free economy, Canada, as such, has not been subject to direct government direction. However, for the last 30 years, governments have engaged in reorganizing the ecumenes; but it is no longer a question of guiding expansion, but rather of replanning the marginal regions. In the Abitibi area of the Québec Clay Belt, for example, the trend has been to consolidate parishes rather than to develop new ones, and on the margins of the Eastern Townships in Québec recently opened areas are already in the process of abandonment. The principal federal legislation has been in the form of two Acts. The first, the Prairie Farm Rehabilitation Act of 1935 (PFRA), was concerned chiefly with the problems of western agriculture; the second, the Agricultural Rehabilitation Development Act (ARDA) of 1961, is oriented to identifying and resolving rural development problems throughout Canada. In Eastern Canada alone, one of the goals of ARDA is to bring about an adjustment in land use on more than 5 million acres of marginal land. In Québec, former agricultural land will be returned to forest and will become part of the renewable ecumene of exploitation; this type of rational politics will bring about a retreat of the frontiers of settlement. It appears the ecumene will tend to become "planned," and will no longer be a reflection of individual initiative; a *rational ecumene* will replace an empirical one. The characteristics of the Canadian ecumene seem well suited to a certain amount of state intervention. The dispersion of the habitat necessitates the establishment of efficient methods of communication which will meet economic, political and social requirements; a powerful infrastructure in conjunction with enlightened planning should be able to combat the discordant tendencies of the ecumene.

Conclusion

The application of the notion of the ecumene to Canada not only involves research in regional and descriptive geography, but it is also the study of an element which has deeply influenced the destiny of man on this continent. Throughout North America,

the spatial dimension has physically molded the country, punctuated the advance of settlement, affected human attitudes towards the mobility of manpower and the waste of land, and developed a taste for types of pioneer life. Gildore of Felix-Antoine Savard is only one of a large number of pilgrims who travelled the "vastness" described by Gabrielle Roy. Despite its burden on Canada, space, as such, has been studied very little, and generally the vision that we create for ourselves of the country is based mostly on historical linking, on ethnic components, and on economic and political structures. The only exceptions to this are the studies done by a few historians who, by extending Turner's theme of American frontierism, have examined the influence of environment in pioneer zones. Hopefully, these exploratory remarks have indicated certain landmarks in understanding the problem of Canadian space and have systematized the general notion of the ecumene, of "man and his world."

BIBLIOGRAPHY

Biays, Pierre, *Les Marges de l'oekoumène dans l'Est du Canada (Partie orientale du Bouclier canadien et Ile de Terre-Neuve).* Presses de l'Université Laval, Québec, 1964. Collection *Travaux et Documents* du Centre d'Etudes Nordiques, no. 2.

Clibbon, P. B., *Les relevés sur l'utilisation du sol.* Conseil d'Orientation Economique, Québec, 1963.

De Smet, R. E., "Degré de concentration de la population," *Revue Belge de Géographie,* Vol. 86 (2), 1962, pp. 38-66.

Gajda, R., "The Canadian Ecumene—Inhabited and Uninhabited Areas," *Geog. Bulletin,* Ottawa, No. 15, 1960, pp. 5-18.

Hamelin, Louis-Edmond, "Direction Nord," Presentation to the Royal Society of Canada, No. 17, 1962, pp. 19-25.

Institute of Current World Affairs, *Newsletter,* (New York), 1966.

Institut de Géographie, Université Laval, Québec. *Répartition de la population, 1961, Québec, Canada.* Map prepared under the direction of Louis Trotier. Conseil d'Orientation Economique, Québec, 1966.

Jefferson, M., "The Problem of the Ecumene. The Case of Canada," *Geografiska Annaler,* Vol. 16, 1934, pp. 146-159.

Juillard, E., *L'économie du Canada.* Que Sais-je?, Paris, 1964, pp. 11-16.

Morrissette, H., *et al., Report on the Study of Oecumen in Canada.* Direction de la Géographie, Ottawa, 1959, (unpublished).

Nicholson, N. L. and Z. W. Sametz, "Regions of Canada and the Regional Concept," *Resources for Tomorrow,* Conference Background Papers, Vol. 1, pp. 367-383.

Siegfried, André, *Le Canada. Puissance internationale.* Colin, Paris, 1947.

Sorre, M., *Les fondements de la géographie humaine.* Tome I, 1951; Tome II, 1948 and 1950; Tome III, 1952; Index to the ecumene, pp. 497-498 in Tome III.

Stamp, L. D., (ed.), *A Glossary of Geographical Terms.* London, 1961, Oekumene, p. 344.

Stone, K. H., *Human Geographic Research in the North American Northern Lands.* In Arctic Research, Arctic Institute of North America, Montreal, 1955, pp. 209-223.

Stone, K. H., "The Development of a Focus for the Geography of Settlement," *Economic Geography,* Vol. 41, 1965, pp. 346-355.

Watson, J. W., *North America.* London, 1963, pp. 286 and 460, 1 map.

The Expansion of Settlement in Canada: A Contribution to the Discussion of the American Frontier

Eckart Ehlers

THE DEVELOPMENT OF THE FRONTIER IN THE U.S.A. AND CANADA TO 1900

The history of North American colonization is closely linked with the concept of the "frontier" — that transitional zone lying between the land occupied by European colonists on one side and Indians on the other side. Wherever the advance of colonization pushed the white man's frontier forward, a battleground emerged wherein the settler had to stand the test: fighting against nature, the Indians, or other settlers. Historically, the frontier is the setting of the so-called "Wild West" where adventurers, speculators and "outlaws" tried their luck. Although the border area of settlement shifted with the opening up of the North American continent, its structure remained the same. By 1891, when most parts of the continental United States had been settled, the frontier ceased to exist.

The nature of the American frontier owes its interpretation to F. J. Turner who saw the frontier as a battlefield in which the European intruder pushed into the unknown wilderness. In doing so he had to conquer nature, or be conquered by her. Ignoring fur trappers, merchants, and a few gold prospectors, it was the farmer-colonist who pushed the American frontier westward. The classical American frontier — the region between the settled and unsettled land—was characterized by a pattern of evolution that was repeated with each advance of settlement. According to Turner, there were four social and cultural phases by which a frontier evolved:

(a) the first phase is characterized by the presence of hunting or farming Indians and white hunters and explorers;

(b) the second phase is characterized by the presence of merchants and other entrepreneurs who forged ahead into this previously unoccupied land to establish trading posts;

(c) the third phase is considered the true pioneering stage. It consisted of the advance of cattle breeders followed by pioneer farmers who practiced a system of land and crop rotation and used the land fairly intensively. These pioneers formed the basis of the permanent population;

(d) the last stage is characterized by the growth of towns and industrialization, a

● Eckart Ehlers *is Dozent, Geographisches Institut, Universitat Tübingen, Tübingen, West Germany.*

Reprinted from *Geographische Rundschau,* Vol. 18 (9). 1966. pp. 327-337. with permission of the author and the publisher.

process which effectively terminated the pioneer stage.

The settlement of the American West, that is, the colonization movement of the 19th century, began with the advance of the Appalachian frontier. By the end of the 18th century, settlers had made a breakthrough via the Cumberland Gap to Kentucky, thereby gaining access to the West. Railway construction, promoted by Lincoln, permitted rapid development of the Middle West and Great Plains from the east. But with the admission of California into the Union in 1848, the unsettled region became hemmed in also from the west. Hence, the last 50 years of the American frontier were marked by the expansion of two pioneer regions: one from the West, the other from the East. Witness for example that the completion of the first transcontinental railway in the U.S.A. did not take place on the west coast, but near Ogden, Utah, where in 1869, the tracks of the Central Pacific Railway, pushed forward from California eastwards, and the Union Pacific Railway, built westwards from Nebraska, were joined. Moreover, the fact that the last great farming frontier was not in the Far West, but rather in the Great Plains (and in Oklahoma, which was Indian territory until 1889), shows that the opening up of the continent resulted not only from the spread of settlement toward the west, but also in later years from both the east and the west.

The end of the United States frontier was by no means coincident with the end of the North American frontier. By the middle of the 19th century, the dwindling availability of land in the United States led to a small flow of settlers into Canada. This emigration pattern, combined with an influx of European settlers, was intensified after 1870. The Canadian Prairie Provinces of Manitoba, Saskatchewan and Alberta, with their extensive grasslands suited to farming and cattle ranching, had a strong appeal to the land-hungry colonists.

In 1870, there were less than 45,000 people in the area known today as the Prairie Provinces, and it was only by 1886 that a narrow strip of land between Winnipeg, Regina and Calgary was settled. A few years later, however, the entire grassland of the three Prairie Provinces had been brought under cultivation. The population rose to over 1.3 million by 1911, and tilled land increased from approximately 500,000 acres (1886) to more than 5 million acres (1911). By this mass immigration to the Canadian provinces, which changed the predominantly westward direction of the American frontier to a northerly one, the land-hungry settlers opened up the last large reserves of land in North America within a few years. With the outbreak of World War I settlement of the great agricultural frontier on the northern edge of the Canadian Prairies came to a stand-still.

The expansion of settlement in Canada since the beginning of this century, the opening up of the northern forest areas, and the present methods and aims of colonization differ considerably from the earlier development phases. Three factors are responsible for this: (1) the application of technology, especially in transportation media as, for example, the airplane and railway to the opening up of new areas; (2) the absence of good agricultural land in the forest area, bounded by the agricultural areas in the South and the sterile Canadian Shield in the North; and, (3) the different aims of colonization, which formerly were directed toward the development of new agricultural lands; today, however, the emphasis is on the development of exploitable mineral resources. Typical of the Canadian frontier in the 20th century is the fact that an agricultural area and an industrial region often exist side by side. This is a fundamental difference from Turner's classical American frontier which separates agriculture and industry in time as well as in space.

THE CANADIAN AGRICULTURAL FRONTIER

The Canadian agricultural frontier is that area in which the ecumene is expanding toward the North (Figure 1). Following the occupation of the favourable agricultural land in the East, especially in Ontario and Quebec, in the 18th and early 19th centuries,

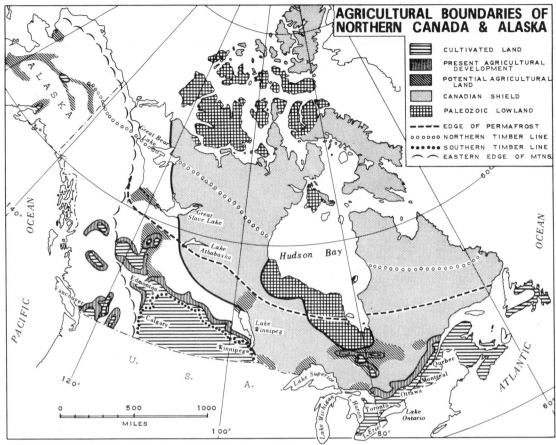

AGRICULTURAL BOUNDARIES OF NORTHERN CANADA & ALASKA

CULTIVATED LAND
PRESENT AGRICULTURAL DEVELOPMENT
POTENTIAL AGRICULTURAL LAND
CANADIAN SHIELD
PALEOZOIC LOWLAND
--- EDGE OF PERMAFROST
ooooo NORTHERN TIMBER LINE
••••• SOUTHERN TIMBER LINE
⌒⌒ EASTERN EDGE OF MTNS.

(AFTER VANDERHILL, 1962; EXTENT OF PERMAFROST AFTER TAYLOR, 1957)

Figure 1

and of the farming and grazing lands on the Prairies of western Canada before World War I, the flow of European immigrants and American farmers diminished. The few settlers who acquired land and established their own farms after this date, were pushed to the northern edge of the Prairies. Here, during the 20th century, two different areas of agricultural settlement developed: (1) a border area in the West stretching in a narrow band between the developed areas in the South and the unopened forest regions in the North; and, (2) large isolated areas of land cleared in the midst of the North Canadian forest region, such as the Peace River Country in the West, and the Great Clay Belt in eastern Canada.

The Western Border Areas

The settlement fringes in the North lie between 53° and 55° N. latitude and form a broad belt in which the cultural and natural elements of the landscape are closely linked. The colonization of this territory began in the years following World War I and reached a peak between 1930 and 1940. The heavy flow of colonists was due, in part, to high world prices for agricultural products, which attracted many unemployed immigrants and farmers from elsewhere in Canada. Another significant factor was the severe and prolonged drought of the "dirty Thirties" that turned large parts of the southern Prairies into unproductive farming areas. Many farmers, either of their own accord or under government sponsored schemes, moved from the southern Prairies into the northern frontier areas. Since this time, however, the pioneer fringe on the northern edge of the prairies has stagnated. In fact, these fringe areas have experienced a great loss in population and today they are marked by the deteriorating remains of many former agricultural operations. The

32

comparatively small number of recently established farms is being surpassed by the abandonment of older ones. For the most part, however, the cleared and arable fields of the deserted farms have been absorbed by neighbouring farmers so that, despite farm abandonment, the cultivated lands are preserved and the remaining farms are enlarging their operations significantly.

As a result of a lack of interest in the free acquisition of lands, land development along the edge of the settled areas remains under the control of the provincial governments. In Saskatchewan and Manitoba there are 13 areas of closed homestead land which have been developed with government assistance since the end of World War II. Even though the amount of development land is restricted, there is still land available. Despite numerous financial and technical assistance measures, only a few settlers are prepared to settle and bring new land under cultivation.

The Peace River Country and the Ontario-Quebec Clay Belt

On the other hand, there are two especially active pioneer areas of agricultural settlement in northern Canada: the Peace River Country in northern Alberta, and the Great Clay Belt in Ontario and Quebec, which stand as agricultural islands in the midst of the northern forest. In contrast to the Prairie border areas, these areas are large; they possess large reserves of potential agricultural land; and they are growing in terms of settlers and population.

Peace River Country—Today the Peace River Country, which extends into British Columbia, is the largest pioneering zone in Canada (Figures 2 and 3). As a result of its northerly position (between 55° and 59° N. latitude), and its somewhat "stormy" development, people like to consider it the last offshoot of the American frontier. The settlement of the Peace River Country

A township in the Peace River Country, which was settled between 1912-1916. The township is typical of the North American Great Plains: the division of the land into sections; isolated farms, and scattered central places. Note the abandoned farm buildings.

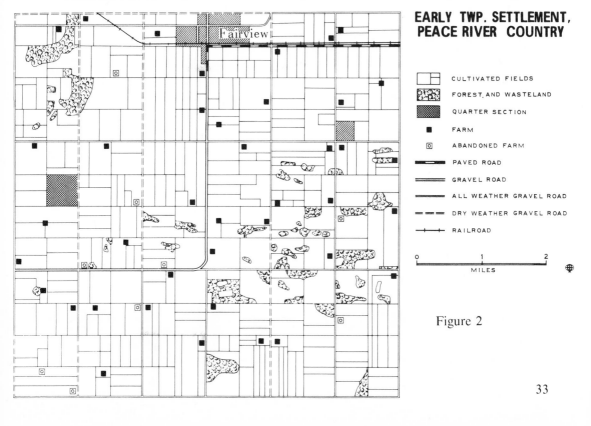

EARLY TWP. SETTLEMENT, PEACE RIVER COUNTRY

CULTIVATED FIELDS
FOREST, AND WASTELAND
QUARTER SECTION
FARM
ABANDONED FARM
PAVED ROAD
GRAVEL ROAD
ALL WEATHER GRAVEL ROAD
DRY WEATHER GRAVEL ROAD
RAILROAD

0   1   2
MILES

Figure 2

33

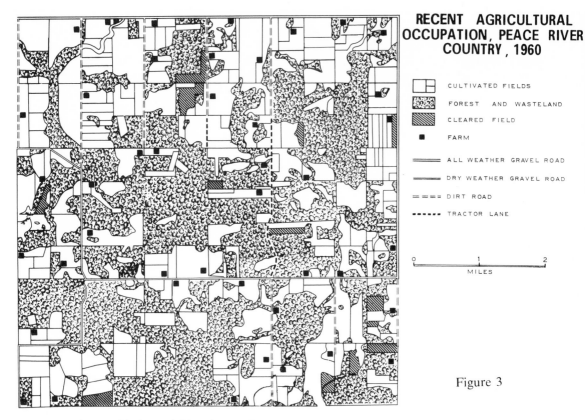

RECENT AGRICULTURAL
OCCUPATION, PEACE RIVER
COUNTRY, 1960

CULTIVATED FIELDS

FOREST   AND   WASTELAND

CLEARED   FIELD

FARM

ALL WEATHER GRAVEL ROAD

DRY WEATHER GRAVEL ROAD

DIRT ROAD

TRACTOR LANE

MILES

Figure 3

A township located 31 miles northwest of the settled township in Figure 2. The Town-
ship is in a recent stage about (1960) of settlement and clearing. The systematic sectional
survey is discernible in the northern part of the township.

started at the beginning of this century, when free land in the southern Prairies was no longer available for settlement and the late-comers were forced into the northern forest areas. The inter-mixture of grassland and parkland, combined with rich degraded black soil, were advantageous for successful farming. These prospects were further enhanced by the construction of the railway into northern Alberta which ensured the movement of farm products to market. The population climbed from only 1,200 in 1911 to 12,000 in 1921 to nearly 30,000 people by 1931. Today about 120,000 people, almost all of whom are engaged in farming, reside in this 135,000 square mile area.

The expansion of settlement in the Peace River Country is progressing at an undiminished rate, especially in the northern part between the town of Peace River and Fort Vermilion. Recent development has been influenced by the completion of the Great Slave Lake Railway in 1964, built to carry the valuable lead and zinc ores out of the Northwest Territories. This railroad runs north-south through the northern Peace River Country and has had a marked impact in attracting settlement. Grain elevators at the newly established railway station of High Level, the most northerly on the American continent, bear witness to this extensive new development. In spite of comparatively unfavourable settlement conditions, new settlers continue to arrive. While financial assistance is offered in Saskatchewan and Manitoba, the settlers in Alberta have to establish their farming enterprises on their own financial resources. It is only after several years of breaking and clearing and after the construction of a house and farm buildings that the settlers receive the title to their 320-acre or sometimes even 480-acre farms.

34

Ontario-Quebec Clay Belt—Modern agricultural development in the Great Clay Belt of eastern Canada is of similar proportions. The moderately fertile soil, which has developed on former glacial lake deposits, and the land, which is comparatively easy to clear, give promise of successful farming in this area, 250 to 300 miles north of the agricultural core of southern Ontario and Quebec. As in the Peace River Country, the completion of the railroad in 1913 was a motivating factor in the settlement of this area. The acquisition of land proceeded in much the same way in Ontario and Quebec, but has become distinctly different since 1930 when the government placed all federal land under the control of provincial administration. As a result of different land policies, the two parts of the Great Clay Belt show divergent trends along the frontier of settlement. In Ontario, for the most part, there is a large number of abandoned farms and the appearance of the cultivated land bears strong resemblance to the stagnant areas of western Canada. Many of the former settlers are migrating to the industrial centres of the province. In the Great Clay Belt in Quebec, with its largely French-speaking population, new farms and well planned towns are being established. The population tripled between 1931 and 1956 and is still increasing. The success of agricultural development and expansion in this area can be explained only by the assistance given to the settlers by the Church and the government, while in Ontario settlers are given very little assistance.

Thus, there are remarkable differences in regard to the extent and the success of colonization along the northern Canadian agricultural fringe. However, the methods of colonization and the economic structure of the fringe areas reveal strong similarities. In contrast to the settlement of the grasslands of the classical American frontier, the clearing of forests is a pre-condition of farm establishment along the agricultural frontier of northern Canada. At the beginning of this century, the axe, saw and the horse were the most important of the settler's land-clearing implements. Since World War II these have been largely replaced by mechanized equipment. Today the costly process of forest clearing proceeds in four steps: (1) clearing, which is done by bulldozers; (2) separation of the useful stumps from brushwood, which is dried and burned; (3) deep plowing of the virgin soil to cut off the remaining roots and bring them to the surface; and, (4) removal of rocks and stones, an operation that is usually done by hand. Because of mechanization, it is now possible to clear much larger areas in less time. In the past, the pioneers could bring only 2-5 acres at the most under cultivation each year; nowadays, with the use of a bulldozer, it is possible to bring as much as 300 acres or even more into cultivation.

The farming structure of the border areas of settlement exhibits widespread homogeneity. The great danger from frost, accompanied by long, cold winters, as well as dryness in the summer, and relatively infertile soils are typical of all the border areas. In addition to poor physical conditions, distance from market and isolation seriously hinder settlement. Moreover, many institutions such as schools and hospitals are absent. Even today settlements in these border locations are reminiscent of classical pioneer life. This is especially true where government assistance is not substantial, for example, in the Peace River Country, in northern Ontario and in the frontiers of the Prairies. In many cases, settlers often live in simple wooden shacks, block houses, sheds, trailers and corrugated iron huts. On the other hand, in the government homestead areas of Saskatchewan and Manitoba, and in Quebec's Great Clay Belt, modern homes were provided for the pioneers.

Agricultural land use in these border areas reflects the harsh physical environment. Except in the central Peace River Country, where wheat can be raised successfully, the growing of coarse grains, especially barley which matures quickly and is frost resistant, prevails in all other pioneer areas. In addition, the cultivation of grasses and alfalfa is widespread, and they form the basis of a flourishing beef cattle industry, except in Quebec, where cattle are kept chiefly for milk production. Furthermore, successful grass production has led to specialization in seed production as well as to an increase in the production of oil plants, flax and rape, which can tolerate the subarctic climate and the soil conditions.

THE CANADIAN INDUSTRIAL AND MINING FRONTIER

The second type of Canadian pioneering frontier, the mining and industrial frontier, is, in contrast to the agricultural frontier, a product of the 20th century technological age. Characteristic of this frontier is the development of small and isolated mining and industrial settlements. Unlike the agricultural border areas, the isolated mining and industrial settlements result in little, if any, areal expansion of the settled areas. In many cases, they are so isolated that they can be reached only by plane. In the mining and industrial frontier several developmental phases are apparent with each phase being

dependent upon new technological developments, particularly with respect to transportation techniques and prospecting methods.

The emergence of an industrial and mining frontier in Canada began in the 19th century when isolated settlements associated with the Gold Rush were established in the Yukon and Alaska. The Klondike gold rush of 1898 and its centre, Dawson City, are an example of this early development. The real industrial and mining frontier began, however, with the building of the transcontinental railway, especially in those places where the railway crossed the Canadian Shield, which until then had been dotted with scattered encampments but had never been adequately explored. In Ontario and Quebec, the Canadian Shield pushes far into the south and covers most of the two provinces. It is, therefore, not surprising that ores such as copper, gold and nickel were discovered here in the early part of the 20th century. Simultaneously, entrepreneurs began to open up the previously undisturbed

Figure 4

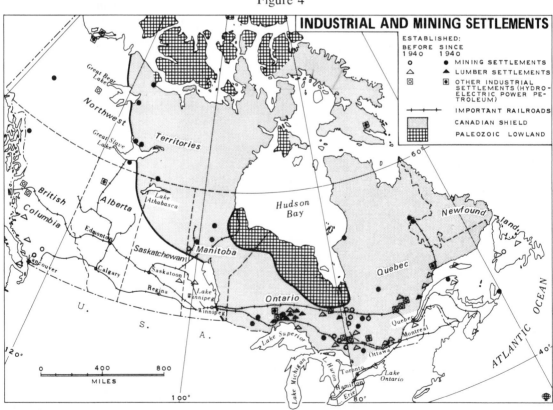

forest reserves in the southern part of the Shield, and numerous sawmills and wood-processing industries sprang up along large rivers and along navigable lakes. Most of the early mining and industrial settlements were adjacent to their chief means of transportation, the railway. Of 66 lumbering and mining settlements established between the turn of the century and 1940 in Canada (Figure 4), 40 were in northern Ontario and Quebec. Of these, 23 were based on the extraction of gold, silver, nickel, copper, and other base metals, and 17 for the storing and processing of wood. Outliers of the forest and mining frontier developed in northern Manitoba, which had been opened up early by the railroad, and in the Rocky Mountains of British Columbia for the extraction of gold and copper. Nine of these settlements were based on wood storage and processing. Disregarding these developments, it is apparent that the core of the industrial and mining frontier was concentrated in eastern Canada, especially in those areas which had been opened up by the railroad.

The second expansionist phase of the industrial and mining frontier began about 1940 and was based clearly upon technological advances, particularly in transportation. These developments have made possible a more detailed exploration of and penetration into the northern and relatively unknown forest. Of special importance was the development of aerial photography and the application of the magnetometer to geological exploration and mapping. The result has been the discovery of numerous exploitable metal bodies widely scattered over the whole Canadian Shield.

Between 1940 and 1958, forty-five new mining and lumbering settlements were established. While the newly founded lumber settlements (12) have been restricted to the southern edge of the Canadian Shield, mining settlements have spread across all of northern Canada and lie, with the exception of those in British Columbia, in the Canadian Shield; ten are situated in the northern part of Quebec; nine in Ontario; and four each in the northern Prairie Provinces and

British Columbia. There are six new mining settlements in the Yukon and Northwest Territories, such as Pine Point on Great Slave Lake.

Figure 4 reveals the isolation and wide distribution of the post-World War II mining and industrial settlements. Almost all the land which has been shown to have metal deposits by magnometric survey lies so far north on the Canadian Shield that it can only be reached by aircraft. Because of the high transportation and development costs, only the most valuable sources of metal have been developed. Since 1940, these developments have included gold (10), copper (8), lead-zinc (6), and iron ore (5). For many of these mining settlements, it was necessary to fly in all construction materials, drilling equipment and operational machinery. And in many cases, the concentrated mining products must be flown to the processing centres in the South.

In spite of the differences in age and location, the old and new border areas of the industrial and mining frontiers are very similar. These similarities are valid both for the appearance of the cultural landscape and for their economic structure. Characteristic of both the old and the new is their rigorous adjustment to their resource base. Hence these settlements have also been labelled "single enterprise communities." Their development and size mirror the importance of the industrial company to which they owe their existence. Headframes and refineries associated with the industrial settlements, the open-cut workings, the administration and work centres, and the living quarters of the workers and employees (some have small gardens) represent the entire developed area, which appears island-like and insecure in the middle of the forest land. The lumbering settlements are somewhat larger in area, but even so, they are definitely restricted in their physiognomic effects. Besides sawmills and associated processing industries, there is generally an extensive network of roads for the opening up of the useful forest reserves, the transport of the lumber and for reforestation activities.

On the whole, the industrial and mining frontier of northern Canada stands in contrast to the farming frontier. Whereas the farming frontier is characterized by a spatial expansion of agriculturally used land, the industrial and mining frontiers are isolated point developments. The farming frontiers, however, are not uniform in their physiognomic appearance. The many small settlement patches of northern Ontario and Quebec form a distinct, although only secondary, settlement fringe, which is especially well developed adjacent to the railway lines. Development is more pronounced where farming operations have been established to supply milk, potatoes, meat, and vegetables to the nearby settlements. In contrast to this, the visual impact of the new, isolated industrial settlements is not nearly as strong This in part is because they are spread irregularly over such a large area that they appear to be very small and insignificant cells of settlement in the middle of the forest. In addition, because they are situated so far north, cultivation is almost impossible; hence, they are dependent upon the south for their food supply.

In only a few instances has the modern industrial and mining frontier of northern Canada contributed to the expansion of the transportation network, even though the development of the first frontier was totally dependent on the building of the railroad. But, since World War II only two railroads have been built to exploit the mineral wealth of the Canadian Shield: (1) the railroad completed in 1953 from the harbour at Seven Islands on the St. Lawrence to the vast ore body at Schefferville/Knob Lake in Labrador; and, (2) the Great Slave Lake Railway, (opened officially in 1964), which was built to transport the rich lead-zinc deposits from Pine Point, on the southern shore of Great Slave Lake, to the refinery at Trail, British Columbia. Major highways in the North have also influenced the industrial and mining frontier: the famous Alaska Highway, built during the War for strategic reasons, and the Mackenzie Highway, extending from the Peace River Country to Yellowknife on the north shore of Great Slave Lake, are important examples. For the most part, however, these transportation developments have not resulted in a marked expansion of the settled area. Nonetheless, they are of major importance because they link the northern outposts of settlement in the sub-ecumene with the developed area to the south. Only here and there along these routes are there permanent settlements, which function as communication posts or small service centres.

In addition to the newly developed ore-mining and lumbering areas, there are other settlements in the North American industrial and mining frontier whose development is dependent on the opening up of other raw materials. Only in a marginal way do they belong to the border region of settlement since, with the exception of the petroleum communities in the Mackenzie Valley and the new oil settlements in the Peace River Country, most of them are in the midst of developed areas. The same applies to the communities adjacent to dams and power plants, which are usually characterized by power stations or large aluminum industries (Kitimat, Arvida).

A new type of frontier is to be seen in the military communities established in the Canadian Arctic and Subarctic since the War, and which collectively could be called a new military frontier with radar stations, air bases, supply depots, and spacious living quarters. To this group belong the numerous weather stations of the Canadian Arctic, as well as the last remaining trading posts of the Hudson's Bay Company on the coasts of the Arctic Ocean, the Labrador Sea and Hudson Bay. Common to all of these settlements, even more than to the towns of the new mining frontier, is their dependence on the aircraft for supplies and other services.

SUMMARY AND PROSPECTS

The characteristics of the Canadian frontier—the expansion of settlement especially in the 19th and beginning of the 20th century in a northerly direction, and the division of the pioneer borders into two types—agriculture and mining-industrial—are the most important contrasts to the classical American frontier of which the westward expansion by farming settlers was typical. The question remains, however, as to whether and to what extent the Canadian pioneer border is an extension of the classical American frontier or if it implies a complete change and is, therefore, of a different kind. It is clear that the agricultural frontier shifted gradually from the U.S.A. into Canada, and that the last active offshoots are the two forementioned points of settlement in eastern and western Canada. Hence, we can conclude that the present agricultural border of settlement in northern Canada is the last stage of the classical American frontier because of its continuation in time and space. This statment is justified also if one takes into account the progress of technology and the changed conditions for the colonists in the 20th century. These changes have doubtlessly facilitated the colonization of virgin lands, but they have only slightly changed the physiognomic appearance of the modern settlement fringes.

In contrast, the North American industrial and mining frontier is a truly Canadian form of the North American frontier and can be explained only by the technological advances of the 20th century. A temporary frontier grew up in the U.S.A. in the 19th century as well and nowadays there are frontier conditions in Alaska similar to those in northern Canada. Nevertheless, the United States mining frontier differs from the Canadian one in several respects. The present mining and forest frontier in Alaska, for example, is similar to the one in northern Canada in that it is confined to a small area and can be considered more as an offshoot of the northern Canadian frontier than as an independent American frontier. The American mining settlements of the 19th century in Colorado, California, the Great Basin, Idaho, and Montana are, in contrast to those of northern Canada today, different because they were based on the mining of gold and silver, first grade metals, and all lie within an area of agricultural potential. These mining settlements of the 19th century were also capable of surviving without help from the

outside and the environment was favourable for agriculture. Nowadays, the search for and refining of valuable minerals has become so remunerative and includes all kinds of metals and has expanded into areas where, owing to adverse physical conditions, large-scale settlements are almost impossible to establish. The development of the post World War II industrial and mining frontier is not related to the expansion of agriculturally useful land, but only with the opening up of new resource development areas. Expansion into the sub-ecumene has led to the rise of settlements that are dependent upon new transportation techniques, and whose existence is dependent upon the importation of goods from the south.

Future settlement in northern Canada is conditioned by two factors: (1) the amount and suitability of land available for agricultural settlement, and, (2) the potential impact of modern technology and research. The agricultural land potential in most of the Canadian provinces is exhausted so that expansion of the ecumene is being confined to smaller areas in which agriculturally useful reserves of land are available. Among these areas are parts of the Hudson Bay lowlands and the Mackenzie Valley, the Great Clay Belt and the Peace River Country. There are many millions of acres of land available for cultivation, but only a small portion will be developed in the next few decades since Canada is a surplus agricultural producer and the costs of clearing and developing new farm land are high. In accordance, the border of close settlement will, in general, move only slightly farther north in the near future.

In contrast, important mining and industrial developments may be expected in northern Canada. Throughout northern Canada, especially within the Canadian Shield, geological investigations of new mineral deposits and development of new and economic refining techniques are in progress. All this will hardly result in an expansion of the border regions of settlement, but will definitely cause a greater density of settlement within the Shield. The development of the Athabaska oil sands (whose reserves are greater than those of all the known oil sites put together), the search for petroleum in northern Alberta and in the Northwest Territories, the opening up of undeveloped forest resources, and the establishment of more military posts and research centres will bring the unsettled regions of the Canadian Arctic and Subarctic closer to the settled areas. However, these settlements will remain "exotic" because of the physical environment and they will still be dependent upon the importation of food supplies from the south. Thus, in the course of future expansion, large areas of the sub-ecumene will become semi-ecumene, but the ecumene itself will expand only insignificantly toward the north.

BIBLIOGRAPHY

Bartz, F., *Alaska* (Stuttgart, 1950).

Billington, R. A., *Westward Expansion: A History of the American Frontier* (New York, 2nd ed., 1963).

Blanchard, R., "Etudes Canadiennes, IV: L'Abitibi-Temiscamingue," *Révue Géographie Alpine*, Vol. 35, 1949, pp. 421-551.

Czajka, W., *Lebensformen und Pionierarbeit an der Siedlungsgrenze* (Hanover, 1953).

Ehlers, E., *Das nördliche Peace River Country, Alberta, Kanada: Genese und Struktur eines Pionierraumes in borealen Waldland Nordamerikas*, Tübinger Geogr. Studien 18, 1965.

Krenzlin, A., *Die Agrarlandschaften an der Nordgrenze der Besiedlung im intermontanen British-Columbia*, Frankfurter Geogr., Heft 40, 1965.

Lenz, K., *Die Prärieprovinzen Kanadas: Der Wandel der Kulturlandschaft von der Kolonisation bis zur Gegenwart unter dem Einfluss der Industrie*, Marburger Geogr. Schriften, Heft 21, 1965.

Lower, A. R. and Innis. H. A., *Settlement and the Forest and Mining Frontiers, Canadian Frontiers of Settlement*, Vol. 9, (Toronto, 1936).

McDermott, G. L., "Frontiers of Settlement in the Great Clay Belt, Ontario and Quebec," *Annals Association of American Geographers*, Vol. 51, 1961, pp. 261-273.

Murchie, R. W., Allen, W., and Booth, J. F., *Agricultural Progress on the Prairie Frontier, Canadian Frontiers of Settlement*, Vol. 5, (Toronto, 1936).

Putnam, D. F., *Canadian Regions: A Geography of Canada* (Toronto, 5th ed., 1961).

Riegel, R. E. and Athearn, R. A., *America Moves West* (New York, 5th ed., 1964).

Robinson, I. M., *New Industrial Towns on Canada's Resource Frontier*, University of Chicago Research Papers, No. 73, 1962.

Schott, C., *Landnahme und Kolonisation in Kanada am Beispiel Südontarios*, Schriften Geogr. Institut Kiel, Vol. 6, 1936.

Schott, C., "Die Erschliessung des nordkanadischen Waldlandes," *Zeitschrift für Erdkunde*, Vol. 5, 1937, pp. 554-563.

Schott, C., "Die Entwicklung Nordkanadas unter dem Einfluss der modernen Technik," *Petermanns Geographische Mitteilungen*, Vol. 98, 1954, pp. 295-301.

Taylor, G., *Canada: A Study of Cool Continental Environments and Their Effect on British and French Settlement* (London, 1957).

Turner, F. J., *The Frontier in American History* (New York, 1920).

Vanderhill, B. G., "The Farming Frontier of Western Canada, 1950-1960," *Journal of Geography*, Vol. 61, 1962, pp. 13-20.

Frontiers of Settlement in the Great Clay Belt of Ontario and Quebec

GEORGE L. McDERMOTT

At the beginning of the second quarter of the 20th century, only two large areas of potentially arable land in Canada remained open for settlement. These two areas, the Peace River Country of western Alberta and the Great Clay Belt of northeastern Ontario and northwestern Quebec, are enclaves beyond the zone of continuous agricultural settlement. The Great Clay Belt is of special interest to the geographer for it is shared by two provinces with differences in cultural environment, religion, and philosophy of colonization that have resulted in strong contrasts in population numbers and distribution, method of settlement, and rate of settlement growth.

Many of the pioneer settlement studies initiated or inspired by Isaiah Bowman were concerned with the economy of the fringe settlements, hence the term "pioneer" was used. The term "frontier" seems more appropriate in this study, for it refers to the area or zone between the settled and unsettled or used and unused land.[1] It is in this context that the term frontier is used here.

In his preface to *The Pioneer Fringe*, Bowman states that "settlement habitually advances and retreats on the outer fringe of ocupation."[2] Since Bowman's writing, Stone has been the only geographer to refer to advancing and retreating frontiers of settlement, which he has shown cartographically for Anglo-America.[3] This paper is concerned with the simultaneous advance and retreat of the agricultural frontier in the physically homogeneous Great Clay Belt.

The Great Clay Belt lies almost entirely within two political divisions: the District of Cochrane, Ontario, and Abitibi County,

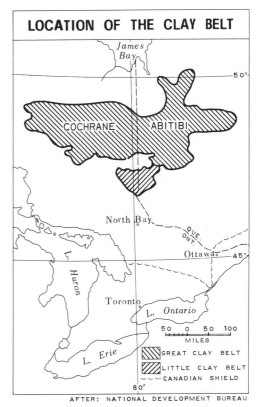

LOCATION OF THE CLAY BELT

AFTER: NATIONAL DEVELOPMENT BUREAU

Figure 1

Quebec (Figure 1). The grey clay, which was laid down in a temporary glacial lake, is estimated to cover sixteen million acres in northern Ontario and thirteen million acres in northern Quebec.[4] However, only 3 percent of this total is improved farmland. Even within one of the oldest and most thickly settled rural areas near the village of Cochrane, forest and brush occupy two-thirds of the land (Figure 2).

● GEORGE L. McDERMOTT *is Chairman, Geography Department, State University of New York at Cortland, Cortland, New York.*

Reprinted from the *Annals of the Association of American Geographers*, Vol. 51 (3), 1961, pp. 261-273, with the permission of the author and the editor.

41

LAND USE IN COCHRANE AREA, 1955

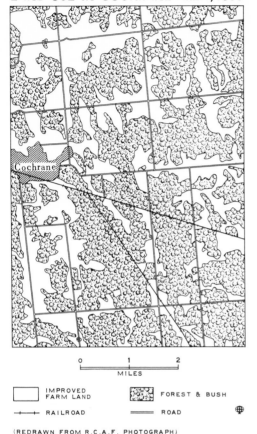

```
        0        1        2
        └────────┴────────┘
             MILES
```

☐ IMPROVED
   FARM LAND   ▦ FOREST & BUSH

+—+ RAILROAD   ══ ROAD   ✛

(REDRAWN FROM R.C.A.F. PHOTOGRAPH)

Figure 2

POPULATION AND SETTLEMENT

The first agricultural settlements in the Great Clay Belt formed a ribbon-like pattern along the Timiskaming and Northern Ontario Railroad beginning south of Matheson and extending northward toward the village of Cochrane (Figure 3A). Even before the railroad was completed to Cochrane, settlers had begun to establish farms in the area,[5] and at least a year before the Canadian National Railroad reached the site of Amos, settlers had begun making clearings in its vicinity[6] (Figure 3A).

When the Canadian National Railroad was completed across the Clay Belt in 1913, settlement spread discontinuously along it from the village of Cochrane to Hearst (Figure 3B). In Abitibi, settlement spread along the railroad from a nucleus at Amos to the present eastern and western limits of settlement in less than ten years (Figures 3A, 3B, and 3D). Even from the beginning, the area of settlement has been more compact in Abitibi than in Cochrane. By 1931, the areas of settlement in Cochrane had reached their present proportions, except for some rather small areas of retreat along the peripheries. The unoccupied land adjacent to the railroads in Cochrane is either a timber concession of a large paper company and therefore not open to settlement, such as the area between the village of Cochrane and the Quebec border, or the land is physically unfit for settlement, as is the case between Matheson and Timmins. By 1931, settlement in Abitibi had spread several miles deep along either side of the railroad from the Ontario border to Senneterre, the present eastern margin of agricultural settlement. The several finger-like northward and southward projections from the main axis of settlement became enlarged by the establishment of twenty-eight new parishes in the quarter-century following 1931 (Figure 3D).

The railroads have set the major axis of the settlement pattern, but the types of land survey have determined its detailed alignments. The differences inherent in the three types of survey used in the Great Clay Belt account for the details of the pattern of rural population distribution, and in part, for the local density (Figure 4). In the southern part of Cochrane District, the townships are six miles square and consist of thirty-six one-square-mile sections. Each section is surveyed into four parcels of approximately one hundred and sixty acres each. In the northern part of Cochrane District, including the area of settlement from the village of Cochrane to Hearst, the townships are nine miles square. This system of survey has contributed to the wider dispersion of rural population, for neighbouring farms are seldom closer than one-quarter mile and frequently as much as four miles of forest, swamp or brush separates adjacent farms.

42

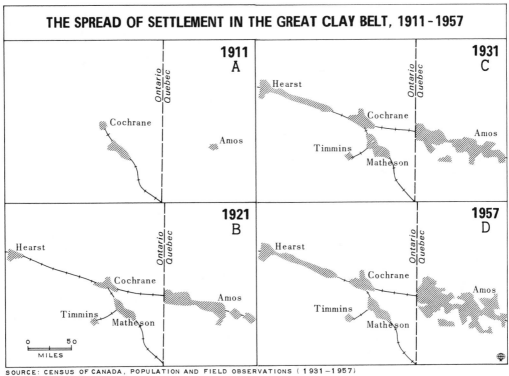

SOURCE: CENSUS OF CANADA, POPULATION AND FIELD OBSERVATIONS (1931–1957)

Figure 3

EFFECT OF DIFFERENT SURVEY SYSTEMS ON PATTERNS OF RURAL SETTLEMENT

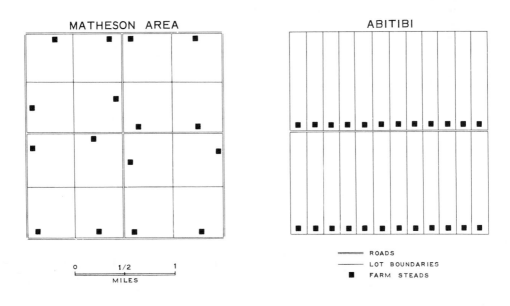

Figure 4

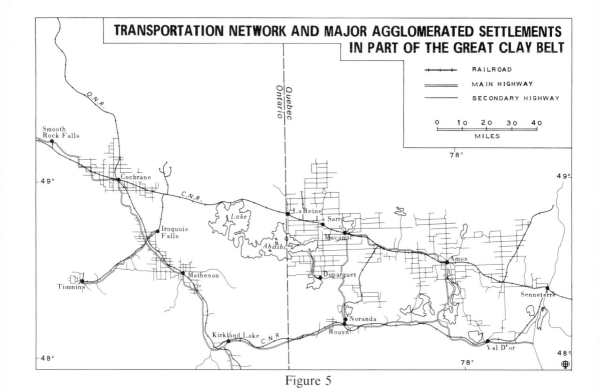

Figure 5

Widespread farm abandonment has led to even wider dispersion of rural population in Cochrane District.

The land survey system used in Abitibi has laid the framework for a pattern of rural population distribution that contrasts sharply with that of Cochrane. The ten-mile-square townships are surveyed into ranges one mile wide that extend east-west at two-mile intervals and the individual farm lots are one mile deep north-south with approximately 880 feet frontage on the road (Figure 4). Farm houses are so closely spaced that Randall has compared the settlement along a range road to the *strassendorf* development.[7] Although the settlement pattern is linear along any one of the east-west roads, the gross pattern is more compact and less ribbon-like than in Cochrane (Figure 5).

Both Cochrane and Abitibi were opened for settlement about 1910, and until 1931 the farm population, the number of farms, the amount of land in farms, and the amount of improved land on farms remained about the same for each area. A quarter of a

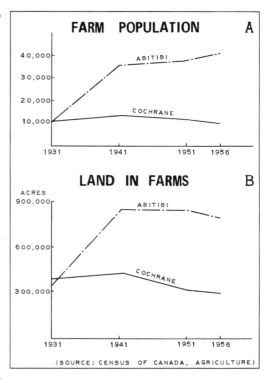

(SOURCE: CENSUS OF CANADA, AGRICULTURE)

Figure 6

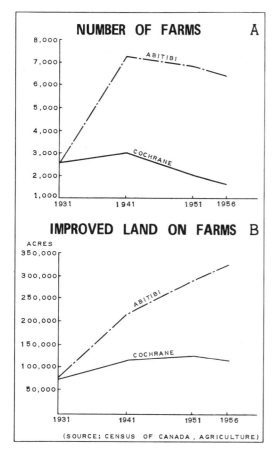

NUMBER OF FARMS A

ABITIBI

COCHRANE

IMPROVED LAND ON FARMS B

ACRES

ABITIBI

COCHRANE

(SOURCE: CENSUS OF CANADA, AGRICULTURE)

Figure 7

century later the contrasts between Cochrane and Abitibi in these four areas are indeed striking (Figures 6 and 7). During this period, Abitibi's farm population more than tripled, while Cochrane's rose to a peak in 1941, then declined again to about the 1931 level (Figure 6A). The total number of farms increased more than two and one-half times in Abitibi, but decreased by one-third in Cochrane (Figure 7A). As the number of farms increased in Abitibi, the amount of land in farms increased proportionately, and as the number of farms declined in Cochrane, the amount of land in farms decreased over 24 percent, in contrast to an increase of 224 percent in Abitibi. Between 1951 and 1956 the amount of improved land in Cochrane declined over 8,000 acres, while during the same period Abitibi's improved

land increased more than 39,000 acres (Figure 7B).[8]

The decrease in farm population and the number of farms in Cochrane is not a cause for alarm, for such decreases have occurred over most of the United States and Canada during the last half-century. However, unlike most of Canada, the amount of land in farms in Cochrane has also decreased (Figure 6B).[9] This is of particular significance, for little of the abandoned land has been absorbed into the remaining farms, as has been the case in most other areas in Canada. Thus, in Cochrane the withdrawal of people from the frontier has resulted in the complete abandonment of farm land.

Analysis of the census statistics reveals that settlement has been advancing in Abitibi and at the same time retreating in Cochrane. The censuses, however, do not give data for units small enough to permit accurate mapping of areas of advance and retreat within each area. Extensive field work was required to determine where the frontier had advanced and retreated in the quarter-century prior to 1957.

THE RETREATING FRONTIER

The abandonment of land, ostensibly taken up for agriculture, has been one of the major problems associated with the settlement of the Ontario Clay Belt. Inasmuch as some abandonment took place very early in the history of settlement, it is necessary to distinguish between abandonment that occurred prior to and since 1931, the beginning date for this study.[10] Only those abandoned lots with standing buildings or evidences of former buildings were mapped, for on these lots the settler had made an attempt to comply with the agricultural settlement duties which required the construction of a habitable house.[11]

Although farm abandonment is rather widespread throughout the Great Clay Belt, it is much more concentrated in Cochrane than in Abitibi. One very striking difference in the pattern of abandonment occurs between Cochrane and Abitibi; in Abitibi the abandoned farms are scattered throughout the settled areas, and although the same is true in Cochrane, there is a marked concentration of abandonment at the ends of the roads and along the peripheral roads in the nodes of settlement, such as around the villages of Cochrane and Matheson. In Cochrane many abandoned farms are several hundred yards removed from the nearest road, but in Abitibi the abandonment is adjacent to the roads for the settlers were permitted to buy land only along existing roads.[12]

In spite of government inducements to the French-Canadians to settle in Abitibi, about two-thirds of the rural population in Cochrane District is French.[13] Comparison of the number of French and English householders served by thirty-four rural post offices in Cochrane District for the years 1931 and 1956 reveals that English Canadians account for over 70 percent of the total decline.[14] Inasmuch as the post offices serving the villages, pulp mill towns and the mining settlements were excluded from this comparison, it can be safely asserted that the English and other non-French-Canadians have accounted for the greatest proportion of the farm abandonments.

There are relatively few areas of settlement in Cochrane larger than two square miles where some abandonment has not occurred since 1931. Only three areas are outstanding for their relative lack of abandonment: (1) Montjoy Township, northwest of Timmins, (2) the area immediately northeast of the village of Cochrane, and (3) the area west of the Cochrane-Matheson road in the vicinity of Iroquois Falls.

A retreating frontier is defined, for the purposes of this study, as an area in which more than 60 percent of the farms are abandoned and the land is not used by neighbouring farmers. Empirical evidence indicates that where 60 percent of the farms are abandoned, the whole landscape takes on a derelict, reverting-to-nature appearance and agricultural operations are minimal. In the areas of retreat, many of the occupied farms are on the brink of abandonment and the care of a garden or a cow is the extent of the settlers' agricultural endeavour.

Eleven areas have a sufficiently high percentage of abandonment to be classed as a retreating frontier. In Cochrane, they are: (1) northeast of Matheson, 75 percent, (2) west of Matheson, 62 percent, (3) southeast of Cochrane, 80 percent, (4) east of Cochrane, 70 percent, (5) northwest of Cochrane, 66 percent, (6) southwest of Cochrane, 63 percent, (7) south of Kapuskasing, 81 percent, (8) east of Hearst, 91 percent, (9) northwest of Hearst, 80 percent, and (10) south of Hearst, 67 percent. Only

Figure 8

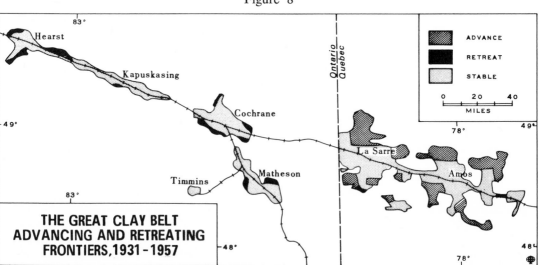

THE GREAT CLAY BELT
ADVANCING AND RETREATING
FRONTIERS, 1931-1957

one area of concentrated abandonment was found in Abitibi, a rather small area on the southern fringe of settlement twenty miles south of La Sarre, 77 percent (Figure 8). Actually, overall land abandonment is somewhat higher than these figures indicate for they are for the period since 1931 and some land abandonment took place before that date.

The overall distribution of retreat forms a spotty pattern peripheral to existing settlement and removed from the main east-west highway and railroad. On Figure 8, the areas of settlement are outlined and those that have shown no appreciable change during the period 1931-1957 are designated as "stable."

The retreating frontier displays certain observable characteristics, the combination of which aid in its identification in the field. The recently abandoned fields, guarded only by broken fences, are choked with weeds, the first echelon of the more formidable army of alder and willow, birch and poplar that will soon follow. Abandoned houses and barns in varying stages of decadence are a common sight on the retreating frontier. Many of the buildings, particularly those with glass windows, have become the prey of vandals and thus have aged prematurely. Some settlers have even left machinery standing in the fields where it was last used. A log or clapboard barn falling down around a rusting hayrake is an ephemeral monument to man's attempt, and failure, to cope with the numerous problems that beset the settler on the Clay Belt Frontier.

As man withdrew from the frontier, the institutions and commercial establishments that served him also were abandoned. Unoccupied schools and churches and closed country stores are common on the retreating frontier. Although these cultural features have been slowly vanishing from the rural scene in Anglo-America since the automobile became common, in the Clay Belt their abandonment is almost exclusively confined to the zones of retreat.

The roads on the retreating frontier show few signs of use. Double-lane, graded gravel roads, which are common throughout most of the Clay Belt, give way to single lane, ungraded, and often dirt-surfaced roads on the retreating frontier. Some township roads are so overgrown with grass that even the wheel tracks have been obliterated.

Also characteristic of the retreating frontier is the absence of new clearings, good pasture, and cattle on the occupied farms. Many of the occupied farms on the retreating frontier appear to be on the brink of abandonment. Those settlers who remain derive very little of their income from farming, but work at jobs elsewhere while continuing to live on their farm lot.[15] Scattered here and there on the retreating frontier are a few settlers earnestly trying to succeed at farming. However, these settlers are quite discouraged, for the value of their land declines more rapidly with each abandonment in the neighbourhood than it can be increased by improvements brought about by hard labor and investment. Abandonment breeds abandonment.

In the areas where more than half of the farms are still occupied—those areas designated as the stable frontier—settlement seems to be on firmer ground (Figure 8). On the stable frontier, most of the farms have cattle, the pastures are relatively free from brush, the fences are in good repair, some new but small clearings are observed on the back lots, and the roads are well traveled. Here, man appears to have a certain measure of control over the physical environment, whereas on the retreating frontier, man's works are slowly being obliterated by nature.

THE ADVANCING FRONTIER

The advance of the frontier has received more attention in the literature than the retreating frontier, perhaps because it is more common. Certainly it is more pleasant to report on man's achievements than his defeats. In *Pioneer Settlement*, twenty-six authors discussed the problems attendant upon the advance of settlement in seventeen of the world's major frontier areas.[16] In

47

few, if any, of these regions have the principles of scientific settlement advocated by Bowman,[17] been so vigorously applied as in the Quebec portion of the Great Clay Belt, where the frontier has advanced with remarkable rapidity during the last quarter of a century.

An advancing frontier is defined, for the purposes of this study, as an area in which the farms have been established since 1931, and in which more than 60 percent of the total number of farms is occupied. Data on the dates of establishment were obtained from old maps, interviews, Annual Reports of the Ontario Department of Lands and Forests, the Quebec Ministry of Colonization, and from parish histories.

The gross distributional pattern of advance, as well as retreat is clearly divided along provincial boundaries (Figure 8). All the areas of advancing frontier are in Abitibi and are larger in extent than the areas of the retreating frontier in Cochrane. This is congruent with the census data which indicate great growth in Abitibi and modest decline in Cochrane (Figures 6 and 7).

In detail, the distribution of areas of advance is peripheral to the older established areas of settlement. Inasmuch as the land adjacent to the railroad and main east-west highway was settled quite early, the advancing frontier is several miles—in some places more than thirty miles—removed from the main axis of settlement. The advance of the frontier has taken the form of broad, finger-like projections into the forest both northward and southward from the railroad. Some of these areas, particularly those southwest, south, southeast, and northeast of La Sarre, are individual parishes, while each of the other areas of advance consists of three to six contiguous parishes.

Although small, ephemeral advances of the frontier were made in Cochrane—especially west of Matheson, southwest of village of Cochrane, and south of Hearst—by the Relief Land Settlers during the early 1930's, nearly complete abandonment of these areas has subsequently occurred. Thus, no area larger than a few widely scattered individual farms has advanced the frontier

in Cochrane during the last quarter of a century. Not only did the frontier in Cochrane fail to advance during this period, but it also failed to maintain the limits once established.

The advancing frontier may be identified in the field by the combination of several observable characteristics. The most striking features are the new, small clearings flanked by forest on one, two, or three sides. The land, newly exposed to sunlight, is covered with a reddish brown peat, which has a raw, cold appearance, and when plowed becomes mottled as the peat and grey clay are interwoven.

In Abitibi, the newness and style of the homes and farm buildings are indicative of the advancing frontier. In the four parishes established since the Second World War, the houses are characteristically square, one and a half stories high and covered with red or blue asphalt shingle siding. These houses were built according to specifications of the Quebec Ministry of Colonization and are identical in every outward detail. Houses built prior to the Second World War are quite varied in style, although some form of the two-story type prevails. There is an obvious lack of such refinements as surfaced driveways, sidewalks, lawns, flowers, and general landscaping around the houses.

Because the Province of Quebec has given substantial financial aid to each new parish for the construction of a central school and parish church, these buildings are more permanent and more attractive than those commonly found on the advancing frontier in the Prairie Provinces.[18] The amount of landscaping is frequently the only observable difference in the exterior appearance of the churches and schools in the stable areas and on the advancing frontiers.

Three mechanical symbols of the advancing frontier have recently appeared in Abitibi: (1) the bulldozer which is used to remove tree stumps after the merchantable timber has been cut, (2) the giant plow, pulled by a twenty-four-ton caterpillar-type tractor, which penetrates the soil to a depth of three feet or more, and (3) the power shovel which is used to excavate deep ditches to drain marshy land. These three govern-ment-owned implements have done much to hasten the process of establishing farms, and in many instances have rendered fit for cultivation land that was too poorly drained or too thickly covered with peat to be uti-lized under previous land-breaking tech-niques.

Great mounds of brush and peat in the fields are no longer symbolic of the advanc-ing frontier in Abitibi, for the use of deep plows permits the turning under of small saplings and branches, up to four inches in diameter, and several inches of peat, thus obviating the necessity of piling peat and stacking brush to be burned. As a result, the incidence of forest fires in Abitibi is much less than in Cochrane where brush burning is still common.[19]

The roadsides in front of the newly estab-lished settlers' homes give testimony of the changing landscape for they are lined on either side with shoulder-high stacks of pulp-wood logs. Yet, despite the settlers' valiant efforts to make farms from forests, the pro-cess is a slow one; the clearings will remain small (Figure 2) and logs will line the road-sides for many years. Forest dominates the landscape on both the retreating and advanc-ing frontiers. On the retreating frontier, fields are giving way to brush and forest, while on the advancing frontier the forest is giving way to fields.

CAUSES OF RETREAT

A large number of the people who attempted to settle in the Ontario Clay Belt had neither experience in nor familiarity with forestry or agriculture. The settlers' unfamiliarity with the conditions of the Clay Belt as well as the techniques of their newly adopted means of livelihood caused dis-couragement and ultimate abandonment. For example, the settlers who were sent to Cochrane under the Relief Land Settlement Plan in 1932 and 1933 were skilled and semi-skilled tradesmen from southern On-tario cities.[20] In the Kapuskasing and Hearst area, 70 percent of the relief settlers had abandoned their lots within seven years.[21]

By 1957, eight of the original 124 relief settlers in the Matheson area remained on the land, but only two were attempting to farm.[22] The zone of retreat west of Mathe-son was formerly occupied by these relief settlers (Figure 8). Only two of the original 93 relief settlers who took up lots southwest of the village of Cochrane remained in 1957.[23] The 105 relief settlers in the Kapus-kasing-Hearst area had also dwindled to only two by 1957.[24] These settlers originally occupied the area of retreat south of Kapus-kasing and south of Hearst (Figure 8).

Frontier settlement in the mid-twentieth century is more subject to human failure than a century ago for the gap between life on the frontier and life in urban communi-ties or well-established farming districts is far greater than at any time in the past. Careful screening of the settlers would help to reduce suffering and the misuse of land due to the presence of misfits on the frontier. Settler-screening programs have been em-ployed in Quebec[25] and Argentina[26] for sev-eral years, and officials express general satis-faction with the criteria used, which include the following: the prospective settlers must have (1) experience in or aptitude for farm-ing, (2) enough capital or equipment to begin modest farming operations (in Que-bec, the government gives this kind of aid to the settler), and (3) a good reputation and be in good health. It would be to the advantage of the settler and the government alike if only those persons who have the greatest possibilities for success were per-mitted to purchase a lot on the frontier.

In order for an Ontario settler to obtain a letter of patent or title to his land he must build a house of 320 square feet and reside in it for six months of every year and clear and cultivate a minimum of two acres of land each year until fifteen acres are under cultivation.[27] These settlement duties, how-ever, have not been rigidly enforced, with the result that many settlers removed the timber from their land but made little or no effort to farm.

None of the land in the Ontario Clay Belt had been surveyed prior to settlement to determine its agricultural capability.[28] The settler, after selecting a lot, is required to inspect it thoroughly and signify that the lot is 50 percent cultivable.[29] The large number of poorly drained lots taken up for farming and then abandoned, gives cause to wonder if the settler even inspected the land before the purchase was made. Binns considers a land-use capability survey an essential preliminary step to settlement.[30]

The major factor affecting the retreat of the frontier is the low financial return from the "bush farm" in comparison with other occupations. Of the scores of settlers, ex-settlers, and officials interviewed, each listed low farm income in proportion to labor expended as the principal cause of farm abandonment in both Cochrane and Abitibi. The estimated average farm income for Cochrane in 1955 was $1,160, which was only one-third the average farm income for the Province of Ontario.[31] The median farm income in Cochrane for that year was less than $250. For the same year the average income of the Cochrane forestry worker was estimated to be over $4,000 or nearly four times the average farm income in the same area.[32] Since the demand for newsprint has steadily risen, employment with the pulp-wood companies has been extended to ten or eleven months a year, the slack period coming during the spring thaw when movement in the woods becomes difficult. During the active period, jobs usually exceed the local labor supply and in 1955 over one and a half million dollars was paid in wages to Quebec residents for working in the forests and mills of the Cochrane District.[33]

A large number of settlers have given up serious attempts to farm, but still remain on the land. These settlers have at one time cleared and cultivated enough land, about fifteen acres, to obtain a patent, but now keep only a cow or two and cultivate a garden, and although they are essentially non-agricultural workers their land is classed as a farm.[34] In Cochrane, 70 percent of the farms are classed as part-time and small scale farms with an annual farm income of less than $250.[35] Thus, the majority of the farmers obtain most of their income from work off the farm.

Mining, as well as forestry, has attracted settlers from their farms in at least two areas of the Clay Belt. Some of the settlers in the zone of retreat northeast of Matheson have gained employment in the asbestos mine eleven miles east of Matheson. The only area of retreat in Abitibi lies twenty miles south of La Sarre on the southern edge of the developed area and one mile north of the gold mines at Duparquet (Figures 5 and 8). Here, as at Matheson, farms with considerable improved acreage have been forsaken for higher paying jobs in the mines.[36] The mines in this case have served as a magnet, not drawing farm produce from the local area as Innis suggests,[37] but rather drawing the farmer himself. Inasmuch as only three areas of agricultural settlement adjacent to mining centres were studied—the two mentioned and the area near Timmins—it is difficult to draw conclusions concerning the effect of the mines on farm abandonment. However, it is significant that only the Timmins area has little abandonment while the other two areas are classed as parts of the retreating frontier. More research needs to be conducted in the Canadian Shield to test the theory that mining stabilizes agriculture.

Providing for the basic necessities of life is the fundamental problem facing the new settler during the first few years on the frontier. The gap between the economic return of the bush farm and the wages offered by other occupations is basic in explaining why three-fourths of the settlers in both Cochrane and Abitibi have found it necessary to couple farming with forestry or some other occupation, and why so many have given up farming entirely. The present part-time farming-forestry occupance of the Great Clay Belt is viewed by Ontario officials as a temporary, yet essential, phase in the sequence of occupation leading to full-time farming. Quebec officials, on the other hand, view part-time forestry as a short circuit en route to full-time farming, and an attempt is being made by financial grants and numerous services to shorten the period of dependence on part-time employment off the farm. However, it would appear that part-

50

time farming is the final stage for many, if not most, of the present generation of settlers in both Cochrane and Abitibi.

CAUSE OF ADVANCE

For years the Province of Quebec has been faced with the problem of a rapidly growing population on a limited agricultural land base. This is particularly serious in Quebec, where both the church and state subscribe to the philosophy that large numbers of people should live on the land, and especially that those raised on farms should be encouraged to go into farming on their own. Only small parcels of land suitable for agriculture remain unoccupied in Old Quebec; and for a half-century or more fathers have had difficulty in establishing their sons on farms. As a result there has been a drift of people to the cities or to the United States. In an attempt to stem this flow, the Province of Quebec launched a colonization program in 1923[38] that has become more elaborate and generous with each passing year.

The Quebec Department of Colonization is dedicated to the task of assisting the settler in becoming established on the frontier. The government assistance is manifest in many forms, but may be discussed under three broad headings: (1) pre-settlement planning, (2) financial assistance, and (3) technical assistance.

Pre-settlement Planning

Each new Abitibi parish is carefully surveyed, mapped, and planned before the settlers arrive. Soil surveys are made lot by lot to determine the capability of the land for agriculture, and land that needs extensive improvement will be temporarily withheld from settlement and land unfit for agriculture will remain in forest.[39] Upon completion of the land capability maps, the location of roads, farm lots, houses, and the parish centre is planned. The latter are centrally located and usually consist of a church, a *presbytère,* a general store, an elementary school, and a few houses. Compactness of settlement is desired by the Department of Colonization and this is fostered by the long

lot survey (Figure 4), and the department policy that settlers should be no more than five miles from the parish centre and that every physically suitable lot should be occupied.[40]

Abitibi is prepared for a future advance of the frontier, for twenty-seven projected parishes, mostly north of the present area of settlement, have been surveyed, mapped, and planned.[41] These parishes will not be opened for settlement until all the suitable vacant lots in the older parishes have been filled, unless a large group of people wish to settle as a unit.

The settlement duties in Quebec are more demanding than those in Ontario. In order to obtain a letter of patent, the Abitibi settler must (1) clear and cultivate a minimum of three acres of new land each year so that within ten years thirty acres or 30 percent of the lot is under cultivation, (2) build a habitable house at least twenty by twenty-four feet and occupy it until the letter of patent has been issued, and (3) build a barn thirty-four by thirty-two feet. The one-hundred-acre lots sell for thirty cents an acre, payable in five equal payments, interest free, which are deductible from government grants.[42]

Financial Assistance

To encourage and assist the settler in establishing an economic farm unit of about thirty acres as rapidly as possible, the Quebec Department of Colonization has been paying statutory premiums for clearing and plowing since 1923. The first premiums were $4 an acre on the first five acres put into cultivation.[43] Since 1947, the settler has received $40 an acre for the first forty acres cleared and plowed.[44]

In addition to offering premiums for work performed, the Department of Colonization has initiated provincial, or participated in federal, settlement schemes. The Relief Land Settlement Plan of 1932-1934, a joint federal, provincial, and municipal project, previously described for Ontario, was responsible for the establishment of four parishes,[45] whose combined population was 3,650 in 1952.[46] Blanchard states that 28 percent of these settlers abandoned their land by the end of the second year,[47] yet there are very few unoccupied farms in these four parishes because the Department of Colonization makes every attempt to place another settler on the land as soon as an abandonment occurs. This policy protects the investment in the land and the buildings and takes advantage of any improvement made by earlier settlers.

The Vautrin Colonization Plan, in force during 1935 and 1936, aided the establishment of group settlements of more than fifty families by providing transportation for the head of the family, a small grant for home construction, and a small relief allowance during the first year. In addition, the settler was entitled to clearing and plowing premiums of $10 per acre for the first twenty acres cleared and $10 per acre for the first ten acres plowed.[48] During the two years of its existence, the Vautrin Plan helped to establish fourteen parishes,[49] whose combined population in 1952 was over 7,200.[50]

In 1937, the Roger-Auger Colonization Plan, another joint federal, provincial, and municipal project, was inaugurated to help establish urban families on relief as settlers on the Abitibi frontier. Grants for living expenses and construction totaling over $1,200 were spread over a four-year period, at which time the federal and some municipal governments withdrew support, but Quebec kept the plan in force during the Second World War.[51] The six parishes[52] established under this plan had a combined population of over 3,600 in 1952.[53]

The present subsidy plan, adopted in 1947, is more generous and embraces more economic aspects of frontier settlement than the previous plans. Under this plan, the statutory premiums total $40 per acre for the first forty acres cleared and plowed, and in addition, a maximum of $1,050 is granted in credit for the purchase of livestock and farm equipment.[54] Thus, by the time forty acres are under cultivation, the settler will have received $2,650 in bonuses for working his own land for which he originally paid thirty cents an acre.

Other subsidies designed to increase the permanency of settlement include: $30 a month subsistence allowance for the first six months of settlement; a grant of $600 and a loan of an additional $600 for the construction of a house, with the materials furnished at cost; up to $250 for well drilling; $75 for house and barn wiring; a $400 grant toward the construction of a barn; and $40 for planting a garden.[55] By the time the Abitibi settler, established since 1947, has 40 percent of his farm acreage under cultivation, he will have received a maximum of $4,345 in grants and subsidies from the Department of Colonization. Some of the subsidies are for work performed by the settler in establishing a farm on the frontier; others are direct grants in an effort to make settlement more permanent.

Since the inception of the present subsidy program in 1947, four parishes have been established,[56] which in 1952 had a combined population of approximately 2,100 on 370 lots.[57] By 1957, 18 percent of the lots in these parishes had been abandoned, and the process of land clearing and building construction seemed to be progressing at a slow rate. During this period the Department of Colonization has placed its greatest emphasis on settling vacant lots in existing parishes rather than creating new ones. Over 4,600 settlers were assigned partially improved lots in older parishes that had been vacated, or new lots that had recently become suitable for cultivation by the completion of a drainage or deep-plowing project.[58]

The four successive settlement plans that have been in operation continuously since 1932 have had a marked effect on the geography of Abitibi. Of the present sixty agricultural parishes in Abitibi, twenty-eight were established with the aid of one or more of these plans. The 1952 population of these newer parishes was more than 15,000 persons. None of the newer parishes has been disbanded.

Technical Assistance

In addition to pre-settlement planning and the granting of subsidies, the Department of Colonization provides the settler with a variety of technical assistance. After a settler has proved his good intentions to farm by clearing and breaking a portion of his land, he may request the assistance of the Department of Colonization's heavy machinery in further clearing and plowing. Approximately half of the cost of this operation is charged against the settler's premiums.[59] The advice and assistance of drainage engineers, foresters, agronomists, and building engineers is available on request. A settler within each parish, who has business ability and who has developed his own farm to a high degree, is employed to serve as an Inspector. His duties involve advising the settlers on clearing, planting, and other essentials to good farming, and keeping a record of the settlers' performance of the prescribed settlement duties as a basis for the payment of premiums.

The Catholic church also makes a significant contribution to settlement in Abitibi. The Department of Colonization depends primarily upon the priests of the Colonization Societies to handle publicity concerning settlement possibilities on the frontier and to secure recruits. The church parish, and not the township, is the effective unit of settlement. Each parish has a resident curé, who not only serves as a spiritual advisor but also as a financial and agricultural advisor. The strength and success of the parish is intimately related to the drive, enthusiasm, and skill of the parish priest.[60] Strong leadership is especially needed during the formative years of settlement.

SUMMARY

Although there are many factors involved in the movement of frontier in the Great Clay Belt, the role played by the respective provincial governments is of major importance. Ontario has followed a laissez faire policy toward the settler, who has employed the same trial and error methods characteristic of the frontier a century ago. Quebec has taken a more paternalistic approach to settlement by aiding the settler in every possible way through the application of rational principles to settlement planning and a willingness to spend any necessary amount of money with the result that the developed area of Abitibi has doubled in the last quarter of a century. In the final analysis, it appears that the government concerned must decide whether the advance of the frontier is desirable and whether it is willing to pay the price to achieve that goal.

REFERENCES

[1] Kirk H. Stone, "Human Geographic Research in the North American Northern Lands," *Arctic Research,* Special Publication No. 2 of the Arctic Institute of North America, 1956, p. 218.

[2] Bowman, Isaiah, (ed.), *The Pioneer Fringe* (New York: American Geographical Society, 1931), p. v.

[3] Stone *op. cit.,* p. 210.

[4] Gosselin, A. and Boucher, G. P., *Settlement Problems in Northwestern Quebec and Northeastern Ontario,* Publication No. 758, Dominion of Canada Department of Agriculture, 1944, p. 8.

[5] Marwick, A., *Northland Post* (Oshawa, Ontario, The Maracle Printing Company, 1950), p. 129.

[6] Ouellet, G. R., *L'Abitibi* (Quebec: Ministère de la Colonisation, 1952), pp. 31 and 59.

[7] Randall, J. R., "Settlement of the Great Clay Belt of Northern Ontario and Quebec," *Bulletin of the Geographical Society of Philadelphia,* Vol. 36 (1939), p. 58.

[8] All statistical data are from the *Census of Canada, Agriculture,* 1931, 1951, and 1956.

[9] Since 1931, Canada's farm land has increased by eleven million acres, although there has been some decrease in the Maritime Provinces and in small areas of other provinces. Since 1951, the total land in farms in Abitibi has also declined slightly.

[10] The method used for dating abandonment was, in the absence of records, subjective. In each major area of settlement, dates of abandonment of specific properties were determined through interview, then notations were made on the size and kind of natural vegetation that

had encroached upon the cleared land. Then comparison was made with the vegetation on the lot whose date of abandonment was not known. Actually most abandoned lots could be clearly classified in one category or another and very few arbitrary decisions were necessary. However, this method is subject to numerous errors, not the least of which are differential growths of vegetation due to slight differences of soil, differences in individual plants, and the condition of the land when abandoned.

[11] This distinction was made because many of the "timber pirates" lived in tents or in town while they stripped the timber from their lots.

[12] Interview with Clovis Meloncon, Chief of Pre-Settlement Survey and Parish Planning, Department of Colonization, June 25, 1957.

[13] *Number of Householders Served from Rural Post Offices and Rural Routes in the Province of Ontario* (Public Relations Division, Post Office Department, April, 1956), pp. 10 and 55.

[14] *Ibid.*, and personal communications with W. M. Griffiths, Director of Comptroller's Branch of Post Office Department, on the 1931 data.

[15] The major sources of farm income in the Great Clay Belt are the sale of (1) dairy products, (2) cattle, and (3) hay, in the order listed. For a detailed discussion of agriculture in the Clay Belt see John R. Randall, "Agriculture in the Great Clay Belt of Canada," *Scottish Geographical Magazine*, Vol. 56 (1940), pp. 12-28.

[16] Joerg, W. L. G., (ed.), *Pioneer Settlement* (New York: American Geographical Society, Special Publication No. 14, 1932).

[17] Bowman, *op. cit.*

[18] Dawson, C. A., *Settlement of The Peace River Country*, Vol. 9 of *Canadian Frontiers of Settlement* (Toronto: Macmillan, 1934), p. 9. For a more recent discussion see Burke G. Vanderhill, "Observations in the Pioneer Fringe of Western Canada," *Journal of Geography*, Vol. 57 (1958), pp. 431-441.

[19] Hess, Quimby, "Land Use Planning and Resources Development in the Northern Region," Progress Report No. 1, mimeographed, Cochrane, Ontario, March 20, 1956, p. 30.

[20] *Report of the Minister of Lands and Forests,* Province of Ontario, 1941, p. 12.

[21] *Report of the Minister of Lands and Forests,* Province of Ontario, 1940, p. 11.

[22] Interview with L. H. Hanlan, Agricultural Representative, Matheson, July 19, 1957.

[23] Interview with Raoul Portelance, Agricultural Representative, Cochrane, July 24, 1957.

[24] Interview with T. J. Murphy, Lands Agent, Kapuskasing, July 26, 1957.

[25] Interview with Henri Fortier, Chief of the Establishment Service, Department of Colonization, Quebec, August 8, 1957.

[26] Taylor, Carl C., *Rural Life in Argentina* (Baton Rouge: Louisiana State University Press, 1948), pp. 354-355.

[27] *Lands for Settlement,* Ontario Department of Lands and Forests, n. d., pp. 3-4.

[28] In 1960, the Ontario Department of Lands and Forests published *The Glackmeyer Report of Multiple Land-Use Planning,* which includes a detailed map of agricultural use capability of the settled area north of the village of Cochrane.

[29] *Ibid.*, p. 3.

[30] Binns, Sir Bernard O., *Land Settlement for Agriculture* (Rome: Food and Agricultural Organization of the United Nations, Development Paper No. 9, 1951), p. 3.

[31] *Economic Survey of Ontario*, Office of Provincial Economist and Bureau of Statistics and Research, Treasury Department, Province of Ontario, 1955, p. C-6.

[32] Interview with Quimby Hess, Regional Forester, Cochrane, June 18, 1955.

[33] Hess, Quimby, "Progress Report No. 1," Land Use Planning and Resources Development Committee, Cochrane, March 20, 1956, p. 46.

[34] A farm is considered as any holding of three acres on which agricultural operations are carried out.

[35] *Census of Canada, Agriculture,* 1951.

[36] Interviews with André Chabot, Chief of Settler Establishment Service at La Sarre, July 12, 1957 and Gaston Lavoie, Deputy Chief of Colonization at Rouyn, July 10, 1957.

[37] Innis, Harold A., *Settlement and the Mining Frontier*, Vol. VII of *Canadian Frontiers of Settlement* (Toronto: Macmillan, 1936), p. 373.

[38] *Statistical Yearbook of the Province of Quebec*, 1924, p. 142.

[39] Personal communication from J. B. Pouliot, Chief of Land Classification, Department of Colonization, Quebec, March 15, 1957.

[40] Interview with Clovis Meloncon, Chief of Pre-Settlement Surveys and Parish Planning, Department of Colonization, June 25, 1957.

[41] Interview with Romeo Lalande, Deputy Minister of Colonization, Quebec, August 9, 1957.

[42] Lalande, Romeo, "Settler Assistance, In View of Promoting Colonization Through Progressive and Rational Methods," 1956, Department of Colonization, Quebec, mimeographed, pp. 3 and 5.

[43] *Statistical Yearbook of the Province of Quebec*, 1924, p. 142.

[44] Lalande, *op. cit.*, p. 4.

[45] Ste. Gertrude de Villemontel, Laferte, Ste. Germaine de Palmarolle, and Roquemaure.

[46] Ouellet, G. R., *Un Royaume Vous Attend: L'Abitibi* (Quebec: Ministère de la Colonisation, 1952), pp. 56, 74, 81, and 82.

[47] Blanchard, Raoul, "L'Abitibi-Temiscaminque," *Revue de Géographie Alpine*, Vol. 37 (1949), p. 490.

[48] *Statistical Yearbook of the Province of Quebec*, 1935, p. 165.

[49] Rapide Danseur, Destor, Preissac, La Corne, Vassan, Villebois, Beaucanton, Val St. Giles, St. Vital, Berry, St. Dominique du Rosarie, Lac Castagnier, Castagnier, and Rochebaucort.

[50] Compiled from Ouellet, *op. cit.*

[51] *Statistical Yearbook of the Province of Quebec*, 1944, p. 75.

[52] Val Paradis, Manneville, Champsneufs, Val Senneville, St. Edmond, and Ile Nepawa.

[53] Compiled from Ouellet, *op. cit.*

[54] Lalande, *op. cit.*, p. 4.

[55] Lalande, *op. cit.*, pp. 2-4.

[56] St. Eugene, Languedoc, Guyenne, and Despinassy.

[57] Ouellet, *op. cit.*, pp. 45, 50, 59, and 79.

[58] Compiled from the *Rapport du Ministre de la Colonisation de la Province de Quebec*, for the years 1947 through 1956, except for 1951 and 1952 when comparable data were not reported.

[59] Interview with Al Leuzon, Chief of Colonization Works Service, La Sarre, Quebec, July 4, 1957.

[60] Interview with Romeo Lalande, Deputy Minister of Colonization, Quebec, August 9, 1957.

Manitoba Settlement Patterns

JOHN WARKENTIN

Surveyors' lines in themselves consist of no more than momentary sightings, quick chaining, and posts pounded into the ground at intervals; yet these lines help stamp a tangible pattern on the land. Taken together, survey systems and land settlement policies are potent instruments in controlling the arrangements of farmsteads, fields, and roads and in encouraging dispersed or nucleated settlement. Thus they critically influence man's imprint on the land and help shape the economic and social life of an agricultural area.

This paper sets out to describe briefly the systems of land division employed in Manitoba, and to indicate how suitable each pattern is for settlement purposes.

I

The first formal land survey in the Prairies was made in Lord Selkirk's Colony in 1813 by Peter Fidler, a Hudson's Bay Company trader and surveyor.[1] The site selected for the colony, near the junction of the Red and the Assiniboine Rivers, lends itself to the river front pattern, and the two rivers became the bases for the land division. Miles Macdonell, the first Governor of the Colony, was familiar with the river lot system of Lower Canada, and it served as the model for the Red River survey, but the lots were made a third wider for the sake of convenience. On 17 July, 1813, Macdonell wrote to Lord Selkirk:

> There being no time in the spring to lay out regular lots, I gave as much ground on this point to the settlers as they could manage, & as much seed as could be spared. I have since laid out lots of 100 acres, of 4 acres front in the river, according to the annexed rough sketch (No. 18). On these

lots they are now at work preparing to build etc. The farms in lower Canada are of 3 acres front, & the first settlers in U. Canada had only the same—but they found it afterwards too narrow, which induced me to add one acre additional to the breadth of our lots here. This is sufficient for any common farmer.[2]

Lord Selkirk, far away from Red River and concerned about the defense of the colony, recommended a nucleated type of settlement that would afford some protection for the settlers. His initial instructions to Macdonell, prepared in 1811, contain some suggestions about the land division:

> They [the lots] may be laid out from 50 to 100 acres to each man, & should as far as possible combine wood & plain in every lot. If however the Indians should appear disposed to be troublesome so as to excite any apprehensions for scattered settlers, small lots of 5 or 10 acres may be laid out more closely adjoining to the Fort and assigned to the men as a temporary tenure to cultivate till they can safely take possession of their full lots.[3]

But Lord Selkirk was more explicit in a letter he wrote on March 30, 1816 to Colin Robertson, who had rallied the colony in 1815 after the Nor'Westers had severely harassed it.

> I am anxious that the land in cultivation should for the present be concentrated in the vicinity of the principal settlement: & that the settlers should not be scattered at a distance along the River as they were last year. To encourage them to remain together I would wish a compact village to be laid out in building lots of about half an acre, to each of which I would annex a lot of 8 or 10 acres of land as near as possible to the village. These small allotments I mean to give as a gratuitous allowance over and above the regular farm lots, but under the conditions that every settler who receives one, will build

● JOHN WARKENTIN *is Professor of Geography, Department of Geography, York University.*

Reprinted from *Transactions*, Historical and Scientific Society of Manitoba, Series III, No. 16, 1961, pp. 62-77, with permission of the author and the Executive of the Manitoba Historical Society. The paper has been slightly amended.

and reside in the village, till he has brought the whole of his extra allotment into complete cultivation. The people will thus be kept together for the first 2 or 3 years, which will not only tend to their actual safety, but will give them a feeling of security of great importance in tranquillizing their minds.

If the N.W. Co. are still in a condition to be troublesome by their intrigues among the settlers, the village should be placed a little down the River from the Government House but if the fort at the —————————— in your possession I should suppose that a better situation would be found between that and the old houses of the R.R.S. laying out the village on some such plan as this—the lots being about 100 feet in front, and 200 deep— the street 80 or 100 feet broad reserving breadth of 4 or 500 yards along the river & laying out the rear lots behind that.[4]

An enclosed sketch shows that the proposed village was to be laid out in a gridiron plan, not just as a simple *Strassendorf* or street village. It appears that the village form of settlement was only meant to be used until it was safe for the farmers to move on their river lots.

The compact nucleated settlements recommended by Lord Selkirk were never established, and the river lot survey prevailed. Indeed, when Lord Selkirk visited the Colony in 1817, he did not attempt to change the mode of settlement, but extinguished the Indian's title to the land for only two miles on either side of the Red and the Assiniboine Rivers (there were wider tracts at the Forts), and ordered more lots to be laid out beyond Fidler's survey in 1813. Most lots were ten chains wide and extended ninety chains westward from the river to make ninety acres, besides which each settler had a ten acre wood lot on the opposite bank of the river.[5]

By 1870, river lots extended for a distance of over forty miles along the Red, both above and below the original lots, and along the Assiniboine as far as Portage la Prairie. The majority of the lots were laid out with a twelve chain river frontage, but by the time the Colony was transferred to Canada in 1870, the lots varied from a chain to half a mile in width.[6] William Pearce, a distinguished land surveyor of the West who was engaged in surveying the river lots for the

Canadian government in the 1870's, said about the lots:

In 1874, I found a party who owned and resided upon, occupied and cultivated, in St. Andrew's parish, a strip of land one chain in width on both sides of the River. He had in all 64 acres of land, but it took a journey of 8 miles to pass from one end of it to the other. In the same year the average width of lots in the parish of St. Andrew's extending along the east bank of the river some 7 miles, was less than 3 chains.[7]

Probably the river front survey was the best land division that could have been devised for the pioneer colony, which was not a specialized agricultural settlement. The settlers farmed the lots in the hope of supplying the Company with produce, but they were also engaged in hunting and fishing, in trading and in working for the Company, so that the river at their front door represented something more than a convenient base for surveys. It was as essential an element of the settlement as the very land they tilled, and therefore it was natural that everyone should desire to live along it. The houses were sufficiently close to each other along both rivers to produce tightly knit settlements which ensured protection, facilitated transportation by both water and land, and facilitated mutual aid in all activities. There was an intangible strength in settlements like this, which can be attributed to the fact that everyone was always aware of what was happening in the entire community, since people in passing by cart or boat along the lots in the ordinary course of their affairs could see what was transpiring. Thus the river lots provided an essential element of cohesion in a pioneer community, where there was always the danger that hard won achievements would be dangerously dissipated over too wide an area. At the same time the river lots did not impose the rigidity often found in European agricultural villages where farming operations often had to be co-ordinated. In the Red River settlement there was room for the flexibility which the varied activities of the settlers demanded.

Only the front of each lot was generally cultivated; the rear was used for pasture and hay. Unlike the land division in Quebec, no second range of long lots was ever laid out behind the river lots. This has been attributed to the fact that the Hudson's Bay Company had extinguished the Indian's claims only for the first two miles on either side of the two rivers and therefore could not grant the land to colonists.[8] This may have ben partly responsible, but of more direct importance was the fact that the river was the favoured place of settlement as the narrowing of some lots through the years indicates. In 1870, there was still much land vacant on the Red, since the lots only extended about ten miles above Fort Garry, so that there was no need for the settlers to go inland. Furthermore, the land immediately behind the river lots was poorly drained, yet it could be used to some extent, because each settler had a "hay privilege"— the exclusive right to cut hay on the outer two miles immediately to the rear of the river lots. On these outer limits some cultivation was occasionally undertaken on better drained lands on what were called "park claims,"[9] but no settlements were founded, though buildings were occasionally erected on the claims. This brought the recognized agricultural limits of the settlers up to a distance of four miles from the river.

The river lot survey started by Lord Selkirk has considerable significance in the settlement geography of the West, because it remains the basis of the land division in the most densely settled part of the prairies, and also it was carried to outlying parts of the region by the métis. A number of cities, for instance Winnipeg and Edmonton, even owe their basic city plans to this survey. It is an outstanding example of the way in which a land division can be designed to take care of the needs of settlers: this was Lord Selkirk's concern from the beginning.

II

In 1869, while the Canadian government was still negotiating for the transfer of Rupert's Land, it sent a surveyor, Colonel J. S. Dennis, to inspect the land in the vicinity of Fort Garry with instructions to select suitable areas for the survey of townships for immediate settlement, and to devise a suitable system of survey for the prairie region. Despite the fact that he was an Ontario Provincial Land Surveyor, Dennis made little use of the survey systems of that province, which was unfortunate because the Ontario townships were divided into long narrow lots which provided some of the same settlement advantages as the river lots.

It is natural that Canadians should investigate the public land surveys of the American prairie states in their attempt to find the ideal scheme for the plains of Western Canada. In discussions held in Ontario before Dennis left for the West, it is evident that the American survey scheme was already being considered as the prototype for the prairie survey. On July 10, 1869 William McDougall, the Minister of Public Works, wrote to Dennis in Toronto.

> The American system of survey is that which appears best suited to the country, except as to the area of the sections. The first emigrants, and the most desirable, will probably go from Canada, and it will, therefore, be advisable to offer them lots of a size to which they have been accustomed.[10]

On his way to Red River Dennis consulted with the Surveyor-General of Minnesota and other "leading and intelligent Americans"[11] who supplied him with full details of the American system of surveying public lands. He arrived at Red River on August 20, 1869, and only eight days later, after some hurried trips within the Red River plain, he sent McDougall his plan, including two maps, for surveying the West. Dennis adopted the American method but made alterations where he thought it could be improved.

> It was generally conceded that the American system is faulty in making no appropriation for public roads, which are subsequently taken from the settler out of the net area of land for which he may actually have paid the Government.
>
> I think further that the townships are unnecessarily small. In a prairie country, where the facilities of communication are greatly in excess of those in a broken or heavily-wooded country, the Townships may well be larger, thus tending to economy in the administration of Municipal Affairs.[12]

In an appendix to this letter Dennis presented his proposed scheme, which was later adopted as the official method of survey by Order-in-Council of September 23, 1869.

Proposed Method for the Survey of the Public Land in the North-West Territory.

1. The system to be rectangular, all Townships to be east and west, or north and south.

2. The Townships to number northerly from the 49th parallel, and the ranges of townships to number east and west from a given meridian. This meridian to be drawn from the 49th parallel at a point say ten miles west of Pembina, and to be called the Winnipeg Meridian.

3. The Townships to consist of 64 squares, of 800 acres each, and to contain, in addition, 40 acres, or five per cent in area in each section, as an allowance for public highways.

4. The Townships on the Red and Assiniboine Rivers, where the same have had ranges of farm lots laid out by the Company to be surveyed, the broken sections butting against the rear limits of such ranges, so as to leave the same intact as independent grants.[13]

Point four, it is worth noting, definitely establishes that the pre-existing surveys were to be retained. Trouble developed when Major A. C. Webb ran the latitudinal line east from the Winnipeg Meridian toward Oak Point (St. Anne Des Chenes), which involved carrying the line across the river lots. He never got there. The entry in his diary for Thursday the 11th of October 1869 reads:

Start on Bearing of N 89°55'20"E from Tp. cor. surveying one section and chained half a section when stoped [sic] by 18 Half Breeds, found it impossible to proceed further. Went into camp within four miles of Winnipeg, and sent Report in and await Col. Dennis [sic] orders to proceed. Changeable weather with slight fall of snow.[14]

But the Riel Uprising of 1869-70 forced a suspension of the field surveys until 1871 and during the interval the original scheme was modified. Lieutenant-Governor Archibald thought that the townships were too large, believing that they should be more akin to the American system. Archibald's

reasons are given in a report dated December 20, 1870:

Now as regards the system of surveys, I take it for granted that the general principle sanctioned by the Government in the Minute of Council of the 23rd September 1869 . . . will be retained. The general principle I take to be that the lands shall be surveyed in rectangular blocks, numbered consecutively, with subdivisions also numbered consecutively, from 1 upwards in each block.

But I cannot help thinking that Col. Dennis has not acted judiciously in recommending a deviation from the system of 6 mile townships.

In the year 1796, now three-quarters of a century ago, when the United States Congress passed their first law on the subject of the surveys of the lands in their Territories, they adopted the system of 6 miles square, subdivided into 36 square miles, each of these again subdivided into 4 square lots of 160 acres each, and these again in certain cases into four ultimate square lots of 40 acres each. . . . This system is known all over the world to the Emigrant classes. A lot of 160 acres is the acknowledged extent of an Emigrant's requirements for farm purposes. The system has been adopted by the most practical people in the world, and after 70 years experience remains unchanged. Why should we change it?

In laying out the boundless prairies to the west of us there might, perhaps, be some justification for liberality in the amount given for a farm lot. But with the limited Territory of Manitoba, with Hudson's Bay Grants and Hudson's Bay Reserves, with squatters' rights and Half Breed rights, and Indian Reserves, our little Territory is already to some extent forestalled, and this does not seem to me to be the occasion nor Manitoba the Province, where we should set the first example of prodigality in the allotment of lands.

But with 5,250,000 acres to grant, and the question before us whether we shall lay it off in lots of 210 or in lots of 160 acres, there is just this difference in the decision to which you come, that the system I should recommend would furnish 32,800 homesteads, while the other would give but 25,000. In effect, therefore, you have by one system 8,000 more farms than by the other, and in a Province limited as ours, I consider this a matter of great importance. 160 acres is quite sufficient for an ordinary farm lot.[15]

By Order-in-Council, April 25, 1871, the townships were changed to contain thirty-six sections, each approximately one mile square, and with a road allowance of one

59

and a half chains (ninety-nine feet). Thus the township measured six miles on each side plus the road allowances. This is known as the first system of survey. In 1881, a different procedure was adopted in subdividing the townships so that the eastern boundary of each section would be a true meridian (the second system), and later that year, it was decided to make the road allowances one chain wide, and place them only on each alternate east and west line, though retaining them on all north and south lines (the third system). The latter changes were designed to save land for arable use and to eliminate the needless building of roads. The first system covers most of Southern Manitoba and extends west to just beyond the 2nd Meridian (which lies 180 miles west of the Principal Meridian), and the third system covers most of the rest of the prairies.

When the sectional survey was adopted for Western Canada, the needs of the state (and the surveyor) were placed above those of the individual settler, although of course, the two need not necessarily be incompatible. Dennis's son, who prepared an official history of the survey, said:

> The primary consideration, having in view the future welfare of the country, was to devise a system under which the country would be rapidly and accurately divided into farm holdings.[16]

The survey certainly accomplished this. A large area had to be surveyed; therefore it was desirable to have a single survey based on astronomical base lines that could be continuously applied over a wide area. In fact, the survey became an index system, because with the scheme adopted of designating townships, range lines, and sections it was very simple to locate any given plot of land.

This survey was not a gradual advance, moving along with settlement, but consisted of the rapid superimposing of a rigid pattern without any regard to topography. No attempt was made to run lines that would conform to terrain and vegetation to ensure that the maximum benefit would be secured for the greatest number of settlers within an area. The surveyors themselves were not completely satisfied with the settlement results,[17] but there was little opportunity for experimenting in ways of dividing the land. There was a note of urgency about the survey, stemming probably from the belief that the West would be rapidly occupied. Thus, throughout almost the entire prairie region, there was laid out an easy to survey and simple to administer, but nevertheless stereotyped pattern of squares. In all fairness I will, however, quote Mr. H .E. Beresford's comments on the survey system, since the survey certainly accomplished what it was designed by Dennis to do:

> For over eighty years, Crown Lands in the western provinces have, with minor changes, been surveyed under [the rectangular] system —a system which has received the highest praise wherever known, a system of survey which has been the greatest single factor in the successful development of Western Canada and one that has caused perhaps less litigation over land boundaries than any other in the world.[18]

Admittedly the survey was not consciously designed to create the settlement pattern of the West, but in lieu of any other settlement planning, the survey, in conjunction with the government's land policy, could not help but be a critical force.

III

The sectional survey was to all intents a *fait accompli* before farmers moved in; it now remains to examine the type of rural settlement it produced, the reactions of critics to this pattern and the few attempts by settlers to make adjustments to the survey.

The sectional survey did not eliminate the river lots along the Red and the Assiniboine, and these were resurveyed in the years 1871-78. By Orders-in-Council April 3 and April 17, 1874, the hay privileges in the "outer two miles" were assigned to the owners of the appropriate river lots, and the boundaries of each parish were produced to the extension of the two mile belt, the rear line being a boundary of the sectional survey.

Road allowances were left between the Inner and Outer two miles, paralleling the rivers, and houses have subsequently been erected along them in a few places, so that a semblance of the second range of river lots, found in Quebec, has developed. Surprisingly enough, the river lot survey was also adopted in the Rainy River country at the suggesting of Dennis himself, when that clay belt was surveyed by the Dominion Government in 1876. By Order-in-Council, June 27, 1876, the land on the river was ordered laid out in river lots, ten chains wide, with the land beyond measured out in the regular sections. At the time, the territory was in dispute between Manitoba and Ontario, and it fell to the Dominion Government to survey the lands. It is the only example in Ontario of the sectional Dominion Lands Survey, although the American sectional survey had been partially adopted by Ontario surveyors along the north shore of Lake Huron.

Nowhere in the prairies was any legally recognized departure permitted from the regular pattern in the first few decades of the survey, except in some districts inhabited by métis. Consequently the sectional survey, together with the Dominion Lands Act of 1872 and its system of alternate homesteads and residence requirements on the homesteads, ensured that there would be dispersed rural settlements through most of the West. This dispersal aroused some criticism shortly after the survey's inception, particularly among visiting Britishers, who apparently were more conscious than Ontario-Canadians of the social implications of scattering people from fairly densely settled areas widely over the prairie. For example, in 1883, Professor Henry Tanner of the South Kensington Institute of Agriculture argued that during the pioneer years, at least, close settlement was essential.[19] W. H. Barneby, a well-informed and observant traveller, voiced a similar opinion in 1884.[20] Further expression was given to this view in 1900 when Professor Mavor of the University of Toronto, in a study commissioned by the Department of the Interior, directed a blast at the failure of the administration to take the idiosyncrasies of people into account in planning settlements. Some immigrants, he argued, would prefer to settle in an isolated way, others in groups. He went so far as to say that:

> . . . it is now generally recognized that the alternate section system was a mistake and that the number of persons who prefer the isolated life to which it induces has often all been shown to be very small, and that the isolation has on the whole been proved especially in the second generation of the settlers to be an immense disadvantage.[21]

I am not sure but that Professor Mavor may have overstated his case. There was, however, a basis for his criticisms and those of Tanner during the early years of settlement.

Planners, geographers, sociologists, and Royal Commissioners have also been critical of the sectional survey. Besides condemning the social implications of the dispersed settlements, they have emphasized the fact that the survey failed to take account of any variations in topography. Probably the severest criticsm that has ever been directed against the sectional survey in Canada was made by an English planner, Thomas Adams, who was brought to Canada by the Dominion Commission of Conservation in 1914. He objected to the fact that:

> . . . a survey designed for one purpose is used as a plan for another purpose, that it is so used without regard being paid to the soil, topography and future development, and that farms are divided and roads located without any properly conceived development scheme being perpared for the areas in which they are situated.[22]

Elsewhere he stated:

> What is needed is the establishment of well planned agricultural villages on good and accessible land. They must be planned in such a way that there will not be an entire absence of facilities for social intercourse, cooperation, transportation and ready means for marketing.[23]

This was written in 1916 when many people were favourably disposed towards constructive planning, because it was felt that many returning soldiers would go to the land, and that they would not take kindly to isolated settlements. Three decades later, a brilliant examination of the section as an ecological unit was undertaken by a geographer, Professor Hildegard Johnson of Minnesota. In her article she supplies striking examples of the practical farming difficulties that may arise when the sectional survey is applied to dissected country, and how farmers have attempted to adjust to the survey within the limits of the land regulations.[24] In 1955, the Saskatchewan Royal Commission on Agriculture and Rural Life published its findings, and in its report on Rural Roads and Local Government records the socially deleterious effects of the sectional survey in Saskatchewan.[25]

Despite the shortcomings of the official survey, very few changes resulted in the land division, although admittedly this is a very difficult thing to do once a survey has been established. The failure of the sectional survey to "fit the land," as it were, is not quite as serious a fault as might be expected. There are many districts where the section is utterly unsuited to the terrain (the Manitoba Escarpment for instance), but in general there was sufficient relatively flat land available to farmers in the prairies to enable them to select suitable homesteads even if the survey took no notice of the topography. And in the days of animal power, it was desirable to live as close as possible to the cultivated land. There is no doubt, when the social welfare of the settlers is taken into account, that the criticism of the isolated life that resulted from this settlement pattern is valid.

But what did the settlers themselves think of the life? When farmers came to the West, they came to an area where land was already measured according to a particular survey, and hence they tended to accept the situation, especially since this division seemed to offer opportunities for economic progress. To them the sectional survey, along with the homesteads, preëmptions and railway lands for sale offered an opportunity for individual expansion. This was a commercial farming economy; land was something to be used, to be bought and sold. Here there was little place for the love of the soil

of a definite locality that characterizes peasant life. Abe Spalding in Frederick Philip Grove's novel, *Fruits of the Earth,* exemplifies the climate of the time. Right from the start Abe selected his land with a view to expansion—and the accumulation of wealth. And most farmers, not soured by a lack of success, quickly accepted the isolation, though the complaints were numerous before the appearance and comfort of the farmsteads were improved, by planting shelter belts and building better farm homes.

IV

It is always very difficult to change an established survey and settlement pattern; hence there were only a few attempts to find alternatives to the sectional survey and the scattering of farmers it had caused. Some attempts were the result of the previous experiences and traditions of a group of settlers, other were devised by planners who were seeking the best agricultural adjustments to local conditions. But at least, in one way or another, they were all efforts to overcome the dispersion resulting from the section survey.

Many métis were quite adamant about retaining the river-front lots in new locations, so that the Government was often forced to recognize their claims. In Ste. Anne, thirty miles southeast of Winnipeg, there was a métis settlement on the Seine River which was surveyed in 1872 prior to the sectional survey, so that the sectional survey had to conform to it. Thus the outlines of Ste. Anne parish (Figure 1) appear today like an open handkerchief casually dropped in the mist of a checkerboard pattern; it is unique in this regard in the prairies. Close by, more métis farmers squatted on river lots along the banks of the Rat and the Seine Rivers *after* the sectional survey was completed. They refused to conform to the sectional division, so finally in 1884, the Government was forced to resurvey their claims, but unlike Ste. Anne, here the rear boundaries of the lots were adjusted to the sectional survey lines. Other métis settlements, which existed prior to the sectional survey, such as those on the

Figure 1

River lots on the Seine River at Ste. Anne, Manitoba, meet the sectional survey. Farms are small in this part of Manitoba and there is a farmstead on nearly every quarter section. Note the linear development of farmsteads along the east-west section road in the lower part of the photograph. Roads, drainage ditches and a railroad cut across the long lots. (Aerial photograph A 15911-9, Air Photo Division, Dept. of Energy, Mines and Resources.)

Saskatchewan River at Prince Albert and Carlton, and on the Sturgeon River near Edmonton, were also surveyed as river lots.

But there were also some schemes to introduce nucleated settlements into the midst of the official survey. Professor Tanner maintained in 1883 that village settlement was essential, and that villages of working men, each with a small holding, should be placed in the midst of farms held by men of capital, who would require labour.[26] His concern about providing employment is well borne out by the fact that in the pioneer years the menfolk normally worked in railway camps, lumber camps and for other farmers to earn much-needed money. Tanner went so far as to prepare plans of village settlements that would fit into the township pattern.

Thomas Adams also worked out a number of village settlement schemes designed to secure the following advantages: the close settlement of farm buildings, convenience and directness of access to the town area and railway station, reduction in the length of roads, and use of swampy and rocky land for timber reserves.[27]

These settlement schemes were not considered practical by surveyors. In 1941, F. H. Peters, Surveyor-General of Canada at the time, said in referring to these and other proposals:

> . . . when it came to the final analysis, none of them seemed to be particularly attractive. It was very difficult to find any way of laying out a system of roads that would be as economical as the system which is followed in the Western provinces.[28]

This conclusion has been challenged by many—most recently by the Saskatchewan Royal Commission of Agriculture and Rural Life.

One rather startling suggestion that was brought before the Select Standing Committee on Immigration and Colonization of the House of Commons in 1883, by Mr. Robert Romains, merits mention.[29] He proposed that agricultural villages be established every few miles along the main railway lines, with the farm lands to be operated from the villages stretching away on either side of the track. The fields were to be reached by tramways extending twenty-five to seventy miles from the villages. This meant that the main railway lines would only be spaced from fifty to one hundred miles apart, with tramways perpendicular to them about every four miles. This scheme, of course, presupposes grain farming, and does solve some social problems, but it would certainly raise a number of new difficulties in farm operations.

The Mennonites of Southern Manitoba were the first people who successfully managed to establish nucleated settlements within the framework of the sectional survey (Figure 2). This was made possible because by Orders-in-Council of March 3, 1873, and April 2, 1876 they were assigned Reserves on the Red River plain for their exclusive use. Within these reserves both odd and even-numbered sections were allocated to the Mennonites as homesteads. An important amendment of the Dominion Lands Act in 1876,[30] granted them (along with the Icelanders) the privilege of settling in villages, thus circumventing the regulation that every farmer had to build a house and improve the land on his *own* homestead. A group of perhaps twenty farmers would pool their homesteads into what then became the village land, select a central site for a village, and divide the land into arable, meadow, pasture and woodland, and further subdivide the arable fields into strips for individual use. The sectional survey and homestead regulations only controlled the pattern to the extent that they determined the size of the village land since each homesteader received a quarter section, and the outside borders, which was a line of the sectional survey. Over 120 villages were established during a period of two decades after 1874, but not more than eighty were fully functioning "open field" villages. Strip farming was not abandoned until the decade of 1920 in a few of the villages. Today, there are still seventeen nucleated settlements left, inhabited by farmers who go out to operate quarter section fields, or parts or multiples thereof.

These Mennonite villages point up the

Figure 2

Hochfeld: a Mennonite farm-operator village in the midst of the sectional survey. Some of the ridges (running north-south and formed by blowing dust) separating the old cultivated strips are still visible north of Hochfeld. The ridges are gradually being plowed down and in time will be obliterated. They are only barely apparent on the field east of the north-south highway. Note the absence of individual farmsteads on the quarter sections in contrast to the Ste. Anne area. (Aerial photograph A 16586-80, Air Photo Division, Dept. of Energy, Mines and Resources.)

FIELD PATTERNS OF THE MENNONITE VILLAGE OF NEUHORST, MANITOBA

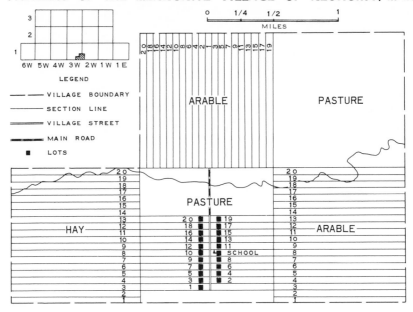

The total land area of the village (the Fleur*) was divided into large fields* (Gewanne)*. Each* Gewann *was subdivided into the same number of strips as there were lots in the village. In this way each farmer had relatively easy access to his arable and hay land. Land that was not suited to cultivation was generally reserved for common use (pasture).*

advantages and disadvantages of nucleated settlements in the prairies, especially since they had an Achilles heel which made them susceptible to any centrifugal influence. Despite the new settlement pattern and land division adopted by the Mennonites, the quarter section remained the legal unit of land, so that if a farmer was dissatisfied no one could stop him from taking his land out of the open fields of the village. This frequently led to the break-up of the entire village. There were various reasons for leaving the villages. Farm management was not as efficient on the strips as on the quarter section. The individual farmers' maximum distance to his fields was rarely more than two miles, but it was a nuisance to have to bring enough water and feed for the horses into the field to do for an entire day. Even more burdensome was the fragmentation of fields, which forced excessive wasteful travelling. The many narrow arable fields separated by grass strips raised some special management problems. Much land was wasted in the grass strips, which often

collected dust and turned into excessively wide ridges. Shrubs and weeds grew on these ridges and often infested the fields. Each farmer had individual access to his fields, yet a certain amount of coordination in farm operations was still essential, and might even prove restrictive to a progressive farmer. One of the main drawbacks of the villages in a North American economy was the limit they placed on the expansion of holdings. Many Mennonites were soon caught up in the North American commercial farming fever, with its tendency to expand operations, and the willingness to move to other seemingly more desirable holdings. Many farmers found village farming confining; it was difficult to acquire more land; it was not so easy to adopt new techniques in the small fragmented holdings —in general, farming went on at a more leisurely pace that did not suit the ambitious. Thus there was a movement right from the first year to the individual farmstead. Since this movement usually went on in the face of very strong religious disapproval from the village community, it is obvious that the dispersed pattern held great attractions, and it is more than a coincidence that the best farmers among the Mennonites were generally men who had left the villages. This does not mean that the quarter section made a better farmer of a man, but rather than the better, more ambitious farmers were the ones who found the quarter section attractive and could use it to advantage.

Only in such a group of people as the Hutterites, where individual enterprise as distinct from "colony" enterprise is unknown, have nucleated settlements proved unyielding in Manitoba.[31] And the somewhat similar cooperative farms of Saskatchewan provide similar evidence.[32] But these settlements are not even the exceptions that prove the rule, because fundamentally each is a single large unified farming enterprise under one management.

Though settlers have shown that they prefer compact fields and dispersed settlements to dispersed fields and compact settlements. a compromise is now being attained through the adoption of the long lot. In a few areas the planned conversion of the squares of the sectional survey into long lots has proved successful. In the sectional survey the section, one mile square, is divided into four quarter sections each a half mile square, with the farms fronting on at least two roads. In the adaptation of the sectional survey to the long lot, each quarter section consists of a rectangle one mile long and one quarter mile wide, with all the farms fronting on one road. Thus the cost of services is reduced, and the farms are brought closer together. After the turn of the century, river lots were laid out within the outlines of the sectional survey on the alluvial soils of the Birch and Whitemouth Rivers in Southeastern Manitoba and along the Carrot River in Northwestern Manitoba. And at present, there is a limited movement, both planned and unplanned, towards the adoption of the long lot, along trunk roads as well as along rivers. The Pasquia Land Settlement Project is making use of the long lot,[33] and Professor W. B. Baker reports that there is an adjustment in farm residence patterns in Saskatchewan, where farmers are either moving into town or relocating on main market roads where possible.[34] An interesting readjustment has been occurring in Southeastern Manitoba since 1940 within the milkshed of Winnipeg. Farmers are establishing themselves on relatively small lots along main market roads, in order to be close to facilities such as feed mills, poultry dressing plants and cheese factories. This, of course, is the result of a change in land use. In these lots the farming advantages of compact holdings and the economies in public services of the linear settlement are secured. And the farms do not have the excessive length or narrowness of some river lots, which can be a handicap.

V

This brief review shows that every land division in the West has originated outside the region — from Eastern Canada, the United States and Europe. The sectional survey, derived from the U.S. survey, is easily of most importance. In the pioneer stage the section is not ideal (even on suitable terrain), mainly because its advantages for large-scale, efficient commercial farming are not required at a stage when only a few farmers at best are able to operate large acreages competently. Consequently from 1871 to 1900 the social disadvantages of the dispersed settlement produced by the sectional survey and the land regulations were especially apparent and the advantages were obscured. But with the introduction of new agricultural techniques and the resultant greater productivity per man-hour, the sectional survey came into its own as a suitable land division for the enterprising farmers of the West . (Witness the break-up of many of the Mennonite agricultural villages after 1885.) People did become accustomed to the dispersion, despite the early complaints and the doubts of various academic critics, and the quarter sections became as accepted a part of the environment of the West as the spacious prairie scene. And when the nature of the topography, the grain growing economy, the increase in size of farm holdings, and the improvements in communications are taken into account, this must be regarded as a fairly sound system of land division for many areas today.

Social disadvantages still exist in the settlement pattern produced by the sectional survey, but they can be partially overcome by concentrating farmsteads along market roads, that is, by adopting the long lot. If the long lot had been incorporated into the original survey, all the highly-valued technical and administrative strengths of the sectional survey would have been retained, with the added advantage of a reduction in the dispersal of the farmsteads and in the cost of municipal services. This survey would have required no great leap of the imagination, or even experimentation, on the part of the surveyors because it had been already introduced in Ontario. Now an adjustment towards the long lot within the general limits of the original survey is beginning rather belatedly and in a haphazard and imperfect manner. This is accomplished on an individual basis by farmers as the opporunity arises to change the locations of their farmsteads, but it is possible that within two generations there may be a change in the landscape of Western Canada as the line settlement develops that is almost as great as that accompanying the enclosure movement in England. It would appear that the main stimuli towards achieving this pattern are to make the public conscious of the economic savings and the social advantages of the long lot, so that as farmers buy and sell land they will tend to move their farm headquarters to the main market roads. The careful location of market roads by municipal authorities and planning boards so that they provide an incentive to farmers to move their homes to the roads will also help. Naturally this involves very long range regional planning.[35]

What of the other survey systems? The few areas laid out in other divisions practically disappear in the great expanse of townships and sections. Even when one passes through them, the river lots cannot readily be distinguished from the quarter sections on the flat prairie terrain, although they can be made out on the banks of the Red River. But many of the lands laid out in river lots are now being urbanized. In Southeastern Manitoba, the area is quite distinctive where seventeen Mennonite villages still exist, because the villages are marked by long rows of cottonwoods in the midst of a country empty of farmsteads. With fast-moving modern farm implements the villages have turned out to be an ideal settlement pattern for Manitoba's special crops area, yet they still possess an old-world atmosphere of comfort and serenity. What can be termed town or sidewalk farming is also coming into existence. In the Great Plains states and in Saskatchewan many farmers now live in towns from where they operate their farms. Naturally this is the maximum in social welfare that a farmer can achieve, even surpassing the line settlement, which provides more service efficiency than social efficiency. This mode of farming became possible with cars and trucks for grain farmers, but you will recall that Robert Romaine in 1883 envisaged tramways leading from central places to farms.

The quarter section and the accompanying dispersal of farms has definitely marked the West. Probably the only quick changes in settlement patterns will come in new special argicultural areas; whether they are carefully planned as in the Pasquia, Bow River or Saskatchewan River projects, or whether they develop spontaneously within the sectional survey in a rapidly intensifying agricultural economy as in Southeastern Manitoba. In both instances, there is an adjustment toward a settlement pattern that satisfies the needs of the people; it is noteworthy that the long lot, which is occasionally emerging, is the equivalent of the survey that Miles Macdonell introduced to Red River in 1813.

REFERENCES

[1] Morton, A. S., *A History of the Canadian West* (London, n.d.), p. 553.

² P.A.M., Selkirk Papers, microfilm copy, Miles Macdonell to Lord Selkirk, 17 July 1883, p. 787.

³ *Ibid.,* Instruction to Miles Macdonell, 1811, p. 180.

⁴ *Ibid.,* Lord Selkirk to Colin Robertson, 30 March, 1816, p. 1894.

⁵ Martin, Archer, *The Hudson's Bay Company's Land Tenures* (London, 1898), p. 11.

⁶ Beresford, H. E., "Early Surveys in Manitoba," *Papers Read Before the Historical and Scientific Society of Manitoba,* Series III, No. 9, Winnipeg, 1954, p. 12.

⁷ Quoted in Beresford, p. 12.

⁸ Martin, p. 99.

⁹ Beresford, p. 13.

¹⁰ Canada Sessional Papers, 1870, V, No. 12, Correspondence and Papers Connected with Recent Occurrences in the North-West Territories, William McDougall to Lieut. Col. J. S. Dennis, July 10, 1869, p. 2.

¹¹ *Ibid.,* Dennis to McDougall. August 28, 1869, p. 6.

¹² *Ibid.,* Dennis to McDougall. August 28, 1869, p. 7.

¹³ *Ibid.,* Dennis to McDougall. August 28, 1869, p. 8.

¹⁴ Surveys Office, Manitoba Department of Mines and Natural Resources, A. E. Webb, Diary of Survey.

¹⁵ Quoted in Sebert, L. M., "The History of the Rectangular Township of the Canadian Prairies," *The Canadian Surveyor,* Vol. 17, 1963, pp. 383-84.

¹⁶ C.S.P., 1892, XXV, No. 13, *Annual Report of the Department of the Interior for the year 1891,* "A Short History of the Surveys Performed Under the Dominion Lands System 1869 to 1889," by J. S. Dennis, p. 31.

¹⁷ See the comments by Peters, F. H., on a paper by Rousseau, L. Z., "Surveys and Land-Use Planning in the Province of Quebec," *Proceedings of Thirty-Fourth Annual Meeting of the Canadian Institute of Surveying,* Ottawa, 1941. Also see comments by Gorman, A. O., on a paper by Beresford, H. E., "Manitoba Surveys," *Ibid.,* p. 58.

¹⁸ Beresford, p. 10.

¹⁹ Tanner, Henry, *Successful Emigration to Canada* (Ottawa, 1883), p. 29.

²⁰ Barneby, W. H., *Life and Labour in the Far, Far West* (London, 1884), p. 358.

²¹ C.S.P., XXXIV No. 10, *Annual Report of the Department of the Interior, 1900,* "Report of Professor Mavor," p. 234.

²² Adams, Thomas, *Rural Planning and Development* (Ottawa, 1917), p. 53.

²³ Adams, Thomas, "Report on Town Planning, Housing and Public Health," *Report of the Seventh Annual Meeting, Commission of Conservation, Canada* (Ottawa, 1916), p. 121.

²⁴ Johnson, H. B., "Rational and Ecological Aspects of the Quarter Section," *Geographical Review,* Vol. 47, No. 3, 1957, pp. 330-348.

²⁵ Saskatchewan Royal Commission on Agriculture and Rural Life, Report No. 41, *Rural Roads and Local Government* (Regina, 1955), p. 59.

²⁶ Tanner, p. 29.

²⁷ Adams, *Rural Planning and Development,* plates facing pages 62 and 63.

²⁸ Comments by F. H. Peters, p. 31.

²⁹ *Canada, Journals of the House of Commons,* 1883, Report of Select Standing Committee on Immigration and Colonization, Appendix 6, pp. 129-133.

³⁰ *Canada Statutes,* 39 Vict., Chap. 19.

³¹ Peters, V. J., *All Things Common, The Hutterians of Manitoba,* unpublished M.A. Dissertation, University of Manitoba, 1958.

³² Wright, Jim, "Co-operative Farming in Saskatchewan," *Canadian Geographical Journal,* August, 1949, pp. 69-90.

³³ Ellis, J. H., *The Pasquia Land Settlement Project Interim Report No. 3, Reclamation of the Pasquia Area, Drainage and Roads,* (Winnipeg, 1957), see map in back pocket.

³⁴ Baker, W. B., 'Changing Community Patterns in Saskatchewan," *Canadian Geographical Journal,* February, 1958, pp. 44-56.

³⁵ For more detailed discussions of this, see T. Lynn Smith, *The Sociology of Rural Life* (New York, 1953), pp. 268-273, and Saskatchewan Royal Commission on Agriculture and Rural Life, Report No. 41, pp. 56-84.

The Rang – Pattern of Rural Settlement in French Canada

PIERRE DEFFONTAINES

When the first French settlers came to Canada, they found everywhere unending forests broken only by the occasional river or an infrequent moss-covered peat bog or *savane* as it was called: there was no previous type of permanent settlement to serve as guide; the natives were wanderers of the forest whose only home was a tent or a makeshift hut. The French Canadian therefore had to plan his own type of settlement and devise his dwelling, knowing little of the country, its geographical features, its waterways, its natural routes, its soil varieties. In a word, he had to people and exploit a country, to initiate a process which in the old countries of Europe took many centuries and constantly required numerous modifications.

DEVELOPMENT OF SETTLEMENTS ALONG THE WATER COASTS

The first settlements were [thus] found along the river coasts; possibly no other colonization was so linked with a country's waterways, which literally served as guides. A veritable river civilization opened up; first it was along the south shore of the St. Lawrence as far up the river as the first rapids at Lachine just beyond Montreal. By the mid-17th century, a second river region was being colonized, that of the tributaries of the St. Lawrence beginning with the Richelieu on the south shore. Steep river banks or high cliffs were no impediment to colonization; at Petite-Rivière, and St. Joseph-de-la-

Rive, homes were studded on to the one-hundredth-metre-high sloping banks. In spite of dangerous rock falls from the cliffs, a community developed at Les Eboulements, while at Cap Rouge, steep steps led down to the waterfront. Nor indeed did muddy shores, dotted with reefs, inhibit settlement.

Toward the end of the French régime, one had only to take a canoe trip along the St. Lawrence and the Richelieu Rivers for an almost complete enumeration of Canadian homes; the waterfront or shoreline was synonymous with settled country.

Along the St. Lawrence, more and more farm lands were staked out, and dwellings sprang up in almost uninterrupted succession along the entire shoreline. On the south shore, however, the tempo of colonization moderated during the 17th century because of the threat of wars with the Iroquois. But with the peace of 1701 came renewed activity that soon matched the pace of development on the north shore. Indeed in the region of Montmagny, for a distance of some twenty-one miles west of the town, one finds today, at every five hundred metres or less, a succession of farmsteads most of which date back to the early days.

A FEUDAL COLONIZATION SYSTEM

The success of this land settlement may be attributed to the particular type of colonization that prevailed. During the 17th and 18th centuries, colonization was modelled on the old French feudal system with this

● PIERRE DEFFONTAINES, *prior to his retirement was Professeur de géographie, Université de Barcelone, and Directeur honaire de l'Institut français de Barcelone.*

Translation reproduced by permission from: Marcel Rioux and Yves Martin (eds.) *French Canadian Society, Vol. I* (Toronto: McClelland and Stewart Ltd., 1964), pp. 2-19. The original article, "Le rang, type de peuplement rural du Canada francais," was printed in *Publications de l'Institut d'Histoire et de Géographie*, No. 5, 1953. Permission to reprint this translation has been received from the author and the publishers of the original article, Les Presses de l'Université Laval.

difference: the Canadian seigniory was not a political institution, as was that of the European Middle Ages, but rather a particular system of settlement and of distribution of new lands (Langlois).

The feudal estate was granted to the seignior, whose duty it was to apportion lots thereof to the settlers, known as *censitaires* because of the *cens* or rent which they had to pay under a bilateral agreement of feudal privileges and duties. The seigniories, naturally situated along the shores, varied in dimensions but usually had a water frontage of about one league, while, in the early period, the depth towards the interior was indeterminate.

Relatively few of the seigniors took up residence on their estates. Most preferred to subdivide the land and parcel out the lots to the settlers—they became veritable colonizing agents spurred on by the prospect of accrued rents. Regulations favouring the new settlers were enforced. This was in sharp contrast to some of the abuses prevalent in the new colonies where large estates specialized in luxury production and slavery provided the labour. The Canadian seigniorial system with its individual family holdings made way for the success of the small farmstead.

Each seignior was obliged to assure the occupancy of his lands, otherwise he risked confiscation of the estate. He thus became a very effective medium of colonization. Though the revenue from the *cens* was not exorbitant, nevertheless he could count on a fixed amount at definite intervals. Intendant Talon was a strong promoter of the seigniorial system to settle the land. Though seemingly antiquated, it was indeed a very efficient system, far superior to the land speculation which later prevailed, and a potent factor in the colonization of French Canada.

In order to assure occupancy of his land, the seignior would actively recruit settlers; army officers of the regiments based in Canada were often granted fiefs and in turn encouraged their soldiers to settle there. Others returned to France to their home provinces, to find occupants for their lands,

as did Giffard who lured many of his fellow countrymen of Mortagne to his seigniory at Beaupré, and was responsible for an important movement of emigration from Perche.

The *censitaire* would be granted a parcel of land on condition that he build his home and grub the land. His rent, initially just a few *sols,* was later twenty *sols* per *arpent* [about 65 metres] of frontage plus fringe payments in kind (a capon, butter, a ham). The seignior had to erect a mill where the *censitaires* were supposed to bring their grain, and also an oven for baking; if he did not fulfill this obligation, the oven and mill could be built by anyone else. The settler further consented to a certain amount of feudal service (the *corvées*), road work in particular. It was often stipulated that any fine oak trees found on the estate should be reserved for the king's navy.

To ensure a certain measure of stability, the seignior had the power to control undesirable transfers of property by withdrawing the tenant's title to the land. He in turn was subject to restrictions with regard to the disposal of his seigniory, and was duty bound to assure the distribution and occupancy of his lands. Regular inspections were carried out by the State, and delinquents were dispossesed of their domains.

In 1640 the population of French Canada was only 340; by 1665 it numbered 2,500 inhabitants. From 1660 on, the King sent an average of some three hundred emigrants annually. During the régime of Intendant Talon most of the officers of the Carignan Regiment were granted estates on which they settled their soldiers, and almost four hundred new farm lands were established in this manner; these soldiers left in the Canadian genealogy names such as Jolicoeur, Laframboise, Brindavoine and Sans Souci. On several occasions (in the period 1672-1711) those seigniories which had been mismanaged were confiscated by the Intendant acting in the name of the King, and were re-allocated. There were eighty-three seigniories in 1683 and some one hundred and ten at the time of the British conquest.

The seigniorial régime was one which

afforded the new settlers security and freedom from isolation. Admittedly the seigniors themselves became an almost extinct class after the British conquest. Many who were in military or civil service returned to their homeland; others had never even set foot on Canadian soil. But the seigniory and its land division were here to stay. The tenants continued to pay the *cens*. Eventually the government made provision for the purchase of the farms by the tenants, although in some instances, notably in Lotbinière, payment of the *cens* was still being exacted in recent years.

THE RIVER LOT

That part of the property bordering the river was known as the river lot. The water frontage was its most important aspect, being usually of narrow dimensions, and the river limits of the property were determined by the highest water line of the equinoctial tides. It was called the waterfront (*fronteau*, or *base* or *devanture*). Here was the starting point for land clearings and on this river lot was built the homestead. Bounded at one extremity by the river, it extended in a long narrow strip to the hinterland.

Farm production was diversified according to the quality of the soil and the state of progress in clearing the land. Near the river in proximity to the house was the vegetable garden; then the cultivated land known as the tilth (*clos*); further up were the meadows and pasture lands dotted with tree stumps. These grubbed spaces, sometimes referred to as loosened or cleared regions made up the open terrain. Beyond this was the reserve of timber, and though a small section of this area was cleared annually, a considerable expanse of wooded land was preserved to provide for present and future requirements of fuel and building material.

The seigniors divided their land into lots of uniform water frontage, the river lot usually being about three or four *arpents* (200 to 250 metres) wide. The surveyor, at three-acre intervals along the shore, marked off the land divisions with a stone under which he put pieces of crockery or

glass, or even charcoal, for identification. This procedure was known as the frontal division of the lands. In the early days one referred to the size of a property by its front acreage only. Its depth to the interior was more often than not indeterminate; sometimes the length was specified as being, ten, twenty or forty acres "more or less," terms which appear frequently in the early acts.

The seignior who granted the land invariably gave good measure with regard to depth of a property. While it was important that the width be specified because this determined the amount of the *cens* or rent, the length mattered little, with the result that land occupancy usually exceeded the limits specified in the deed, a varied terminology being applied to these property surplusses: *allouance, fort mesure, tour du bâton, robinet*.

A few of these farmlands of indefinite length still exist. For example, in Château-Richer, there are properties extending some four and a half miles from the St. Lawrence into the high forests of the Laurentians. Their frontage, however, is but one and a half *arpents* (100 metres) as compared with the customary three *arpents*.

Though indifferent with regard to measurement of depth, the surveyor did pay particular attention to the lie of the lots. As a general rule they were set at right angles to the river front in a northwest by southeasterly direction along the St. Lawrence, and north-south direction along the Ottawa. At times, determining the proper angle was a problem because of the winding character of the river. In these instances the sun served as guide, the direction line being determined by either the 9 o'clock, and 10 o'clock, or mid-day sun.

The Development of the Second Rang

By the end of the French régime, the waterfront lots were practically all occupied along the St. Lawrence and up its main tributaries. For a long time, this remained the one and only rang. In 1668, there were no more available riverside holdings at Sillery, Cap Rouge, and St. Augustin, while at Lauzon by 1681 almost the entire coastline had been settled. With the rapid increase in population over the years it eventually became necessary to move to the interior.

The second-line settlements were begun during the eighteenth century. At St. André-de-Tilly, the first of these were granted in 1722. In subsequent years there appeared third-line farm holdings; during the military rule at Trois-Rivières (1760-64) there were already double-line seigniories and a few of triple depth.[1] At St. Roch-des-Aulnaies, April 8, 1728, was entered in the annals as a day of note; that was the day surveyor Noël Beaupré measured off the first four acres of second-line farms. At Ste-Anne-de-la Pocatière, all farmlands in 1721 were along the riverside but by 1732 some twenty-four back lots were settled.

Hamelin has established evidence of settlement along the St. Lawrence in the diocese of Joliette as early as the seventeenth century; along the smaller rivers such as L'Assomption, farm holdings were established in the late eighteenth century, but colonization of the interior region between rivers did not get underway until the turn of the century and as late as 1820. The first settlers in Beauce County, some three leagues distant from the St. Lawrence, came to the region from the Island of Orleans in 1737 via the Etchemin and Chaudière rivers, but another fifty years were to pass before farm properties were developed inland.

This inland settlement (second rang development) was an important milestone in the history of Canadian colonization; as the new homes were no longer adjacent to the water, a new means of communication had to be devised. Thus began the era of the dirt road. Roads were cleared along the front line (still called *fronteau* although there was no longer contact with the water) of these new inland settlements. With the building of roads, it became imperative to delineate the boundaries of the waterfront properties and to equalize their longitudinal dimension, in the same manner as it had been necessary to determine the width of early settlements. Undoubtedly it was the inland settlement that finally imposed the

second dimension to the original farmlands. Forty *arpents* became the accepted standard but this was later changed to one mile, which gave a length-to-width ratio of ten to one. At L'Islet, for example, the length varies from thirty to forty-two arpents and the width from two to four.

The variations in the lengths of the properties, due either to the meandering of the river or to good measure at the time of distribution, were bought back and served as long narrow strips for laying out the inland road. If these reclaimed abutments exceeded the road requirements, they were generally granted as bonuses to the proprietors of the land to which they were adjacent.[2]

With the resulting uniformity in depth of properties, the inland road ran parallel to the river, adapting itself to the winding of the waterway along its course. Later when a third rang was established, similar parallel roads were to be provided and the settlement pattern took on a peculiar geometric design.

With the introduction of surveying regulations this design became more systematized and crystallized. The newer *rangs,* being further removed from the original river front, were more and more symmetrical. Symmetry became an obsession with the inhabitants and they could conceive a property layout only in this manner. Everywhere the régime of the rang prevailed and we find territories inevitably sectioned off in bands divided into narrow lots whose length corresponds to the width of the rang (generally 30 to 40 *arpents*) and whose width remains 3 or 4 *arpents*.

From the Waterfront to the Roadfront

Accompanying the progress of inland settlement and the appearance of dirt roads is a change in the location of homes and buildings. The first parishes and chapels had been established close to the shore. Later, with the opening of roads, many were rebuilt inland—at Les Eboulements a new church was erected on the plateau in 1803. St. Antoine-de-Tilly had heralded this trend back in 1720. Most of the homes of the first rang by the waterfront were abandoned for new structures built nearer the road, usually referred to as *le chemin du Roi* [the King's way]. From 1704 on, maintenance of the lower river road at St. Augustin (Cap Rouge) was abandoned with the transfer of the church and homes a mile inland where a new road had been opened up. With the substitution of horse and buggy for boat and canoe as means of transportation, home sites at St. Pierre on the Island of Orleans were gradually displaced from the shore toward the interior, close to the road. Only the stone foundations of the early homes and the water well testify to the existence of former habitation. To this day the displacement phenomenon is still being carried out at such places as Montmagny and L'Islet.

With the development of the road, intersettlement communication by waterway became a thing of the past. At one time the St. Lawrence had been the highway or main street of Canada. The *habitants* plied the river from shore to shore and entertained neighbourly relations and friendly rivalries with residents of communities on the other side of the river. Today, with the introduction of highways and the attendant severance of crossriver communication, the nicknames bestowed upon the north shore *habitants* by those of the south shore and vice versa have fallen into disuse. The St. Lawrence has become, as it were, a frontier between north and south shores, and the *habitants* no longer participate in the busy river activity of today.

The first waterfront settlements were often located on low lands. Living near the shore had its advantages with regard to fishing and communication, but these were offset by the cold and damp climate of the lowlands, and the difficulty of access to the arable plateaus in those regions where the river was flanked by high banks or cliffs.

Single and Double Rangs

The newly acquired importance of the road entailed still further adaptations. Initially the homes were located on one side only of the road while on the other side were the extremities known as *trécarrés*, of the next line of properties generally still virgin forest. This early system of the *rang simple* meant that homes were face to face with woodland.

It was not an ideal situation, this confrontation of habitation with an intractable foe, the forest, and for this reason a modification was effected in the pattern of settlement. Under the new plan, homes faced on opposite sides of the road to form what was and still is known as the *rang double*. Since their extremities or *trécarrés* abutted, it made for economy in road construction. The neighbourhood became more dense and the settlement began to take on the characteristics of an elongated village (*stassendorf*).

All recent colonization is of the double-rang type. With a less extended network of highways, it is easier and less costly to provide better road maintenance, and parishes are easier to set up. At Boileau and Ferland the first three rangs that were opened up were of the single type while the last three were double.

There are some instances of single rangs being transformed into *rangs doubles*. Home owners gradually cleared land and moved their houses to the other end of their property so as to be facing the *fronteau* of the next rang. Such a displacement was effected at St. Hilarion near La Malbaie in 1860 as the habitants sought to avoid isolation. The entire procedure was carried out in less than a year, and the road of the first rang was closed since it no longer served a need. On the other hand, there are some regions such as La Mauriceie where the people refuse to adopt the double-rang type of settlement.

ADVANTAGES OF THE RANG

The success of the *rang* stemmed from the facilities it provided. This system had two advantages that cannot easily be obtained simultaneously: it allowed the settlers to possess large tracts of land and at the same time, because of the layout of the lots, it afforded protection from isolation.

In European countries, to benefit from

one of these advantages automatically precluded the other; as a general rule the more the property is concentrated, the more the houses are scattered and isolated, and conversely, the more the fields are distant from the homestead the more effectively may the latter be grouped together. In Canada the home and its attendant land stand in one block and yet, because of this disposition, the *habitants* enjoy a certain proximity to their neighbours.

Thus a strong sense of solidarity permeated relations among neighbours. This was an indispensable factor in a new country where dangers lurked incessantly: Indians, home or forest fires, famine, sickness and the long cold winters. It was to the advantage of the seignior, responsible for the protection of his tenants, to have them close to one another for mutual aid in time of need. Needless to say, he favoured and promoted the rang type of settlement.

This mutual protection policy, however, stayed within certain bounds with regard to settlement. During the wars with the Iroquois, Intendant Talon, on orders from Colbert, had tried to impose a village type of settlement where the homes were brought together in a sort of borough and the respective properties radiated outward in widening wedges. A few villages north of Quebec City were so constituted: Bourg-Royal, Charlesbourg, Village St. Claude, Village St. Joseph, Village St. Bernard, and Petite-Auvergne, but the plan met with no further success. The river lot triumphed over all royal edicts; linear settlement prevailed over the central development type.

On the other hand, the settlers constantly avoided isolation; neighbourly reciprocity was too vital to the Canadian way of life. There was a custom introduced from the French provinces which took on important dimension in Canada, that of the *premier voisin* or the first and nearest neighbour. To the *habitant,* the *premier* voisin comes even before a relative; he is invited to all family gatherings, is consulted about important decisions, and helps out in all large projects.

When a pig is slaughtered, he is given a piece, and when the bread is taken from the oven, a loaf is sent over; if, on the other hand some need arises, it is to the *premier voisin* that one has recourse. Rural living without him is inconceivable.

Thus the narrowness of the *devanture*, or lot frontage, and the fact that the houses faced at first the river and later the road for inter-communication, allowed for relatively easy neighbourly contact, an important consideration in the struggle against an excessively closed economy. Though the road was indispensable to the settler, it was also for him a heavy burden, for it was his responsibility to keep in repair that section which bordered his property. He had to "clear," in other words, to ensure that the *devanture* road was passable. In winter, it was a particularly arduous task, but quite necessary if he did not wish to suffer isolation during many months. So after each snow fall he would take the plough to open up his portion.

During the river-lot era, roads were not a problem, since the river provided all necessary means of communication. Possibly this explains the delay in opening up the second rang. Later, when it became imperative to move to the interior, the narrowness of the properties acquired still more importance to the settler, in that it kept to a minimum his share of the road maintenance. It would have been disastrous had he been over-burdened with too great a distance to clear. With the heavy snowfalls during winters in French Canada, the best system, to have the least distance to maintain and at the same time take advantage of the means of communication, was the lengthy lot with a short side facing the road.

The rang has likewise offered practical advantages in its adaptability to modern-day rural living, notably with regard to mail delivery (mail boxes set on posts along the road at the front of each farm), to the installation of electricity, telephone and bus services; likewise it has facilitated winter school attendance.

THE DISADVANTAGES OF
THE RANG

It must be noted that the rang type of settlement was not without drawbacks. Farming is more difficult because of the length of the lot, which sometimes exceeds a mile. Much time is wasted in carrying out the various farm tasks: the cattle must be led to the meadows, which are usually the more remote lots; a great deal of time and energy is spent getting to and from the pastures for the twice-daily milking chore. This is tiring for the animals and wastes the time of the herdsman. Whenever there is a heavy rain, the road to the back fields is muddy and sometimes impassable; in the fall, only the freeze-up and a good snowfall can remedy the situation so that the farmer can transport heavy loads such as manure, logs for fuel and building, large stones, etc. Travel within the domain is as difficult and inconvenient as road travel is practicable.

The same problem is encountered when it comes to fencing. In Europe the long fields are not fenced off from the open countryside but rather they merge into the landscape; in Canada where the long narrow properties are disposed adjacently, it seemed imperative to mark off the boundaries with fences. In fact, there was a veritable run on pickets and cedar lists (*thuyas*), be it for the straight or serpent type (*clôture à serpent*). A great deal of wood was used in the building of fences due to the method of double construction: either the horizontal bars or lists were held within two straight pickets or these latter were crossed to form a fence known as *clôture à jambette*. In either case, the purpose was to avoid the use of nails, a scarce item in those early days, and also to facilitate the demolition of the fences in winter time for unimpeded travel by sledge or snow-shoe. In certain regions where stone rubble was cleared from the fields in sufficient quantity, the fence would be built of this material, thus "killing two birds with one stone." Today, however, most fences are of wire.

Within the domain, more fences separated pastures from garden and cultivated fields. Not infrequently, the cow path leading from the barn to the fields, via the tilth was also fenced off to prevent the cattle from wandering.

The profusion of fences is typical of French-Canadian countryside: it is not so in English Canada. One sometimes wonders at this preoccupation with fencing, a sort of obsession as it were, to encompass each domain. Perhaps because in those days a property was characterized by its *abattis* or cleared land and by its fences. Thus we find that in all contracts the question of fences is spelled out in minute detail. Perhaps also it is a throwback to a peculiarity of the peasants of western France, ancestors of the first settlers, whose fields were bordered by groves. The fact of the matter that fencing in those long narrow properties was not a practical measure.

The most significant drawback of this elongated form of land holding lies in the variations of geographic structure encountered. Soil differences and surface irregularities are not taken into consideration when the land is marked off. Rigid property limits indifferently cut across highland and lowland, hills, streams, steep gullies and ravines, and over countless jutting rock formations which had been left behind by glaciers of another age and for which the inhabitants devised a rich and colourful toponymy: *crans, caps, ronds de fesse, dos de moutons*. Not infrequently, access to one part of the domain was possible only by making a long detour, often over the neighbour's property.

SOCIAL AUTONOMY OF THE RANG

The rang was the fundamental social unit; "the closeness of the houses along the same road resulted in a tradition of mutual aid" (Falardeau). The location of the rang determines the social classification of its occupants: as a general rule, those close to the shore are the most highly considered, whereas those of the interior, disdainfully referred to as "the concessions" which have been more recently settled, have an inferior status. Some parishes claim as many as twelve rangs in succession, some scarcely broken into, while others are progressively being abandoned.

Each rang has its own organization: there is the rang council, the rang school, the rang chapel and *calvaire* where prayer meetings are held when the parish church is too far away, rang butter and cheese dairies, and the rural women's circle of the rang. Over and above all is the unifying spirit of brotherhood: the poor are cared for—house-to-house collections known as *guignolées* are made to gather food and provisions for the less fortunate—and fires are fought by every able-bodied inhabitant of the rang. As is often the case in the country, the fire victims have little or no insurance, so the entire community comes to the aid of the unfortunate ones: one neighbour supplies the framing, another the boards, still others the labour, until the family is once again in possession of a home. Work bees or *corvées* are frequently organized to help one or the other whether it be for moving or wood cutting, harvesting or land clearing, hackling or wool spinning.

To say of a person that he was brought up on the rangs is testimony to his spirit of brotherhood. Habitants of the rangs and those of villages do not come under the same census categories; some 28 percent of the population of Quebec live in rangs. Villagers will often say the rangs are not their responsibility, particularly when it comes to road maintenance.

THE RANG AND THE PARISH

The establishment of the rang always preceded that of the parish. Under the French régime, the seignior, who was paid according to the frontage of his tenants' lots, had no interest in seeing the land subdivided into small urban holdings. Likewise, but for quite different reasons, the government favoured agricultural colonization while discountenancing commerce and industry which would compete with production in the mother country. The authorities disapproved of settling on lots which were too small for successful farming. A ruling was passed in 1745 whereby it was forbidden to build houses on lots with a frontage of less than one and a half *arpents;* some *habitants* of

the Island of Orleans were prosecuted in 1752 for having disregarded the order.

The type of fraternal relations which characterized the first settlements was not conducive to the development of villages. As was the custom in Brittany, the *habitant* who needed assistance obtained it through the *corvées* rather than by hiring a tradesman from town. The French-Canadian settler was a remarkable and resourceful jack-of-all-trades who managed to make most of what he needed. During the long winters, he spent many hours in his *boutique* tinkering away at various projects and tasks which he had reserved especially for this slack period. For this reason, there was little or no call in rural areas for that class of individual sometimes considered supernumerary and parasitic—the village craftsman.

The uniformity of the lots due to their particular geometric layout made for an unusual brand of social equality; there was no noticeable hierarchy among the *habitants* in their undiversified environment; all were on an equal footing—an important factor in the realization of their mutual assistance policies. The ascendancy of the large estate so often experienced in new countries was never a problem in French Canada.

Another factor which inhibited the development of towns was the absence of trading among the settlers; each family unit provided for most of its daily requirements so that little was purchased from the exterior.

Concentrated settlement around the church developed only after the original land settlement in rangs was completed, and this regardless of the pastor's sustaining influence in promoting colonization. The village finally evolved from the unpredictable over-expansion of the population which resulted in property sharing and concentration of homes.

Here is an account of the establishment of the first church in the parish of L'Ange-Gardien: it was in 1664 that a property with a two-*arpent* river frontage was bought for this purpose. In 1667, the first curé was installed and a church built: the excess acreage was rented to a farmer who agreed to clear two *arpents* of land yearly on the understanding that the parish would build him a home and a shed. In the meantime, he was permitted to store his grain in the garret of the church. Thus the church was considered just another domain, and no provision was made for the development of the village. In 1852, at St. Hilarion near La Malbaie, the proprietor of the *canton* or district donated lot No. 10 of the third rang as a site for the erection of a church; the lot was cleared and the settlers worked in *corvées* to build it. On the church property it was the custom to build shelters to house the horses in winter while the parishioners were attending church service. These small winter stables constituted, as it were, the embryo of the parish village.

Later, the old folks, living off their life savings, took up residence close to the churches; they had passed on their farms to the younger generation so as to enjoy the serenity of the religious environment; they were called *emplacitaires* because they had chosen an *emplacement* or a small plot of land from an old property near the church. They eventually became a class apart from the people living in the rangs. In order to allow for the establishment of these "senior citizens," it was necessary to subdivide land into small plots.

There was so little provision made for the village around the church that often when the church was moved, the entire village was displaced with it; in [French] Canada most churches changed sites at least once, sometimes more than once (twice at Lotbinière and Les Eboulements, and at Ste. Anne-de-la-Pocatière and St. André-de-Tilly three times). At Maskinongé and at Upton the moving of the church resulted in a schism.

Considerable time elapsed before the church village became a stable element. It was a slow process, gaining ground with the years and with the increase in the number of old folks; parishes acquired a certain maturity as their cemetery population became more numerous. A new colony in Abitibi that claimed only three deceased was spoken

of with commiseration. It was the departed who determined the status of the living—the people were proud of a well-filled cemetery, evidence of their long establishment. Such were the beginnings of the village, an agglomeration of small abodes either built or rolled from their original sites to a location close to the church.

Later the professionals—doctors, lawyers, notaries—and the trades people would settle around the borough which finally became a centre apart from the rang. The village spread out in one way or another across the lots, along a single street: a monumental church and presbytery, often a school directed by Sisters or Brothers, likewise of monumental proportions, the general store for either groceries, dry goods (in other words, any merchandise that was not edible), restaurants studded with advertisements and posters where everything was sold: soft drinks, etc.

French Canadians did not take to centralized settlement, which explains the slow development of cities, a situation rarely met with in America. The market for farm produce thus being very limited, agriculture could provide only a low standard of living for the *habitants*; in this semi-closed economy, the settlers had to rely upon complementary activities such as wood cutting and hunting in order to secure ready cash. With this in mind large sections of virgin forest were reserved as a source of supplementary earnings. As the terrace or the mountain had been to the Mediterranean gardener, so was the forest to the Canadian settler—a complementary area that must be respected for the maintenance of equilibrium. At one time, trapping for furs had been a threat to colonization. The *coureurs de bois* constituted an important element of the population; during the eighteenth century, about one-fifth of the male labour force lived in and off the forest. Lumbering, in fact, is still a threat to farming, particularly in certain regions. Land and forest each must make its contribution; what matters is to make an equitable compromise and remain within the respective limits. It must be added that colonization did not progress along a continuous front, but rather in zones or districts situated between vast expanses of countryside. Beauce and the Eastern Townships at one time were such zones, as are the recently developed districts of Lac St. Jean, and Aroostook in Maine. Still more recent is the colonization of the Abitibi region.

With today's massive developments of industry and towns, a new phenomenon is changing the long-established pattern; the new communities which are burgeoning in the vicinity of waterfalls, mines, paper or saw mills, did not start as extensions of rangs but rather with the building of a factory or a mill and of workmen's homes on adjacent small lots. Such was the case at Joliette (originally called L'Industrie), L'Assomption, Forestville, or the lumbering centres of La Tuque, Mattawa, North Bay and Les Piles. Sometimes towns are owned outright by a company; in such "closed towns" one cannot locate without the consent of the company (Baie Comeau, Clarke City, Arvida, and Noranda are examples of these).

EXTENSION OF THE RANGS

Wherever French Canadians migrated over the American continent, there they introduced, without appreciable regional variation, this unusual type of strictly rural settlement which was for a long time without urban support. We find it in eastern Canada on the southwest coast of Newfoundland on the outskirts of Port-aux-Basques, and throughout the countryside of former French Acadia. We find it again at the other extremity of the St. Lawrence on the southern tip of Ontario, around the outposts established by Lamoth-Cadillac, at Lake St. Clair near Windsor, and also in certain parishes along Georgian Bay. In the city of Detroit, the lower wards bear marks of the French rangs that plans for urban development could not wipe out.

Along the Ottawa River, the Rivière-aux-Français, and the Great Lakes, we find the rang moving out of its customary forest habitat. It accompanied the migration of French Canadians to the Prairie Provinces: in Manitoba, the Assiniboine River is bordered with rangs as far as Portage la Prairie, they dot the shores of the Red River district above and below Winnipeg (St. Norbert, Morris, Ste. Agathe); on the Seine (a distinctively French name) they form wedges, within the townships, at Lorette and St. Anne. There are many other islands of rangs: at St. Laurent on Lake Manitoba, at Fort Alexander on Lake Winnipeg, at Rivière-du-Rat (St. Malo) along the South Saskatchewan River (St. Laurent, Batoche, St. Louis) and the North Saskatchewan in the vicinity of Prince Albert and even northward in the Le Pas district.

Farther west, in Alberta, there are river lots scattered in the regions of Lake Labiche, at St. Albert and as far north as Peace River, which is being settled by French Canadians in ever increasing numbers. Nor did the rang settlement stay within Canadian boundaries: along the valley of the Mississippi where long ago French explorers pioneered there are, among others, developments at Prairie-du-Chien where the Wisconsin River flows into the Mississippi, and as far south as Bâton Rouge and in the region of New Orleans.

Here, conditions of settlement are quite different: there is less forest land and consequently less preliminary clearing to be done; lakes and rivers lose some of their importance. Nevertheless, even in regions as dissimilar as these, the Canadians could conceive of no other form of property. It was, in a sense, their trademark.

REFERENCES

[1] Trudel, Marcel, *Le Régime militaire dans le Gouvernement des Trois-Rivières, 1760-1764.* (Trois-Rivières, 1952), p. 24. At Lauzon, a fourth rang was started in 1738.

[2] Provost, Honorius, "En parlant de colonisation seigneuriale," *Revue de l'Université Laval,* III (Avril 1949), p. 672.

BIBLIOGRAPHY

Blanchard, Raoul, *L'Est du Canada français, "Province de Québec"*. Montréal, Beauchemin, 1935, 2 vol.

Blanchard, Raoul, *Le centre du Canada français, "Province de Québec"*. Montréal, Beauchemin, 1947.

Caron, l'abbé Ivanhoe, *La colonisation du Canada sous la domination français. Précis historique*. Québec, 1916.

Caron, l'abbé Ivanhoe, *La colonisation de la province de Québec. Débuts du régime anglais, 1760-1791*. Québec, 1923.

Caron, l'abbé Ivanhoe, *La colonisation de la province de Québec. Les cantons d l'Est, 1791-1815*. Québec, 1927.

Dauvergne, R., "Les anciens plans muraux des colonies françaises," *Revue d'histoire des colonies. Paris*, 1948, pp. 231-269.

Dawson, C. A. and Murchie, R. W., *The Settlement of the Peace River County*. Toronto, Macmillan, 1936.

Falardeau, Jean-Charles, "Paroisses de France et de Nouvelle-France au XVIIIème sièle," *Cahiers de l'Ecole des Sciences sociales*, Université Laval. Québec, 1948.

Gérin, Léon, "L'habitant de Saint-Justin. Contribution à la géographie sociale du Canada," *Memoires de la Société royale du Canada*, 1898, section I, pp. 139-216.

Gérin, Léon, *Le type économique et social des Canadiens. Milieux agricoles et tradition française*. Montréal, 1938.

Hamelin, Louis-Edmond, *La marche du peuplement à l'intérieur du diocèce de Joliette*, Joliette, 1951.

Hamelin, Louis-Edmond, "Le Rang à Saint-Didace de Maskinongé," *Notes de Géographie,* No. 3, Publications de l'Institut d'histoire et de géographie, Université Laval, 1953.

Kalm, Peter, "Voyages de Kalm en Amérique, analysé et traduit par L. W. Marchand," *Mémoires de la Société historique de Montréal*. Montréal, 1880.

Langlois, Georges, *Histoire de la population canadienne-français*. Montréal, 1935.

Le Chevalier, Jules, o.m.i., *Batoche*. Montréal, Oeuvre de Presse dominicaine, 1941.

Ledoux, Burton, "French-Canada. A modern feudal State," *Virginia Quarterly Review*, 1941, pp. 206-222.

Lewis, H. Harry, "Population of the Quebec Province," *Economic Geography*, 1940.

Mackintosh, W. A., *Prairie Settlement: The Geographical Setting. Canadian Frontiers of Settlement*. Toronto, Macmillan, 1936.

Noiseux, D. C., *Dix années de colonisation de Sainte-Anne de Roquemaure*. Québec, Ministère de la Colonisation, 1943.

Provost, l'abbé Honorius, "En parlant de colonisation seigneuriale," *Revue de l'Université Laval*, avril, 1949, pp. 672-678.

Rousseau, L.-Zéphirin, "Surveys and Land-use Planning in the Province of Quebec," *The Canadian Surveyor*, février, 1941, pp. 24-32.

Roy, Joseph-Edmond, *Histoire de la seigneurie de Lauzon*. Lévis, 1897-1904, 5 vol.

Roy, Léon, *Les terres de la Grande-Anse-des-Aulnaies et du Port-Joly*. Lévis, 1951.

Salone, Emile, *La colonisation de la Nouvelle-France. Etude sur les origines de la nation canadienne-française*. Paris, 1906.

Taylor, Griffith, *Canada. A Study of Cool Continental Environments and their Effect on British and French Settlement*. Londres, Methuen, 1947.

Trudel, Marcel, *Le régime militaire dans le gouvernement des Trois-Rivières, 1760-1764*. Trois-Rivières, 1952.

Population Growth and Distribution in Canada, 1921-1961

Population growth in Canada has been rapid during the past several decades and has been accompanied by major movements of population within the country. Redistribution has occurred between farm and non-farm areas and between rural and urban areas, and it has occurred geographically among the provinces.

Since the beginning of the twentieth century Canadian population growth has been consistently more rapid than growth in the United States (Table 1). Although the rate of net immigration[1] differed in the two countries, more significant differences existed in the rate of natural increase.[2] In both countries there was an acceleration in growth during the 1940's and 1950's after a low level during the 1930's, but it was greater in Canada than in the United States, and as a result, the gap between the percentage rates of growth widened.

There were, of course, different regional rates of population growth within each country[3] (Figure 1). Generally, the levels of growth in Canada have been higher than in the United States even though in the Atlantic Region for the first three decades and in the Prairies during the 1930's and 1940's growth was lower than in any region in North America. The highest level of growth in the United States was in the West, where the pattern of change over time was similar to the pattern for British Columbia although the level was slightly lower in the West. Underlying the regional patterns of growth in Canada shown in Figure 1, there were notable differences in the patterns for individual provinces in the Atlantic Regions[4] and in the Prairies. Growth in the former provinces was low throughout the sixty-year period, but the levels were relatively high in Nova Scotia and New Brunswick during

Table 1

PERCENTAGE INCREASE IN THE POPULATION IN CANADA AND THE UNITED STATES BY CENSUS DECADES

Canada		United States	
1901—11	34	1900—10	21
1911—21	22	1910—20	15
1921—31	18	1920—30	16
1931—41	11	1930—40	7
1941—51	19	1940—50	14
1951—61	30[1]	1950—60	18
1901—61	239[1]	1900—60	136

[1] Includes Newfoundland in the 1951—61 change.

Source: Data for Canada are based on *Census of Canada*. Data for United States are from *Population Redistribution and Economic Growth, United States, 1870—1950*, Vol. III, by Hope T. Eldridge and Dorothy Thomas, Philadelphia: The American Philosophical Society, 1964, p. 11, and *Statistical Abstract of the United States, 1965*, (86th edition), U.S. Bureau of the Census, Washington, D.C., 1965, p. 13.

● ISABEL B. ANDERSON *is Assistant Professor, Department of Economics and Political Science, University of Saskatchewan.*

Reprinted from *Internal Migration in Canada, 1921-1961*, Economic Council of Canada, Staff Study No. 13 (Ottawa: Queen's Printer, 1966), pp. 7-17, with permission of the author, the Director of the Council, and Information Canada.

81

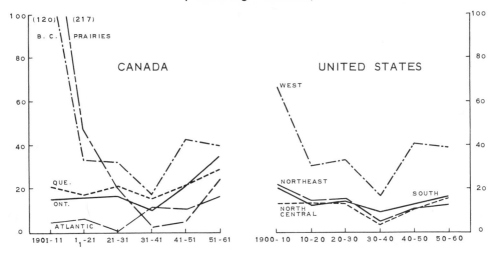

POPULATION INCREASE IN CANADA AND THE UNITED STATES, BY REGION, BETWEEN CENSUS YEARS
(Percentage Increase)

NOTE: ATLANTIC INCLUDES NEWFOUNDLAND FOR 1951-61 ONLY.   SOURCE: CENSUS OF CANADA

Figure 1

Table 2

DISTRIBUTION OF THE CANADIAN POPULATION, BY PROVINCE,
AT CENSUS DATES, 1901 TO 1961
(Percentage share) (1)

	1901	1911	1921	1931	1941	1951	1956	1961
Canada................	100	100	100	100	100	100	100	100
Newfoundland..........	—	—	—	—	—	2	2	2
Prince Edward Island.....	2	1	1	1	1	1	1	1
Nova Scotia............	9	7	6	5	5	4	4	4
New Brunswick.........	6	5	4	4	4	4	3	3
Atlantic................	17	13	11	10	10	12	11	10
Quebec................	31	28	27	28	29	29	29	29
Ontario................	41	35	33	33	33	33	34	34
Manitoba..............	5	6	7	7	6	5	5	5
Saskatchewan...........	2	7	9	9	8	6	5	5
Alberta................	1	5	7	7	7	7	7	7
Prairies................	8	18	22	23	21	18	18	17
British Columbia........	3	5	6	7	7	8	9	9
Yukon and Northwest Territories.............	1	*	*	*	*	*	*	*

(1) Columns may not add to the totals shown because of rounding.
*Less than one percent.
Source: *Census of Canada.*

82

the 1931-41 decade. Within the Prairie Region there were diverse provincial patterns of growth. During the first decade of the century, growth in Saskatchewan was higher than in any other province, but it fell continuously until 1951 and there were absolute declines in the population during the 1930's and 1940's. By the 1920's, growth in Alberta exceeded growth in all the other provinces except British Columbia; it fell below growth in British Columbia, Quebec and Ontario during the 1930's and 1940's, but during the 1950's it exceeded growth in British Columbia by two percentage points. During the same decade, growth in Saskatchewan exceeded growth in only one province—Prince Edward Island.

Despite the substantial differences in regional and provincial rates of population growth, the Canadian population has remained unevenly distributed and the provincial shares have not changed swiftly or substantially, at least since 1911 (Table 2).

Throughout the period since 1911, Ontario and Quebec, together, have contained over 60 percent of the population, the Prairie Region approximately 20 percent, the Atlantic Region 10 percent and British Columbia has contained less than 10 percent of the population. However, even a small change in the provincial shares implies a substantial shift of the population geographically. Quebec, Ontario, Alberta and British Columbia had relatively high rates of population growth and their share of the population increased. Nova Scotia, New Brunswick, Manitoba and Saskatchewan had low rates of population growth and their share of the population decreased.

Underlying the patterns of growth and provincial redistribution, there have been significant changes in the rural-urban and farm-nonfarm structure of the population. Between 1901 and 1961 the urban share of the population in Canada rose from 35 percent to 58 percent (Table 3). In every

Table 3

URBAN[1] SHARE OF THE POPULATION IN EACH PROVINCE,
AT CENSUS DATES, 1901 TO 1961
(Percentage share)

	1901	1911	1921	1931	1941	1951	1956	1961
Canada..............	35	42	45	50	51	54	55	58
Newfoundland..........	—	—	—	—	—	27	32	36
Prince Edward Island.....	14	16	19	19	22	25	30	32
Nova Scotia............	28	37	42	43	45	46	45	47
New Brunswick.........	23	27	31	31	31	32	35	38
Atlantic..............	24	31	36	36	38	36	38	41
Quebec..............	36	44	51	59	60	64	67	72
Ontario..............	40	50	56	59	60	58	56	57
Manitoba.............	25	39	39	42	41	46	50	56
Saskatchewan..........	6	16	17	20	21	30	36	43
Alberta..............	16	29	30	31	31	46	54	62
Prairies..............	19	28	28	30	31	41	47	55
British Columbia........	46	51	46	55	53	51	49	47
Yukon and Northwest Territories............	19	20	—	—	6	10	8	23

[1] Urban population includes people living in incorporated cities, towns and villages of 1,000 and over.

Source: *Census of Canada.*

Table 4

NONFARM SHARE OF THE RURAL POPULATION IN EACH PROVINCE,
AT CENSUS DATES, 1921 TO 1961

(Percentage share)

	1921	1931	1941	1951	1956	1961
Canada...............................	32	37	44	56	63	70
Newfoundland.........................	—	—	—	94	96	94
Prince Edward Island....................	15	22	31	37	38	47
Nova Scotia...........................	27	39	54	68	74	79
New Brunswick........................	28	36	48	58	65	73
Atlantic..............................	26	36	49	69	75	79
Quebec...............................	32	35	38	47	51	56
Ontario...............................	34	43	54	64	72	80
Manitoba.............................	30	37	42	49	52	57
Saskatchewan.........................	24	23	27	31	36	42
Alberta...............................	23	26	30	33	36	43
Prairies..............................	25	27	32	37	41	46
British Columbia......................	65	67	74	81	85	89

Source: *Census of Canada.*

Table 5

TOTAL NONFARM SHARE OF THE POPULATION IN EACH PROVINCE,
AT CENSUS DATES, 1921 TO 1961

(Percentage share)

	1921	1931	1941	1951	1956	1961
Canada...............................	63	68	72	80	83	87
Newfoundland.........................	—	—	—	96	98	96
Prince Edward Island....................	31	37	46	52	57	64
Nova Scotia...........................	58	65	75	82	86	89
New Brunswick........................	50	56	64	72	77	83
Atlantic..............................	52	59	68	80	84	87
Quebec...............................	67	73	75	81	84	88
Ontario...............................	71	77	81	85	88	91
Manitoba.............................	58	63	66	72	76	81
Saskatchewan.........................	37	39	42	52	59	67
Alberta...............................	46	49	52	64	71	78
Prairies..............................	46	49	53	63	69	76
British Columbia......................	81	85	87	91	93	94

Source: *Census of Canada.*

province except British Columbia, the share increased significantly over the sixty-year period. In some provinces and regions, the change was very dramatic. For example, the urban share of the population in Quebec doubled (from 36 to 72 percent), and in the Prairie Region it almost tripled (from 19 to 55 percent). Conversely, the rural share (including farm and rural nonfarm) of the population declined during the period. In 1921 one third of Canada's rural population was classified as nonfarm; this proportion had increased to 70 percent in 1961. (Table 4).

The substantial population shift from farm to nonfarm areas during the 1921-1961 period is shown in Table 5. By 1961, 87 percent of the Canadian population was classified as nonfarm. This shift was general throughout the country, but it was most pronounced in the Atlantic and Prairie Provinces and less pronounced in Quebec, Ontario and British Columbia where, even in 1921, a relatively large portion of the population was classified as nonfarm. In 1961, however, Prince Edward Island and Saskatchewan still had significant portions of the population living in farm areas.

These structural changes in the Canadian population reflect the fact that the urban population has grown much faster than the total population, that the rural population has grown much more slowly than the urban

and nonfarm population, and that the farm population has actually declined. Since the beginning of the twentieth century the rates of urban growth in Canada and the United States have exceeded the rate of total population growth (Table 6).[5] At the same time, not only has total population grown faster in Canada than in the United States, but urban growth has been consistently much faster in Canada.

Percentage increments in urban and rural population in each province during each census decade from 1901 to 1961 are shown in Table 7. Urban growth has been much more rapid than rural growth in every province and in every decade, with the exceptions of Ontario after 1941 and British Columbia after 1931. The rural population, at least after 1921, grew as a result of rapid growth in the population of rural nonfarm areas.[6] During the period from 1921 to 1961, growth in the rural nonfarm population was particularly rapid in Ontario in each of the four decades and in British Columbia in the last two decades (Table 8). The lowest rates of growth have been in the Prairie Provinces. In contrast, the rural farm population—that is, the farm component of the rural population—declined absolutely during the period from 1921 to 1961, almost without exception. At the same time, the total nonfarm population grew rapidly as a result of the combined effect of rapid growth in the urban

Table 6

PERCENTAGE INCREASE IN THE URBAN POPULATION IN CANADA AND THE UNITED STATES, BY CENSUS DECADES

Canada		United States	
1901—11	61	1900—10	39
1911—21	32	1910—20	29
1921—31	30	1920—30	27
1931—41	13	1930—40	8
1941—51	27	1940—50	20
1951—61	42[1]	1950—60	30
1901—61	469[1]	1900—60	315

[1] Includes Newfoundland in the 1951—61 change.

Note: Although there are differences in the definition of urban population, the broad inference from this comparison seems to be valid.

Source: *Census of Canada.* Data for the United States are from *Population Redistribution and Economic Growth, United States, 1870—1950*, Vol. III, *op. cit.*, p. 218, and *Statistical Abstract of the United States, 1965, op. cit.*, p. 16.

Table 7

PERCENTAGE INCREASE IN THE URBAN AND RURAL POPULATION IN EACH PROVINCE, BY CENSUS DECADES, 1901 TO 1961

	1901—11		1911—21		1921—31		1931—41		1941—51		1951—61(1)	
	Urban	Rural	Urban	Rural	Urban	Rural	Urban	Rural	Urban	Rural	Urban	Rural
Canada.........	61	20	32	15	30	8	13	8	27	10	42	17
Newfoundland...	—	—	—	—	—	—	—	—	—	—	68	11
Prince Edward Island........	*	—11	11	— 9	3	— 2	22	4	18	—*	37	— 4
Nova Scotia.....	42	— 6	22	— 2	2	— 5	18	9	13	10	16	14
New Brunswick..	23	1	28	4	6	5	12	12	17	11	37	6
Atlantic.........	32	— 4	23	— 1	3	— *	16	10	13	9	32	9
Quebec.........	50	6	35	4	40	3	18	13	31	8	45	2
Ontario.........	42	— 2	30	2	24	9	13	7	18	26	33	40
Manitoba.......	185	46	32	32	23	9	2	6	20	— 3	45	— 4
Saskatchewan...1,322		382	60	52	47	17	2	— 4	32	—18	58	— 9
Alberta.........	828	332	61	56	29	22	10	8	72	— 7	93	— 1
Prairies.........	357	183	47	48	30	16	4	2	40	—10	68	— 5
British Columbia.	141	101	21	47	59	9	12	25	38	48	28	52
Yukon & North-west Territories	—67	—69	—	2	—	10	—	17	149	42	238	28

(1) Includes Newfoundland in the 1951—61 change for Canada and the Atlantic.
*Less than one percent.
Source: *Census of Canada*.

population and very rapid growth in the rural nonfarm population.

There have been significant provincial differences in the rates of growth of urban, rural, nonfarm and farm population. Urban and nonfarm growth—rapid throughout the country—was particularly rapid in Saskatchewan and in Alberta during the 1951-61 decade and was consistently rapid over the whole period in Quebec, Ontario and British Columbia. In contrast, rural growth was very low and there were absolute declines in the rural population in the Atlantic Provinces before 1931, and in the Prairie Provinces after 1931. The decline in the farm population, which was general throughout the country, was particularly rapid in Saskatchewan and in the Atlantic Provinces, but it has also been rapid in Ontario.

Throughout the period 1921 to 1961 Ontario and Quebec, together, contained about half of the farm population in Canada, the Prairie Region about one third, the Atlantic Region about 10 percent and British Columbia 4 percent (Table 9). However, there were some notable changes. The shares for Alberta and Quebec increased because the rate of decline in farm population was less rapid than elsewhere. The shares for the Atlantic Provinces, Saskatchewan, and in recent years, Ontario decreased because the decline in farm population was more rapid than elsewhere. Approximately 65 percent of the total nonfarm population in Canada was in Quebec and Ontario during the forty-year period. Another 15 percent was in the Prairies, and the Atlantic Region and British Columbia each contained about 10 percent. Alberta and British Columbia both had increasing shares because they had the most rapid growth in total nonfarm population.

Table 8

PERCENTAGE INCREASE IN THE RURAL NONFARM, TOTAL NONFARM AND FARM POPULATION IN EACH PROVINCE, BY CENSUS DECADES, 1921 TO 1961

	1921—31			1931—41			1941—51			1951—61 [1]		
	Rural Non-farm	Total Non-farm	Farm	Rural Non-farm	Total Non-farm	Farm	Rural Non-farm	Total Non-farm	Farm	Rural Non-farm	Total Non-farm	Farm
Canada.........	25	29	1	30	18	− 4	37	30	−11	45	43	−19
Newfoundland...	—	—	—	—	—	—	—	—	—	11	27	13
Prince Edward Island........	40	18	− 9	50	35	− 8	17	18	− 8	23	30	−20
Nova Scotia.....	35	11	−20	54	30	−19	36	22	−22	33	23	−26
New Brunswick..	37	17	− 7	51	29	− 9	34	26	−11	33	35	−32
Atlantic.........	36	14	−13	52	30	−13	34	23	−15	25	28	−26
Quebec.........	11	33	− 1	22	19	8	36	32	− 9	20	40	−15
Ontario.........	39	27	− 7	32	17	−12	51	27	− 4	74	46	−21
Manitoba.......	33	26	− 1	21	8	− 3	12	17	−14	13	33	−20
Saskatchewan....	12	28	18	12	7	− 9	− 4	14	−23	21	43	−23
Alberta.........	37	32	18	26	16	2	5	45	−12	28	74	−15
Prairies.........	25	28	14	19	10	− 4	4	25	−17	20	51	−19
British Columbia.	12	39	4	37	21	*	62	48	7	68	46	−15

[1] Includes Newfoundland in the 1951—61 change for Canada and the Atlantic.

*Less than one percent.

Source: *Census of Canada.*

The shares for the Atlantic Region, Quebec, Manitoba and Saskatchewan declined because growth in the population was lower. The share for Ontario varied and growth in total nonfarm population was moderately rapid.

Provincial differences in population growth, changes in the provincial distribution, and changes in the rural-urban and farm-non-farm structure of the population can be explained, to a significant degree, by the patterns of migration within Canada. Small, but significant, changes in provincial distribution reflect, in part, different rates of net migration and imply internal migratory flows between provinces, as well as preferences of international migrants for particular provinces. Similarly, changes in the distribution between rural and urban and between farm and nonfarm areas reflect, in part, different rates of net migration in these areas, including both internal and inter-national migration. The statistics presented in this paper do not reveal the origin and destination of migrants.

Although Canada has had one of the highest rates of net immigration of any country in the World in the 20th century, internal net migration has been far greater than inter-national net migration during the forty-year period from 1921 to 1960.[7] Table 10 shows that net immigration was 1.4 million while the net movement out of farm areas was 3.2 million people and the net movement into non-farm areas was 4.6 million; over 60 percent of the latter movement was into incorporated centres of 1,000 persons and over—that is, into urban areas. In other words, for the full forty-year period, internal net migration was more than twice as large as the net movement into Canada. However, it is worth noting that the magnitude of net internal, relative to net international, migration varied substantially between decades.

Table 9

DISTRIBUTION OF THE FARM AND TOTAL NONFARM POPULATION IN CANADA
BY PROVINCE, AT CENSUS DATES, 1921 TO 1961

(Percentage share) [1]

	Farm						Nonfarm					
	1921	1931	1941	1951	1956	1961	1921	1931	1941	1951	1956	1961
Canada.........	100	100	100	100	100	100	100	100	100	100	100	100
Newfoundland...	—	—	—	1	*	1	—	—	—	3	3	3
Prince Edward Island........	2	2	2	2	2	2	*	*	*	*	*	*
Nova Scotia.....	7	5	4	4	4	4	5	5	5	5	4	4
New Brunswick..	6	5	5	5	5	4	4	3	4	3	3	3
Atlantic.........	14	12	11	11	10	10	9	8	9	12	11	10
Quebec.........	24	24	27	27	28	28	29	30	30	29	29	29
Ontario.........	26	24	22	24	25	23	38	37	37	35	35	36
Manitoba.......	8	8	8	8	8	8	6	6	6	5	5	5
Saskatchewan....	15	17	16	14	13	13	5	5	5	4	4	4
Alberta.........	10	11	12	12	12	13	5	5	5	5	6	6
Prairies.........	32	36	36	34	33	34	16	16	15	14	15	15
British Columbia.	3	3	3	4	4	4	8	8	9	9	10	10

[1] Columns may not add to the totals because of rounding.

*Less than one percent.

Source: *Census of Canada.*

Table 10

ESTIMATED INTERNATIONAL NET MIGRATION
AND INTERNAL FARM AND NONFARM NET MIGRATION FOR CANADA,[1]
CALENDAR-YEAR INTERCENSAL INTERVALS, 1921 TO 1960

(Thousands of people) [2]

	1921—30	1931—40	1941—50	1951—60[3]	1921—60[3]
International[4]..............	286	— 73	104	1,083	1,400
Internal[5]					
Farm-Nonfarm					
Farm.....................	−472	−592	−925	−1,171	−3,160
Nonfarm.................	758	519	1,029	2,254	4,560
Urban................	621	215	578	1,467	2,882
Rural Nonfarm.........	137	304	450	787	1,678

[1] Excludes Yukon and Northwest Territories.

[2] Columns may not add to totals shown because of rounding.

[3] Includes Newfoundland in 1951-60.

[4] This is net immigration if it is positive and net emigration if it is negative.

[5] This is net in-migration (including international migration) if it is positive and it is net out-migration (including international migration) if it is negative.

Table 11

RATE OF INTERNATIONAL AND INTERNAL NET MIGRATION FOR CANADA,[1] CALENDAR-YEAR INTERCENSAL INTERVALS, 1921 TO 1960
(Number of migrants per decade per 1,000 population) [2]

	1921—30	1931—40	1941—50	1951—60[3]
International	30	− 7	8	68
Internal				
Urban	137	39	88	163
Rural	− 68	− 53	− 81	− 55
Rural Nonfarm	80	139	154	178
Rural Farm	−145	−185	−312	−462
Total Nonfarm	122	68	108	168

[1] Excludes Yukon and Northwest Territories.

[2] This is a decennial crude rate. The population base is the average of the population at the beginning and the end of the interval.

[3] Includes Newfoundland in 1951-60.

Source: *Census of Canada.*

Internal net migration was less than twice as large during the 1920's and the 1950's, but it was more than six times larger than international net migration during the 1930's and the 1940's. Furthermore, over three quarters of the net immigration during the forty-year period occurred in the 1951-60 decade and two-thirds of the internal net movement occurred during the two decades after 1941. Thus the amount of internal net migration was large, even though the amount of international net migration was small.

Internal net migration has been rapid compared with international net migration (Table 11). During the four decades from 1921 to 1960 net movements into urban and non-farm areas have been consistently rapid, even with a significant decline in the 1931-40 decade, and the rate has increased since 1941. Net migration out of rural and farm areas has also been consistently rapid and the rate of net migration out of farm areas has increased very substantially.

REFERENCES

[1] Throughout this paper immigration and emigration refer specifically to international migratory flows—that is, to migratory flows into and out of the country. In contrast, in-migration refers to the movement of people into a province, or a region, and includes immigration and in-movements from other provinces. Similarly, out-migration refers to movements out of a province, including emigration as well as movements to other provinces.

[2] The rate of natural increase is the excess of births over deaths per decade per 1,000 of the population. The population base is the average of the resident population at the beginning and the end of the decade.

[3] The validity of comparisons of regional growth in the two countries depends upon the comparability of the regional units.

[4] The pattern of population growth over time in the Atlantic Region is affected by the inclusion of Newfoundland for 1951-61.

[5] Compare these figures with the corresponding figures in Table 1.

[6] In some provinces growth in the rural nonfarm population primarily involved growth in suburban or urbanized areas, but in other provinces it included significant growth in the areas of resource development.

[7] In Canada, since 1911, the census enumeration has been made on, or near, June 1. In this paper, intervals which begin and end on June 1 of census years are usually referred to

as intercensal intervals and are documented as, for example, the 1951-61 decade. However, the net migration estimates are based on annual, calendar-year vital statistics and on estimated population at the beginning of each census year. The relevant interval is, for example, from January 1, 1951 to December 31, 1960. In this paper it is usually referred to as a "calendar-year intercensal interval" and it is documented as, for example, the 1951-1960 decade.

Changing Community Patterns in Saskatchewan

W. B. BAKER

"Humanity is passing through an age of tumult and upheaval. Nowhere is this felt more acutely than among those engaged in agriculture." This statement was made a few years ago by a Saskatchewan farm housewife. She was directing attention to a transformation going on behind the clamour of Canada's industrial and urban expansion. To understand what she had in mind, one could visit almost any Saskatchewan rural community. Talk to old John Jones who broke trail from Ontario in 1900. Call on young Ted who is trying to make a start in farming. Look in on Jake Simmon at the village general store. Question Bill Miner who has become a town farmer. Each of them will tell you in his own way that trends are emerging to work fundamental changes in agriculture and rural life. Because Saskatchewan agriculture is so well adapted to modern farm technology, what they have to say will eventually apply in varying degree to all of Canada's rural areas.

DEVELOPMENT AS AN AGRICULTURAL ECONOMY

While the province's history extends back over the centuries, its development as an agricultural economy has been brief and rapid. The period of peak settlement (1897-1920) is well within the memory of many living pioneers. By the late 1930's, immigration had dropped to a trickle of the former flow. The first three decades established a pattern just before the outlines of modern prairie agriculture were becoming apparent. The Homestead Act encouraged farmers to occupy small quarter-section and half-section farm units. It also required settlers to reside on their land. The "team haul" made necessary the location of villages and towns at regular six-to-seven-mile intervals along the railroad. Early optimism about the productivity of the land and population concentrations prompted the establishment of one-roomed country schools, country churches, stores and post offices. These, in turn, became the centres of social neighbourhoods where farm families could find relief from the trials of a pioneer existence.

The tractor and related mechanical equipment appeared in the province during the early decades of the twentieth century. While the stage was set for full mechanization by the late Twenties, the depression intervened and blocked any persistent trend. The Second World War both limited and

● *The late* W. B. BAKER *was President of the Canadian Centre for Community Studies, Ottawa.*

Reprinted from the *Canadian Geographical Journal,* Vol. LVI (2), 1958, pp. 44-56 with permission of the author and publisher.

stimulated the mechanization process. The shortage of steel for agricultural implements restricted available supply. The mobilization effort reduced available farm labour and encouraged the economies of machine operation. In the late Forties the first major breakthrough began and equipment flowed out to Saskatchewan farms in increasing volume. In 1946 Saskatchewan farmers purchased just over $24,000,000 worth of new machinery; by 1953 they were investing just under one billion dollars. Since then the slump in farm income has sharply reduced new investment. During the same period other aspects of farm technology encouraged the emergence of "science conscious" farmers. Improved plant and animal breeding practices, more accurate knowledge of soil productivity, chemical weed control, fertilizers, farm management and a host of other advances were made available to the farmer through the research of government, university and industry.

With the establishment of a major trend toward the application of mechanical and scientific methods to farm production, a subtle transformation started. Today, it is challenging and reforming the pioneer patterns on a broad front. It is this transformation that the housewife had in mind when she made the statement which introduced this article. She was pointing to trends which can now be identified and which invite interpretation. Anyone interested in the probable future pattern of rural Canada will want to have some understanding of its dynamics.

THE SMALL-FAMILY FARM

What, for example, is happening to the small-family farm, long held to be the ideal for a sound rural economy? The blunt answer is that it is on the way out, a victim of modern technological forces. Its place is being taken by the larger, highly mechanized, and commercialized operation. Farming as a way of life is giving way to farming as a business. The rapidity of this change is demonstrated by the disappearance between 1936 and 1956 of almost 50,000 of the pioneer 160-acre and 320-acre farms. For every

five of these units there is now in their place one farm 640 acres or larger in size. Extensive capital investments are involved in the operation of the new units. In general the farmers who operate them are able to take advantage of all advances in modern technology. The resulting higher incomes permit living standards comparable to or above those of the average urban dweller. They represent the latest design in "family" farms, for they are not the "factory" farms often imagined. While these farms do tend to employ more farm labour than their smaller counterparts, by and large the chief source of labour remains within the family. Actually, the hired farm labour force in Saskatchewan has shown sharp decreases since the advent of mechanization.

The process of adjusting to farm technology means vastly different things to those who have been unable to take full advantage of the trends. Well over one-half of the 103,000 farms reported for Saskatchewan in 1956 are far removed from the picture book farms so often found in magazines. The situations on these farms is highly complex. More research will be required before their dilemma can be fully assessed. In general they seem to be stalled on the lower rungs of the development ladder. Some of them have been able to obtain credit for mechanical equipment but can obtain no credit for land purchase to fully utilize their equipment investment. Often they find themselves competing for available land, whether for sale or for rent, with larger operators better able to finance. As a result, such farms are often top-heavy with equipment investment; their profit margins narrow or even negative.

While lack of an adequate amount and a proper form of credit seems to be the central difficulty, there are other deep-rooted social and psychological aspects. Operators on these farms are not inclined to make use of farm science; many could not afford to do so even if they were interested. A recent study of small farms indicates a suspicion of credit; yet credit is an essential feature of any modern business. Many such farmers find their social situation satisfying, even though their incomes are inadequate by any economic standard. Frequently the insecurity of leaving the farm for an urban occupation is greater than the insecurity of remaining on the farm. Generally, the children on such farms are not encouraged to complete a high-school education. Since they also tend to belong to families larger than the average, sooner or later most of them migrate to the city. When they do, they suffer the consequences of an inadequate education. Further complications develop as the parents find themselves unable to bear the tax burden necessary to finance modern standards of public service.

In a very real sense, the families on small farms represent a group of citizens caught in one of the pressure spots of a rapidly changing society. Many, if not most of them, are subject to forces which they can neither understand nor control. They belong to what might well be a chronically depressed group in modern agriculture. Few agricultural policies are at present in effect which would relieve their situation. Parity prices are assumed by many farm leaders to be the solution. This is difficult to demonstrate in the light of the high production costs and other complex factors associated with such units. A number of studies suggest that up to one-third of these families would migrate out of agriculture if they could afford it or if they could acquire skills which would qualify them for urban occupations. Many of those who are forced to migrate should not leave agriculture, to which they are best adapted by both interest and skill. In such instances, migration represents a distinct loss to the agricultural industry.

Even the farmers who have adjusted to the modern farm technology are still faced with insecurities. The wheat surplus of the past few years has restricted annual income. At the same time, the costs of production have been mounting. Saskatchewan also experiences greater yield variability than any other province in Canada. These three situations continue to make farming a high-risk operation calling for management ability of the highest order. Significantly, almost all of the young men attending the University of Saskatchewan's School of Practical Agriculture come from the modern family farms.

FARM POPULATION

Population dynamics reflect, in part at least, the basic adjustment in farm production. In only one census year since 1901 has there been any increase in the proportion of the total population of Saskatchewan classified as rural-farm. That was in 1936 when urban jobs were not available and the flow of migrants was reversed. Under normal conditions farm regions always produce surplus populations. Declining economic opportunity in farming and attractive employment in urban centres have accentuated this surplus. In 1951, rural farm population in the province for the first time made up less than one-half of the total.

In the past, the chief destination of farm migrants has been the heavily industrialized areas of Ontario and British Columbia. In recent years, Saskatchewan's own industrial and urban development has been absorbing a growing proportion. Indeed, it can be said that much of Canada's present urban industrial growth would have been impossible without drawing upon the rural labour supply. The only other large potential source would be sharply increased immigration from other countries. It would be short-sighted if this were done in the face of necessary reductions in the rural labour force.

The bulk of the rural migrants is in the work-productive age group. Both males and females tend to leave the home community in greatest numbers during their early Twenties. A conservative estimate would be that from sixty to seventy percent of the young people at present born and living on the farm will eventually settle in an urban centre. Few would argue with the need for this transfer if improved opportunities were available for attaining an acceptable standard of living. When those who must move complete a high school or college education, gains are realized both by the migrants and by the economy as a whole.

As with the adjustment in the pioneer farm unit, there is a negative as well as a positive side to the transfer of farm population. Migration is always a highly personal matter. The decision to leave the community in which one has grown up is seldom an easy matter, particularly when the whole family is involved. The donor communities are depleted of population. In many Saskatchewan communities the extensive migration of young people has left a concentration of the very young, the middle-aged and the very old. Rural communities in particular invest heavily in the development of youngsters, only to have them leave when they are ready to contribute to community development. The receiving communities in turn should be concerned about the quality of the migrants. If they are ill-equipped, then a penalty is inflicted on both migrants and community. In a society that literally breeds demand for skilled persons, poorly-educated citizens represent a distinct social cost.

In general it is the low-income areas that have too many people in agriculture in relation to land resources, if desired income levels are to be attained. Farm families in the higher income areas seem more aware of the need or are better able to make an early assessment of economic opportunities and to prepare for alternatives. It may seem strange that surplus youth tends to be most immobile in areas where migration would result in sharp gains for the remaining families. As indicated, freedom of movement out of these areas is restricted by a number of complex factors. In Saskatchewan, only limited alternative employment is available on a part-time basis. For those who remain, increase in size of holdings and efficiency of operation are not readily attained. Inadequate educational preparation further restricts alternatives. Other families are handicapped through age, illness, or debt burdens.

Whatever the future may hold in store for the farm population, it is fairly certain that, barring any major depressions, the movement to urban industrial centres will continue. The trend to mechanized and commercialized family farms is by no means completed. Urban industrial occupations are liable to continue to be attractive. Despite marked advances in rural levels of living,

differences persist for many farm families. If movement of people from rural to urban areas is to continue, we should be more concerned about the quality of both those who migrate and those who remain in agriculture. Every profession in Canada needs well-educated recruits. An increasing number of technological stations must be manned in our expanding economy. A major human resource is at present being neglected in our rural areas. Similarly Canadian agriculture cannot lag behind the rest of the economy; areas in a state of chronic depression despite sincere effort require various forms of assistance to permit potentially competent farm operators to increase their efficiency and to remain on the farm.

THE RURAL RESIDENCE PATTERN

Changes in farming operation and urban migration influence the number and closeness of neighbours in the countryside. Pioneer expectations were that fairly dense settlement would prevail. Census data show that this expectation was being realized until the 1930's. Since that time the number of persons per square mile has been decreasing (Figure 1). By 1951, Saskatchewan's population density was less than three in the prairie region and just under five in the park region. This trend has a direct bearing on the rural residence pattern (Figure 2). Not only does density determine the distance separating neighbours, it also determines the cost of bringing expensive public services to farm families; social space has a price tag on it.

Most agricultural peoples of the world live in villages and go out to the fields each day. In Saskatchewan, as in other Great Plains states, an isolated rural residence pattern has prevailed. It would be difficult to imagine many ways in which a more expensive pattern could be designed. Recent trends are making it even more expensive. In most instances, each of the 50,000 quarter-section and half-section farms which have disappeared represented a farmstead. In many Saskatchewan districts this shows up in a reduction as high as sixty percent of the

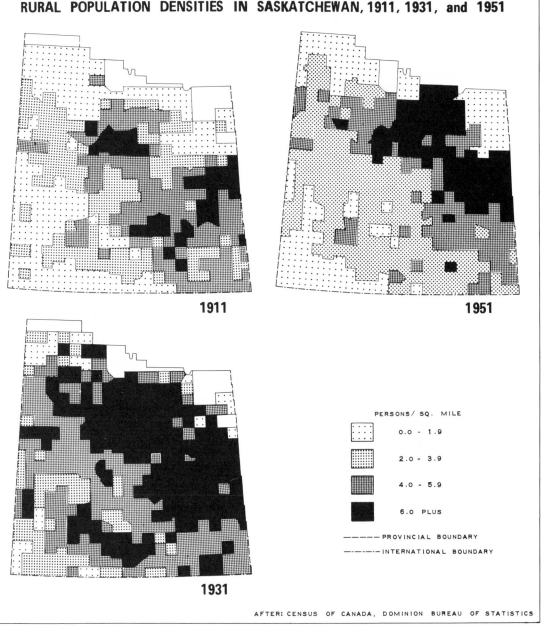

RURAL POPULATION DENSITIES IN SASKATCHEWAN, 1911, 1931, and 1951

1911

1951

1931

PERSONS/ SQ. MILE

0.0 - 1.9

2.0 - 3.9

4.0 - 5.9

6.0 PLUS

————— PROVINCIAL BOUNDARY

—·—·— INTERNATIONAL BOUNDARY

AFTER: CENSUS OF CANADA, DOMINION BUREAU OF STATISTICS

Figure 1

originally occupied homes per township. This contraction is the converse picture of the more dramatic suburban expansion.

To this should be added the emerging trend toward town-resident farmers. A major revision of farm-residence patterns can be identified. There is evidence that most of the town farmers move to a village or town within ten miles of their farms. But as early as 1951, this group included twenty percent of all farmers in the prairie region. The 1956 census shows that it has since increased to over twenty percent for the province as a whole. While many of those who relocate express a desire to return to farm residence at a later date, it is highly improbable that

FARM RESIDENCE PATTERNS IN SASKATCHEWAN, 1953

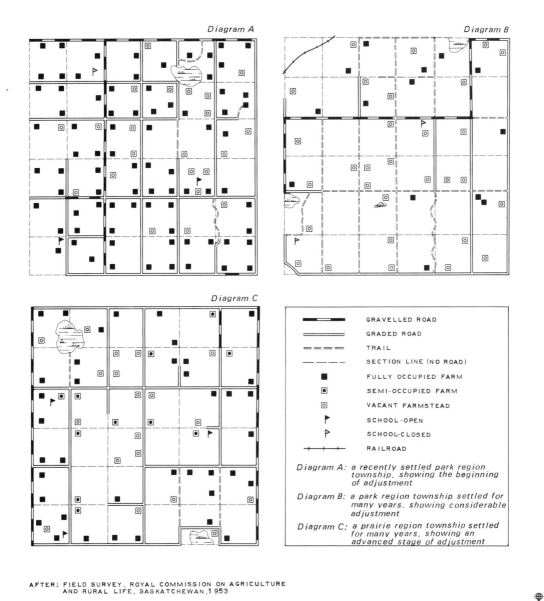

Diagram A

Diagram B

Diagram C

GRAVELLED ROAD
GRADED ROAD
TRAIL
SECTION LINE (NO ROAD)
■ FULLY OCCUPIED FARM
◉ SEMI-OCCUPIED FARM
◎ VACANT FARMSTEAD
⊳ SCHOOL-OPEN
⊳ SCHOOL-CLOSED
RAILROAD

Diagram A: a recently settled park region township, showing the beginning of adjustment

Diagram B: a park region township settled for many years, showing considerable adjustment

Diagram C: a prairie region township settled for many years, showing an advanced stage of adjustment

AFTER: FIELD SURVEY, ROYAL COMMISSION ON AGRICULTURE AND RURAL LIFE, SASKATCHEWAN, 1953

Figure 2

they will do so. Parents and children establish many personal ties with town families. They become involved in the organized life of the urban centre. In the meantime, the farm home may be neglected or moved into town. The coming of television, rural electrification, centralized schools, and all-weather roads may slow down or even reverse the tendency to relocate. There are areas, however, where all of these conditions already exist but the more recent pattern continues to develop. More subtle community factors seem to be involved.

For those who, either by choice or of necessity, remain on the farm, the problems of distance and costly public service will continue. Much could be done to relieve the situation by adoption of the French Canadian pattern of line settlement. Main market roads could be selected and developed. Farm families could be assisted to relocate along such trunk lines. Not only would service costs be reduced but some of the earlier "neighbouring" would be re-established. There are practical difficulties in such an adjustment but a few communities are showing imaginative leadership in this direction.

RURAL TRADE CENTRES

Both pioneer patterns of residence and trade distribution are being affected by an accumulation of other changes. There are at least 800,000 fewer horses in Saskatchewan today than in 1926. The automobile and truck have helped to send old Dobbin to the pastures or to the canning factories. In the days of the "horse haul," distance was measured in terms of miles. Today it is measured in time. New conditions of travel have brought about a shrinking of both time and space. There are very few farms today without an automobile, a truck, or both.

Even though the farm population has become sparse, in terms of time families may be as close to their neighbours as they ever were. As rural roads are improved they will be brought even closer. Beyond this, greater ease of travel has extended the farm family's

Figure 3

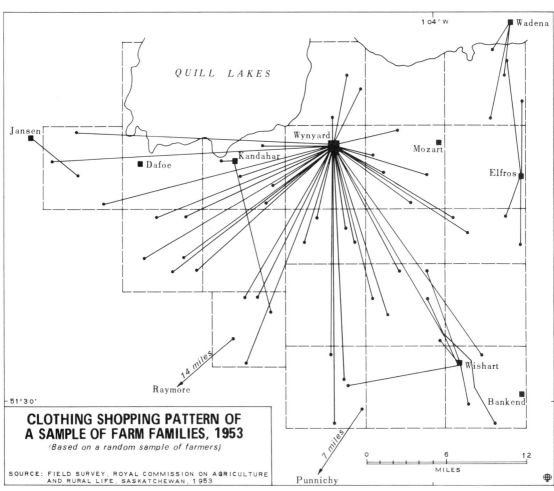

CLOTHING SHOPPING PATTERN OF A SAMPLE OF FARM FAMILIES, 1953
(Based on a random sample of farmers)

SOURCE: FIELD SURVEY, ROYAL COMMISSION ON AGRICULTURE AND RURAL LIFE, SASKATCHEWAN, 1953

shopping horizon. Rising rural living levels, radio, the press and television have contributed to a more critical and selective purchasing pattern. The village of an earlier day with its often gloomy general store has been unable to satisfy this demand. As a result, farmer patronage has been shifting to the larger, more favourably-situated centres (Figure 3). As the favoured centres draw upon an expanded trading population, increased specialization becomes economically feasible. The grocery, hardware and women's wearing apparel stores take the place of the general store. In the course of time, commercial recreation services expand to complement the retail trade. Professional personnel—doctors, lawyers, dentists, teachers —quite clearly prefer these larger centres and tend increasingly to concentrate in them. Provincial and local government administrative centres gradually shift in the same direction.

This tendency toward the reorganization of the rural trade pattern produces a number of curious consequences. When the pioneer quarter-section and half-section farms become outmoded under modern technology, they can be absorbed into balanced farm units. Obviously, this cannot occur in the instance of the pioneer villages except over long periods of time. The total number of incorporated and unincorporated centres of all sizes in the province approaches 1,500. Less than one-third of these centres were large enough to merit individual listing in the 1956 census. Saskatchewan was credited with eight cities, ninety-eight towns and 377 villages. No other Canadian province has as many small centres or as high a proportion of its population living in them.

Despite the apparent trend in farm-family patronage, the surprising fact is that the proportion of centres in each population size class has remained almost constant in the 1911-56 period. This suggests that there are a number of systematic relationships between classes of centres and that these relationships tend to persist. It also suggests that centres differ, not only in size, but also in the functions which they perform.

In the light of this evidence, it is difficult to conclude that in the foreseeable future all or even most of the small villages now suffering a decline in farmer patronage will disappear. What seems more likely is that some reduction will occur in the number of retail outlets of the small villages, but they will continue to meet farm family needs for convenient services: groceries, mail, gasoline and oil, public school, church, curling and so on. On the other hand, the larger centres will provide specialized modern merchandising as well as professional, recreation and administrative amenities. To this extent the large and small centres have complementary functions to perform.

While there is reason for this optimistic assessment of the future of the small village, those whose livelihood depends upon the volume of trade are facing difficult times. Competition between the village merchants is certain to increase. This basic fact leads to much bitterness not only between merchants living in the same village but also between village merchants and the mobile farm family. In numerous instances, villages are losing confidence in their future. Not having access to any factual analysis, a situation of low morale can prevail. Sooner or later this is reflected in many subtle social and psychological tensions. Again it is a case of a special group of citizens caught in one of the pressure spots of rural social and economic change.

LOCAL GOVERNMENT

Most of the trends thus far identified are based largely upon the decisions of individual farm families. The farmer need consult with no one other than perhaps his family and banker when he decides to double the size of his farm by buying out a neighbour. Similarly, he can change from horses to tractor or shift his patronage from village to town. When individual decisions of many farmers living in the same area are added together, however, a number of consequences emerge which require collective adjustments. No farmer, in deciding to double the size of his farm, consciously decides at the same time to close his local school. Yet in the aggregate this is just what does happen.

When Saskatchewan was first settled, thousands of one-roomed country schools were established on the initiative of the parents. Even by 1922, almost five percent of the original units were no longer operating. By 1953-54, over thirty percent of the 5,221 school districts established by that time were no longer in operation. In 1944, provision was made for the establishment of some sixty larger school units of administration, each including approximately eighty of the original districts. While the trend toward the closing of country schools was initiated well before the larger unit, it has since been accelerated. Population decline, the teacher shortage and rising educational standards are the primary motivating factors. There is little doubt that in the next decade the majority of the one-roomed schools will be replaced by centralized schools. This has already happened in the neighbouring province of Alberta. An important aspect of the regrouping of schools is that town and country schools are being integrated. Farm and town families are losing many of their former differences in experience and responsibility.

Other local government units have been going through the same process. Hospital districts were established in the early twenties. In the last two decades, larger health units have been established over much of the province. Agricultural representative districts have been set up which include a number of municipalities. The one type of local administrative unit which has continued to resist adjustment is represented by approximately 300 nine-township municipalities established in 1909. It is fairly certain that their days are numbered. The development of a motorized agricultural industry has created widespread demand for a system of all-weather roads. Since Saskatchewan has more miles of road than any other province in Canada, heavy public investments will be involved. The nine-township unit established to satisfy the needs of a simpler society cannot provide the necessary standard of modern service. Discussion is now under way on the merits of municipal reorganization. The county

system, or some modification of it, is expected to take its place.

One of the costly consequences of social and economic change is the lag which persists between individual and collective decision-making. Local government reorganization proceeds on a painful, piecemeal basis over the years. Changes are resisted until the majority of farm families are convinced of the urgency of the need. Aside from the delay in providing costly modern public services, a growing confusion is apparent in the relationship between the various administrative units which have some claim on the farmer's tax dollar. As the confusion mounts, the farm family finds itself increasingly less able to exercise democratic control over those to whom responsibility is delegated. Indifference and hostility to the provision of needed services is one result. Indeed, there is evidence that the lag in rural public service is one important factor in pushing farm families off the land into other occupations or into urban residence.

THE RURAL COMMUNITY

Agriculture and rural life are traditionally associated with strength in community and small group experiences. The growing concern for the social problems rooted in the mass society of urban concentrations tends to confirm these sentiments. At the heart of the small rural community are the myriad voluntary organizations which give it meaning. Today these organizations find themselves crowded together in contracting communities. Competition for available leadership is often extreme. One small village of 600 population reported some 150 organizations with accounts in the local bank. There is every indication that voluntary associations are shifting in the direction of a larger community. In it, the close personal contacts and neighbourliness of an earlier day will be much more difficult to develop and retain.

RURAL VALUES

Finally, a word must be said about trends in rural attitudes and values. It is now apparent that social and economic adjustment means radical changes in almost every aspect of rural living. It is one thing to identify the direction of change; it is quite another matter to evaluate the desirability of change. Is it enough to point out that the trend to urbanization promotes higher levels of living or that centralization means more and better services? There are serious students, and many of them farm operators and housewives, who value those close personal relationships which give depth and meaning to living. Others place high value on the opportunity for intimate association with nature. How to maintain these values in an era of modern technology presents a problem which is not to be taken lightly.

CONCLUSION

Humanity is passing through an age of turmoil and upheaval, and rural folk are feeling this most acutely. While no one can foresee in detail the nature of the agriculture and rural life which shall prevail in the future, the broad outlines are already apparent. The changes anticipated are both good and bad from the viewpoint of the man on the land. Those able to take advantage of the trends are likely to enjoy satisfaction in their occupation. Those, who for various reasons find themselves unable to adapt, may be placed at a relative disadvantage. Progress is always a potentiality and not an inevitable consequence of change. Special programmes will be necessary to soften the pains of technological advance. To attempt to stop or even to slow down the tide of change would be too costly for farmers in the long run.

Changing Settlement Patterns Among the Mackenzie Eskimos of the Canadian Northwestern Arctic

M. R. HARGRAVE

Following 1576, when Martin Frobisher encountered Eskimos on Baffin Island, intermittent contact was kept up by explorers and nineteenth century American and Scottish whalers until the establishment of Hudson's Bay Company posts in the present century. Yet Bisset's paper,[1] sketching present conditions in the life of the Melville Peninsula Eskimos, indicates that in spite of the superimposition of trapping, the proximity of "DEW Line" radar stations, the introduction of firearms and trade goods, in spite of schools, medical and other facilities, the pattern of life and settlement in the 1960's is not radically different to the traditional one. A relatively long navigation season and wide approaches to Baffin and Hudson's Bays, and to Foxe Basin, proximity to markets and home ports permitted many of the whalers to complete their hunt and return to civilization in the same season. The pattern of Eskimo settlement was scattered, so that what depredation by undisciplined whaling crews did occur had little effect on the population in general. The eastern and central Eskimos were also shielded from southern influences by a broad buffer zone of barren lands.

A very different situation has prevailed in the Canadian Western Arctic, where more

● M. R. HARGRAVE *is coordinator, Central Region Recreation Sector, Canada Land Inventory, A.R.D.A., Department of Regional Economic Expansion, Ottawa.*

Reprinted from *The Albertan Geographer*, No. 2, 1965, pp. 25-30, with permission of the author and the Editorial Board Chairman.

recent and more intensive contact has profoundly influenced Eskimo society. Different geographic conditions have contributed to the destruction not only of traditional society but also of almost the entire original population. First, the Beaufort Sea had large numbers of bowhead whales but the configuration of the "peninsula" of Alaska, together with the long distance from markets and bases made it imperative that the commercial whalers winter in the Arctic. Secondly, a different settlement pattern obtained from that of the eastern Eskimos: the concentration of the Mackenzie Eskimos in large villages facilitated cultural contact and ensured rapid transmission through the entire group. Thirdly, the Mackenzie River fostered the entry of southern goods and influences. Lastly, the proximity of the tree-line and the presence of a rich muskrat resource in the wooded Mackenzie Delta encouraged white settlement, and offered the Eskimos an alternative economy to that of the coast.

The nineteenth century explorers were generally unanimous in their admiration of the Mackenzie Eskimos as a robust and self-reliant group. Lieutenant Pullen, R.N., during his search for Sir John Franklin in 1850, had this comment to make on the Cape Bathurst villagers:

> I cannot help saying here, that I never saw a finer body of men, nay women too, than I found the natives here, and I have often thought what a glorious expedition it would be to introduce the blessed Gospel among them.[2]

In contrast, though some of the same attributes noted by the explorers are still discernible today, serious social and economic problems are evident, and what may be called a "métis society" has developed.

This paper endeavours to sketch the traditional life and settlement patterns of the Mackenzie Eskimo and contrast it with that of today. In accounting for changes which have taken place, the writer pretends to no significant original research; rather he aims to point to the need for research that might shed light on causes that have led to present distributions and problems.

TRADITIONAL SETTLEMENT PATTERNS AND ECONOMY

From the journals and writings of the explorers, especially of Sir John Franklin (1826),[3] Sir John Richardson (1826 and 1848),[4] and of Oblate Father E. Petitot (1865),[5] together with later information gathered by Vilhjalmur Stefansson during his travels in the Western Arctic (1906-18),[6] it is possible to reconstruct the pattern of traditional life before modern influences disturbed the equilibrium.

It is commonly thought that the Eskimos of Arctic Canada were a nomadic people whose environment compelled them to disperse in small groups. This was not so in the case of the Mackenzie Eskimos. Franklin and Richardson report meeting numerous groups along the coast from Herschel Island to Cape Bathurst, but settlement appears to have been concentrated at five locations: (a) Kittigazuit, at the mouth of the easterly channel of the Mackenzie Delta; (b) Atkinson Point, almost seventy miles east of the Delta; (c) Baillie Island and adjoining Cape Bathurst; (d) at three locations along the southerly coast of Herschel Island; and, (e) at the mouth of the Anderson River (Figure 1). Several sub-groups of the Mackenzie Eskimo have been recognized, but older natives refer to the whole group as the Tariumiut (the sea-dwellers).

Richardson reports that in 1848 about 200 kayaks propelled by generally younger mature males (constituting probably the "home-defense force") surrounded his boats at Kittigazuit. Franklin's and Richardson's journals indicate that the average family consisted of no more than four or five individuals; therefore, if one able-bodied male from each family met Richardson, it would appear that the population of Kittigazuit was between 800 and 1,000 individuals. There were also villages at Shingle, Kay and King Points, and three villages on Herschel Island—all to the west of the Mackenzie Delta. Stefansson estimates the population of Atkinson Point and Baillie Island-Cape Bathurst at 500 each, and suggests a total population at 4,000. Petitot estimates the total population at 2,000 which

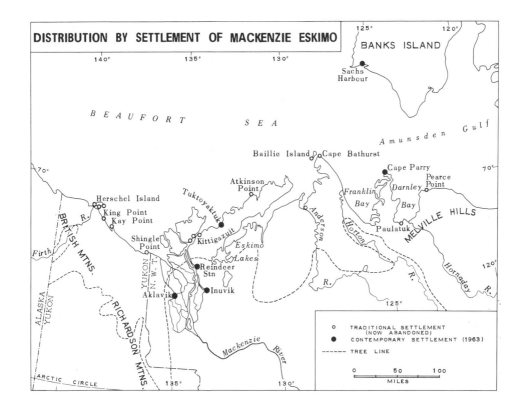

Figure 1

figure seems to be more in accordance with the journals of Franklin and Richardson.

The traditional economy was largely based on the resources of the sea. In the immediate area of the Delta the sea is shallow, silty and only faintly brackish, owing to the discharge of the river. The sea off Atkinson Point, Herschel Island and Cape Bathurst is less influenced by river water, and is deeper and clearer. The large bowhead whale and seal were hunted at the eastern and western extremities of the region, which, together with six (whitefish, anadromous herring and arctic char) caught with nets of caribou sinew, were the basis of life. The more numerous Kittigazuit people, and the whaling village of Shingle Point subsisted on fish (mainly whitefish and herring) and on the smaller white whales (beluga) which migrate from the west to enter the shallow estuaries of the lower Delta in July and August to bear their young in the warmer Mackenzie waters. The Anderson River

Eskimos primarily depended on a rich fishery resource, sealing (many large bearded seal, or "ugruk," can be seen on the ice basking in the early summer sun near the mouth of the river), and caribou hunting. The journals of Franklin and Richardson make almost daily reference to caribou along the coast, and this was an important auxiliary resource, particularly for its value in furnishing clothing and sinew for sewing and fish nets. Of lesser importance were wildfowl, ptarmigan and the collection of berries on the tundra.

Trade was carried on with the Eskimos of North Alaska, exchanging furs for Russian-made knives, and with the Copper Eskimos of Coronation Gulf, but it appears to have terminated prior to 1850, probably due to the development of trade with the neighbouring Indians. Petitot refers to the commencement of trade in 1849 with the Hare Indians of the Fort Hope area, via the Anderson River, and with the Loucheux

101

Indians of the Fort McPherson area, via the Mackenzie River. At that time the Indians became the middlemen between the Hudson's Bay Company posts and the coast-dwellers. In return for furs the Indians introduced firearms, tobacco, beads, and basic hardware items such as files, matches, cooking utensils, knives and traps.

Yet in spite of the rudimentary beginning of trade in furs, the pattern of settlement and economy remained basically unchanged. The continued dependence on the bowhead and beluga whales necessitated community co-operation in large villages.

PRESENT SETTLEMENT PATTERNS AND ECONOMY

The "Eskimo Identification Disc List," District W-3, 1963 edition,[7] listing all Eskimos to January, 1963, enumerated some 1,560 resident in the region. The composition of the group, the settlement pattern and economy were radically different, however, to those of a hundred years ago.

Of this number some 400[8] lived in Tuktoyaktuk, the terminus of the Mackenzie River transportation system, and trans-shipment point for all water freight for DEW Line (Distant Early Warning radar stations) stations and settlements in the Western and Central Arctic as far east as Spence Bay. At Sachs Harbour, Banks Island, the centre of what is probably the Arctic's richest white fox trapping area, there lived 85 Eskimos. A similar number lived at Cape Parry, in which area numerous caribou are still hunted, and where the rare musk-oxen seem to be regaining a foothold.[9] Fifty lived at Reindeer Station, the headquarters of the Canadian government reindeer herd. About 40 were living at DEW Line stations where the family heads were employed. The remainder, about one thousand, lived in the wooded southerly half of the Mackenzie Delta, mainly within the settlements of Aklavik (total population, Eskimo, Indian and White, in 1965 estimated at 600-700), the former regional "capital" and center of muskrat trapping, and in Inuvik (total population estimated at 2,200-2,300 in 1965), the modern community completed in 1960 to serve as adminis-

trative centre for many thousands of square miles of the northwestern Mackenzie District. Clairemont[10] states that in 1961, 106 "bush" Eskimos lived in log cabin camps close to Aklavik.

A hunting economy based on caribou and seal, supplemented by white fox trapping, and the sale of a few polar bearskins, prevails at Cape Parry. At Sachs Harbour the economy is based almost wholly on the white fox which is trapped in large numbers by an elite group of aggressive Eskimo trappers. At Reindeer Station the population is supported entirely by salaried reindeer herders. Wage labour, divided about equally between permanent and casual, maintains the Eskimo population at Inuvik, and is also the predominant source of income at Tuktoyaktuk and Aklavik. In Aklavik, of 29 settlement Eskimos between the ages of 16 and 29 only two earned over $400 by trapping in the 1960-61 season, and what in the early 1950's was referred to as the "million dollar business" (referring to muskrat trapping and shooting which has for many years accounted for 95 percent of the wooded delta's fur catch) was in 1961 referred to as the "$100,000 activity". Most of the more than fifty trappers at Tuktoyaktuk set their traps along the sea coast between Atkinson Point and the outer edge of the Delta islands, and even in the best years trapping income is inadequate owing to the proliferation of short traplines. With the availability of some wage labour, trapping as a way of life appears to have been rejected not only by many younger men but also by some of the older Eskimos, especially in Inuvik, Aklavik and Tuktoyaktuk.

The Delta and surrounding waters are rich in anadromous herring and whitefish, and throughout there is heavy dependence on subsistence fishing, not only for the table but also for the numerous dog teams.

The traditional occupation of hunting the beluga (white whale) is still carried on in July and early August at Shingle Point, Kendall Island, and Whitefish Station by a few Delta Eskimos, and from Tuktoyaktuk, but the hunt is no longer a communal pursuit but an individual chase by a few old and small schooners and river craft purchased during the rich trapping years of the 1920's, by small motor craft and even outboard canoes. The beluga is trailed by following the slight swell it makes in the shallow water, shot when it surfaces, and harpooned before it sinks.

Of the almost 1,600 Eskimos in 1963 probably less than twenty percent could be traced back to the original Tariumiut—and that only by virtue of generous infusions of white and immigrant Alaskan Eskimo blood. With the death in 1964 of an old Tuktoyaktuk man, who was able to recall hunting with bow and arrow as a boy on Herschel Island prior to the arrival of the whalers in 1889, it is likely that no pure-blood Tariumiut remains. The other significant fact with regard to the present Eskimo population is its alarming fecundity. Clairemont has noted that roughly fifty percent of the population of Aklavik was under fifteen years of age in 1961, compared to a corresponding figure for Canada of thirty-four percent. Ferguson[11] reported 55.4 percent below the age of fifteen in Tuktoyaktuk in 1957, and in the same year a natural increase rate of 5.59 percent as compared with 2.02 percent for Canada in 1955. The writer calculated a natural increase of approximately six percent in 1962. Truly the term "population explosion" is valid here!

THE CHANGING PATTERN OF LIFE

A number of factors can be recognized as having affected or influenced the changes which have taken place since the traditional stage of Mackenzie Eskimo life. Some are obvious and important; others are tenuous and their importance is less evident, or not as yet measured. Certainly three factors can be seen as crucial, and chronologically it is possible to divide the history of change into three periods: the commercial whaling period, 1889-1912; the trapping period, 1912-1954; and the modern period, 1954 to the present.

The Commercial Whaling Period,
1889-1912

The New England whalers, who had penetrated the Bering Sea by 1850, slowly increased the range of their operations until in 1889 seven steam whalers reached Herschel Island. From that year to 1910 huge profits were made from the harvesting of baleen, used in the manufacture of ladies' corsets and horse whips. By 1912 substitutes and the development of the automobile made whaling in the Beaufort Sea unprofitable. In 1894-95 fifteen whaling ships wintered at Herschel Island; Bodfish[12] reports that as many as six hundred men wintered there some seasons. Ships also wintered at Baillie Island and in southern Franklin Bay.

The contact with the whalers was drastic in its effects. Eskimos flocked to Herschel Island from the Delta, and Nunatamiut (land dwellers), Eskimo caribou hunters from interior Alaska, immigrated into Canada to enter the service of the whalers as professional hunters. Bodfish indicates that during two winters, 1897-99, his ship, the "Beluga," received 47,000 pounds of venison, and that during the whaling period the professional hunters killed all the musk-oxen "within an area of one hundred and fifty miles."[13] The material culture of the Mackenzie Eskimos was to all effects destroyed; the whale boat replaced the kayak and umiak almost overnight. The bowhead whale resource was decimated and the Eskimos were debauched, and exposed to diseases to which they had no immunity. "The R.C.M.P. census of 1911 showed only forty descendants of the local people, although there were also in the country considerably over a hundred immigrants, and since this census of two years ago six of the forty have died and three have gone permanently insane."[14]

The Trapping Period, 1912-1954

The year 1912 marked the visit of the last whaling ship and the establishment of the first Hudson's Bay Company post close to the present site of Aklavik. The high prices for white fox during World War I and during the 1920's led to fierce competition between free traders and the Hudson's Bay Company, and the Western Arctic proliferated with trading posts. The competition was encouraged when Eskimos purchased schooners and whaleboats from the proceeds of their trapping and were able to choose the best posts at which to buy or sell. In 1924 the Eskimo fleet at Aklavik totalled thirty-nine schooners, nineteen with auxiliary power, and twenty-eight whaleboats. These were valued at $128,000, and had been bought within the previous five years.[15] The price of white fox skins soared up to as high as $70.00 in 1928, and a good silver fox could be exchanged for a whaleboat. In the Mackenzie Delta, muskrat prices rose from fifty cents in 1914 to an average of $1.31 for the period 1921 to 1929. The price plunged to 31 cents in the early depression years, but by 1935 had risen again to $1.00.

The expansion of the fur trade after 1912 brought about the dispersion of the Eskimo population along the coast as far east as Pearce Point, and few settlements had more than three or four families. The competition between the free traders and the Hudson's Bay Company ended with the purchase by the latter of some of the free traders' posts and by declining prices when other traders, having reaped rich profits, left the country. High muskrat prices in the Delta during the 1920's attracted new immigrants from nearby Alaska, and white trappers from the south. The rising price in 1935, following the earlier decline, brought a new wave of Alaskan immigration: reference to the W-3 Eskimo Identification Disc List indicates that no less than 29 present family heads born in Alaska arrived in the Delta area between 1935 and 1946. The population of Aklavik and immediate area in 1931 was 411, composed of 180 Indians, 140 Eskimos and 91 whites. By 1958 it had risen to 1,500, comprising 384 whites, 242 Indians and 883 Eskimos.

After 1939, and the abandonment of the Hudson's Bay Company posts at Herschel and Baillie Islands, the small populations there moved to Tuktoyaktuk where a store and transportation centre had been set up, and where there existed limited opportunity for wage employment during the busy summer months. Two small groups remained east of Tuktoyaktuk at Stanton, near the mouth of the Anderson River, and at Paula-tuk, in Darnley Bay east of Cape Parry. The Eskimo had lost his independence: the location of trading stores governed his distribution. In the absence of commercial posts the Roman Catholic missions at Paulatuk and Stanton engaged in trading until 1954 when their two functions were found to be incompatible. The Stanton residents, discouraged by low fur prices, moved to Tuktoyaktuk while the Hudson's Bay Company was prevailed upon to re-open a store at Lettie Harbour to cater to the needs of the Cape Parry area Eskimos. The possibilities of sales to Dew Line Eskimo and other employees together with the attractions of casual employment resulted in the moving of the store to Cape Parry in 1959, close to the large radar station there.

By 1954 the present settlement pattern had largely emerged. Only Sachs Harbour remained a viable trapping community. The low prices for fur in the late 1940's and 1950's were accompanied by higher food and equipment prices. The muskrat population remained high but apart from cyclical fluctuations in the numbers of white fox there appears to have been an actual downward trend in the numbers of these animals along the mainland coast. Three reasons have been proposed. First, at the end of the whaling period many whole carcasses were strewn along the storm beaches, and, the rate of decay being very slow in the Arctic, the carcasses were for many years the source of food for the foxes—and foxes for the trapper. Second, the introduction of white fox trapping in Banks Island in the late 1920's and the high catches which have prevailed there since may have reduced the number of animals crossing to the mainland after a westerly migration through the Arctic Islands. Third, the radar stations along the coast have been accused of disturbing the ecology of the white fox. It would appear that at least the first two reasons have some validity.

The Modern Period, 1954 —

Trapping continues to be an important feature of the Eskimo economy in Aklavik and Tuktoyaktuk, but for some years wage employment has surpassed trapping as the major economic activity. The decline in trapping coincided with an intensification of the "Cold War" between Russia and the West, an awareness of the need for hemispheric defence against possible transpolar air attack, and a national awareness of the social and economic needs of the indigenous people of the North. A surge of economic activity started in 1955 when the construction phase of the Dew Line began, followed by the construction of government health, educational, administrative and other facilities throughout the North. It was decided that a modern town was to be built in the Delta when the site of Aklavik proved subject to flooding and unsuitable for further expansion. The construction of Inuvik in the late 1950's created many casual and permanent jobs as did the construction and manning of the Dew Line stations.

There is evidence of a levelling off in construction activity in the North. In the Mackenzie Delta area some Eskimos are finding employment and developing new skills, but opportunities are limited, and there is a high rate of unemployment and under-employment.

In the absence of research information other more tenuous factors may be best treated by asking questions rather than by hazarding answers. What has been the influence of the missions, medical facilities, schools, public assistance and other community facilities on population movement and settlement patterns? Can the decline of trapping be explained in purely economic terms, or are other factors involved?

Certainly the structure and functions of northern settlements have changed radically during the last few years, and new institutions—good in themselves—militate against traditional pursuits and the development of a stable economy. The problem is recognized and alternative pursuits have been fostered by the federal government. As long ago as 1935 reindeer herding was introduced to the Mackenzie Delta Eskimos, and since 1960 there have been fishing and whaling projects, and small fur garment shops (each employing about twelve Eskimo women and making fur parkas and sportswear) have been set up in Aklavik and Tuktoyaktuk.

With the exception of the latter, success has been limited. While the long range aim is assimilation, there remain short term problems referred to above which are perhaps equally important.

CONCLUSION

As opposed to the Canadian Eastern Arctic, the influence of the first sustained white contact was violent to the extent that it virtually destroyed a numerous and self-reliant group. Under the influence of the southern material culture and the vagaries of the fur industry—itself governed by changing fashions and cyclical fluctuations of fur-bearing animals—settlement patterns in the Western Arctic have "revolved" from the large traditional villages based on communal hunting of sea mammals, to dispersed and mobile "camps," and back again to the present nucleated settlements. There is, however, a basic difference. Under the prevailing physical and social conditions, the traditional village was an economically viable unit, and with an apparently low birth rate[16] the Eskimo lived in relative harmony with his environment. On the other hand, the present large settlements are not economically viable. The rejection of dependence on the traditional diet, the factors which militate against dispersion and more thorough harvesting of resources in areas not presently utilized, the restricted employment opportunities, and last but not least, the fecundity of the population, together comprise a serious problem.[17]

The changing patterns of settlement and the concomitant social and economic changes engendered are not unique to the Mackenzie Delta. Similar patterns have developed, are developing, or will develop elsewhere. The process is merely more developed there than in most of Arctic Canada (Labrador must be excepted). It would appear, then, that the area is an excellent research laboratory for the social scientist—a laboratory in which the anthropologist, sociologist and the geographer can pursue not only academic studies but also pragmatic solutions. To reconstruct the initial, and to report the present stage is not enough; the process, too, is important. The principle of uniformitarianism, that "the present is the key to the past," has long been held in geology. Similarly in human geography, the past (the dynamic past) can be the key to the present—and of assistance in planning for the future.

REFERENCES

[1] Bisset, D., "Recent Changes in the Life of the Iglulik Eskimos," *The Albertan Geographer,* No. 1, 1964–65, pp. 12–16.

[2] Pullen, W. J. S., "Pullen in Search of Franklin," *Beaver,* Outfit 278, 1947, p. 25.

[3] Franklin, Sir John, *Narrative of a Second Expedition to the Shores of the Polar Sea in the Years 1825–26 & 27,* London, 1828.

[4] Richardson, Sir John, *Journal of a boat voyage through Rupert's Land and the Arctic Sea, in search of the Discovery Ships under the command of Sir John Franklin,* Longmans, Brown, Green & Longmans, London, 1851, 2 V.

[5] Petitot, E. F-S., *Monographie des Esquimaux Tchliglit du Mackenzie, et de l'Anderson,* Librairie de la Société Asiatique de l'Ecole des langues Orientales Vivantes, de la Société Philologique des Sociétés Asiatiques de Calcutta, de Shanghai, de New-Haven, etc., Paris, 1876.

[6] Stefansson, V., *My Life with the Eskimo,* Macmillan, New York, 1913; "The Distribution of Human and Animal Life in Western Arctic America," *Geogr. Journal,* Vol. 41, 1913, pp. 449–459.

[7] Compiled by the R.C.M. Police, the northern representatives of the Registrar of Vital Statistics, Northwest Territories.

[8] Author's calculations and estimates. The figures for Tuktoyaktuk, Sachs Harbour and Cape Parry are accurate. The Eskimo population is mobile, and small fluctuations are a constant feature.

[9] In 1962 a Cape Parry Eskimo reported seeing fifty musk-oxen in the Melville Hills, seventy miles southeast of Cape Parry.

[10] Clairemont, D. H. J., *Deviance Among Indians and Eskimos in Aklavik, N.W.T.,* Northern Co-ordination and Research Centre, Department of Northern Affairs and National Resources, Ottawa, 1963.

[11] Ferguson, J. D., *The Human Ecology and Social Economic Change in the Community of Tuktoyaktuk, N.W.T.,* Northern Co-ordination and Research Centre, Dept. of Northern Affairs and National Resources, Ottawa, 1961.

[12] Bodfish, H. H., *Chasing the Bowhead Whale,* Harvard University Press, Cambridge, Mass., 1936.

[13] *Ibid.,* p. 186.

[14] Stefansson, V., *The Distribution of Human and Animal Life in Western Arctic America, op. cit.,* p. 453.

[15] Innis, H. A., *The Fur Trade in Canada,* University of Toronto Press, Toronto, 1956.

[16] Older informants were questioned closely with regard to the possibility of high infant mortality or infanticide but no evidence was adduced for either as the governing factor.

[17] In an introductory paragraph it was suggested that the Mackenzie River was a cultural corridor. It was then a factor in the breakdown of the Mackenzie Eskimo society. It may well be that the Mackenzie system will change its role to that of a highway of progress. Whether the movement will be south or north is as yet too soon for conjecture.

POSTSCRIPT: Many changes have occurred in the Canadian Northwestern Arctic since 1963. Control of the federal government reindeer herds has been transferred to Inuvik, and Reindeer Station lies abandoned. Cape Parry Eskimos have moved back to the southerly shore of Darnley Bay where a relatively flourishing community, centred at Paulatuk, lives on the resources of land and sea from Pierce Point to as far west as the Anderson River. Oil — in as yet unknown quantity — has been discovered at Atkinson Point, and the entire region is a hive of oil exploration activity, and controversy with regard to the ecological implications of that activity. Significant progress has been made in the provision of low-cost and low rental housing, community oil supply, health and other services, as well as community development in terms of participation by Eskimo and Indian in local government.

The construction activity, referred to above, has not yet levelled off, thanks to government housing programmes and extension of community services, and to increased freight movements subsequent to the commencement of oil exploration. Labour costs, however, are not recognized as a major cost item in either oil exploration or development, and indeed, the *direct* effect of exploration on employment of the local people has not been great. In short, the present reliance on wage labour in Aklavik, Inuvik and Tuktoyaktuk rests on an insecure and ephemeral base. In the meantime, the "population explosion" prevails, as does what is described above as a "métis society," with its related problems.

One hope for the future is that oil discoveries and methods of transporting it to markets will be such as to absorb as many local people — trained for that purpose — as is possible.

Changing Patterns of Indian Trapping in the Canadian Subarctic

The importance of trapping to the Indians of the Canadian Arctic and Subarctic is a matter of historical fact, and the changes brought about in the traditional Indian way of life by the introduction and development of a trapping-trading economy have been well documented for many tribes throughout Canada by historians and anthropologists. The author's field work at Snowdrift, a Chipewyan community at the eastern end of Great Slave Lake in the Northwest Territories, provided information about recent acculturative factors affecting the trapping pattern. The purpose of this paper is to show that these factors are not peculiar to Snowdrift alone, but are widespread and appear to be altering the significance of trapping in the present-day economy of peoples throughout the eastern and western Subarctic.

SNOWDRIFT

The community of Snowdrift is located on the southeastern shore of Great Slave Lake in a region that is entirely within the area of Precambrian rocks (Figure 1). The eastern end of the lake has an extremely intricate shoreline with large numbers of bays and innumerable islands. The country around the village is characterized by wooded, rolling hills from 500 to 1,000 feet above sea-level; many lakes of various sizes dot the area; the vegetation and fauna are essentially subarctic in character.

Snowdrift, with a population of approximately 150 persons in 1961, is a very recent village whose physical existence in its present form goes back no more than 10 years. However, the area has been a focal point for residents of the surrounding region since 1925, when the Hudson's Bay Company established a post at the site of the present community. Prior to that time the population at the eastern end of Great Slave Lake consisted of an unknown number of Chipewyan families who hunted, fished, and trapped throughout the area and moved about the country as single families or groups of families. Most of these families traded at Fort Resolution, a long-established post on the southwestern shore of the lake, and considered that community to be their trading centre. When the Snowdrift post was established many of these families, together with some who had traded at posts to the south and southwest on Lake Athabasca, shifted their centre of activity to the new post. Factors responsible for the recent concentration of a permanent population at Snowdrift are not, however, specifically connected with the fur trade. They include: (1) the increase in government services that have reduced reliance on income derived from trapping; (2) the recent establishment of a federal school in the village; (3) improved housing; and, (4) wage-employment.

The yearly round of subsistence activities at Snowdrift includes fall fishing near the village, winter and spring trapping, a certain amount of hunting throughout the year for moose and caribou, and some wage-employment during the summer. Winter and spring trapping keeps men away from the village for varying periods of time, usually not more than 2 weeks at a time, and the late summer and early fall run of caribou usually takes many families to the extreme northeastern end of the lake for a period of about

● JAMES W. VANSTONE *is Curator, Department of Anthropology, Field Museum of Natural History, Chicago, Illinois.*

Reprinted from *Arctic*, Vol. 16(3), 1963, pp. 158–174, with permission of the author and editor.

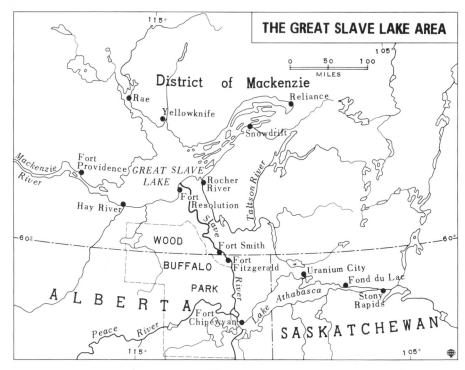

Figure 1

2 weeks to 1 month during late August and early September. Opportunities for wage-employment in the form of commercial fishing and tourist guiding are of growing importance, and there is also a government sponsored road-way clearing project that takes many young men away from the village during January or February.

Trapping begins officially on the first of November, when the season for most fur bearing animals opens. The area that has been used by Snowdrift trappers is bounded on the north by Walmsley and Clinton-Colden lakes, on the east by Whitefish Lake, on the south by the southern Taltson River-Nonacho Lake area and on the west by the Doubling-Meander Lake region (Figure 2). This is a large area, but the majority of trappers have not trapped to the peripheries of the region in recent years. In 1961, for example, only one man trapped north of Narrow Lake in the Doubling Lake-Meander Lake area and only occasionally do trappers extend their activities as far north as Clinton-Colden Lake or as far east

as Whitefish Lake. In fact, the number of trappers who operate beyond the tree line is very small indeed. During the winter of 1959-60, when fox prices were relatively high and the animals few in the wooded country, only three men trapped in the Barren Grounds. Most men do a large part of their trapping within a radius of from 65 to 80 miles from the village and the Taltson River-Nonacho Lake area is the centre of concentration.

The length of time that the trappers stay away from the village varies considerably. A very few men go out in the late fall, return for Christmas and stay several weeks, then go out again and do not return until late in the spring. However, a much larger number return frequently to trade furs and obtain supplies. Most informants maintained that they seldom stayed out more than 2 weeks before returning to trade. Thus a great deal of time is spent in travelling to and from the trap lines. A man seldom comes to the village without staying at least a week.

108

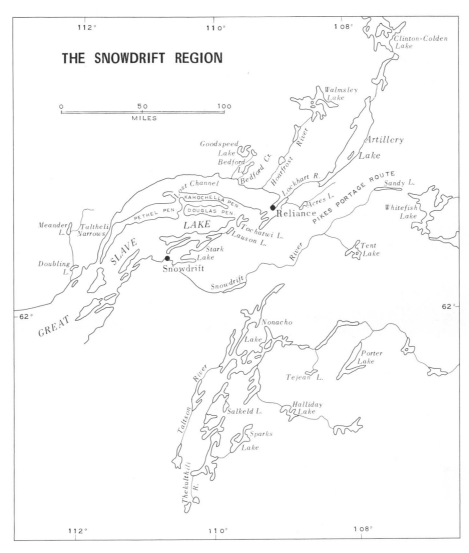

Figure 2

Trappers always have to do a certain amount of hunting because the supplies they take with them from the village usually consist of little more than staples, regardless of the amount of credit they have received from the Hudson's Bay Company manager. The most important hunting during the winter and spring is for caribou. Some men feel very strongly that their dogs should be fed on caribou meat during the winter, especially when they are being used on the trap line; thus, there are trappers who will not set nets while they are on the line, but will spend time hunting instead of trapping. When they are unsuccessful, they must feed their dogs flour or cereal from their own food supply, and therefore, run out of these staples rapidly and must return to the village. It seems certain that for many trappers, looking for caribou is the most important thing they do on the trap line and always takes precedence over trapping. Some men are even unwilling to leave the village if no caribou have been reported in the area where they trap.

In this connection, it is worth comment-

109

ing on a statement made about 30 years ago by a manager of the Hudson's Bay Company store at Snowdrift. He believed that the success of the Snowdrift Indians as trappers was directly related to the presence or absence of caribou in the trapping areas. At the time this statement was written into the records, fur prices, particularly for white fox, were very high, but unless caribou were reported in the Barren Grounds, the Indians would not go there to trap foxes. Thus, trapping seems, to some extent, to have been incidental to hunting even as late as 1925 or 1930. The same is true today, even though caribou meat is no longer quite as essential a food item as it was 30 years ago.

The existence of clearly defined family trapping territories among the Chipewyan has been the subject of some dispute. At least one early observer believed that these territories did exist,[1] while another has denied their existence.[2] It must be admitted that the evidence collected at Snowdrift does not support the theory of family trapping territories. Most informants stated that they had learned to trap in one of the areas trapped by their fathers and afterwards often continued trapping in that area. However, few trappers could be found who had trapped more than 5 years in any single area. Two or three years was the more usual length of time, depending on how successful the trapping was in a particular locality. Some trappers had trapped at one time or another throughout most of the area within a 100-mile radius from the village. During the winter of 1961, one trapper relocated his trap line in the middle of the season because he felt that the snow was too deep in the region where he had originally set his traps. Consistent opposition to trap line registration on the part of the villagers is a further indication that they desire freedom to determine for themselves where the trapping is best.

There can be no doubt that the total area trapped by Snowdrift residents has been shrinking steadily in recent years, particularly since the Indians began to live permanently around the trading post. The factors primarily responsible for this situation have already been mentioned. However, it is worth discussing them in more detail, particularly as they relate to the changing trapping pattern.

The payment of family allowance on a monthly basis provides an additional form of income that can be obtained at regular intervals. This has meant that trappers, even when they normally trap without their families, are reluctant to be away from the post when there is a cheque to be cashed. Other government services such as rations and old age pensions have a similar affect and the income from these sources reduces the reliance on income derived from trapping.

The recent establishment of a federal school in the village must be regarded as a stabilizing feature whose effects will mainly be felt in the future. However, even in the winter of 1961, when the school was only in its second year of operation, many families were beginning to realize and accept the fact that their mobility during the trapping season would be reduced. Several families who had previously spent most of the winter in the bush on their trap lines have been forced to change their pattern of operation. Now only the heads of these families go out on the line and the tendency is for them to return to the village at frequent intervals. It should be emphasized that it is the most vigorous, and therefore usually the most successful, trappers who take their families with them on the trap line; it is the activities of these men, usually middle-aged with large families, that have been most affected by the construction of the school.

Improved housing should probably be considered as being closely associated with population stability since the people appear to be interested in improving houses that they are going to live in all or most of the year. The Canadian government has encouraged this by instituting home improvement and home building programs. Thus, it is possible to note a greater contrast between the difficult, uncomfortable life on the trap line and the comfortable, gregarious life in the village. It is probably also safe to say that the discomforts of a bush trapper's camp are more difficult to tolerate when the trapper is alone or with another man rather than with his family. It is little wonder, then, that the trappers tend to become easily discouraged and desire to return frequently to the comforts of their homes in the village.

Wage-employment is, as yet, not of much importance to the people of Snowdrift during the winter months. However, in recent years the government has introduced a roadway-clearing project at the northwestern end of Great Slave Lake. This project, which involves Indians from every community in the Great Slave Lake area, takes many men away from Snowdrift during January or February. Government policies for the Indians in the Great Slave Lake area include developing an alternative means of livelihood to trapping and the roadway-clearing project is thought of as a step in that direction. So far, the Indians of Snowdrift have only participated in the project after Christmas when the main trapping period has ended. Thus, its direct effect on trapping has not been great. An indirect effect, however, must be taken into consideration. During December, 1961, the author noted that many trappers were not particularly worried about their lack of success and adopted a definitely desultory attitude toward the whole trapping procedure. The feeling was that, since a comparatively large amount of money was to be earned clearing roadways in January, it was not necessary to worry about the number of mink or marten trapped in November and December. Thus, we must think of this government project, as well as others more extensive that are likely to follow in the years to come, as being another factor that greatly influences the Snowdrift trapping pattern.

If one is prepared to say that the total area trapped by Snowdrift men is shrinking, then it is equally certain that the region around the village is not being trapped as efficiently or effectively as it might be. At first glance there appears to be a contradiction here. If the Indians no longer trap to the peripheries of their areas because of their desire to visit the village frequently, why is it that they do not trap more intensively than formerly in the area close to the village? The answer to that is that the factors that are responsible for the shrinking of the total Snowdrift trapping area also operate to reduce trapping effectiveness and, in short, to reduce the number of hours spent by the men on their trap lines.

Keeping these factors in mind, it is not difficult to understand why trapping has become increasingly unpopular with the Snowdrift Indians. Nevertheless, one would suspect that enthusiasm for the activity might have remained relatively high if definite and predictable benefits continued to be derived from it. This, however, has not been so. Not only is life on the trap line very hard and carried out under extreme climatic conditions, but the rewards are unpredictable. It is this last point that is of particular importance. As Honigmann has pointed out for the Attawapiskat Cree, "the enthusiasm which is an important motivating factor in the Indian hunter and trapper is keenest when his efforts promise to bring him not a few but many animals."[3] This is also true for the Indians of Snowdrift and this enthusiasm is also observable when the prices paid for furs are high. Honigmann further points out that in some societies relative scarcity of resources means that people will work harder. At Attawapiskat, and also at Snowdrift, the reverse is true. It is the author's opinion that at Snowdrift it is the extreme fluctuation of fur prices as much as anything else that has had a discouraging effect upon the Indians' interest in trapping as a means of making a living.[4] This fact, along with others previously mentioned, has simply meant that the men, with few exceptions, are spending less time on their trap lines regardless of how far these lines are from the village. During the main trapping period, which extends from about the first of November until Christmas, most of the village trappers are on their lines for no longer than 4 weeks. In recent years there has been only a moderate amount of trapping after Christmas. The value of mink

111

and marten pelts drops rapidly after the first of the year and only white-fox pelts continue to be in prime condition. Thus, largely because of the unwillingness of most Snowdrift trappers to go into the Barren Grounds to trap white foxes, there is relatively little trapping activity until muskrat and beaver trapping begins in the spring.

CHANGING PATTERNS IN THE SUBARCTIC

The material on contemporary trapping presented in the preceding pages leaves the distinct impression that this activity is a changing aspect of the Indian economy and that, particularly in recent years, this change has been rapid, extensive, and influenced by factors that are characteristic of Indian acculturation throughout much of Canada, particularly among subarctic peoples in the east and west. Therefore it seems probable that the changing pattern of trapping at Snowdrift can be duplicated in other parts of subarctic Canada. Whether and to what extent this actually is so will now be examined.

Although neglected for many years by anthropologists, both the eastern and western Subarctic of Canada have recently received more attention, particularly with regard to problems of social and cultural change. Much of this work is still unpublished, and unfortunately, even the available reports do not deal with trapping in such a way as to make them comparable with the information collected at Snowdrift. This is, of course, not the fault of the investigators but stems from the particular emphasis of the various research designs. Nevertheless, enough information does exist so that some statements can be made about similarities in trapping patterns, particularly with regard to the problem of area utilization as it was described for the Snowdrift Chipewyan.

Turning first to other studies in the western Subarctic, it is found that Helm,[5] in describing trapping activities of the people at "Lynx Point," a Slave Indian community on the Mackenzie River near Fort Simpson, notes that the time spent on the trap lines varies considerably. Most of the men have short lines and seldom spend more than 1 or 2 nights on the trail. Those with longer lines may set up tents and spend as much as 2 weeks in the bush. Only one old man in the community has a winter trapping camp in the bush and stays away from the community as long as 2 months at a time. Helm mentions that in spite of the definite economic advantages of such an arrangement, most "Lynx Point" trappers would prefer more frequent visits to the community because they do not like the social isolation of trapping-camp life. They prefer the comforts of their own homes where children can be more easily tended, and they like to be near the store. Although not specifically stated by Helm, one gets the definite impression that there are as many factors that draw the trappers of "Lynx Point" closer to the community as in Snowdrift, and that it is only the older trappers who think in terms of long periods of intensive trapping with infrequent visits to the post.

A similar situation can be said to exist among the Dogrib Indians of Lac la Martre. Here, according to Helm and Lurie,[6] a trap line is set only an overnight distance from the village and even an extended trapping tour undertaken by several men together seldom lasts more than 2 weeks. The authors also mention that an important factor affecting the fur take at Lac la Martre is the increasing access to other sources of income, "both in terms of providing less arduous income activities and of cutting down on time-energy remaining for trapping."

In the eastern Subarctic the picture is somewhat less clear and is complicated by the Indians' trapping in many areas in government apportioned and registered trapping territories. This means, for one thing, that a trapper may construct a log cabin at a convenient point on his trap line and generally will not hesitate to improve his line, both by using it better and by providing greater comfort. A Snowdrift trapper, on the other hand, seldom traps more than 4 or 5 years in the same area and for that reason hesitates to make elaborate preparations or improvements on any one trap line. Thus, the trapper in the eastern Subarctic, if he is working a registered trap line, presumably finds that there is less contrast between life in the bush and in the village, at least as far as physical comfort is concerned. His cabin and other improvements are also likely to commit him as far as intensity of trapping and time away from the village is concerned. Therefore, he would be less likely to abandon all or part of a line because of distance from the community or fluctuations in fur prices. Thus the trapper in the eastern Subarctic would seem to be less susceptible to some of the influences that are affecting trapping areas in parts of the west.

Although the situation appears, then, to be more complex in the east than in the west, it is nevertheless possible to isolate the factors under discussion. At Northwest River in the Melville Lake region of Labrador, for example, there was, during the early 1950's, a growing tendency for the Montagnais Indians to spend more time near the trading post. McGee[7] indicates that with greater dependence on the Hudson's Bay Company store, the summer camp has increasingly become a base of operations from which the Indians move out and to which they return frequently. Wives and children are usually left at the camp, which discourages the trappers from staying away for long periods of time. In emphasizing the importance of pensions and relief payments as part of the total picture of changing trapping patterns in the Melville Lake region, McGee further points out that more Indians now have a basis for credit at the Hudson's Bay Company store that is not connected with the trading of fur. It is possible for a family to subsist more or less entirely on relief allotments and other forms of unearned income.[8] "Because of the availability of welfare money, and because of the resident missionary with a church and school, the number of individuals who do not go to trapping grounds has grown rapidly to encompass at least 25 percent of the (Northwest River) band."[9] It should be emphasized that those families living permanently in the village have given up trapping entirely and are living on welfare payments and some

wage-employment during the summer. Trapping areas close to the Northwest River have been taken over by whites so that it is impossible for an Indian to operate a trap line from the village. Since these trends described by McGee were already well established at the time of his field work in 1951 and 1952, it is probable that they have continued and become more significant during the past 10 years.

Moving west into northern Ontario, we find that at Attawapiskat many inland trappers leave the village shortly after the beginning of the new year and do not return until Easter. However, some remain in the community until the end of February. A large number of trap lines are located relatively close to the coast and the trappers who work these lines leave their families in the village, and consequently, return frequently during the trapping season. Men who trap beaver on Akimiski Island also leave their families at the post and make a number of trips to the island by sled. These men remain away from the village for about 10 days at a time and then return, each visit to the post lasting about 1 week.[10] Thus, one receives the distinct impression of diminishing use of trapping areas and the gradual development of a more sedentary community life. Honigmann's field work was done in 1947 and 1948 and it should be emphasized that although trends that point toward changes similar to those documented for the west are observable, traditional trapping patterns were still of considerable importance. Thus, we learn that in the two seasons, 1944-45 and 1945-46, 43 percent of the listed Attawapiskat trappers operated within 60 miles of the post and that these men earned 36 percent of the trapping income for these years. On the other hand, 57 percent of the trappers operated at a distance greater than 60 miles from the village and earned 64 percent of the income. It is obvious from these figures that greater proportionate earnings are achieved by those men who travel farther from the post into the area where fur bearing animals are more plentiful.[11] It is possible too, in the light of previous statements, that these trappers work harder because of the greater potential rewards awaiting their efforts. At any rate, it will be seen that there were still many factors that in 1948 encouraged Attawapiskat trappers to maintain a trapping routine involving long absences from the community.

In describing field work among the northern Ojibwa at Pekangekum in northwestern Ontario, Dunning[12] has little detailed information to offer concerning changes in trapping over time, but he does document some of those factors of economic and social change that have been seen to influence trapping at Snowdrift. Thus, we learn that government buildings, a school, a nursing station, and houses have been constructed on the reserve within the past 10 years. This, together with a steady increase in government subsidy of the economy until it constituted more than 41 percent of the total income in 1955, has resulted in a change in the pattern of residence in summer "from domestic units spread widely over the trapping territories to a cluster of population at each of two centres."[13] The author points out that this concentration of population "could not have occurred under the condition of the traditional hunting and trapping economy,"[14] and we can assume that such a fundamental demographic change has had a definite effect on the pattern of trapping, very likely in the direction that has been documented here for other communities.

At Winisk, less than 200 miles northwest of Attawapiskat on Hudson Bay, Liebow and Trudeau[15] say that the construction of a radar base between 1955 and 1957 offered the Cree Indians of the area alternative ways of making a living. Prior to the construction of the base, each family group would leave for its own trap line area in late September, return for a week at Christmas or Easter, or both, to sell furs and obtain food supplies, and then go back to the trap line until late May or June. Thus there would be only a little more than 2 months during the summer when all families were in the village. The authors also say that some families had only a few miles to travel to their trap lines and it seems safe to assume that they visited the post more frequently.[16] With the construction of the radar base, the pattern outlined above changed almost immediately to nearly complete dependence on wage labour and year-round residence in what became a community.[17]

It certainly would be difficult to find a more dramatic example of the preference that subarctic Indians have for wage labour over trapping as a means of making a living. The reasons for this preference, documented in some detail for the Winisk Cree by Liebow and Trudeau and for the Snowdrift Chipewyan by the present author,[18] are of no concern in this context. What is of particular interest is the great change in trapping patterns, actually the almost total disappearance of trapping, in response to the sudden introduction of an alternative means of making a living. What is seen here is the rapid culmination of a process that is going on also in other parts of the eastern and western Subarctic.

In a paper discussing changing settlement patterns among the Cree-Ojibwa of northern Ontario, Rogers[19] gives the clearest statement of the series of trends that have already been noted for Snowdrift and other areas. He finds that there has been an increasing tendency for larger groupings of people to come together and remain more sedentary for a longer period of the year than formerly. This change to larger settlements and a more sedentary existence can be mainly attributed to the impact of Euro-Canadian culture in much the same way as can be documented for other parts of northern Canada. With regard to trapping, Rogers documents certain changes in the yearly cycle at Round Lake, Ontario, as follows:

> With the advent of fall the men make ready for the coming season of trapping. When the time arrives, primarily during October, nearly half the population of the village departs for their winter camps, located as a rule within the boundaries of their trapping territories where they will reside until just before Christmas. During this period some of the men will occasionally return to the village to sell their furs and obtain more supplies. Some trappers take their families with them, others leave them in the village. Those men who exploit the area in the immediate vicinity of the settlement do not

establish winter camps in the bush but rather operate directly from their homes in the village.[20]

This pattern continues with the return of those trappers, who have been in winter camps, to the village for Christmas. After Christmas, trapping is again resumed, but some men do not return to their bush camps; rather, they "borrow" a territory, or, more likely, secure permission to trap in a territory near the village.

It appears, then, that at least half the Round Lake trappers trap near the village. Some have done so since the post was first established, but for others this form of trapping represents a break with the old tradition of winter trapping camps. In his Round Lake study, Rogers[21] states that in spite of the growing tendency for trappers to exploit areas close to the village, there has been no noticeable lowering of the yield per trapper. It is probable, however, that this will occur in the future. What Rogers does not state, and what would be of particular interest in this connection, is just how frequently those trappers who do stay in winter camps before Christmas return to the village to sell furs and obtain supplies. Another point of interest would be whether the number of trappers who return to the winter camps after the holiday festivities is declining or remaining approximately the same. One would suspect that the number of men staying in the village after Christmas is growing.

At the risk of reading more into Rogers' information than it actually contains, it could be said that among the Round Lake Ojibwa the trapping pattern is undergoing changes similar to those noted for Snowdrift and other communities in the eastern and western Subarctic. Namely, there is a general decline in the efficiency with which the total available trapping area is used. Less time is being spent on the trap lines and more either in the community or going to and from the community. These are general statements that seem to apply, to a greater or lesser degree, wherever furbearing animals are being trapped today in subarctic Canada.

It would, of course, be naive and over simplistic to suggest that the factors responsible for this situation are everywhere exactly the same or are taking place at the same rate. However, it is probable that lack of uniform documentation over the entire area prevents, to some extent, the easy recognition of those similarities and parallel developments that do exist. One factor that complicates this situation and about which little has been said here is time. Even such information as has been presented indicates that changes in trapping patterns have not taken place at an equal rate. Thus, it is the opinion of one experienced field worker that in many areas, particularly in some parts of the eastern Subarctic, the Indians trap more today than they did a hundred years ago.[22] This may be true in the west as well since the proliferation of trading posts and the availability of consumer goods on a large scale, both of which have encouraged trapping, have occurred during the past 30 or 40 years. Similarly, although it has been possible to make general statements about the decline in trapping effectiveness and land use, which apply over the entire area of subarctic Canada, it is undoubtedly true that statements of this kind must be tempered by the recognition that the concentration of Indian populations in permanent communities has sometimes resulted in greater use of the trapping area near the community as the trappers have increasingly withdrawn from the peripheries. This is not true at Snowdrift, but it is at Round Lake, and there are indications that it may also be true of other communities in the eastern Subarctic.

All this appears to suggest that in spite of conditions that are working to the detriment of continued interest in trapping, there are still a number of important factors that tend to insure that it will be a long time before this means of livelihood entirely disappears. Undoubtedly the most important of these, at least in the western Subarctic, is that trapping remains the only source of income besides uncertain wage-labour, relief, and welfare. It is unlikely that very many Indian groups will desire, or be permitted, to live entirely on unearned income, and it has been only under unusual circumstances, such as at Winisk, that wage-employment opportunities in any part of the Subarctic have existed in sufficient abundance and permanence to enable the Indians to make a complete change in their means of subsistence. Even if job opportunities were to increase considerably, it is by no means certain that trapping would rapidly disappear. There exist, in many areas, factors that encourage a continued interest in trapping. In the east, the logging and mining industries, as well as tourist guiding, have tended to keep the Indians in the bush and helped to maintain their interest in and close association with the environment they know thoroughly. Commercial fishing and the growing importance of tourist guiding has had, and is likely to continue to have, a similar effect in the west. It must also be remembered that trapping has been important in the eastern Subarctic for more than 300 years and in the west for nearly as long. Time is an important factor in consolidating adaptive forms of culture and it required perhaps a dozen or more generations for patterns of socialization, family holdings, and marriage preferences based on trapping to be built up and strongly consolidated in both the east and the west. Indeed, it is the strength of these developed patterns that is likely to determine the reaction of any one group of Indians to the introduction of the various factors mentioned above that have the potential to inhibit trapping.

In this connection one point in particular must be examined; that is the presence or absence among subarctic Indians of normative pressures to hunt and trap. Although information on this point is not as complete as it might be, the author feels that this is not a significant factor among Indians in the western Subarctic. At Snowdrift, for example, the men seemed to take very little pride in their trapping skill, and the author never heard comments about the pleasures or compensations of bush living. In the Mackenzie-Great Slave Lake area the field worker will hear a great deal about how the Indian is most happy and content when he is in the bush on his trap line, but these comments always seem to originate with local whites and never with Indians. At Snowdrift all informants maintained that they would give up trapping at once if an opportunity for steady wage-employment presented itself.[23] It has been previously noted that the "Lynx Point" trappers prefer the comforts and conveniences of the village to the social isolation of the trapping camp.

The extent to which this is also true in the east can only be surmised since, with one exception, the authors cited in this paper have not discussed the matter. However, it seems clear that there were definitely more strongly developed normative pressures to hunt and trap than have been documented for the west. For example, in speaking of the more or less permanent Indian population living in the vicinity of Seven Islands on the coast of the St. Lawrence, Speck notes that for the period of his field work, between 1915 and 1925, these people could claim no prestige through their close contact with whites and assimilation of white values. Both social and financial prestige lay with the interior hunters and trappers. The coast people are spoken of as being glad to give up their precarious employment and restriction of freedom, should the opportunity arise, for "the adventure and possible greater profit of furs of the big woods."[24] With reference to general aspects of Montagnais-Naskapi economy, Lips[25] makes a similar statement. The Naskapi have always been hunters and trappers and they wish always to remain so. McGee, writing about the people in the Northwest River region of Labrador is more explicit concerning the compensation of bush living as opposed to the more comfortable life in the village, and his statements have added significance because he is dealing with the more or less contemporary scene. First of all, trappers are close to good caribou territory and are thus in a position to satisfy their need for meat. Second, old people who might have difficulty getting on alone at Northwest River have relatives who will keep them supplied with meat and firewood on the trapping grounds; in other words, it is not so easy to ignore old people in a trapper's camp. Finally, and perhaps most important of all, in the bush, for young and old alike, there are no non-Indian sanctions on conduct to worry about.[26]

It cannot be denied, therefore, that factors exist throughout the Subarctic, although perhaps most notably in the east, that work to maintain interest in and dependence on trapping as a means of livelihood. These are essentially conservative factors and they arise from a tendency to associate trapping with the old Indian way of life that is right and good and as a reaction against the insecurities and uncertainties of the newly developing community life and increased contact with whites. It seems, nevertheless, to be true that in most if not all parts of the Subarctic, factors favouring the persistence of trapping as a major economic undertaking are in rapid decline. That is why the trends and changes described in this paper have, for the most part, become significant within the past 15 years, or even much more recently than that.

By way of summary and conclusion, then, it can be said that in most subarctic communities there are, at the present time, factors operative which tend to reduce the reliance on income derived from trapping. These factors, which for the most part involve additional sources of income, have also been effective in developing tendencies toward sedentary community life. The Indians are turning increasingly toward wage employment and dependence on various forms of government assistance as they attempt to achieve a higher standard of living. It is not surprising, therefore, that both Indians and white administrators are agreed that one of the most important problems facing the peoples of subarctic communities today is the need to achieve financial stability and to be free from the uncertainties that are characteristic of an economy based on trapping.

REFERENCES

[1] Seton, E. T., *The Arctic Prairies: A Canoe Journey of 200 Miles in Search of the Caribou.* London: Constable, 1920 (new rev. ed.), pp. 150–151.

[2] Penard, J. M., "Land Ownership and Chieftaincy Among the Chipewyan and Caribou-Eaters," *Primitive Man,* Vol. 2, 1929, p. 21.

[3] Honigmann, J. J., *Folkways in a Muskeg Community. An Anthropological Report on the Attawapiskat Indians.* Ottawa, Dept. of Northern Affairs and National Resources, NCRC–62-1, 1962, pp. 89–90.

[4] VanStone, J. W., *The Economy of a Frontier Community, A Preliminary Statement.* Ottawa, Dept. of Northern Affairs and National Resources, NCRC–61-4, 1961, pp. 17–18.

[5] Helm, J., *The Lynx Point People: The Dynamics of a Northern Athapaskan Band.* Ottawa, National Museum of Canada, Bulletin 176, 1961, p. 24.

[6] Helm, J., and N. O. Lurie, *The Subsistence Economy of the Dogrib Indians of Lac la Martre in the Mackenzie District of the N.W.T.* Ottawa, Dept. of Northern Affairs and National Resources, NCRC–61-3, 1961, pp. 40–41.

[7] McGee, J. T., *Cultural Stability and Change Among the Montagnais ·Indians of the Lake Melville region of Labrador.* Catholic University of America, Anthropology Series No. 19, 1961, p. 56.

[8] *Ibid.*, pp. 56–57.

[9] *Ibid.*, p. 57.

[10] Honigmann, J. J., p. 89.

[11] *Ibid.*, p. 121.

[12] Dunning, R. W., "Some Implications of Economic Change in Northern Ojibwa Social Structure," *Can. Jour. of Econ. and Pol. Sci.,* Vol. 24, 1958, pp. 562-566.

[13] *Ibid.*, p. 565.

[14] *Ibid.*

[15] Liebow, E., and J. Trudeau, "A Preliminary Study of Acculturation Among the Cree Indians of Winisk, Ontario," *Arctic,* Vol. 15, 1962, pp. 190-204.

[16] *Ibid.*, 193-94.

[17] *Ibid.* p. 195.

[18] VanStone, J. W., *The Changing Culture of the Snowdrift Chipewyan.* Ottawa, National Museum of Canada, Bulletin 209, 1965.

[19] Rogers, E. S., "Changing Residence Patterns Among the Cree–Ojibwa of Northern Ontario," *Southwestern Journal of Anthropology*, Vol. 19, 1963, pp. 64–88.

[20] *Ibid.*, pp. 75–76.

[21] Rogers, E. S., *The Round Lake Ojibwa.* Royal Ontario Museum, Art and Archeology Division, Occasional Paper No. 5, 1962, p. C36.

[22] Personal communication from E. S. Rogers.

[23] VanStone, J. W., *The Economy of a Frontier Community, A Preliminary Statement*, p. 19. Ottawa, Dept. of Northern Affairs and National Resources, NC RC-61-4, 1961, 34 pp.

[24] Speck, F. G., and L. C. Eiseley, "Montagnais–Naskapi Bands and Family Hunting Districts of the Central and Southeastern Labrador Peninsula," *Proc. of the Am. Phil. Soc.*, Vol. 85, 1942, p. 222.

[25] Lips, J. E., "Naskapi Law (Lake St. John and Lake Mistassini Bands). Law and Order in a Hunting Society," *Trans., Am. Phil. Soc.*, Vol. 37, 1947, p. 387.

[26] McGee, J. T., p. 58.

America's Eskimos: Can They Survive?

DIAMOND JENNESS

Quite recently, we have discovered that in our American Arctic, near the Alaska-Canada boundary, Eskimos, or a people resembling Eskimos, were resisting the unbroken midsummer daylight and the unbroken midwinter night, snow and ice and temperatures of 60° below zero Fahrenheit, centuries before the Romans founded their city on the Tiber. How long before or how many centuries even then these Eskimos, or proto-Eskimos, had been living in our Arctic we do not know; nor do we know from where they came or for what reason they remained there: the only race, white, yellow or black, that has ever wrestled unaided with the polar ice and won the struggle. No other people has settled in that region more than a few days or weeks without constant assistance and support from the warmer and richer world to the south.

Why was it that only the Eskimos succeeded in overcoming the Arctic environment? What advantages did they possess over other races? What qualities, physical or mental, that other people lacked?

With some qualities, we know, they were not endowed, although other living things that dwell in the Arctic were fitted for survival. The Eskimos were not shielded from the bitter cold as were the seals and the polar bears with an abnormally thick layer of fat under their skins. The Eskimos were not protected with a dense warm coat of hair or fur such as is wrapped around the caribou and the fox. The Eskimo does, or did, wear a coat of fur, but it was an *ersatz* one that he stole from animals such as the seal and the bear; the Eskimo himself is more hairless than a white man and seldom

● *The late* DIAMOND JENNESS, *prior to his retirement, was Chief of the Division of Anthropology, National Museum of Canada, Ottawa.*

This paper, which has been slightly abridged, was delivered at the 1962–63 Waterloo Lutheran University Lecture Series "Canada and the Emerging Peoples and Nations," and is reprinted with the permission of the author and Waterloo Lutheran University.

grows even a thin beard to protect — or endanger—his chin.

Nevertheless, the Eskimos were—and are —a little different from other people, even though they have often been mistaken for Chinese. The long arms of a Negro reach to the knees and occasionally lower; a white man's arms rarely touch his knees; and an Eskimo's extend only about halfway down his thigh. His hands and feet, too, are smaller than ours, as you would quickly discover if you tried to put on his gloves or sealskin slippers. Is the reason that, in a very cold climate, it is easier for the heart to pump warm blood into short limbs than into long ones? Then, again, the nose cavities through which an Eskimo breathes are smaller than the nose cavities in other races. Is that to protect his lungs from freezing when the wind is howling and the thermometer register 50° below zero? . . .

I am unable to answer these questions. However, I do know, that, although the Eskimo's hands and feet will freeze almost as quickly as my own, he is a tougher individual than I am and will struggle on without a groan. I know, too, that he has an unusually sensitive funny-bone and will plod doggedly along a seemingly endless trail until he drops from sheer exhaustion, provided I keep tickling that funny-bone with remarks that he finds amusing. He is a loyal and cheerful companion on a hard journey.

He has needed more than toughness, however, and a lusty sense of humour, to win a livelihood and to raise a family in the Arctic, generation after generation, without any help from the outside world. It has required great adaptability and inventiveness to cope with the strange and exceedingly difficult environment, where no trees grow because the ground is forever frozen below the top few inches, where berries seldom ripen on the rare bushes because the summer is too short, where all birds except three or four species —one of them the useless raven—flee the region for eight months of every year, and where fewer animals roam the land than hide in the sea. . . .

In many, many ways the Eskimos have revealed unusual adaptability and ingenuity.

It was from them that we Europeans learned to speed over the frozen sea with dog-sleds, and to capture great whales and walruses with hand harpoons. The light fold-boat in which some holiday-seekers explore the inland waters of Europe, and even a few of our North American rivers, is merely an imperfect copy of the Eskimos' skin-covered *kayak* from which he harpooned seals in the ocean and speared caribou as they swam the rivers and the lakes during their spring and autumn migrations.

For Arctic living, nevertheless, the Eskimos needed more than short limbs and narrow noses, and more than adaptability and resourcefulness. Their stern environment where, throughout nearly half the year, cold and darkness confined them for sixteen to eighteen hours each day to one-roomed dwellings into which every neighbour entered at will and into which every stranger or chance visitor squeezed for shelter and rest—an environment whose innumerable dangers, foreseen and unforeseen, limited the average span of life to less than twenty-five years, forcing the Eskimos to cling to one another for safety, to share every hardship of fishing and the chase, and to hold all food in common — that environment also imposed on them a deep social consciousness: strong social bonds that elsewhere man has generally worn more loosely, and only the ants, the bees, cattle, and several other forms of life that share this earth with us, wear inescapably tight. The Eskimos had to abrogate all privacy, to submerge—yet at the same time to preserve—their separate individualities, and they had to acquire an endurance of one another's company, a tolerance of one another's idiosyncrasies, far beyond the capacity of most Europeans. . . .

I have said that the Eskimos possessed (and, I believe, still possess) special qualities, some hereditary, others the product of their environment, that enabled them to adapt to the rigours of Arctic existence and to maintain themselves in that forbidding region for many thousands of years. Let us recall how, not so long ago, Sir John Franklin and more than a hundred hardy English seamen perished from starvation and exposure in the area around the Magnetic Pole, a place where Eskimos have been living for centuries and where they still struggle to preserve a foothold so that they may pass on the torch of life to their children. Then let us ask ourselves, "Should we permit so remarkable a people to fade away and disappear from the face of the earth because their outward appearance is a little different from ours and because many of them have not yet learned to speak our language or to think and act exactly as we do?"

"Everything is in motion," said the ancient Greek philosopher; and the Eskimos, together with the environment in which they live, have changed and are changing. We Europeans have brought them to a crossroads in life's journey, and their path ahead is obscure. Their heredity is changing. The blood of white men now flows in the veins of most of them, modifying their genes but not, we hope, destroying the qualities that make for survival and progress. Missionaries, traders, and government officials have submerged their culture, and many of their ancient patterns of thinking and living have vanished, or are vanishing now. Even their physical environment has changed since the coming of the white man, so that it is no longer possible for them to live as their ancestors did. The wheel of their operations has been broken, a modern Greek would say, and they must either rebuild that wheel or perish. Unless they can adapt themselves to radically new conditions, they will not survive.

What are these new conditions, this new environment that has enveloped them so suddenly? Except in southern Greenland, the climate of the Arctic has not sensibly changed. Winters are not a day shorter than they were in pre-European times, the summers no longer. Snow and ice still cover the whole landscape from September to the end of May, and the earth and sea have not ceased to support the same animals, the same birds, and the same fish that they nurtured two thousand years ago. In what ways, then, has the Eskimo's environment altered?

Game is still there, but the firearms we whites introduced have decimated their numbers, and only in two or three places can Eskimo hunters now obtain sufficient meat and fish to nourish their families throughout the twelve months of the year, or even enough furs to protect their bodies from the intense cold. The walrus has disappeared from many districts, the erratic movements of the seals have heightened the difficulties of capturing them, the musk-oxen are threatened with extinction, and the caribou, estimated late in the nineteenth century to total two and a half million, are now reduced to one-tenth of that number. A century ago, when every Eskimo hunter required six caribou hides to clothe himself, an equal number for his wife, and three, four or five hides for each of his children, he endeavoured to kill in each season at least two dozen of the deer; but today many of his grandsons have never seen a live caribou, and they know neither how to stalk one with bows and arrows as their grandparents did, nor even how to make serviceable bows. The grandparents were completely self-supporting; everything they needed—food, clothing, shelter, cooking utensils, weapons and tools—they obtained by their own exertions or fashioned with their own hands, using only the materials that nature placed at their command. The modern Eskimo has reached the stage when he must obtain most of his food and clothing from the outside world, along with the boat in which he hunts and travels, the timber from which he builds his home, the fuel to make that home habitable, and the furniture he requires, or believes that he requires, to give his household minimum comfort.

Naturally, the outside world does not offer him all these commodities as free gifts. In our commercial civilization, imports must be paid for by exports either of money (or services calculated in money) or of goods. No money circulated in the American Arctic until quite recently—in Canada's far north not before World War II—and almost the only goods our merchants would accept from the Eskimos were the pelts of fur-bearing animals—foxes, polar bears, muskrats, seals, and on the lower Yukon River, beaver and mink. By far the most remunerative were the pelts of foxes. Accordingly, late in the nineteenth century, there sprang up a flourishing fur trade; this trade, which began as simple barter, gradually introduced the Eskimos to our money economy and created the economic base they needed to maintain steady commercial relations with the civilized world.

This new base, however, demanded a revolution in the Eskimo's daily life. Just as the invention of the motorcar has changed the habits of every civilized nation, so the necessity of trapping the small fur-bearing animals upset the traditional routine of Eskimo life. Instead of hunting seals on the frozen ocean from November to April of each year, they were forced now to trap foxes, which could not yield blubber to heat and light their homes and were useless for meat except to starving dogs. The food, the clothing, and the other necessities of life, for which trapping allowed them neither time nor opportunity to acquire themselves —even when they retained the skill—they purchased from the traders with the cash or credits received for their furs. There were years in which an Eskimo was paid $40 or $50—at the peak $65—for a single white-fox fur, $200 for the rarer blue fox, which is a colour variety of the white, and for the black fox, which is a colour variety of the red, a slightly different species, $300 and $400. So their life pattern changed.

In spite of this change, however, all appeared to go well for half a century, because in both America and Europe fine furs continued in strong demand and brought high prices, while the costs of food, clothing and other goods remained relatively low. Civilization's diseases — measles, smallpox, influenza and tuberculosis—committed great ravages, but the morale of the Eskimos held firm, and the apparent stability of the fur trade sheltered them from some of their earlier hardships and perils. Throughout this half-century, however, their economic independence was declining, their sturdy self-sufficiency evaporating. More and more tightly they were being chained to the trad-

ing posts, exposed to the mercy of fickle trade currents that they could neither comprehend nor escape. They could not turn the clock back; they could not revert to the life of their ancestors and recover their lost freedom, any more than we can revive the age of the oxcart and the spinning wheel. The trade current carried them relentlessly along, although for those fifty years its small rapids seemed to threaten no danger.

Then, in the early 1930's, the great depression toppled the world of commerce, and the fragile bark to which the Eskimos had committed their fortunes foundered. The prices that civilization now offered them for their furs slumped heavily, while the costs of flour, tea, bacon, clothing, ammunition, tobacco, and other commodities without which they could no longer survive, skyrocketed. In northern Labrador, the veteran Hudson's Bay Company, which had been operating half a dozen fur-trading posts along the coast, threw up its hands and abandoned to the Newfoundland government the burden of supplying the needs of the region's inhabitants. Everywhere throughout the American Arctic, traffic in the furs of the wild animals became unprofitable, or yielded inadequate returns. Today, an Eskimo, whether he lives in Canada or in Alaska, can no longer support his family for more than a few weeks on the miserable pin-money—from $200 to $500 a year—which is all he can expect to earn from his pelts.

Time has convincingly proved that fox furs never can be a suitable base on which to rest the economy of a whole people. They are luxury articles, subject to all the winds and breezes of fickle fashion. This year, white-fox furs may be the height of style, next year outdated. Moreover, the harvest of furs never holds steady from one season to another, because the population of white foxes, like that of many other Arctic animals, runs in cycles; in the white-fox cycle, it reaches a peak every four or five years and then declines. Finally, man has greatly complicated the fur trade by domesticating some of the wild animals and introducing synthetic furs. For these reasons, no one now looks forward to any return of the old high prices for Arctic furs, or believes that the Eskimos can ever again become prosperous from trapping. If they are to survive and prosper in the Arctic, they must find another economic base, or combination of bases, more secure than the fur trade, on which to rebuild their lives.

Where can they look for such a base? It must be one that will yield each family an income, in cash or in cash and kind, of from $1,500 to $2,000 yearly, for that is the minimum the family must now possess to pay for the necessities and near-necessities that it cannot itself win from the Arctic, but must acquire from southern Canada. What resources does the far north hold that will provide such an income to every Eskimo family?

As far as we know today, the Arctic is extremely poor in resources useful to man, except perhaps mineral resources; and the exploitation of mineral wealth requires more capital than the Eskimos now possess, and also more knowledge. The severe climate, and the grave transportation problems it creates, make the region one of very high costs, and ores must be unusually rich or valuable to overcome that handicap. Hitherto, we have failed to uncover such ores in adequate tonnage, though we have taught the Eskimo to mine a little coal in Greenland and Alaska, and for four years have operated, largely with Eskimo labour, a small nickel mine, now unhappily exhausted, on the west coast of Hudson Bay. Nevertheless, geological structures in northern Labrador and elsewhere favour the presence of valuable minerals, and it would seem to be only a question of time before the world's hunger for metals will lead to more intensive investigations and the discovery of ore bodies that will more than repay their exploitation. When that day comes, development will demand a large labour force, large enough, perhaps, to embrace all or most of the Eskimos who now live in the Arctic, and who prefer to remain there if only they can find the means to support their families and themselves.

Until mining does develop in the Arctic,

however, what can we do, and what are we doing now, to preserve these Eskimos from dire poverty, outright destitution, and perhaps slow extinction?

In a few districts, such as Barrow on the northern tip of Alaska, Iglulik in Hudson Bay, and perhaps Resolute in the Arctic Archipelago, small bodies of Eskimos can still kill enough seals, whales and walruses to keep the wolf of hunger from their doors; in other districts, hunting is more precarious and frequently ends in failure. In any case, hunting and trapping, as we have seen already, no longer bring in the income the Eskimos require to purchase clothing, ammunition, boats and lumber for proper housing and other needs. Authorities in Alaska and Canada are encouraging commercial fishing and handicrafts, but these pursuits, though valuable, have yielded less than trapping and benefit only a very small percentage of the population. Government construction projects in various districts between North Alaska and the Atlantic coast of Labrador, oil explorations on the shores of the Arctic Ocean, and one or two minor enterprises, have given welcome work to a few hundred Eskimos, but many of the projects have since lapsed or are lapsing. Consequently, only wage employment seems capable of providing the solid economic base the Eskimos need to rebuild their lives; but at the present time, the Arctic itself offers very little employment and will probably never offer enough, or even nearly enough, until mining expands into the far north and gains a firm foothold. The region lacks the natural resources to feed and clothe its present population, even at mere subsistence level, and the number of years that must pass before mining and other industries spring up to make it economically viable no one knows.

Meanwhile, in the Canadian Arctic, where the Eskimo working force numbers almost 6,000, nearly 2,000 of them young men between the ages of fifteen and thirty-five, only about five percent are able to find jobs that bring them steady wages. All the rest must continue to fish and to hunt for their daily food, snatch an odd day's work whenever the opportunity comes their way, and trap foxes or muskrats during the winter to provide $200—or if they are lucky $300 —in hard cash. To that pocket-money they may be able to add an equal sum by selling a few fish or handicraft objects; but after that, merely to keep themselves and their families alive, they must look to the government to double their paltry incomes with pensions, family allowances, and straight relief. . . .

Can we keep them healthy in either mind or body now that they have lost the proud self-sufficiency of their forefathers and have become mere appendages, mere hangers-on, of our impatient, hustling, and in many ways ruthless, civilization? I have lived with primitive Eskimos, some of whom were seeing a white man for the first time; I have felt them scrutinize and measure me, and been conscious that they considered themselves the white man's superior. And in their own Arctic, the only world they knew, they were superior, whenever fate stripped the white man of civilization's rifles and other inventions and left him as bare-handed as they were. Ignorant though they were by our standards, they overflowed with health and energy and self-confidence, untouched by the tuberculosis, influenza and virtually all the other diseases that civilization carries in its train. "We are captains of our fate," they seemed to say, though they knew, just as well as we do, that fate weaves her web blindly, and that no man knows the day or the hour when she will snip his strand.

Last summer, in Hudson Bay, I visited some younger and more sophisticated kinsmen of the primitive Eskimos with whom I had lived many years. To outward appearance, they were healthy still, although many hundreds of them had been hospitalized with tuberculosis during the last fifteen years. Inwardly, they had greatly changed. Gone was the sturdy bearing and proud assurance of former years. They seemed distrustful and listless; instead of looking straight into my face, they watched me from the corners of their eyes. They were sick in their souls, sick with that malady we have seen in Euro-

peans who were uprooted from their homes and their work during and after World War II and were herded into makeshift refugee camps, where months and sometimes years of idleness, a humiliating dependence on strangers for their daily bread, and the blankness of their future gradually undermined their morale and sapped their energy and courage.

Fortunately, Eskimos have not changed entirely. These depressed, demoralized Hudson Bay Eskimos, and their kinsmen elsewhere in the American Arctic, whether they hunt and trap foxes for a bare livelihood or sit in their homes and work at handicrafts, have not yet lost that most characteristic trait of their ancestors, the super-sensitive funny-bone. Their faces can still light up at trivial jokes and glow with the old-time animation; and as long as they retain their bubbling sense of humour, they can still bequeath something of value to our modern world. Doubtless, they will vanish as a separate race, different physically from other races on our globe. Survival of that kind would demand a return to their former isolation, and that is no longer possible in our shrinking world. Already, you can see freckled faces and flaming red hair in some Hudson Bay settlements, while many Greenlanders can be distinguished from Scandinavians only by their speech. Eskimo speech, too, may follow their peculiar physical traits into oblivion. But the soul of the people, the hardihood, the courage, the ingenuity, the bubbling humour, these and other sturdy qualities that enabled them to conquer the Arctic, they can still bequeath to future generations, provided we restore their dignity and their self-confidence, help them to stand once more on their own feet economically, and invite them to step side by side with us in civilization's onward march.

Goethe has pointed out the road for them and for us too in the second part of his *Faust*. Man's redemption comes only from work: honest, constructive, rewarding work. If, today, the Arctic cannot offer such work to its Eskimos, we should encourage as many of them as is possible to move into the industrial south and to train for the same jobs as their fellow citizens, whether these be of English or French extraction, Russian or Chinese. We need not bring all of them south, or nearly all, but should recruit mainly the younger and more adaptable volunteers, men and women, between the ages of sixteen and twenty-five, for whom today the north is unable to make adequate provision. And they need not remain forever in our midst. Many, we would hope, will voluntarily return to their homeland as soon as it can offer them a living wage.

It will not be easy to settle America's Eskimos in the south, whether it be in the United States or in southern Canada. Immigrants who come to this hemisphere from Poland, from Greece, or from China immediately find relatives, or at least fellow-countrymen, who stretch out welcoming hands, familiar churches that open wide their doors, and people who speak their tongue ready to help them obtain employment. But the Eskimos will find no one who understands their language, save three or four ex-missionaries, and as yet, very few of our Canadian Eskimos have acquired more than the feeblest smattering of English. These first migrants to the south will be unable to find their way around in our cities, to order a meal at a restaurant, or to engage a room in a hotel, unless we appoint ourselves their guardians and watch over them

during the first months or year of their sojourn. We will have to protect them from civilization's pitfalls, guide them through the mazes of urban life, buoy up their morale and speed up their education and training until they are capable of standing alone. They, for their part, will have to conquer loneliness and frustration, bewilderment and dismay; they must struggle with all their might to fit themselves into a world different from any they or their forefathers could ever have seen in their most imaginative dreams. It will be difficult, very difficult, for them, and difficult for us also, demanding all the patience and wisdom we and they can muster. But is it not better for all of us that they should struggle and work for a better life in our industrial south than rot in idleness in the far north, maintained by government doles?

The survival of the Eskimos, the survival, not of their identity as a separate race—for that is impossible—but of those admirable qualities that their race has evolved through centuries of isolation in the Arctic, rests with ourselves. They are too poor, too inexperienced in our ways, to take the initiative in any change; the initiative must come from us. And time presses. Every year, Canada's expenses on relief in the Arctic grow heavier, and the morale of our Eskimos sinks lower. If we are to save them we must act quickly. Will we?

Bewildered Hunters In The 20th Century

ABE OKPIK

This problem came into being within the last fifteen years, and still continues to grow, and will be in existence for some time yet to come.

The situation, as it existed in time past, was based on the economic resources of the land, which furnished some needs in most communities throughout the north. Trapping was a big factor because it provided income and power to buy equipment such as rifles, boots, outboard motors, tents, toboggans, etc. Hunting comes into this category as well, for if you had a good hunting ground you can picture yourself getting more fur. Fishing was another resource in large parts of the north and still is.

Each individual with his family was the bread winner for his family and sometimes for other unfortunate ones. Sometimes two or three or more families agreed to stay together to hunt and trap in the same area providing that the surrounding country was economically sound. The people did not dwell in the settlement simply because, in times past, the resources did not allow it. So, people in the North never lived all together in one place at all times, because the situation did not allow it. The location of homes was not a problem, because it was the general trend, and accepted that the house they built would be no more than a log house or sod house. To the east, a tent insulated with moss was used. Sometimes the house or tent was no larger than eight by ten feet. Some coastal people who live mainly near the sea use ice houses, or ice frame with a tent inside: a method found mostly in the central Arctic. Still, in a large part of the North, for groups comprised of movers-from-place-to-place, an igloo is most used.

All over the Arctic, this type of housing was pretty well accepted, for the place of dwelling did not matter too much in space, structure, or shape as long as it was a place to keep the body and soul warm and together. In the summer the families moved into a tent. This housing among the people did not change until the last decade when the employment programmes started developing in the North.

The population of each community came into being from the surrounding areas, and the houses grew around the settlement without too much observance by anyone. Instead of being apart and distant neighbours, the people of the area now found themselves cuddled up in and around the settlement. Any type of structure or material was accepted. Although the intention was to stay only for a short while, the people who found jobs began to stay around, bringing more and more of their relatives and people into the village. Some schools were set up in order to teach the children, thus bringing more population into the community. A small settlement, which often consisted of a missionary, a Hudson's Bay store, an R.C.M.P. building, and a few houses found itself developed into a larger community where a school, a hostel, or nursing station and administrators, power plants and welfare housing, and other people entered into the picture. Thus a settlement became a larger community within the last few short years.

I will try to outline the problems involved among the northern communities. Mentioned earlier, the people congregated around the settlement in order to get jobs and an opportunity to work for a short while without any real intentions of staying there. Sometimes

● MR. OKPIK *is with the Government of the Northwest Territories.*

Reprinted from *North,* Volume XIII(4), 1966, pp. 48-50 with permission of the author and the editor.

125

a hunter became a community garbage collector, and sometimes a camp drudge became an assistant-to-the-administrator though often he did not understand the circumstances involved where a person hanging around town could be a valuable helper to the administrator, but in time the hunter sometimes gets the idea of the reasons. Then the hunter and family who occasionally visited the settlement began to come to the settlement more frequently, sometimes even for a short time to work. They would camp in the same place they normally visited for the first few times, then their visits became longer and the children started attending school part time. Eventually the families move into the settlement without being too conscious about what they were really moving into. The hunter could no longer take extended trips to make a successful hunt as he used to.

The usual camping visit to the settlement was now permanent. The tent he sets up is renovated into a temporary house. The family is stuck in the settlement by this trend. The people who lived in the area are now living in settlement either to work or have their children go to school. This accumulation turned the village into a larger community without any planning involved. The planner came later, and this is where concern for poverty started to be seen in the community.

The community grew without too much pressure from the authorities or planner as to how the shacks or houses should be kept, or what type of structure the homes should be. Once an individual hunter who still thinks as a hunter is absorbed, even without his awareness of it, he loses control of his hunting equipment: his dogs, boats, canoes, traps which were at one time his vital possessions for living off the country. He gives this a thought and soon the whole thing flows over: "I'll be all right." So therefore his beginnings are left in the mist, and this new life starts eruption. At first he is without bewilderment but then his first wages are earned, and more and more necessities are required in order to settle the mind. Now, the ideal life is a mystery to him. In his own

opinion the whole dream will be over when he gets out on the land again. Then the expectation of this dream is not fulfilled; it materializes into the facts of a new way of life.

This gives him no point of return, and he loses all livelihood since his life possessions are now gone. As time carries on, he is watching for an opportunity to go somewhere. But where? He finally decides to get work elsewhere, wherever a job was found available. But this does not solve anything because when he leaves his family for an indefinite time he worries about them. This is where his stability is really tested by those strange things that lead from one thing to another. To keep the family together he decides to leave his job and make an attempt to go back to the family and the old environment. But this seldom brings anything new. In fact, he now finds it harder to survive on the land than he used to. Then he finds himself more or less alone, caught in a situation between two great gaps. His mind is bewildered and confused. He pictures the both sides. The outsiders came here to give him an opportunity to change to a new and conformed way of living. But he believes that he cannot really change his ways. His children are learning the other side of conforming since they have been going to school. The hunter realizes that he is too old to take any schooling. For the sake of his children he must try to make some adjustment, but how? First he hears that his cousin was arrested for drinking in a public place and given a light sentence in jail. Then he hears of another cousin or relative who was involved in stealing some money. He then gets a shock by hearing of one of his nieces who got pregnant while going to school. Perhaps by someone in school or some white man who works with the people who are building the houses in town. He then may hear a rumour that a cousin and his friend were arrested by the police for fighting and resisting arrest, and were given a sentence by a local Justice of the Peace. This time each was given three months. To him this seems odd—once the policeman was a friend, but now what has he become towards the people? During these incidents his mind is restless. Once or twice he tries hard to start again on the land, but his children are not happy out there. No one in the family cares to stay out any length of time. So he goes back to the community to try and make enough money to start over on the land. This time when he goes back it seems like all his relatives were really glad to see him. They will come and eat with him bringing their share, just the way it used to be. Pretty soon the country food which he brought back to town with him is gone. He goes on the hunt alone but he can not go far because the equipment and dogs are not as ready and good. Disappointed, he goes back to town and goes over to see the trader he used to know to see if he can be given a grubstake to start for a better trap line. But even the trader has little to offer. He is busy, selling to others who have more money to spend. He could not convince him to give him a grubstake. He wonders what happened to the storekeeper who used to help him with anything he wanted, especially nets, shells. But now the trader does not even trust him. Still he thinks in his mind that everything will turn out better. He goes and sees some friends he knows who can perhaps help him. But they too have changed their attitude and it seems to him they don't care.

He manages to get a little money here and there. Soon another addition to the family arrives. His housing condition is the same. He knows this used to be okay in the bush and there is nothing to be ashamed of as his was better than others. He has nothing to worry about although his home is of tar paper, cardboard and a double tent.

He has been working on part-time jobs off and on but he cannot understand where all his money has been disappearing to.

He then decides to have a drink with friends he knew to ease his mind and talk with them about some things which have been bothering him. This has no effect either. Finally, on the last effort, he goes and has a serious talk with the administrator. But the administrator has very little time to spend with him as he is also very busy. He finally ends up talking to the Welfare Officer. The officer is prepared to listen so he tells him all the things that have been bothering him and why. He tells him about what is happening to his family. The closely-knit group of relatives is falling apart, their words do not have a meaning anymore.

Within a few short years these good people he used to know seem to have deteriorated to a lowly thinking type of people. They do not seem to care. Something else has taken over their responsibility. He does not think of it as bad or poor, he thinks of it as a broken machine which cannot be repaired because it is completely useless or helpless.

Even the Missionary who used to be helpful does not come and visit anymore. It seems more and more teachers and other people are arriving from outside and they are all supposed to be helping the native. Now they say his house is not good as it has no washing facilities, no flush toilet. He wonders, "what do they mean by that? What do they want me to do about it. It used to be all right. Anyway, I am not going to do anything about it until maybe when someone else starts getting new houses."

This is the beginning of poverty—when a person is bewildered and has no way to improve his ways in a completely new environment.

This example in this writing is what has happened to the hunter when he is caught between two cultures: one he knows, and one he guesses at.

This is a beginning and the end of an era which has confused the once leader, hunter, provider and honest man. He is caught between this new influence and the old way of living.

This is what is happening to many people of the North, especially the people who were born before 1945 and who are not going to change. This group is the core who refuses to change.

What is their future?

CANADIAN CITIES

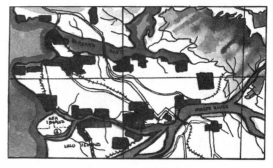

An Introduction

According to the 1966 Census of Canada, approximately 75 percent of the Canadian population live in urban centres of 1,000 or more population, and another 15 percent are in the rural non-farm category. This is almost a reversal of the 1901 distribution pattern when only 35 percent of the population was classified as urban. Of the 14 million urban dwellers in Canada, over one-half reside in the Great Lakes-St. Lawrence Lowlands of Quebec and Ontario. Metropolitan Montreal, with a population of 2.7 million in 1970, and Metropolitan Toronto with 2.6 million residents, account for slightly over 20 percent of Canada's total population. In fact, individually, each of these two cities has a greater population than the total population of any province in Canada other than Quebec or Ontario—a striking testimony to their dominant role in Canada. Metropolitan Vancouver with slightly over one million (1970) population, is their next closest rival. These figures strongly support Hamelin's observation that, as the Canadian ecumene expands, the size of the head (the city) is becoming more disproportionate to the size of the body (the remainder of the country).

The literature on Canadian urban geography is adequate in terms of individual city studies and land use descriptions, but the more formal analysis of Canadian cities and their inter- and intra-regional relationships, and many of the changing internal, social and economic functions are just now being investigated. In this regard, a number of studies are now being initiated by a group at the University of Toronto's Centre for Urban and Community Studies. In this section, however, the focus is on urban growth, the quality of cities, and their characteristics and functions.

In the introductory paper, the process of Canadian urbanization—interpreted in a historical and systems framework, is briefly analyzed by N. H. Lithwick.

In the second paper, J. W. Maxwell develops a functional classification of Canadian cities defined in terms of basic employment. From this analysis two urban Canadas emerge—the heartland cities of the densely settled Great Lakes-St. Lawrence Lowlands where manufacturing functions dominate, and the periphery, which "is characterized by huge sparsely settled areas, giving it almost vassal status to its relationship with the heartland".

E. T. Rashleigh presents a penetrating and sensitive statement on the characteristics and personality of Canadian cities. His observations include regional comparisons on the quality of urban cores; an evaluation of the basic urban amenities which facilitate "the good urban life"; the present state of affairs as reflected in suburbia, downtown, and in apartment living; the functions of open space; and the future prospects, alternatives, and challenges facing the urban planner in creating a better environment for human growth and development.

According to Maxwell's functional classification of Canadian cities, no city, in terms of employment criteria, is sufficiently specialized to merit classification as a financial centre. On the other hand, as Donald Kerr emphasizes in his exploratory article on the geography of finance in Canada, financial activity is a vital function and its magnitude and spatial functional organization is a significant facet of urban activity that has been virtually ignored by urban geographers. His analysis emphasizes the overwhelming dominance of Montreal and Toronto in Canada's financial geography.

The Process of Urbanization in Canada

N. H. LITHWICK

I. THE EARLY HISTORY OF URBANIZATION IN CANADA

The importance of history in determining the basic features of the urban system is apparent in Canada because of the specialized nature of our economic development.

Staple Exports as the Engine of Growth

Canada's economic development has been analyzed by H. A. Innis and synthesized by M. H. Watkins into a *staple theory* of growth.[1] Canada's early economy was based on exogenous demands for a sequence of resource exports, beginning very early with cod, through fur, timber, and wheat in the seventeenth to nineteenth centuries, and finally to minerals and fuel oils in this century. The particular sequencing of resources had a profound impact on the shape of the urban system. With exports of fish and timber, maritime port cities developed, particularly Halifax. As furs grew in importance, continental penetration by the voyageurs led to the focussing of international trade on Montreal. Wheat had a different impact; for, in contrast to the continuous shifting of base of operation for the earlier staples, wheat technology requires residentiary settlement.[2] Inevitably, true central places began to develop in the wheat-growing areas, the prairies. These had to be linked to the grain export ports, first Montreal and then Toronto, and subsequently the railway emerged as the main inter-urban link of the past and present century.

But central places need their intermediate inputs and final outputs from larger centres. At a very early stage of development, these were imported from the United Kingdom and then from the United States. As exports continued, domestic incomes rose; markets reached adequate scale, and domestic production, first concentrated in Montreal and then increasingly, because of its more advantageous location, in Toronto, began to replace imports. This shift from Eastern to Central Canada was abetted by the shift of export markets to the United States as the mineral phase of the staple sequence emerged, and as the share of foreign capital from the States began to displace that from the United Kingdom. Coincident with this shift to the States was the shift of the centre of gravity of the American economy itself to the West. This led to a second north-south line of force operating on Canadian cities which further weakened the East, and which culminated in the phenomenal growth of the west coast in the United States and therefore of Vancouver. In recent years Vancouver has further expanded as a result of Canada's rapidly increasing post-war trade with the Far East—Taiwan, Hong Kong, and particularly Japan.[3]

The growth pattern of the Canadian economy therefore has been affected very directly by these major exogenous forces. This patterning can be clearly seen in Table 1. The critical role of urban Canada in national development is clearly seen in the last line of the above table, where population growth was, in the last half century, almost exclusively in the largest urban centres. Like-

● N. H. LITHWICK *is Professor of Economics, Department of Economics, Carleton University.*

Reprinted from "Urban Canada: Problems and Prospects," a report prepared by N. H. Lithwick for the Honourable Robert Andras, Minister of State for Urban Affairs, Government of Canada, and published by Central Mortgage and Housing Corporation, Ottawa, December, 1970.

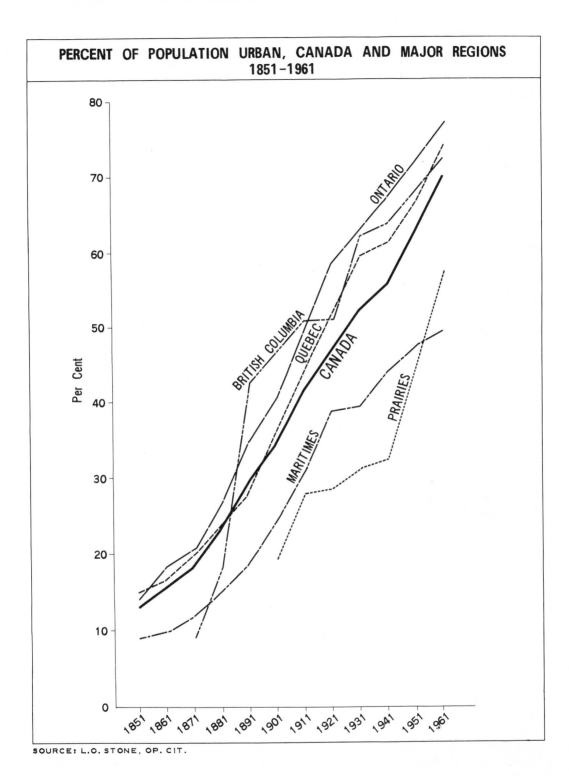

PERCENT OF POPULATION URBAN, CANADA AND MAJOR REGIONS 1851–1961

SOURCE: L.O. STONE, OP. CIT.

Figure 1

Table 1

THE GROWTH OF MAJOR URBAN CENTRES

Historical Increase in Population of Metropolitan Areas (thousands)

Region	City	1650-1750*	1750-1850	1850-1900	1900-1950
East	Charlottetown		7**	5	6
	Saint John		27**	23	46
	Halifax		25**	26	130
Quebec	Quebec	3	32	54	266
	Montreal	8	69	296	1,534
Ontario	Kingston†		8**	10	36
	Ottawa		15**	81	334
	Toronto		31**	241	1,553
	Hamilton		14**	65	304
Prairies	Winnipeg			48	428
	Regina			2	110
	Edmonton			3	324
	Calgary			4	273
West	Vancouver			27	763
TOTAL				858	6,107
CANADA				2,893	8,299
% of Increase in Urban Areas				30	74

* Dates are approximate.

* * Population as of 1850.

† In 1831, Kingston was actually larger than Toronto (then called York).

Source: C. F. Osler, *op. cit.*

wise, the sequencing of major urban growth from East to West is clearly seen. Even within regions, there is the same tendency. Further evidence, involving all urban areas, is summarized in Figure 1. The role of the very early response to staples is omitted because of lack of data, but the differential levels and slopes of the curves do substantiate the importance of these exogenous forces in determining the timing and location of urban growth in Canada.

The Shift to Endogenous Development and the Evolution of Mature Urban Units

Despite this shift in relative importance, the absolute growth of urban centres has continued to be very substantial in all regions (Table 2). Thus, it appears that the inter-industrial and hence inter-urban links were already forged, so that regardless of the locus of initial new export demands, the linkages in the economy led to a system-wide expansion with regional variations. In addition, by 1900 the larger urban centres, espe-

cially Montreal and Toronto, had become sufficiently large and their economies had become so well developed that they can be judged to have achieved the maturity stage of self-sustained growth. The economy itself therefore was on the threshold of such a developmental phase, because so much of the nation's modern economic activity was concentrated in these two centres. Vancouver's recent growth has been extremely rapid, bringing it to the mature state by the start of the 1960's.

Table 3 presents some comparisons of Canadian metropolitan growth performance with that of similar-sized United States metropolitan areas. Toronto and Montreal grew faster, by a substantial margin, than all comparable cities in the States. This finding reflects the higher growth rate in Canada than in the United States over this period, and the higher labour intensity of that growth.[4] It may also reflect the greater polarization of the two largest Canadian cities, whose counterparts are substantially

Table 2

POPULATION FOR THE PRINCIPAL REGIONS OF
METROPOLITAN DEVELOPMENT IN CANADA, 1901-1961

Principal Regions of Metropolitan Development	1901	1911	1921	1931	1941	1951	1961
	Population (in thousands)						
Halifax	51	58	75	79	99	134	184
Montreal	415	616	796	1,086	1,216	1,504	2,156
Quebec	117	133	158	207	241	297	383
Hamilton	79	112	154	190	207	266	359
London	52	61	74	87	97	129	181
Ottawa	103	133	168	197	236	296	436
Toronto	303	478	686	901	1,002	1,264	1,942
Windsor	22	32	66	117	129	163	192
Winnipeg	48	157	229	295	302	357	476
Calgary	8	56	78	103	112	156	290
Edmonton	15	48	87	116	136	211	374
Vancouver	—	—	224	338	394	562	790
CANADA	5,324	7,192	8,776	10,363	11,490	13,623	17,743
	Percent change in population since last census						
Halifax	—	13.4	30.6	4.1	25.5	35.8	37.3
Montreal	—	48.6	29.2	36.3	12.1	23.6	43.3
Quebec	—	13.9	18.5	30.7	16.7	23.0	29.2
Hamilton	—	40.6	37.5	23.7	8.8	28.7	34.9
London	—	18.5	20.9	18.0	11.6	32.7	40.6
Ottawa	—	28.4	26.4	17.6	19.9	25.3	47.2
Toronto	—	57.7	43.6	31.3	11.2	26.2	53.6
Windsor	—	44.7	103.4	77.2	10.6	26.1	17.6
Winnipeg	—	223.7	46.0	28.7	2.4	18.1	33.4
Calgary	—	570.2	39.5	33.0	8.1	39.7	85.8
Edmonton	—	215.5	83.0	32.4	17.9	55.1	77.1
Vancouver	—	—	—	51.3	16.5	42.7	40.6
CANADA	—	35.1	22.0	18.1	10.9	18.6	30.2

Source: L. O. Stone, *op. cit.,* Table L. 5, p. 278 and L. 1, p. 269.

smaller than the major centres in the States by a factor of up to four. The table also reveals the western drift of urban growth in the States that has paralleled the situation in Canada.

Thus, we can observe the close relationship between national economic development and the urbanization of these leading cities. Early exogenously determined national development led to the highly specific patterning of urban growth. The fact that this development was sustained through a fortunate sequence of export expansion permitted the basic conditions for urban development to emerge: incomes rose, markets expanded, capital and labour accumulated, appropriate structural changes were forthcoming, and these were capable of launching a sustained growth process. But the very success of this urban development entailed the transformation of the economy itself, so that Canada soon was launched on its current development path.

These trends are readily seen in the changing structure of the economy. In 1870, over half of economic activity was primary, and therefore non-urban. By the 1920's, this proportion had fallen to one-quarter, and currently it stands at just over one-tenth.[5] Thus, the role of staples as the engine of growth

Table 3

COMPARATIVE GROWTH RATES:
METROPOLITAN AREAS IN CANADA AND THE UNITED STATES

Metropolitan Area[a]	Population 1960	Percent Change 1950-1960
Metropolitan Toronto[b]		
Municipality of Metropolitan Toronto	1,618,787	44.9
Census Metropolitan Area	1,824,481	50.7
Other Census Metropolitan Areas in Canada with More than 500,000 Population in 1961		
Montreal	2,109,509	43.3
Vancouver	790,165	40.6
Metropolitan Areas of the Same Size Range in the United States[c]		
St. Louis, Mo.-Ill.	2,104,669	19.9
Washington, D.C.-Md.-Va.	2,001,897	36.7
Cleveland	1,909,483	24.6
Baltimore	1,727,023	22.9
Newark, N.J.	1,689,420	15.0
Minneapolis-St. Paul	1,482,030	28.8
Buffalo	1,306,957	20.0
Selected Metropolitan Areas in Western United States		
Los Angeles-Long Beach	6,038,601	45.5
Houston	1,243,158	54.1
Dallas	1,083,601	45.7
Denver	929,383	51.8
Fort Worth	573,215	46.0

[a] Population figures for the United States' cities are for Standard Metropolitan Statistical Areas 1960.
[b] Metropolitan Toronto population figures are for 1951 and 1961.
[c] Includes Standard Metropolitan Statistical Areas with between 1.3 million and 2.3 million population in 1960.

Source: Robert A. Murdie, *Factorial Ecology of Metropolitan Toronto, 1951-1961*, Department of Geography, Research Paper No. 116, University of Chicago, 1969, Table 1, p. 3.

Table 4

COMMODITY COMPOSITION OF TRADE, CANADA, 1957-1968

	1957		1968	
	Exports	*Imports*	*Exports*	*Imports*
Grain	8%	—	5%	—
Industrial Materials	53%	29%	43%	21%
Highly Manufactured Products	6%	31%	25%	44%
Services	25%	31%	21%	29%
Other	8%	8%	7%	7%

Source: Economic Council of Canada, *Sixth Annual Review*, September 1969, Table 5-5, p. 89.

has been continually reduced, and endogenous urban growth has taken over. This does not mean that the economy has become autarchic; indeed, foreign trade still accounts for one-quarter of Canada's GNP. But this trade increasingly is based on industrial specialization on a continental scale, rather than on primary products exclusively (Table 4). Such specialization is intimately connected with the urbanization process, for the source of most of our exports and the destination of most of our imports is increasingly the urban system.

The link between the development of the Canadian and the American urban systems has not yet been analyzed in detail. One study of capital flows suggests very strong systemic links.[6] Other material, such as emigration, transport, and information and commodity flows, would probably substantiate the major impact of the American urban system on the Canadian. This interdependence undoubtedly has accelerated the economic development of the major Canadian urban centres, but the close linking has also contributed to an increasing loss of Canadian identity and possibly independence[7] for these centres, particularly if

polarization is occurring on a continental scale. For some Canadians at least, this effect represents an additional cost of our current urbanization process that warrants some additional thought.

II. THE DIMENSIONS OF MODERN URBANIZATION IN CANADA

Economic Transformation

Despite the lack of economic data on urban Canada, there can be little doubt about the economic maturity of our major urban centres, Montreal and Toronto. The high rates of growth and shares in the key macro-variables—labour force, income, and capital—and their growing structural complexities have been well documented.[8] In Table 5, we present a rough synopsis of these developments.

A parallel development of smaller metropolitan areas has been observed, although only Vancouver has attained a degree of sophistication approximating that of Montreal and Toronto. A number of centres, however, have particular features that indicate that they are on a sustained growth path, because of rapidly growing markets (Oshawa, Oakville), because of a highly

Table 5

ASPECTS OF URBAN ECONOMIC GROWTH AND STRUCTURE

Metropolitan Area	No. in Area 1964	% Inc. 1951-1964	Retail Sales Inc. 1951-1964	% Managers and Professionals 1961	Income (Perc. Disp.) % of Can. 1962	% Inc. 1951-1962	No. of Mfg. Establishments (1964)	Indust'l Energy Consumption MKWH 1963	Value of New Construction Million $ (1964)	Total Value Added % of Canada 1965*
Montreal	3,032	57	85	20.5	13.6	107	5,144	4,847	540	16
Toronto	3,038	72	78	21.2	15.4	102	5,115	3,011	603	18
Vancouver	1,019	57	55	21.5	5.9	95	1,701	619	188	4
Ottawa	778	48	76	23.9	3.1	124	333	349	148	1
Winnipeg	631	41	52	18.9	3.3	89	1,008	277	97	2
Hamilton	572	71	71	17.8	3.0	88	713	1,897	91	6
Quebec	556	34	101	21.5	1.6	158	570	17	64	1
Edmonton	940	60	112	22.0	2.4	144	519	343	114	1
Calgary	617	75	59	23.1	2.2	142	412	377	96	1
London	707	33	110	20.5	1.3	108	319	263	52	1
Halifax	597	23	68	19.3	1.1	126	144	59	26	—
Windsor	366	24	21	19.6	1.1	45	391	326	44	2
Regina	215	44	26	22.1	0.8	109	122	92	36	n.a.
CANADA		39	87	19.1	100.0	106	33,630	73,311	5,379	100

* DBS, *Canada Yearbook*, 1968, Table 16, p. 714.

Source: Canadian Daily Newspaper Publishers Association, *Canadian Markets Data Manual, 1965.*

Table 6

EMPLOYMENT STRUCTURE IN CITIES, 1931-1961
SHARE IN NATIONAL EMPLOYMENT BY SECTOR

City		Primary Industries	Manu-factur-ing	Con-struc-tion	Utilities	Trans-porta-tion	Trade	Finance et al.	Services	Govern-ment	N.E.S.	Total
Quebec	1931	—	2	2	2	1	2	1	2	3	1	1
	1961	—	1	1	1	1	1	1	1	3	1	1
Montreal	1931	—	15	15	11	12	13	15	11	12	8	9
	1961	—	11	7	6	10	8	11	8	5	9	8
Ottawa	1931	—	1	1	2	1	2	2	2	10	1	1
	1961	—	1	2	1	1	1	2	2	8	2	2
Toronto	1931	1	13	9	13	8	13	17	10	10	4	7
	1961	—	6	5	5	5	5	9	7	5	6	5
Hamilton	1931	—	4	2	2	1	2	2	2	2	1	2
	1961	—	3	2	1	1	1	1	1	1	1	2
London	1931	—	1	1	1	1	1	2	1	1	1	1
	1961	—	1	1	1	1	1	2	1	1	1	1
Windsor	1931	—	1	1	1	1	1	1	1	1	—	1
	1961	—	1	1	1	1	1	1	1	—	1	1
Winnipeg	1931	—	3	3	4	3	5	5	4	4	3	2
	1961	—	2	2	2	3	2	3	2	2	2	2
Edmonton	1931	—	1	1	—	1	1	1	1	2	1	1
	1961	—	1	2	2	2	2	2	2	3	2	2
Calgary	1931	—	1	1	1	1	2	2	1	2	1	1
	1961	1	1	2	2	2	2	2	2	2	2	2
Vancouver	1931	1	3	4	2	5	5	5	4	4	4	3
	1961	—	2	2	2	3	4	4	3	2	3	3

Table 7

EMPLOYMENT STRUCTURE IN METROPOLITAN AREAS, 1941-1961
SHARE OF NATIONAL EMPLOYMENT BY SECTOR

Metropolitan Area		Primary Industries	Manu-factur-ing	Con-struc-tion	Utili-ties*	Trans-porta-tion	Trade	Finance et al.	Services	Govern-ment	N.E.S.	Total
Quebec	1941	—	2	3	—	2	2	2	3	5	3	2
	1961	—	2	2	1	2	2	2	3	4	2	2
Montreal	1941	—	18	15	11	15	16	21	14	12	20	11
	1961	1	18	13	10	16	13	18	13	9	14	12
Ottawa	1941	—	2	2	2	2	3	4	3	17	6	2
	1961	—	1	3	2	2	2	3	3	12	3	3
Toronto	1941	—	16	11	13	10	17	23	11	11	15	10
	1961	1	17	12	14	11	15	23	13	9	11	12
Hamilton	1941	—	4	2	2	1	2	2	2	1	2	2
	1961	—	4	2	2	1	2	2	2	1	2	2
London	1941											
	1961	—	1	1	1	1	1	2	1	1	1	1
Windsor	1941	—	3	1	1	1	1	1	1	1	2	1
	1961	—	2	1	1	1	1	1	1	1	1	1
Winnipeg	1941	—	3	4	3	6	6	6	4	3	5	3
	1961	—	4	3	4	5	4	4	3	3	3	3
Edmonton	1941											
	1961	—	1	3	2	3	3	2	2	3	2	2
Calgary	1941											
	1961	1	1	2	2	2	2	2	2	2	2	2
Vancouver	1941	1	4	5	2	6	6	7	5	4	6	4
	1961	1	4	5	4	6	6	7	6	4	5	5

* Only city totals available for 1941.

developed industrial structure (London, Winnipeg, Calgary), or because they have a real advantage in selected areas (Edmonton, oil; Hamilton, steel; Ottawa, Quebec, and Halifax, public administration).[9] Because the latter are key inputs into the economic system, all these areas are highly responsive to the rapid economic growth of the economically leading areas, and the prospects for completing their economic development appear to be very good.

In addition to this selected material, we have assembled comprehensive information on one key aspect of the evolving urban economic structure, and that is the structure of the labour force in the key urban areas (Tables 6 and 7).

The major city share in overall employment clearly has declined somewhat in the East and has tended to rise in the West, a trend we have already noted with regard to population. The role of primary industry, however, which is essentially non-urban, tends to distort this picture. Over the period 1931-1961, the share of primary industry in the national labour force declined from over one-third to under one-sixth. Thus, part of the reason for the weakness of the decline in total urban shares in the labour force is the shift out of primary industry, which necessarily increases the importance of urban activities. If we examine the various sectors individually, we observe a systematic and much sharper *decline* in the relative position of the Western cities. This finding tends to conflict with our stress on economic polarization.

The reason for this emerges in Table 7 when we examine entire metropolitan areas, rather than the cities proper. If the process of flight from the core is indeed taking place, our wider-area data should reveal this. The evidence indicates that this is in fact what has been happening. Overall major metropolitan area shares in the labour force have tended to increase, as have shares in most of the individual industrial sectors. Polarization has indeed taken place therefore, but it has been subject to the constraints of the micro-urban process of suburban drift of persons and jobs.

Demographic Responses

The mechanisms that have led to the development of the urban economic system have had a direct impact on the demographic structure of urban Canada, and hence on the nation as a whole. With economic growth increasingly concentrated in the larger urban centres, their relative attractiveness—particularly in terms of higher wages for labour and more employment for the highly skilled, but also because of the increased variety of consumer goods and services[10]—leads to an increasing concentration of population. The evidence on incomes, employment, and expenditure has been provided elsewhere.[11] In this section we trace out the demographic responses to urban concentration.

The increasing concentration of the Canadian population is shown in Table 8. Note that this steady shift to the major urban centres has continued with one very significant deviation, and that is the retardation during the Depression. Over the past century, there has been an eighteen-fold expansion in the number of urban centres over 100,000 in size, a twelve-fold expansion in the number of cities of 30,000-100,000, and a nine-fold increase in the number of units under 30,000. Furthermore, the rate of growth of the largest metropolitan areas has been faster than that of all other urban centres. Urban growth, in other words, has become increasingly polarized in the largest urban centres. This is not to deny that selected smaller urban centres have grown extremely fast, or that selected metropolitan areas have had relatively slower growth rates; but the general pattern remains one of polarization.

The sources of this demographic pattern are natural increase and net migration. In general, the rate of natural increase in metropolitan areas has been lower than in the rest of the country. This means that the potential supply of labour from the cities themselves has been growing at a lesser rate than in the non-urban areas. The very rapid growth of the urban economy has imposed rapidly increasing demands for labour, substantially greater than the urban supply can meet. This

Table 8

POPULATION IN THE PRINCIPAL REGIONS OF METROPOLITAN DEVELOPMENT
(PRMDs), CANADA AND MAJOR REGIONS, 1901-1961

CANADAª AND MAJOR REGIONS	1901	1911	1921	1931	1941	1951	1961
	Population in the PRMDs ('000)ᵇ						
CANADA	1,338	2,076	3,103	4.098	4.615	5.904	8,575
Maritimes	102	111	138	141	170	212	280
Quebecᶜ	532	750	954	1,292	1,457	1,800	2,539
Ontarioᶜ	629	908	1,266	1,640	1,851	2,354	3,452
Prairies	76	307	459	616	657	853	1,352
British Columbia	—	—	287	408	480	684	953
	Percentage of the total population in PRMDs						
CANADA	26.0	30.5	35.4	39.5	40.2	43.3	48.3
Maritimes	11.4	11.9	13.7	14.0	15.0	16.9	19.4
Quebecᶜ	33.1	38.3	41.4	46.0	44.7	45.4	49.5
Ontarioᶜ	28.2	35.3	42.4	46.9	48.0	50.2	54.2
Prairies	18.0	23.1	23.5	26.2	27.1	33.5	42.5
British Columbia	—	—	54.8	58.8	58.7	58.7	58.5

ª Exclusive of Newfoundland, Yukon Territory and Northwest Territories.

ᵇ Regional figures may not add to Canada total due to rounding error.

ᶜ Because Ottawa MA is partly in Ontario and partly in Quebec, Hull County in Quebec has been allocated to the Ontario total.

Source: L. O. Stone, *op. cit.*, Table 6-2, p. 13.

Table 9

STRUCTURE OF POPULATION 14 YEARS OF AGE AND OVER
BY MIGRATION TYPE, CANADA AND REGIONS, 1965

REGION	Population	Migrants 1964-1965			
		Intra-regional	Inter-regional (thousands)	From Abroad	Out Migrants
Atlantic Provinces	1,275	57	13	*	25
Quebec	3,828	142	22	22	29
Ontario	4,649	217	53	48	40
Prairie Provinces	2,222	124	28	14	40
British Columbia	1,252	68	33	12	15
CANADA	13,226	608	149	100	149

* ⟨10,000.

Source: D.B.S., *Geographic Mobility in Canada, Oct. 1964-Oct. 1965*, Special Labour Force Studies. No. 4, April 1967, Table 2-A, p. 8.

had led to a sustained net migration into the city, accounting for over half of the population increase in metropolitan areas over the past 40 years.[12] Much of this increase has occurred at the expense of the rural areas, but in addition a substantial number of migrants from outside the country augment this urban increase.

The relative importance of various migration flows is summarized in Table 9.

The high rate of population mobility is apparent. It is also clear that the bulk of the migrants moved within the same region, no doubt primarily because of the distance factor, but also, especially in the case of Quebec, for social and cultural reasons. The very important role of immigration from abroad in meeting labour demands, particularly in Ontario and Quebec, is obvious. In the following sections we discuss the urban implications of internal and international migration.

Urban Implications of Internal Migration

Migration Patterns—A comprehensive study of migration in Canada has been published recently by D.B.S.[13] This section will draw heavily on their findings.

Inter-regional net migration is highly responsive to the economic pressures we have referred to.

"The provinces that enjoyed the highest levels of income, modernization and economic growth in recent decades (Ontario, British Columbia and Alberta) were the only ones sustaining net gains in the 1956-61 five-year migration; the provinces that had the highest concentrations of work force in primary activities had the sharpest net losses. . . ."[14]

If economic factors are the chief explanatory variables of migration, the foci of economic activity—the major urban areas— should be the main destination of such migration. This was indeed found to be the case.

"Among the selected urban size groups, the 100,000 and over group was most favoured as a destination for five-year emigrants, even after the concentration of 1956 population in this size group is taken into account."[15]

This finding has profound implications for explaining mobility. The relevant spatial unit in the migration flow is the city. It serves to attract migrants because of its employment potential. The role of the province as a spatial unit is largely derivative; it is a function of its urban structure.

". . . the presence of a large urban agglomeration or metropolitan area significantly affects a region's retentive power upon population. . . ."[16]

It is true that intra-provincial migration flows exceed inter-provincial migration (by four to one, 1956-61[17]), but this can be explained largely by the distance factor.[18] Migrants move to jobs but face some distance impediments that naturally bear a relationship to provincial boundaries. Thus, provinces play a relatively passive role in explaining migration flows; the urban structure is central here. Any attempts to analyze and deal with provincial demographic trends that ignores the central role of the urban unit will be based on totally inappropriate assumptions.

As rural areas decline towards some limit, this important source of urban population tends to dry up. The major flows then must be within the urban system.[19] If our hypothesis about the central role of the major centres is valid, the inter-urban flow must be from smaller to larger units.

"For the future, the most significant internal migration flows will be among urban centres, and the streams involving the large urban agglomerations and metropolitan areas will be particularly important."[20]

In other words, our analysis of the urban system *qua* system, inherently linked to the nation's ecomonic structure, appears to be fully substantiated by this evidence. The main flows are to urban centres, and because of their growing relative attractiveness, primarily to the largest centres.

In addition, higher-than-average levels of education and occupational skills are shown for the migration streams involving metropolitan areas, relative to both source and destination populations and alternative flow patterns.[21]

To summarize these conclusions, it is worth quoting Stone at length:

"Over the 1941-51 decade, Canada underwent rapid structural changes high-lighted by the decline of primary activities and the growth of manufacturing, sales and services. The rapidly growing economic sectors were spatially concentrated in certain regions of Canada, and these regions had thus unusually large increases in the economic opportunities that attract migrants . . . the 1951-61 decade saw a continuation of (these) trends. . . . Among the counties or census divisions, the major relevant shifts probably involved the decline of agriculture, and advances in urbanization, manufacturing and tertiary activity. In regard to the urban complexes, the major relevant shifts probably involved the degree of increase in the performance of metropolitan functions, which spurred the demand for a more highly educated and professional work force, and pushed specialization in activities like wholesale trade, business and financial services."[22]

A final intra-metropolitan process, that relates to our concern with the poor, has been documented.

". . . the result of the migration into and out of the MA (Metropolitan Area) was a net loss to the central city and a net gain to to the ring."[23]

But the real concern is less with size than with composition of the population.

". . . the net effect of this redistribution was to raise the levels of educational and occupational skills in the ring and to lower it in the central city . . . the redistribution tended to increase the proportion of married persons in the ring and lower it in the central city."[24]

Mobility and Migrations—Some further analysis of migration and its relationship to employee changes in industry and occupation can be made on the basis of unemployment insurance records (Table 10).

In both periods, about half of all persons in the sample changed jobs *each year*. Obviously most changes were in occupation, largely within the same firm, and hence industry and location. One-fifth of all job changes, however, entailed changes in location, indicating a high degree of responsiveness to economic opportunities despite the substantial costs of moving. The evidence also suggests a relationship of mobility to unemployment: the lower the rate of unemployment, the greater the total mobility. This, of course, reflects the importance of demand conditions. On the supply side, the unemployed show much higher mobility rates than the employed, particularly with respect to locational mobility.[25]

There appears, therefore, to be much potential responsiveness of labour to changing economic conditions. This extends to locational change as well. In line with Stone's findings, moves tend to favour nearer areas rather than farther ones.[26] Further, the young are the most mobile, as are males relative to females.[27] The result is a continual drift of the "best" workers to the best jobs—a result that permits the continual healthy expansion of the national economy but which has serious implications for the losing areas.

Table 10

PERCENTAGE OF JOB CHANGES BY TYPE, 1952-1960

Mobility Variable	1952-1956 %	1956-1960 %
One Variable		
Occupation	42.0	44.3
Industry	15.0	16.4
Location (UIC Local Office)	5.0	6.1
Two Variables		
Occupation and Industry	24.0	19.3
Industry and Location	3.0	3.4
Occupation and Location	3.0	3.0
All Three Variables		
Occupation and Industry and Location	8.0	7.5
TOTAL	100.0	100.0
% Involving Location	19.0	20.0

Source: 1952-1956, D.B.S., *Canadian Statistical Review*, July 1960.
1956-1960, D.B.S., *Canadian Statistical Review*, November 1961.

140

The Urban Situation of Internal Migrants—
The problems of adjustment for migrants to urban areas from elsewhere in Canada have not been analyzed as thoroughly as those of immigrants from abroad. Presumably it is assumed that internal migrants are socially and culturally more similar to their new urban neighbours than are immigrants. This may not be the case, however. Immigrants to Canada must meet certain skill requirements, and so the incoming flow is screened; internal migrants undergo no such screening process. One need only reflect upon the serious urban problems of native peoples, or of rural migrants,[28] to be aware that "citification" may be a much greater ordeal for them than for formerly urbanized, highly educated, skilled immigrants from the United Kingdom, the United States, and other countries.

In their study, Charbonneau and Légaré indicate that internal migrants in Quebec located typically in the east end of Montreal, the lowest income district of the city. Foreigners tended to locate in the downtown area, while migrants from other provinces located relatively more frequently in the west end.[29] In contrast to Toronto, the proportion of foreign migrants was quite small. The effect of internal migration to Montreal thus appears to be an exacerbation of a low-income situation in the poorest areas. These conclusions are, of course, highly speculative, and it is recommended that serious research into the general situation of internal migrants be undertaken, particularly in the light of Stone's general findings on the rapidly changing source of internal migrants.

Urban Implications of International Migration

*The Urban Orientation of Migrants—*Data recently assembled by the Department of Manpower and Immigration permit us, for the first time, to discover the intended urban destination of immigrants to Canada.

In 1968, over one-half of all immigrants intended to settle in the three largest metropolitan areas of Canada—Montreal, Toronto, and Vancouver (Table 11). In the light of the fact that these centres held one-quarter of the Canadian population, the large-city orientation of the migrant flow is readily seen. Indeed, 75% of all immigrants were destined for nine of the largest cities in Canada in that year.

Table 11

IMMIGRATION AND MAJOR URBAN CENTRES, 1968

Census Metropolitan Area	CMA Population (000) 1968	% of Total Pop.	Average Annual Growth Rate 1966-68	Immigrants* Destined for (1968)	% of all Immigrants	Immigrants in 1968 as % of 1968 CMA Pop.
				(000)		
Montreal	2,527	12.2%	1.8%	28.8	15.7%	1.1%
Toronto	2,280	11.0%	2.8%	53.8	29.2%	2.4%
Vancouver	955	4.6%	3.5%	13.4	7.3%	1.4%
Ottawa	518	2.5%	2.3%	4.6	2.5%	0.9%
Hamilton	471	2.3%	2.5%	5.3	2.9%	1.1%
Edmonton	425	2.1%	3.0%	5.1	2.8%	1.2%
TOTAL: 6 cities	7,176	34.7%		111.0	60.4%	1.5%
Rest of Canada	13,522	65.3%		73.0	39.6%	0.5%
TOTAL: Canada	20,698	100.0%		184.0	100.0%	0.9%

* The close correspondence between intended and actual destination for the first three months of 1969 makes the use of these intentions data valid.

The ratios of actual to intended were as follows: Montreal 92%; Toronto 108%; Vancouver 102%; Ottawa 91%; Hamilton 122%; and Edmonton 93%.

This actually adds to the polarization in Toronto of the immigration flow.

Source: Department of Manpower and Immigration, supplementary Table 3.

In addition to the big-city direction of this flow, there is some evidence that the rate of growth of particular cities is related to immigration. Thus, Toronto, which has been growing more rapidly than Montreal, attracted almost twice as many immigrants.

The Impact of Urban Growth—The result of these trends is that immigration has been an important factor in urban demographic growth. Immigration amounted to more than half of the net growth in population in most centres, with the proportion in Toronto and Montreal being particularly high.

In addition to the urban specificity of the overall immigration flow, we observe that the flow to the largest cities is relatively labour intensive, giving them an even greater share of the productive immigrant population. Thus, whereas the three largest metropolitan areas absorbed 52% of total immigration in 1968, they claimed 55% of the immigrant labour force. Since the other large urban centres had about the same share of immigrant population and labour force, by implication the less urbanized areas had relatively more non-productive dependents, further augmenting economic disparities between these areas. From this we can infer that those provinces which lack major urban centres benefitted substantially less from the migration flow than did the wealthier urbanized provinces.

A second set of estimates adds another dimension to the city-selectivity of immigrants. A sample for the first three months of 1969 reveals that the largest metropolitan areas received not only a relatively greater share of the labour force than of total immigration, but within that labour force, a high proportion of clerical and service workers and craftsmen and labourers. For the non-metropolitan areas, there was a high incidence of workers in primary industries. As for the non-labour force component of the immigration flow, the metropolitan areas showed a high incidence of future productive labour, particularly college students, while the non-metropolitan areas contained a disproportionate number of retired persons with zero productive potential .

On the basis of this evidence, we would conclude that the single most important facet of the immigration process is its strong metropolitan orientation. This has had two effects, one a benefit and the other a cost. The benefit is that the quality of this metropolitan flow has been relatively higher than the non-metropolitan flow, measured in terms of labour productivity. This must be qualified by the cost, which is the substantial augmentation to the metropolitan growth rate that this flow has induced, further aggravating the serious urbanization problems. The net outcome for these metropolitan areas cannot be ascertained at present.

The Urban Situation of Immigrants— While we have identified several of the parameters indicative of the impact of immigration on Canada's urban system, it remains for us to examine the urban condition of these immigrants. Regrettably, the amount of information we have is incomplete and precludes cross-city comparisons. Some data have been assembled for Montreal and Toronto, and we include them for two reasons: first, these cities account for almost half of all immigrants; and second, the socio-economic hardships of immigrants in these major metropolitan areas are probably worse than anywhere else.

For Montreal, available information indicates that the densest area of immigrant residency is the central city, that part of the city where the lowest levels of income are found.[30]

For Toronto, there is much more evidence not only on the location of immigrants, but also on their socio-economic status, housing conditions, and so forth.[31]

For housing, it was found that apartment dwelling was most characteristic of immigrants who arrived between 1956 and 1961. Natural born Canadians and pre-war immigrants were most likely to live in single detached dwellings, and immigrants in the central city who arrived between 1946 and 1955 were most likely to be living in single attached dwellings.

As for the quality of housing, a quarter of all post-war immigrant households in To-

Table 12

DWELLING CHARACTERISTICS
IN METROPOLITAN TORONTO, 1961

	Pre-War Imm. %	1946-1961 %	Other Canadians %	Total %
a) *Amenities*				
Running Water	99.57	99.80	99.47	99.56
Flush Toilet	89.89	85.28	91.61	89.83
Bath or Shower	92.23	86.43	94.23	92.06
Furnace	96.87	97.35	97.18	97.25
Refrigerator	97.94	98.79	98.69	98.56
Home Freezer	5.74	4.58	9.87	7.83
Television	91.86	87.61	93.25	91.69
Car	62.62	70.03	77.84	72.95
b) *Type of Housing*				
Single Detached	59.47	43.88	59.11	55.75
Single Attached	19.53	24.60	13.93	17.49
Flat or Apartment	20.99	31.45	26.83	26.67
Owned	77.73	61.87	65.95	67.45
Rented	22.27	38.13	34.05	32.55
Reporting a Mortgage	27.88	32.84	37.97	34.72
No. of Households	99,106	108,943	274,441	482,940

Source: Anthony H. Richmond, *Immigrants and Ethnic Groups in Metropolitan Toronto, op. cit.*, Tables 33 and 34, p. 57, from 1961 Census of Canada.

ronto were sharing bath and toilet facilities. This compares with an overall city average of less than 10%, and suggests that the bulk of low-quality housing was occupied by this group. The fact that Toronto's share of families not maintaining their own household is twice as high as the national average and the metropolitan average further supports this view.[32] Some evidence regarding this is contained in Richmond's statistical study.[33] (Table 12).

The relative lack of flush toilets and bath and shower facilities for post-war immigrants is readily seen, although for other items, such as furnace, refrigerator, and running water, this group scores slightly higher than both pre-war immigrants and other Canadians. One important lack is cars, but even this group is in a better position than pre-war immigrants. As for types of housing, the apartment orientation and consequent rental status is clearly seen.

It is not necessary to draw the conclusion that Dr. Richmond does, however—namely, that "this is indicative of the seriousness of the housing shortage for this section of the population".[34] Sharing of facilities is an economical way for newly arrived immigrant households to minimize their expenditure on housing, thereby enabling them to accumulate savings for later acquisition of their own homes. This transition-easing pattern of behaviour is highly rational, and the all-too-easy inference that those behaving in this way are bearing a heavy toll in terms of the "social effects and human implications"[35] may lead to inappropriate policies. There is some evidence that several urban renewal schemes, responding to the data with little awareness of the process, have destroyed a highly functional component in the urbanization process.

Despite this qualification, Richmond does provide evidence about the housing market in Toronto that has direct application to immigration. His evidence suggests that the housing market in Toronto is not homogeneous but is clearly stratified into a number of largely independent markets based upon the variables of age and family status, economic status, and ethnic origin. This means that there may be a surplus of housing in one sector of the market which is useless in overcoming shortages in another.

Table 13

PERCENTAGE WITH INCOME OF $3,000 OR MORE BY DISTANCE FROM CITY CENTRE, MALES IN LABOUR FORCE, SELECTED MOTHER-TONGUE AND BIRTHPLACE GROUPS, METROPOLITAN TORONTO, 1961

Zone[1]	Total	English	German	Italian	Polish	Ukrainian	Other
			MOTHER TONGUE				
1	41	46	53	32	38	41	32
2	58	64	61	36	57	59	54
3	65	68	77	46	68	67	63
4	74	75	77	57	80	78	73
5	77	77	79	66	79	78	76

Zone[1]	Total	Canada	U.K.	U.S.	Germany	Italy	Poland	U.S.S.R.	Other European	Other
					BIRTHPLACE					
1	41	45	49	51	49	31	39	39	32	26
2	58	63	65	69	59	36	59	61	50	57
3	65	68	69	78	69	46	70	70	60	64
4	74	75	76	83	75	56	76	76	72	74
5	77	77	77	83	79	62	81	39	75	77

[1] To derive zones: Metropolitan Toronto was divided into five distance zones, radiating from the city centre (Queen and Yonge intersection). The first zone consists of that area within a three-mile radius of the centre; the second, third, and fourth zones are each four miles wide; the fifth zone is made up of the rest of the metropolitan area (the most peripheral part).

Source: Wilfred G. Marston, "Social Class Segregation within Ethnic Groups in Toronto," *Canadian Review of Sociology and Anthropology*, 6 (2), 1969, Table VII, p. 77.

Spatial differentiation among immigrant groups is seen in Table 13. where there is a clear concentric variation in income distribution within ethnic groups and clear variations within concentric zones among ethnic groups. In other words, both ethnicity and income influence the housing choice of immigrants, creating complex sub-markets based upon the variables cited above.

CONCLUSION

In this brief excerpt the broad relationships between Canadian economic development and the basic structures of the urban system have been established, and selected dimensions and characteristics of modern Canadian urbanization—particularly the impact of internal and external migration on major urban centres, have been explored. Immigration has been a significant factor in urban growth but it has created problems. Low quality housing, lower incomes, and residential segregation have combined to create dissatisfaction among many immigrants. How much of this situation is part of an adjustment process and how much is likely to persist is not as yet known.

REFERENCES

[1] M. H. Watkins, "A Staple Theory of Economic Growth," *Canadian Journal of Economics and Political Science*, Vol. XXIX, May 1963.

[2] Douglass North, *The Economic Growth of the United States*, 1790-1860, Prentice-Hall, Englewood Cliffs, N.J., 1961.

[3] For a more comprehensive review of this relationship of urbanization to national development patterns, see C. F. Osler, *The Process of Urbanization in Canada*, 1600-1961, M.A. Essay, Department of Economics, Simon Fraser University, March 1968. See also L. O. Stone, *Urban Development in Canada*, D.B.S., 1961 Census Monograph, Ch. 2.

4 Economic Council of Canada, *Sixth Annual Review*, 1969, p. 23. From 1950 to 1962, real output in Canada and the United States grew by 4.8% and 3.3% respectively. Employment contributed 1.5 and 0.9 percentage points to these growth rates.

5 N. H. Lithwick and G. Paquet, eds., *Urban Studies: A Canadian Perspective*, Methuen, Toronto, 1968, Table 2-3, p. 29.

6 D. Michael Ray, "Urban Growth and the Concept of Functional Region," in N. H. Lithwick and G. Paquet, *op. cit.*, pp. 60 ff.

7 The monopolization of information resulting from the technological superiority of the United States in the computer field is already of concern, not only because of possible public policy limitations, but because of the potential for influencing individual Canadians.

8 See, for example, the study of W. G. Gray, M. Jalaluddin, and F. Charbonneau, *A Tale of Two Cities: Economic Growth and Change in Toronto and Montreal*, Economics Division, Policy and Planning Branch (ARDA,) 1968. Also, the analysis of Donald P. Kerr, "Metropolitan Dominance in Canada," in John Warkentin, ed., *Canada—A Geographical Interpretation*, Methuen, Toronto, 1968.

9 L. O. Stone, *op. cit.*, p. 193.

10 D. Michael Ray, in N. H. Lithwick and G. Paquet, *op. cit.*

11 See N. H. Lithwick and G. Oja, *Urban Poverty*, Research Monograph No. 1.

12 L. O. Stone, *op. cit.*, Table 6.4, p. 142 for regional data and Table 9.2, p. 179 for metropolitan area data.

13 L. O. Stone, *Migration in Canada: Regional Aspects*, D.B.S., 1961 Census Monograph, 1969.

14 *Ibid.*, p. 10. See also pp. 14-15.

15 *Ibid.*, p. 11.

16 *Ibid.*, p. 58 and Chapter 8.

17 *Ibid.*, p. 57.

18 *Ibid.*, p. 16.

19 "Rural-urban flows were dwarfed by urban-urban flows in the 1956-61 period." *Ibid.*, p. 66.

20 *Ibid.*, p. 58.

21 *Ibid.*, pp. 12-13.

22 *Ibid.*, p. 19.

23 *Ibid.*, p. 13.

24 *Ibid.*, p. 14.

25 D.B.S., "Unemployment as a Factor in Labour Mobility," *Canadian Statistical Review*, April 1962.

26 D.B.S., *Canadian Statistical Review*, July 1960.

27 *Ibid.*

28 For a limited but in-depth analysis of the problem, see Jane A. Abramson, *Rural to Urban Adjustment*, ARDA Project No. 37003, Ottawa, May 1968.

29 H. Charbonneau and J. Légaré, "L'extrême mobilité de la population urbaine au Canada," *Revue de Géographie de Montréal*, pp. 248-250.

30 H. Charbonneau and J. Légaré, "L'extrême mobilité de la population urbaine au Canada," *Revue de Géographe de Montréal*, pp. 248-250.

31 Much of this work has been conducted by the Institute for Behavioural Research, Ethnic Research Programme, York University, under Anthony H. Richmond.

32 CMHC, *Canadian Housing Statistics, 1968*, Table 101, p. 73.

33 Anthony H. Richmond, *Immigrants and Ethnic Groups in Metropolitan Toronto*, York U., June 1967.

34 *Ibid.*, p. 8.

35 *Ibid.*, p. 10.

The Functional Structure of Canadian Cities: A Classification of Cities

J. W. Maxwell, J. A. Greig and H. G. Meyer

INTRODUCTION

Geographers have long been interested in developing methods for classifying and comparing cities by function and specialization. Dickinson[1] reports that pioneer work was done as early as 1841 on the comparative study of towns as commercial centres. With the advent of modern tabulating machines, which have made feasible the publishing of detailed statistics for very small units including cities of small populations, quantitative methods of classification have come to the fore. Harris[2] lists the many quantitative city classification studies that have been made since he and William-Olsson published the first such studies in 1943.[3] These studies have proven useful at the preliminary stage of investigation since they allow the important functions of many cities to be identified by broad categories in a rapid and inexpensive manner. Such studies "suggest different types of relationships which urban agglomerations may have to one another and to economic regions."[4] They present an overview of the urban scene at the regional or national level. It is the purpose of this study to develop such an overview for the Canadian urban scene by studying the functional structure of the major cities in the nation.

CITIES STUDIED

All of the 180 incorporated urban centres in Canada that had 1961 populations of 10,000 and more were included in the study.

The total number of cities was reduced to 110 by recognizing census metropolitan areas as individual cities and grouping other contiguous municipalities to form single units. The group of adjoining urban municipalities allows a better approximation of geographical centres to be obtained.

Unincorporated urban centres and cities with populations of less than 10,000 in 1961 were omitted because the necessary labour-force statistics are not published for them. Kirkland Lake (1961 population: 15,484) was the only large centre affected by this lack of data. Table 1 gives the population, regional location, and constituent municipalties or census metropolitan area of all cities studied. Figure 1 shows the city locations and the system of regions that was established for analytical purposes.

SOURCE MATERIAL

Canadian census labour-force statistics for place of residence classified by industry were utilized exclusively as source material. As data from the 1971 census were not yet available when the study was done (1971), it was necessary to use 1961 statistics. The use of somewhat dated source material was not thought to be a serious limitation since the emphasis of the project was on methodolgy as well as on fact-finding. The results reveal the character of the functional profiles of the major Canadian cities at a particular point in time.

● J. W. Maxwell *is with the Lands Directorate, Lands, Forest and Wildlife Service, Department of the Environment, Ottawa.*
J. A. Greig *is an undergraduate student in geography at the University of Toronto and* H. G. Meyer *is a graduate in geography from Bishop's University.*

Reprinted from the *Geographical Bulletin*, Vol. 7(2), 1965, pp. 79-104 with permission of the author and Information Canada. The article was revised in 1971 by J. A. Greig and H. G. Meyer.

TABLE 1

City[a]	Population 1961	Location	Component Municipality or Census Metropolitan Area	Cat.
Alma**	13,309	Eastern Periphery	Alma	I
Amherst**	10,788	Eastern Periphery	Amherst	I
Arvida	14,460	Northern Periphery	Arvida	I
Asbestos**	11,083	Heartland	Asbestos	I
Barrie	21,169	Heartland	Barrie*	II
Belleville	30,655	Heartland	Belleville*	III
Brampton**	18,467	Heartland	Brampton	I
Brandon	28,166	Western Periphery	Brandon	II
Brantford	55,201	Heartland	Brantford*	III
Brockville	17,744	Heartland	Brockville*	I
Calgary	279,062	Western Periphery	Calgary C.M.A.	IV
Charlottetown	18,318	Eastern Periphery	Charlottetown*	I
Chatham	29,826	Heartland	Chatham*	II
Chicoutimi	42,886	Northern Periphery	Chicoutimi and Chicoutimi South	III
Cobourg**	10,646	Heartland	Cobourg	I
Corner Brook**	28,185	Eastern Periphery	Corner Brook	II
Cornwall	43,639	Heartland	Cornwall	III
Dawson Creek	10,946	Western Periphery	Dawson Creek	I
Drummondville	27,909	Heartland	Drummondville*	II
Edmonton	337,568	Western Periphery	Edmonton C.M.A.	IV
Edmunston	12,791	Eastern Periphery	Edmunston	I
Flin Flon**	11,104	Western Periphery	Flin Flon	I
Fort William/Port Arthur	90,490	Western Periphery	Fort William, Port Arthur	III
Fredericton	19,683	Eastern Periphery	Fredericton	I
Georgetown**	10,298	Heartland	Georgetown	I
Glace Bay	24,186	Eastern Periphery	Glace Bay	II
Granby	31,463	Heartland	Granby*	III
Grand'Mere	15,806	Heartland	Grand'Mere	I
Guelph	39,838	Heartland	Guelph	III
Halifax	183,946	Eastern Periphery	Halifax C.M.A.	IV
Hamilton	395,189	Heartland	Hamilton C.M.A.	IV
Joliette	18,088	Heartland	Joliette	I
Jonquière	28,588	Northern Periphery	Jonquière*	II
Kamloops**	10,076	Western Periphery	Kamloops	I
Kelowna**	13,188	Western Periphery	Kelowna	I
Kenogami**	11,816	Northern Periphery	Kenogami	I

(continued over page)

Table 1 (continued)

City[a]	Population 1961	Location	Component Municipality or Census Metropolitan Area	Cat.
Kenora**	10,904	Western Periphery	Kenora	I
Kingston	53,526	Heartland	Kingston	III
Kitchener	154,864	Heartland	Kitchener C.M.A.	IV
La Tuque**	13,023	Heartland	La Tuque	I
Lethbridge	35,454	Western Periphery	Lethbridge*	III
Lindsay**	11,399	Heartland	Lindsay	I
London	181,283	Heartland	London C.M.A.	IV
Magog	13,139	Heartland	Magog	I
Medicine Hat	24,484	Western Periphery	Medicine Hat*	II
Moncton	43,840	Eastern Periphery	Moncton*	III
Montreal	2,109,509	Heartland	Montreal C.M.A.	IV
Moose Jaw	33,206	Western Periphery	Moose Jaw*	III
Nanaimo**	14,135	Western Periphery	Nanaimo	I
New Waterford	10,592	Eastern Periphery	New Waterford	I
Niagara Falls	22,351	Heartland	Niagara Falls	II
Noranda**	11,477	Northern Periphery	Noranda	I
North Battleford**	11,230	Western Periphery	North Battleford	I
North Bay	23,781	Northern Periphery	North Bay	II
Orillia	15,345	Heartland	Orillia*	I
Oromocto**	12,170	Eastern Periphery	Oromocto	I
Oshawa	62,415	Heartland	Oshawa	III
Ottawa	429,750	Heartland	Ottawa C.M.A.	IV
Owen Sound	17,421	Heartland	Owen Sound*	I
Pembroke	16,791	Heartland	Pembroke*	I
Penticton	13,859	Western Periphery	Penticton	I
Peterborough	47,185	Heartland	Peterborough	III
Portage La Prairie**	12,388	Western Periphery	Portage La Prairie	I
Port Alberni**	11,560	Western Periphery	Port Alberni	I
Port Colbourne**	14,886	Heartland	Port Colbourne	I
Prince Albert	24,168	Western Periphery	Prince Albert*	II
Prince George**	13,877	Western Periphery	Prince George	I
Prince Rupert**	11,987	Western Periphery	Prince Rupert	I
Quebec	357,568	Heartland	Quebec C.M.A.	IV
Red Deer**	19,612	Western Periphery		I
Regina	112,141	Western Periphery	Regina	IV
Rimouski	17,739	Eastern Periphery	Rimouski	I
Rivière du-Loup**	10,835	Eastern Periphery	Rivière du-Loup	I
Rouyn	18,716	Northern Periphery	Rouyn	I
Saint John*	95,563	Eastern Periphery	Saint John C.M.A.	III
St. Catharines	84,472	Heartland	St. Catharines*	III
St. Hyacinthe	22,354	Heartland	St. Hyacinthe	II

Table 1 (continued)

City[a]	Population 1961	Location	Component Municipality or Census Metropolitan Area	Cat.
St. Jean	26,988	Heartland	St. Jean	II
St. Jérôme	24.546	Heartland	St. Jérôme*	II
St. John's	90,838	Eastern Periphery	St. John's C.M.A.	III
St. Thomas	22,469	Heartland	St. Thomas*	II
Sarnia*	50,976	Heartland	Sarnia	III
Saskatoon	95,526	Western Periphery	Saskatoon	III
Sault Ste. Marie	43,088	Northern Periphery	Sault Ste. Marie*	III
Sept Iles**	14,196	Eastern Periphery	Sept Iles	I
Shawinigan	44,852	Heartland	Shawinigan and Shawinigan South	III
Sherbrooke	66,524	Heartland	Sherbrooke	III
Sorel	17,147	Heartland	Sorel	I
Stratford	20,467	Heartland	Stratford*	II
Sudbury	110,694	Northern Periphery	Sudbury C.M.A.	IV
Swift Current**	12,186	Western Periphery	Swift Current	I
Sydney	33,617	Eastern Periphery	Sydney	III
Thetford Mines	21,618	Heartland	Thetford Mines	II
Timmins	29,270	Northern Periphery	Timmins	II
Toronto	1,824,481	Heartland	Toronto C.M.A.	IV
Trail	11,580	Western Periphery	Trail	I
Trenton	13,183	Heartland	Trenton*	I
Trois Rivières	80,402	Heartland	Trois Rivières Cap-de-la-Madeleine	III
Truro	12,421	Eastern Periphery	Truro	I
Val d'Or**	10,983	Northern Periphery	Val d'Or	I
Valleyfield	27,297	Heartland	Valleyfield*	II
Vancouver	790.165	Western Periphery	Vancouver C.M.A.	IV
Vernon**	10,250	Western Periphery	Vernon	I
Victoria	154.152	Western Periphery	Victoria C.M.A.	IV
Victoriaville	18,720	Heartland	Victoriaville*	I
Welland	36.079	Heartland	Welland*	III
Whitby**	14,685	Heartland	Whitby	I
Windsor	193,365	Heartland	Windsor C.M.A.	IV
Winnipeg	475.989	Western Periphery	Winnipeg C.M.A.	IV
Woodstock	20,486	Heartland	Woodstock	II

[a] Geographical rather than legal cities have been used where possible. Census Metropolitan Areas are considered as single cities, as are clusters of contiguous municipalities.
* Indicates a change in the municipal boundaries since the preceding census.
** Cities having reached 10,000 population since 1951 census.

Note: Certain urban complexes have not been recognized because the component cities have highly individualistic and often contrasting functional profiles. By reporting these cities separately, their outstanding functional characteristics are fully recognized. Such is the case for the cities forming the urban complexes in: (1) the Saguenay Valley (Chicoutimi, Arvida, Jonquière), (2) Cape Breton Island (Sydney, Glace Bay, New Waterford) and (3) the St. Maurice Valley (Shawinigan and Grand Mére).

Figure 1

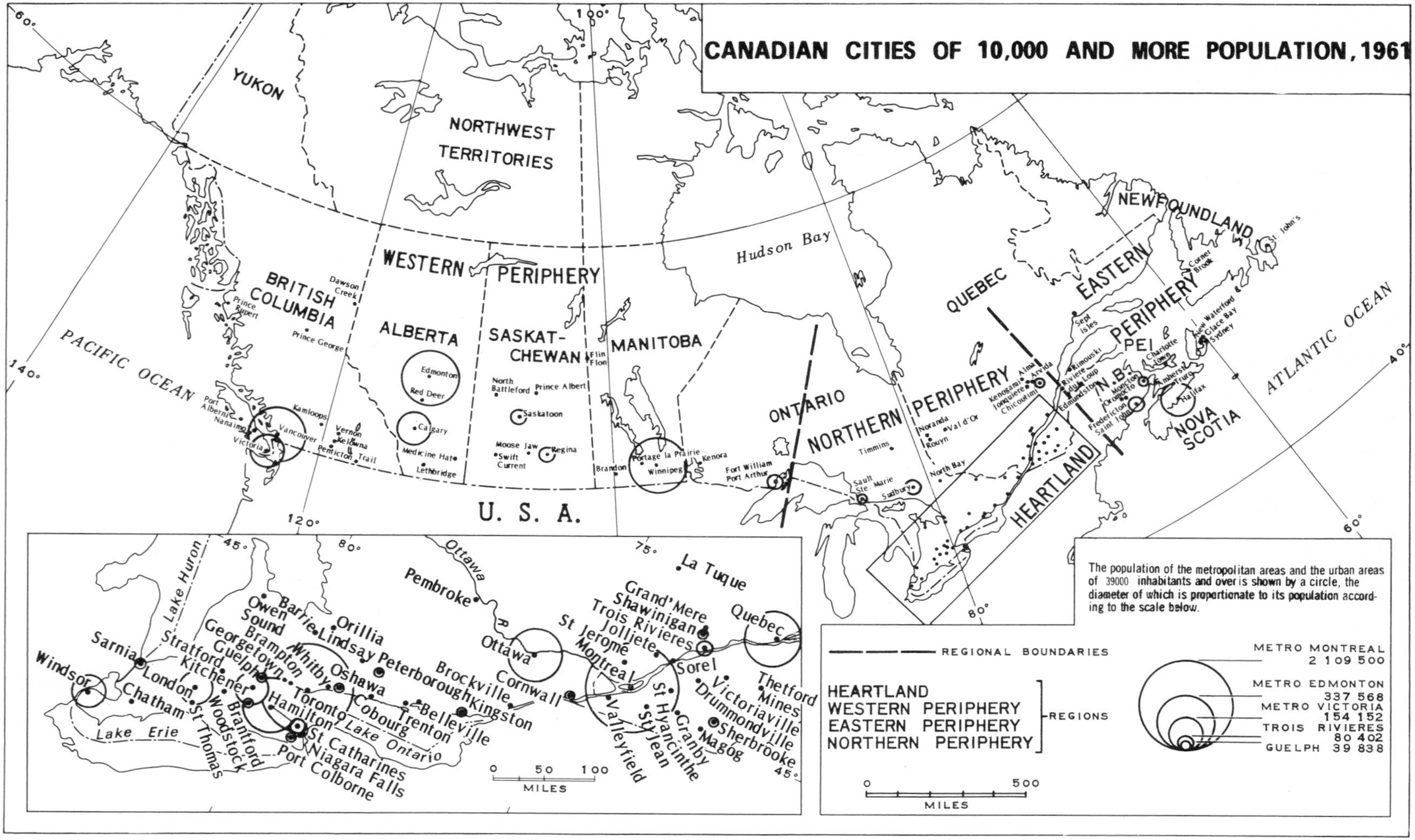

CANADIAN CITIES OF 10,000 AND MORE POPULATION, 1961

ATLANTIC OCEAN

NEWFOUNDLAND

EASTERN PERIPHERY

St John's

QUEBEC

NOVA SCOTIA

P.E.I.

N.B.

HEARTLAND

NORTHERN PERIPHERY

ONTARIO

Hudson Bay

Timmins

Sault Ste Marie
Ste Marie

Fort William
Port Arthur

MANITOBA

Flin Flon

Winnipeg
Portage la Prairie
Brandon

SASKAT-
CHEWAN

Prince Albert
North Battleford
Saskatoon
Moose Jaw
Swift Current
Regina

WESTERN PERIPHERY

NORTHWEST TERRITORIES

ALBERTA

Edmonton
Red Deer
Calgary
Medicine Hat
Lethbridge

BRITISH COLUMBIA

Dawson Creek
Prince George
Kelowna
Vernon
Penticton Trail
Kamloops
Vancouver
Victoria
Nanaimo
Port Alberni
Prince Rupert

YUKON

PACIFIC OCEAN

U. S. A.

The population of the metropolitan areas and the urban areas of 39000 inhabitants and over is shown by a circle, the diameter of which is proportionate to its population according to the scale below.

METRO MONTREAL   2 109 500
METRO EDMONTON   337 568
METRO VICTORIA   154 152
TROIS RIVIERES   80 402
GUELPH   39 838

─ ─ ─ REGIONAL BOUNDARIES

— · — HEARTLAND
        WESTERN PERIPHERY   REGIONS
        EASTERN PERIPHERY
        NORTHERN PERIPHERY

0   500
MILES

(inset — southern Quebec / Ontario)

La Tuque
Grand'Mere
Shawinigan
Trois Rivieres
Jolliette
St Jerome
Montreal
Quebec
Thetford Mines
Victoriaville
Drummondville
Sherbrooke
Magog
Granby
St Hyancinthe
St Jean
Valleyfield
Sorel

Ottawa
Cornwall
Pembroke
Brockville
Kingston
Belleville
Trenton

Orillia
Barrie
Lindsay Peterborough
Owen Sound
Brampton
Georgetown
Guelph
Kitchener
Stratford
London
Woodstock
Brantford
St Thomas
Chatham
Sarnia
Windsor

Whitby
Oshawa
Cobourg
Toronto
Hamilton
St Catharines
Niagara Falls
Port Colborne

Lake Ontario
Lake Huron
Lake Erie

Ottawa R.

0   50   100
MILES

FUNCTIONAL CATEGORIES

City "functional structure" was equated to city "employment structure" as it was thought that the activities in which a city's people earn their living provide good indicators of city functions. The terms "city activity" and "city function" were taken to be synonymous. The city functions recognized consist of groupings of census industrial categories. The 1961 *Census of Canada* breaks down the services performed by the labour force into twelve industry divisions, which are further broken down into major groups. The first column of Table 2 lists all census industry divisions and major groups that were relevant for the study. The second column of the table presents the city functional categories derived from the census categories. The functional classes represent broad groups in the employment structure of the city. Although highly aggregated, these categories are sufficiently detailed to distinguish in reasonably specific terms the functional character of cities.

Table 2

CENSUS INDUSTRIAL CATEGORIES AND CITY FUNCTIONAL CATEGORIES
USED IN THE STUDY

Census Industry Divisions and Major Groups significant to the study[a]	City Functions recognized in the study
Agriculture	Omitted
Forestry	Omitted
Fishing and Trapping	Omitted
Mines, Quarries and Oil Wells	Extraction
Manufacturing	Manufacturing
Construction	Construction
Transportation, Communication and Other Utilities	
Transportation, storage and Communication	Transportation
Electric Power, Gas and Water Utilities	Public Utilities
Trade	
Wholesale Trade	Wholesale Trade
Retail Trade	Retail Trade
Finance, Insurance and Real Estate	Finance, Insurance and Real Estate
Community, Business and Personal Service	
Community Service[b]	Community Service
Recreation[c]	Recreation
Business Service[d]	Business Service
Personal Service[e]	Personal Service
Public Administration and Defence	
Government Service[f]	Government Service
Industry Unspecified or Undefined	Omitted

[a] Dominion Bureau of Statistics, *Tenth Census of Canada: 1961, Labour Force*, Vol. III. Tables 2, 3, 5, and 6 (Ottawa: Queen's Printer, 1963 and 1964).

[b] "Community Service" includes labour force in education and related services, health and welfare services and religious organizations.

[c] "Recreation" includes labour force employed in motion picture and recreational services.

[d] "Business Service" includes labour force employed in accounting, advertising, engineering and scientific services, labour and trade organizations, law and other business services.

[e] "Personal Service" includes labour force employed in shoe repair shops, barber and beauty shops, private households, laundries, cleaners and pressers, hotels, restaurants, and taverns, and lodging houses and residential clubs, funeral directors, dressmaking, and other personal services.

[f] "Government Service" includes labour force employed in work peculiar to public service only. It includes federal, provincial and local administrators as well as other government personnel.

METHOD OF ANALYSIS

Basic Employment

Basic employment—that part of total city employment engaged in activities that produce goods and services for markets outside the city—was used when characterizing city functional structures. It is the basic employment that identifies the *raison d'être* of a city and reveals the functional realtionships that exist between the city and its hinterland.[5] To calculate each city's basic employment in the respective functions, a technique developed by Ullman and Dacey,[6] termed the "minimum requirements method," was used. These analysts were able to produce:

> ". . . a quantitative statement which closely approximates the minimum percentage of a labour force required in various sectors of its economy to maintain the viability of an urban area. The employment in an urban area which is greater than this minimum requirement is called excess employment . . . the excess employment approximated the export or basic employment." (p. 176)

Minimum requirement values for selected functions were determined empirically by Ullman and Dacey for American cities of various size classes. These values were given as percentages of total city labour force. They found, as had earlier investigators[7], that the minimum requirement values varied with city size. As city size increases, the minimum employment in the various functions required to keep the city viable also increases; however the rate of this increase differs among functions. By plotting the empirically derived values for city minimum requirements in functions against the logarithms of the city population values, Ullman and Dacey found that the plotted points closely fit a straight line. Using these data, they were able to develop linear regression equations that allow expected minimum requirement values for functions to be calculated for cities of all sizes. Once a city's minimum requirement values for functions are known, the excess employment can be calculated. This excess employment can be taken as a good approximation of basic employment.

Excess (Basic) Employment in Canadian Cities

The regression equations derived from the Canadian data were developed by calculating the percentage of each city's urban labour force* in each of the functions, and grouping the cities into four arbitrarily determined groups on the basis of population. For every function the minimum percentage figure in each city group was noted (Table 3). These values were considered to be empirical minimum requirements. They were plotted against the logarithms of the appropriate city population values and regression lines were fitted to them. Expected minimum requirement values for cities of any size could then be read off the regression lines. Figure 2 shows the regression lines used in determining the expected minimum requirements for the functions.

Comparison of the regression lines based on Canadian data with the equivalent Ullman-Dacey lines showed that the latter gave lower minimum requirement values at the high-population levels. This situation arises, it is thought, because in Canada there are few very large cities and they are widely dispersed. None of these cities are units in well-integrated "city systems" such as the Manufacturing Belt of the United States, where cities the size of the largest Canadian centres are clustered together, and where exchange among centres reaches large proportions. In the city of the system, concentration on one or few functions is possible. Only the minimum activity necessary to keep the city viable needs to be maintained for other functions. In contrast, the non-system city must be more self-sufficient because it cannot depend to the same degree on other cities for its needs, and because it must be much more concerned with the requirements

---

* Total city labour force excluding the labour force in the "Agriculture," "Forestry and logging," "Fishing and Trapping," and "Not Stated" categories of the 1961 Census of Canada industrial classification. Employment in these categories is very insignificant for urban areas. For practical purposes, urban labour force can be equated to total city labour force.

of its rural hinterland. This latter situation is not conducive to the maintenance of only minimum requirements in functions. Although "city systems" do exist for medium sized cities in Canada—for example, the city cluster focused on Toronto in southern Ontario—they do not exist for the very large cities in the country. For this reason, the possibility of finding only the minimum activity necessary to keep a city viable is remote for these large centres.

CHARACTERIZATION OF CITY

Functional Structures

Three characteristics were used to typify city functional structures: the city's dominant function, its distinctive functions, and its degree of functional specialization. The dominant function of a city is defined as the activity having the largest share of the city's total excess employment. A distinctive function is defined as an activity whose share of

Figure 2

REGRESSION LINES USED TO DETERMINE EXPECTED MINIMUM REQUIREMENTS FOR FUNCTIONS

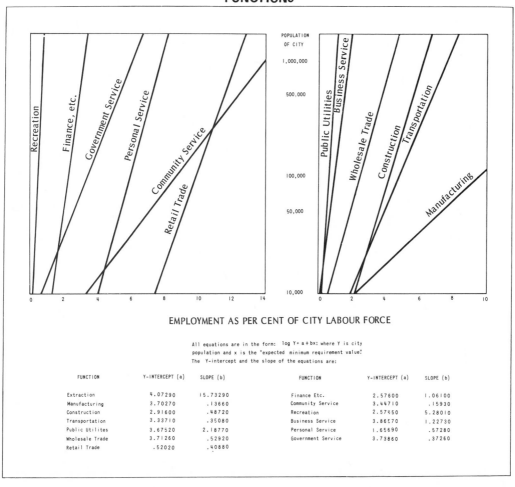

EMPLOYMENT AS PER CENT OF CITY LABOUR FORCE

All equations are in the form: $\log Y = a + bx$; where Y is city population and x is the "expected minimum requirement value". The Y-intercept and the slope of the equations are:

FUNCTION	Y-INTERCEPT (a)	SLOPE (b)	FUNCTION	Y-INTERCEPT (a)	SLOPE (b)
Extraction	4.07290	15.73290	Finance Etc.	2.57600	1.06100
Manufacturing	3.70270	.13660	Community Service	3.44710	.15930
Construction	2.91600	.48720	Recreation	2.57450	5.28010
Transportation	3.33710	.35080	Business Service	3.86570	1.22730
Public Utilites	3.67520	2.18770	Personal Service	1.65690	.57280
Wholesale Trade	3.71260	.52920	Government Service	3.73860	.37260
Retail Trade	.52020	.40880			

Table 3

MINIMUM PERCENTAGES EMPLOYED IN CANADIAN CITIES OF VARYING SIZE CLASSES, 1961

| Function | Cities (population in thousands) | | | |
	10.0-19.9 (50 cities)	20.0-29.9 (20 cities)	30.0-99.9 (24 cities)	100 & over (16 cities)
Extraction	00.0	00.0	00.0	00.0
Manufacturing	02.4	05.6	09.6	10.5
Construction	02.2	03.3	04.1	04.8
Transportation	02.3	04.3	04.3	04.2
Public Utilities	00.2	00.5	00.4	00.8
Wholesale Trade	00.5	01.7	01.8	03.2
Retail Trade	09.4	09.5	11.2	10.5
Finance, Insurance and Real Estate	01.4	01.7	02.1	02.7
Community Service	04.6	08.6	08.9	08.4
Recreation	00.3	00.4	00.3	00.5
Business Service	00.1	00.5	00.7	01.2
Personal Service	04.1	05.6	05.3	05.9
Government Service	02.7	01.2	02.7	03.3
TOTAL	30.2	42.9	51.4	56.0

Source: Calculated from the Tenth *Census of Canada,* 1961.

total excess employment in a city greatly exceeds the share it usually has in most cities. City specialization is given by an index based on the relationship of a city's distribution of total excess employment among functions to its distribution of minimum requirement employment among the functions.

Dominant and Distinctive Functions

By recognizing two types of functions—dominant and distinctive functions—two kinds of functional importance are considered. Functions are rated in terms of their importance in a city's functional structure relative to the importance of the other functions in the city's structure when determining the dominant activity of a city. When identifying a city's distinctive functions, activities are rated in terms of their importance in a city's functional profile relative to their importance in the functional profiles of all cities studied.

When identifying the dominant functions in cities, two classes of manufacturing dominance were recognized: "Manufacturing I" and "Manufacturing II." This was done be-

cause this function has the largest share of excess employment in so many cities. Included in the first class are all cities where 50 percent or more of a city's excess employment is accounted for by manufacturing. In the second class are all cities where manufacturing is dominant, but accounts for less than 50 percent of the city's excess employment.

Four classes of distinctive functions were recognized. The classes were defined in terms of standard deviation values of excess employment above the mean value of excess employment for an activity. For each function, Class I includes all cities whose value of excess employment for the function under consideration is equal to the mean value of excess employment for the function, plus a value of at least two standard deviations above the mean. Class II includes all cities whose values of excess employment for the function exceed the mean by between one and two standard deviations, and Class III includes all cities whose excess employment value for the function is between the mean and one standard deviation above the mean. Class IV includes all cities where excess em-

Table 4

VALUES OF EXCESS EMPLOYMENT FOR DETERMINING
CLASSES OF FUNCTIONAL IMPORTANCE

Function	Mean value of excess employment	Standard deviation of excess employment	Class IV (mean value of excess employment)	Class III (between mean value and one standard deviation above the mean)	Class II (between one and two standard deviations above the mean)	Class I (over two standard deviations above the mean)
				percent of city urban labour force		
Extraction	6.429	18.02	< 6.43	6.44 - 24.44	24.45 - 42.46	42.47+
Manufacturing	33.837	25.83	< 33.84	33.85 - 59.66	59.67 - 85.49	85.50+
Construction	6.051	3.91	< 6.05	6.06 - 9.96	9.97 - 13.87	13.88+
Transportation	9.041	7.34	< 9.04	9.05 - 16.38	16.39 - 23.72	23.73+
Utilities	1.702	2.42	< 1.70	1.71 - 4.12	4.13 - 6.54	6.55+
Wholesale Trade	5.369	4.49	< 5.37	5.38 - 9.85	9.86 - 14.34	14.35+
Retail Trade	5.961	3.72	< 5.96	5.97 - 9.68	9.69 - 13.41	13.42+
Finance, etc.	2.894	2.60	< 2.89	2.90 - 5.50	5.51 - 8.41	8.42+
Community Service	10.285	7.32	< 10.29	10.30 - 17.60	17.61 - 24.92	24.93+
Recreation	.564	.44	< .56	.57 - 1.00	1.01 - 1.44	1.45+
Business Service	1.774	1.23	< 1.77	1.78 - 3.01	3.02 - 4.25	4.26+
Personal Service	4.638	2.93	< 4.64	4.65 - 7.57	7.58 - 10.50	10.51+
Government Service	11.445	16.02	< 11.45	11.46 - 27.46	27.47 - 43.48	43.49+

ployment values for the function are below the mean. Table 4 shows the values determining the classes of importance for each function.

Cities were rated in each function in terms of the classes of "distinctiveness." An activity was considered "distinctive" in the functional structure of a city when the city qualified for one of the first three classes. Cities have as many distinctive activities as they have excess employment values qualifying them as being of Class I, II, or III in "distinctiveness." A function was not considered "distinctive" in a city when the city qualified only for a Class IV rating.

City Functional Specialization

A procedure frequently used in determining the degree of functional specialization of cities is to compare the functional structure of individual cities with an idealized functional profile representing what is thought to be a balanced profile or a situation of minimum specialization[8] The city's degree of specialization is then given by the deviation of its functional profile from the model. The problem has been to develop meaningful models representative of a balanced structure.

In this study a specialization index developed by Ullman and Dacey[9], which utilizes the minimum requirement values for functions, was adopted. These analysts have used the distribution of a city's total minimum requirement employment among the functions as a model of a balanced structure, that is, the case of minimum specialization. If the city's excess employment is distributed among the functions in the same proportions as is the minimum requirement employment, the case of least specialization is considered to exist. City specialization is thought to increase with increasing deviation of the city's excess employment distribution from the distribution of the city's minimum requirement employment among functions.

Ullman and Dacey designed their index so that large deviations of the excess employment distribution from the distribution of minimum requirement employment are accentuated. Their formula for determining the index is:

$$S = \Sigma_i \left[\frac{(P_i\text{-}M_i)^2}{M_i} \right] \div \frac{(\Sigma_i P_i\text{-}\Sigma_i M_i)^2}{\Sigma_i M_i}.$$

where: "S" is the index of specialization, "i" refers to each of the thirteen functions, "P" to the percentage of city labour force employed in each of the "i" functions, "M_i" the expected minimum requirement value for each function, and "Σ_i" the sum of all the functions.

An index value of unity indicates that the excess employment of the city is distributed among the functions in the same proportions as is the city's minimum requirement employment. This is the case of minimum specialization. As the value of the index increases, specialization is considered to increase.

The formula for the index operates satisfactorily as long as the expected minimum requirement value or the excess employment value is not less than zero. In cases where either value registered as zero or less an arbitrary value of .01 was substituted.

THE RESULTS OF ANALYSIS

Maps showing the distributions of functions as dominant and distinctive activities in cities and of city specialization values were constructed. The distribution patterns were examined in terms of the Canadian heartland—the densely populated St. Lawrence Lowlands and southern Ontario—and the periphery of Canada—the remainder of the country. This regional framework is useful when analyzing city functional character at the national level because tthe distribution of cities within the two regions differs so greatly.

The patterns of functional dominance and specialization given in Table 5 and Figure 3 show in a dramatic way the overwhelming importance of manufacturing in city functional structures. This function is dominant in 66 of the 110 cities. Eight other functions register as dominant activities. Extraction is dominant in twelve centres, government service in eleven, transportation in nine, community service in eight, utilities in one, wholesale in one, retail in one, and personal in one.

Although manufacturing is almost ubiquitous as a dominant function, it achieves an especially important position in the functional profiles of heartland cities. Even within the heartland, there is a high concentration of manufacturing dominance. All but three of the heartland cities with ratings of "Manufacturing I" are found in the city clusters focused on Toronto and Montreal respectively. These two nodes make up the manufacturing heart of Canada. Manufacturing in most of the heartland cities between and beyond the two city clusters, while still the dominant activity, has a much lower rate of importance.

Manufacturing generally plays a much less significant role as a dominant function in the profiles of periphery cities than in those of the heartland cities. All but eight of the fifty-two heartland cities have manufacturing as their dominant function. In contrast, just over one-third of the periphery cities are similarly characterized; the balance have transportation, retail trade or some other function as the dominant activity. In the periphery ratings of "Manufacturing I" are infrequent. Only six of the fifty-eight cities in the periphery have such ratings. Further, there is a fundamental difference between "Manufacturing I" cities in the heartland and the periphery.

The heartland cities with ratings of "Manufacturing I" form basic units in fully integrated manufacturing regions where interaction among cities is very significant. High degrees of industrial linkage have developed among industries at both the intracity and intercity levels, and all types of industrial activity occur, from blast furnace activity to electronics manufacturing and food processing. Contrasting markedly with this situation is the position of the periphery "Manufacturing I" cities. They are few in number, have isolated locations, and usually have only one important industrial activity. These manufacturing cities also stand apart from the other periphery cities, being highly specialized and not primarily concerned with servicing a rural hinterland. Although based on different activities, they have many traits in common with the extraction centres such

Figure 3

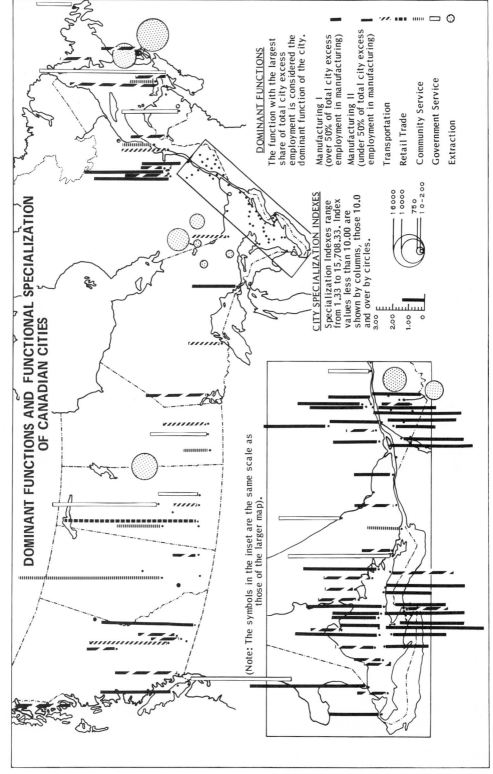

DOMINANT FUNCTIONS AND FUNCTIONAL SPECIALIZATION
OF CANADIAN CITIES

(Note: The symbols in the inset are the same scale as
those of the larger map).

CITY SPECIALIZATION INDEXES

Specialization Indexes range
from 1.33 to 15,708.33. Index
values less than 10.00 are
shown by columns, those 10.0
and over by circles.

3.00

2.00

1.00

0

16000
10000
750
10-200

DOMINANT FUNCTIONS

The function with the largest
share of total city excess
employment is considered the
dominant function of the city.

Manufacturing I
(over 50% of total city excess
employment in manufacturing)

Manufacturing II
(under 50% of total city excess
employment in manufacturing)

Transportation

Retail Trade

Community Service

Government Service

Extraction

157

Table 5

DOMINANT FUNCTIONS AND SPECIALIZATION INDEXES OF CANADIAN CITIES

City Periphery: Heartland	Specialization Index	Dominant Function	City Periphery: Heartland	Specialization Index	Dominant Function
Chicoutimi	1.33	Manufacturing II	Winnipeg	2.67	Transportation
Nanaimo	1.47	Manufacturing II	Woodstock	2.70	Manufacturing I
Moose Jaw	1.49	Transportation	St. Jerome	2.76	Manufacturing I
Charlottetown	1.51	Community Service	Brampton	2.82	Manufacturing I
Medicine Hat	1.53	Manufacturing II	Lindsay	2.82	Manufacturing II
Brandon	1.59	Community Service	Whitby	2.87	Manufacturing I
Lethbridge	1.62	Wholesale Trade	Sorel	3.06	Manufacturing I
St. Thomas	1.64	Manufacturing II	Windsor	3.13	Manufacturing I
Saint John	1.66	Manufacturing II	Victoriaville	3.15	Manufacturing I
Prince Albert	1.68	Community Service	St. Catharines	3.24	Manufacturing I
Saskatoon	1.70	Community Service	Fredericton	3.34	Government Service
Sherbrooke	1.71	Manufacturing II	Brantford	3.35	Manufacturing I
London	1.74	Manufacturing II	Shawinigan	3.38	Manufacturing I
Joliette	1.74	Manufacturing II	Portage la		
Rimouski	1.74	Community Service	Prairie	3.44	Government Service
Chatham	1.75	Manufacturing II	Cobourg	3.69	Manufacturing II
Truro	1.83	Manufacturing II	Prince Rupert	3.71	Manufacturing II
Owen Sound	1.84	Manufacturing II	Kenora	3.75	Manufacturing II
Belleville	1.86	Manufacturing II	St. Jean	3.77	Manufacturing I
Orillia	1.87	Manufacturing II	Pembroke	3.78	Government Service
Kelowna	1.88	Manufacturing II	Drummondville	3.78	Manufacturing I
Prince George	1.89	Manufacturing II	Vancouver	3.83	Manufacturing II
Corner Brook	1.93	Manufacturing II	Grand'mère	3.91	Manufacturing I
Penticton	1.93	Personal Service	La Tuque	3.92	Manufacturing I
Fort William-			Welland	4.13	Manufacturing I
Port Arthur	1.97	Transportation	Granby	4.32	Manufacturing I
Kingston	2.03	Community Service	Port Alberni	4.47	Manufacturing I
Edmunston	2.16	Manufacturing II	Peterborough	4.67	Manufacturing I
North Bay	2.17	Transportation	Kenogami	4.70	Manufacturing I
Moncton	2.20	Manufacturing I	Oshawa	4.83	Manufacturing I
Cornwall	2.21	Manufacturing I	Kamloops	4.92	Transportation
Vernon	2.25	Manufacturing II	Port Colbourne	5.04	Manufacturing I
Sarnia	2.27	Manufacturing I	Arvida	5.09	Manufacturing I
St. John's	2.28	Government Service	Kitchener	5.11	Manufacturing I
Niagara Falls	2.40	Manufacturing II	Magog	5.54	Manufacturing I
Stratford	2.41	Manufacturing II	Amherst	5.97	Manufacturing II
Jonquière	2.45	Manufacturing I	Georgetown	6.12	Manufacturing I
Québec	2.45	Government Service	Victoria	6.20	Government Service
Barrie	2.45	Manufacturing II	Hamilton	6.28	Manufacturing I
Brockville	2.48	Manufacturing II	Trenton	6.40	Government Service
Regina	2.49	Government Service	Edmonton	6.65	Government Service
St. Hyacinthe	2.51	Manufacturing I	Valleyfield	7.24	Public Utilities
Alma	2.53	Manufacturing II	Swift Current	8.29	Retail Trade
Sydney	2.53	Manufacturing II	Halifax	8.61	Government Service
Guelph	2.57	Manufacturing I	Red Deer	8.89	Community Service
North Battleford	2.60	Community Service	Trail	19.21	Manufacturing I
Trois Rivières	2.62	Manufacturing I	Ottawa	19.77	Government Service
Rivière du Loup	2.63	Transportation	Toronto	33.31	Manufacturing I
Sault Ste. Marie	2.65	Manufacturing I	Calgary	36.58	Extraction

Table 5 (continued)

City Periphery: Heartland	Special-ization Index	Dominant Function	City Periphery: Heartland	Special-ization Index	Dominant Function
Oromocto	48.23	Government Service	Glace Bay	708.50	Extraction
Montreal	54.83	Manufacturing I	Thetford Mines	812.45	Extraction
Dawson Creek	101.18	Transportation	Val d'Or	5416.33	Extraction
Sept Iles	201.71	Transportation	Asbestos	7111.85	Extraction
Rouyn	423.31	Extraction	Noranda	10101.76	Extraction
Sudbury	536.00	Extraction	New Waterford	15565.25	Extraction
Timmins	616.80	Extraction	Flin Flon	15708.33	Extraction

as extremely high specialization rates and the tendency to be "one-industry towns." Without exception the periphery cities with "Manufacturing I" ratings are resource-oriented manufacturing centres such as Trail and Arvida.

Many of the cities with "Manufacturing II" ratings, both in the periphery and the heartland, are service or transport centres despite the high rating enjoyed in them by manufacturing as a dominant function. The fact that many of these centres are service centres suggests that the kind of manufacturing activity found in them will differ considerably from that of the "Manufacturing I" cities. It is thought that market-oriented manufacturing industries associated with the final stages of production will take up a proportionately larger share of city manufacturing activity in the service centres than in the specialized manufacturing cities ("Manufacturing I" centres), where industries associated with the earlier stages of production for both final and non-final markets will be relatively more important.

Industries that use local, ubiquitous materials such as water, whose products are highly perishable, where the manufacturing process involves weight gain, or where the size of market for one minimum-sized plant is small, will make up the bulk of the manufacturing activity in the service centres. Such industries as baking, soft drinks, and ice cream manufacturing are representative of this group. These are the kinds of industries that reinforce the central place functions of the city.

Because manufacturing enjoys a position of such overwhelming importance as a dominant activity, it tends to blanket many of the distinct and unique features in city functional profiles. This results in an almost uniform and colourless impression of city functional character. For this reason the distribution patterns of functions as distinctive activities need to be examined to determine the significant functional rôles of many cities. This is especially true when manufacturing is the dominant, but not a distinctive function in a city. This is the case in most cities with ratings of "Manufacturing II." If the most distinctive functions of this group of cities are examined the importance of the central place functions soon becomes apparent. Wholesale trade, finance, recreation and community service rank high as distinctive functions, followed closely by retail trade, and government service. Transportation also ranks high as a distinctive function in some of these cities and serves to isolate those instances where cities owe their existence primarily to transport functions although having manufacturing as their dominant activity. Prince Rupert provides the classic example of a city that has built up an important manufacturing function based on a transport break-in-bulk point.

The relationship between the central place functions and manufacturing in the functional profiles of cities is clearly illustrated by Figure 4. Wholesale trade, a key central place function, is shown to have an almost unbroken high rating of distinctiveness in the periphery. In the heartland it appears distinctive only in eight cities, four of which are outside the manufacturing nodes. Pro-

159

viding a marked contrast to this pattern is the distribution of manufacturing as a distinctive function. Manufacturing distinctiveness is almost completely restricted to the heartland cities of the manufacturing cores. Only in the resource-oriented manufacturing cities does the function appear distinctive in the periphery. Manufacturing, although dominant in many cities, both in the periphery and the heartland, is highly localized as a distinctive function. A major difference is seen to exist between the periphey and heartland in the relationship that wholesale trade and manufacturing have to one another in the functional profiles of the cities.

Further study of Table 5 and Figure 3 shows that a close association exists between city specialization, city location, and dominant functions. The highest rates of functional specialization are found in the extraction centres. These high rates are due primarily to the exclusive priority necessarily placed on resource location for this function. This feature often precludes the possibility of other city-forming activities emerging as important basic activities in extraction centres, resulting in extreme specialization rates. Next to the extraction centres, the most specialized group of cities are the "Manufacturing I" centres. These cities in the two manufacturing nodes of the heartland are highly specialized because the close proximity of cities to one another has allowed a high rate of city interdependence to develop. Concentration on one or a few functions is possible for a city since it can rely on the other cities within the nodes for markets and many of its requirements. The resource-oriented manufacturing centres in the periphery are highly specialized for much the same reason as are the extraction centres. They are not located in areas with densely populated hinterlands, where central place functions are required to a degree that would offset the manufacturing specialization. Further, because their industrial activity is concerned with one or few operations, rather than many as in the case of the cities in the manufacturing regions, the possibility of diversification developing even in the manufacturing component of their functional structures through industrial linkage is small.

Certain government centres also have high rates of specialization. The government function makes the national capital, Ottawa, one of the most specialized cities in the country. Without this activity the city would have a diversified structure and would be a relatively small service centre serving the lower Ottawa Valley. Victoria and Halifax also are highly specialized because of the existence of both administrative and military activities in their profiles. Like Ottawa, they would be relatively small regional centres with diversified structures if the government activity was not present. Most other cities that have important government functions have more diversified structures than the three cities cited because other functions are also important. This has a diversifying effect on their functional structure. Quebec City and Regina are examples of important government centres that have quite diversified profiles.

Functional diversification is greatest among the cities of the periphery, where serving a large hinterland is the *raison d'être* of the city. Rimouski, serving the Richelieu River basin, has the least specialized profile in the country. These periphery cities are closely followed, in terms of specialization, by heartland cities outside the manufacturing regions that act primarily as service centres. Specialization is slightly greater for these cities than their counterparts in the periphery because they are closer to the population centre of the nation and to the manufacturing nodes. The possibility of their receiving special activities such as the national headquarters of large firms is greater than it is for periphery cities. London, for example, is the principal service centre in southwestern Ontario. In addition, it is also an important financial and insurance centre on the national level.

These observations on dominant and distinctive functions, and city functional specialization suggest that a fundamental difference in functional character exists between the cities of the periphery and the heartland. The periphery cities, except for the resource-oriented manufacturing cities and the extraction centres, are less specialized and place a

Figure 4

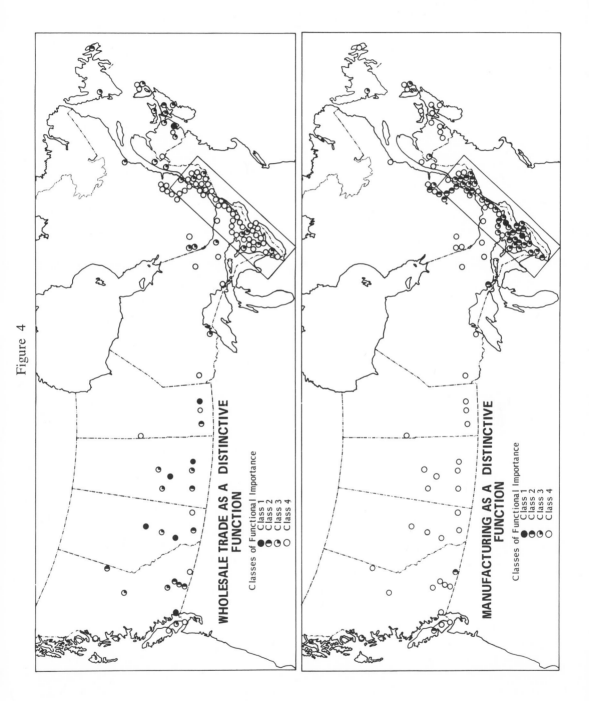

WHOLESALE TRADE AS A DISTINCTIVE
FUNCTION

Classes of Functional Importance
● Class 1
◐ Class 2
◔ Class 3
○ Class 4

MANUFACTURING AS A DISTINCTIVE
FUNCTION

Classes of Functional Importance
● Class 1
◐ Class 2
◔ Class 3
○ Class 4

161

lower emphasis on manufacturing than do the heartland cities. Most periphery cities have an important involvement with functions associated with distance such as wholesale trade and transportation. The difference is illustrated most clearly in Figure 4, where wholesale trade is shown to be almost completely restricted to the periphery as a distinctive function. Manufacturing distinctiveness, in contrast, is shown to be limited to the heartland.

A review of the characteristics of wholesale trade and manufacturing is useful when examining the difference between heartland and periphery cities. Wholesale trade, while requiring the concentration of facilities and people for its performance, is essentially a function of space. It is concerned with the collection, transfer, and distribution of goods. It can be expected, therefore, to reach its highest levels of relative importance in those areas where extensive activities are of prime importance. Such is the situation in the eastern and especially the western periphery. A very large proportion of the total "economic energy" generated in these areas is "consumed" in overcoming the great distances associated with extensive activities such as the wheat-production and ranching functions of the brown and dark-brown soil regions of the Great Plains. The cities in these areas reflect in their functional profiles this involvement with distance, and have the central place functions ranking high in their structures.

The manufacturing function, on the other hand, is focused on adding value to goods through their physical conversion. Economies that accrue through agglomeration and economies of scale tend to promote the concentration of manufacturing activity in areas where large populations and other manufacturing activities are located. Such locations provide easy access to market, allow specialization of production, provide more opportunities for substitution of inputs, and guarantee an ample labour supply. Only when the costs of assembling inputs is high in relation to total costs, when no substitutes are available, or when the weight loss is high

in the manufacturing process will resource or input location factors override those factors favouring market-oriented locations.

Because of good accessibility to both the Old World and the United States, the presence of resources demanded on world markets, and the existence of a resource base favourable to agriculture, the heartland developed early as an area of population concentration. The existence, within a relatively small area, of energy resources, raw materials, and port facilities, together with the population concentration, satisfied the prerequisites required for manufacturing development. The heartland developed early as a manufacturing area in Canada and has become the centre of manufacturing activity for the nation. This is readily seen in the functional profiles of the heartland cities.

CITY TYPES IN CANADA

To achieve a summary view of the Canadian urban milieu and to illustrate further the basic differences existing between the functional positions of cities in the heartland and in the periphery, the cities were grouped into city-type classes. The classes were developed from a synthesis of the materials already presented. Four groups of cities were identified on the basis of the importance of wholesale trade, manufacturing, and degree of specialization in cities. Obviously, other city groups exist, but different criteria are required to identify them. The criteria selected and the groups recognized are considered to be the most significant ones at the macro level where the total Canadian urban milieu is being viewed. The percentages of a city's total excess employment in manufacturing and wholesale trade respectively, were used as measures of importance for the two functions. These values for each city were plotted on an isometric graph (Figure 5), together with values of city population and functional specialization to facilitate the classifying of centres. So that other functions were not completely ignored, special note was made when individual cities within groups were known to have important functional positions that are not identified by the grouping cri-

Figure 5

CITY TYPES IN CANADA

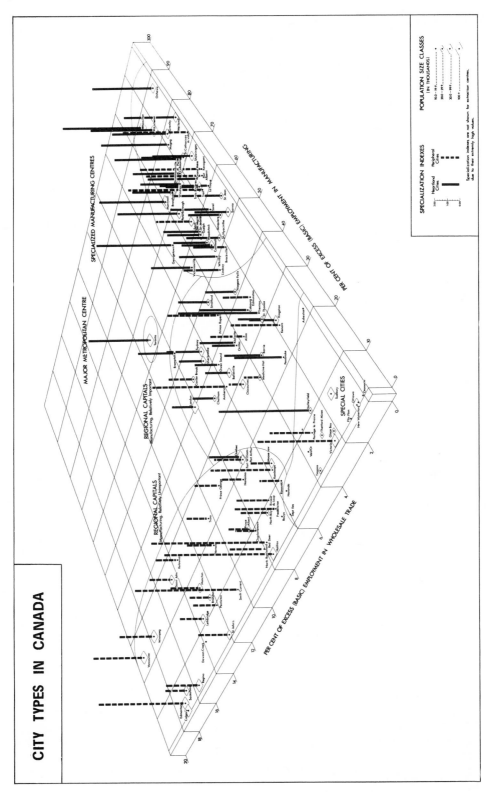

teria. In this way "transportation," "extraction," and "government" centres were identified.

Figure 5 reveals two familiar patterns: specialization generally increases with increases in the importance of manufacturing, and heartland cities are generally more specialized than the periphery cities. It also discloses several fairly distinct groups of centres. First, there are thirty-six cities with high to very high values of manufacturing importance, but with moderately low to low ratings for wholesale trade. Since these cities appear, for the most part, to have nearly minimal trading functions and, as a group, exhibit a high rate of specialization, they have been called "specialized manufacturing centres." All but five of these cities are located in the heartland. The five periphery cities are resource-oriented manufacturing centres where the nature of the manufacturing process, and either: (1) the lack of a large, densely populated hinterland that would promote central place functions, or (2) the assumption of hinterland servicing by a nearby centre, have given rise to high specialization rates and a tendency for one industry to be dominant. In the heartland, it is the "specialized manufacturing centres" that make up the two manufacturing regions. While all these cities exhibit a high specialization in manufacturing, a fair degree of variation exists within the group. Oshawa, and Magog, for example, present extreme cases of manufacturing specialization. These cities are almost juxtaposed to larger centres and are, in a sense, specialized suburbs or components of "dispersed cities." Victoriaville and Trois Rivières, on the other hand, are important as regional capitals, although the central place functions are definitely secondary to manufacturing. The city with the lowest rating of manufacturing importance in the group, Lindsay, is a commercial centre as well as a manufacturing city. This feature is revealed by "distinctive" ratings for both the retail and financial functions in this city.

Secondly, there is a group of fifteen cities that are very highly specialized, but have low rates of importance for both manufacturing and wholesale trade. Other functions, obviously, have the significant rôles in these cities. Because this feature (high specialization rates, coinciding with low values in both manufacturing and wholesale trade) is characteristic of so few cities, these centres have been called "special cities." Included in this group are highly specialized government and extraction centres. The most specialized Canadian city, the mining centre of New Waterford, has the lowest ratings of manufacturing importance in Canada. In some ways it resembles the highly specialized centres of Oshawa and Magog in that it is a component of a "dispersed city." This city, with Sydney, Glace Bay and smaller suburbs, makes up the iron and steel complex of Cape Breton Island. The functional structures of these cities are well integrated with one another. The mining centre of Rouyn presents an interesting case because its value of wholesale trade is higher than for any other "special city." It is an important distribution centre for settlement in the Clay Belt. Here is a case where a highly specialized extraction centre also acts as a service centre for surrounding areas.

A third group of cities is characterized by relatively high values of wholesale trade importance, low values of manufacturing importance, and generally low levels of specialization. Since city functional diversification and high ratings for trading functions are closely associated with hinterland servicing, the seventeen cities of this group have been called "regional capitals" with manufacturing relatively unimportant.

It is in these cities that the central place functions show their greatest strength. Significantly, all the cities of this group (with the exception of Quebec) are located in the periphery. These are the cities most concerned with collecting, transferring, and distributing goods. It is not surprising that six of seven cities where transportation is the dominant function are in this group. Also in the group are the six cities where community service is dominant. Several cities act as political capitals as well as regional ones; included are Edmonton, Regina, Fredericton,

Winnipeg and Halifax. The first four maintain functional profiles typical of regional capitals despite the government function. Halifax, however because of the presence of a large military establishment in addition to other government activities, has a high rate of specialization. This is one of several atypical specialization indexes in the group. Other examples are Dawson Creek, the terminus of the Alaska Highway, and Calgary, with its specialized extractive function.

In 1951 four cities were isolated on the basis of population size and relatively high values for both wholesale trade and manufacturing. These "major metropolitan centres" were Vancouver, Winnipeg, Toronto and Montreal. The development during 1960 suggests that Toronto has emerged as the single most important metropolitan centre in Canada at least in terms of maintaining relatively high values for both the wholesale trade and manufacturing function. Montreal has lost importance, in relative terms, in wholesaling but has increased its manufacturing in relative terms. It now falls within the cluster of specialized manufacturing centres. Winnipeg and Vancouver, located in the periphery, are much more involved, relatively, with wholesale trade (also transportation) than Toronto or Montreal, which are situated in the heartland. Since 1951 both Winnipeg and Vancouver have lost importance, in relative terms, in manufacturing and reflect a functional structure more closely akin to the regional capitals.

The remaining twenty-five cities are "midway," in terms of specialization and importance of wholesale trade and manufacturing, among the specialized manufacturing centres, the regional capitals where manufacturing is relatively unimportant, and the special cities. In this group are found some of the best examples of multi-functional cities. It is perhaps the most heterogeneous group of all because there is no outstanding trait common to it. In the groups previously described, all cities share one distinctive characteristic, be it: (1) a very high degree of specialization, (2) a high degree of manufacturing importance, or (3) a high rate of wholesale trade

importance combined with low ratings for manufacturing and specialization. The lack of any real "distinctiveness" in terms of specialization, wholesale trade or manufacturing means that many must be differentiated on the basis of other functional characteristics if detailed analysis is required. In general terms, the larger centres are regional capitals with a moderate to heavy emphasis on manufacturing, and the smaller cities, while often serving as regional capitals, do not have very significant wholesale trade functions. They are frequently involved with other types of activity such as government service and transportation.

Eight of the twenty-five centres in the group are periphery cities. They are all important regional capitals but also have significant manufacturing functions, which are based on either local resources and/or advantages accruing from transportation factors. For example, the existence of local coking coal is basic to Syndney's iron and steel complex; local clay deposits and natural gas reservoirs provide the principal materials for Medicine Hat's clay products industry; and regional forest resources feed the large pulp and paper operation in Edmundston.

Prince Rupert and Kenora represent cases where transportation factors are of prime importance as underpinnings of the manufacturing function. Local resources are relatively unimportant in the two cities as factors in manufacturing location; the advantages accruing to break-in-bulk-points and the gate-way function, respectively, provide the basis for their manufacturing. These periphery cities are only slightly more specialized than most of their counterparts which serve as regional capitals but where manufacturing is relatively unimportant.

Some functional diversity is exhibited by the heartland cities in this group. For example, Trenton, and Pembroke are closely associated with military camps; St. Thomas is an important railway centre; Kingston is perhaps the closest approximation in Canada to an "institutional" centre, having important educational, military, religious, and penal facilities; Orillia is the site of a large mental hospital; Owen Sound and Barrie are the centres of a large recreational hinterland. Because of this diversity it is difficult to formulate general statements on these cities. Of note, however, is the fact that the majority of heartland cities in this group are not members of the city clusters. Most of them are outside the clusters and are not typified by high specialization indexes as are centres of the manufacturing nodes. Even the cities within the manufacturing cores are more diversified in comparison to other member cities. Consider that Sherbrooke and London are the second and third most diversified cities in the heartland. In essence, most of these heartland cities are basically regional capitals exhibiting functional profiles not very dissimilar from those of their counterparts in the periphery. Their rates of specialization, for example, are closer to those of the periphery service centres than to those of cities within the manufacturing regions. As stated in the preceding section, the propensity for obtaining special functions and moderately high rates of manufacturing activity is greater, for the heartland regional capitals have tended to produce more complex functional structures, because the additional functions are superimposed on the "service-centre type" of functional profile.

An interesting exercise is to isolate and to consider the values for the major metropolitan centres. The positions of these cities in Figure 5 symbolizes the two distinct groupings of Canadian cities: periphery vs. heartland functional types. Winnipeg and Vancouver, located in the periphery are much more involved, relatively, with wholesale trade (also transportation) than Toronto or Montreal, situated in the heartland. On the other hand, the latter two cities emphasize manufacturing to a greater extent than their counterpart in the periphery. The implication is clear; a familiar thesis is again stated: the friction of distance requires urban centres in the periphery to "consume" greater amounts of "economic energy" in achieving their functional objectives than is required by the heartland cities. In keeping with the position of the heartland, its metropolitan centres are the largest cities in the nation and their in-

fluence infiltrates all segments of the national economy. The influence of the two periphery metropolitan cities, in contrast, is restricted to western Canada. The northern and eastern peripheries are essentially appendages of the heartland—as is the western periphery, but to a lower degree—and they do not have any cities that rank as national metropolises. They do have several cities that could be described correctly as "regional metropolises."

CONCLUSION

Although the city groupings developed do not allow an exhaustive analysis of city functional structure to be made, they do provide an overview of urban character in Canada. The distribution patterns studied illustrate the close relationship existing between the functional profiles of cities and the character of the economic regions in which the cities are located. They show that a fundamental difference exists between the functional profiles of the heartland and periphery cities, thus reflecting the major elements of Canadian economic geography.

The heartland, enjoying excellent locational relationships for most manufacturing processes, the key urban function, is the site of urban concentration as well as manufacturing concentration. The region contained 66 percent of Canada's urban population in 1961, yet it occupies only 2.5 percent of the nation's land area; fifty-two of the 110 cities studied are in the heartland. Whereas great concentration of economic activity, financial power, and population exist in the heartland, the periphery is characterized by huge sparsely settled areas, giving it almost vassal status in its relationship with the heartland. The periphery, because of its involvement with the friction of distance, must devote much more its "economic energy" to overcoming this friction. In the heartland the friction of distance is comparatively of little consequence. These fundamental differences in situational characteristics are reflected by the functional profiles of the cities in the two regions.

166

REFERENCES

[1] Dickinson, R. E., "The Scope and Status of Urban Geography," *Land Economics,* Vol. 24 (3), 1948, p. 231.

[2] Harris, C. D., "Methods of Research in Economic Regionalization," *Geographia Polonica,* No. 4, 1964, p. 60.

[3] See Harris, C. D., "A Functional Classification of Cities in the United States," *Geographical Review,* Vol. 33 (1), 1943, pp. 86-99; William-Olsson, W., "Utredning angaende Norrlands naringsliv," *Statens offentlinga utredningar,* Vol. 39, 1943; and Alexandersson, G., *The Industrial Structure of American Cities* (Lincoln: Univ. of Nebraska Press, 1956).

[4] Harris, C. D., "Methods of Research in Economic Regionalization," p. 60.

[5] Alexander, J. W., "The Basic-Nonbasic Concept of Urban Economic Functions," *Econ. Geog.,* Vol. 30 (3), 1954, pp. 246-261.

[6] Ullman, E. L. and Dacey, M. F., "The Minimum Requirements Approach to the Urban Economic Base," *Papers and Proc. Reg. Sc. Assoc.,* Vol. 6, 1960, pp. 175-194.

[7] Morrissett, I., "The Economic Structure of American Cities," *Papers and Proc. Reg. Sc. Assoc.,* Vol. 4, 1958, pp. 239-259.

[8] Florence, P. S., Fritz, W. G., and Gilles, R. C. *Measures of Industrial Distribution, Industrial Location and Natural Resources.* United States National Planning Board. U.S. Government Printing Office, Washington, D.C., 1943; and Rodgers, A., "Some Aspects of Industrial Diversification in the United States," *Econ. Geog.,* Vol. 33 (1), 1957, pp. 16-30.

[9] Ullman, E. L., and Dacey, M. F., p. 189.

Observations on Canadian Cities

E. T. RASHLEIGH

Over the past ten years, any planner working in Canada has been kept constantly aware of the rapid pace of urban growth, particularly in the larger centres. Everyone takes for granted that this growth will continue to typify the expansion of the country, and no one, to be more specific, seemed to question the validity of the predictions for 1980 made by the Royal Commission on Canada's Economic Prospects in 1957. What was surprising was that, if no one questioned the prospects of feverish urban growth over the next 25 years, few seemed very concerned about it. Yet it was my impression, as a working planner, that the larger metropolitan centres, which were expected to more than double their populations in 30 years, had not, so far, demonstrated any remarkable prowess in city building and had ignored several fundamental problems that would become critical as the rapid urban build-up continued. After well over a decade of post-war building, the most we could point to were isolated successes across the country: a good neighbourhood subdivision here, an inoffensive highway approach to the city there, a block of slum clearance somewhere else.

This paper is, in large part, the product of a tour of twelve of the larger Canadian cities. It is a critical commentary on what is on the ground today, but not so much in any city in particular as the larger cities generally.[1]

THE VITAL CORE

It would be wrong to describe the typical form of growth exhibited by the central commercial cores of Canadian cities, without first recognizing that they differ from each other in many respects. By this I refer, not only to such obvious qualities as the old-world charm of Quebec and the maritime orientation of some cities, but to the fact that, for example, Quebec and Ottawa have developed two cores instead of the customary one, each with a different function; that the core of Regina displays an emphasis on serving, not just the city, but the surrounding farm population, whereas Windsor has a relatively small downtown core because Detroit lies just across the river, and because another section of Windsor, once a separate municipality, still maintains a flourishing shopping district; and of course, that each core reflects the character of its original townsite layout.

The Typical Pattern

Nevertheless, one can generalize to a considerable extent on the manner in which all cores are growing at the present time. The traditional heart of downtown still prevails in most cities as a hard core of better retail outlets extending for several blocks along a street that connects directly with the main roads out of town. The less-than-best outlets try to get as close to this centre of gravity—on the same street if possible, or on other main intersecting streets. Thus, the retail area acquires a strip or vertebrae shape. When it grows, it tends to be attracted by the main arteries more than any other single factor. But the medium-sized cities of Canada cannot sustain a hard retail core much more than four or five blocks long without low-rent, low-density gaps appearing, and it is also apparent that a core will stretch only so far along the main street before it loses its "downtown" quality. However, because there are so many streets on

● E. T. RASHLEIGH *is Senior Associate Planner, Greater Vancouver Regional District Planning Department.*

Reprinted from *Plan*, Vol. 3 (2), 1962, pp. 60–77, with permission of the author and the editor.

which the core may deepen, the small businessman does not like the gamble of locating on a side street or of losing a direct connection with the existing retail core; so premium retail development does not usually occur behind the main street unless there is a large office building, parking garage or big store to establish the pattern. The usual deepening process starts on the side streets crossing the heart of the core, which build up until the back street, running parallel to the old main and between these streets, can be built up. This second main street is a fairly recent feature of core development. Edmonton and Calgary have one that built up in the typical piece-meal fashion, while the Wellington Square enclosed mall in London is a deliberate attempt to deepen the core *en bloc*.

Around the hard core are other functional segments of downtown. Specialty stores must be near the heart of the core. But in trying to locate cheap space they help give a new direction to retail expansion. The cheap outlets may locate anywhere on the periphery, or out along a main street, but they may also congregate—as in the retail core of a bygone era, as a skid-row near the railroad station, and so on. Cities with a farmers' market, like London, Windsor and Ottawa, have a special retail dimension to one side of the main street; regrettably, this element has almost disappeared in Hamilton and will probably be removed in Windsor. Most cores contain groups of related businesses, such as wholesalers, banking and insurance companies; this tendency to congregate has been repeated in recent years with the building up of prestige office streets, although rather than establish a completely new pattern, they tend to capitalize upon an existing, well laid out and attractively landscaped street. A prominent exception to this is the complex of new oil company office buildings in Calgary, which is several blocks removed from the old core; here the transformation has been so rapid that a 19-storey building virtually rubs shoulders with a 2-storey frame house. In many cities the retail core has acquired, in recent years, a sprawling backside of parking lots and other automobile services so that it stands isolated in a no-man's land of asphalt, the inevitable result, under present circumstances, when large cities have a high incidence of automobiles.

Influences and Inadequacies

Having generalized on the pattern of the core, I now turn to the effect of this pattern. By and large, most cores possess cacophony and at the same time monotony, a relentless regularity in plan and a chaos in elevation, an anemia of content and a chronic lack of intentional emphasis. Where this description does not apply, it usually means that the major factors shaping the core have been offset. These are a rigid adherence to the past, a jealous retention of property rights, and an absence of a unifying hand.

Literally underlying every development is the pattern of streets and individually owned parcels of land, which were designed in a different era for a completely different purpose. These remain and are extremely resistant to either modification or reorganization. The grid street system creates the overall mold of uniform blocks. Narrow lots break the blocks into small parcels that were laid out with no appreciation of the fact that retail operations require a different layout from residential properties. So stores and office buildings have long since had to make maximum use of the land, leaving the task of circulation wholly to the public rights-of-way. Efforts to improve traffic conditions immediately run into both the expense of acquiring intensively built-up frontage, and the difficulties of changing traffic flows when nearly every street constitutes the only access to a block of properties. The pattern is equally hard for new private construction to break. The assembly of a parcel of adjoining properties is so difficult that instead a patchy, jagged-tooth facade builds up; even where assembly has been possible, the new development is still designed as a two-dimensional facade, because it is so much more difficult to assemble a parcel running through the block between streets. Thus, the old format is reaffirmed and the old errors duplicated at higher densities. Not only do these rigidities inhibit improved site development, they also obstruct the logical expansion of the various functional parts of the core. A stubborn owner, an old business that has become an unsuitable neighbour, or a type of development that creates dead frontage, all can isolate a portion of a block, can prevent an area from redeveloping logically, and can even drive expanding businesses to other parts of the core, or outside the core, or even outside the city.

Forces other than business produce the colourful elements in the core; business is too much a creature of the national and international world, and too much motivated by a self-interest that stops at property lines to create anything more grand than a building, or to add anything unique to the street scene. The municipality is the major co-ordinating influence in the core, but so far, few have acted vigorously, partly because there are so many interests that place a great value on the downtown. But the municipalities must share the blame more directly. Most seem to have a very narrow conception of the core's function: it is expected to move cars and store them, contain stores, offices and factories, and provide pedestrian access to them. It is hard to believe that municipal authorities regard their downtown as a place to enjoy, to relax in, or to look at. So for this reason, too, innovation, experiment, and imagination in new growth is rare; most of the pleasant features, the idiosyncracies, the surprises are inherited from the past. Also, one gets the impression that dollars and cents alone are deciding the fate of the core as the main centre of non-business activities.

This restraint on the part of the municipality manifests itself in a number of shortcomings. In several cities, the future direction of core expansion is not at all clear, as though it is being left to work itself out through the push and pull of individual decisions. Public institutions are scattered through the core and outside it, sometimes in unsuitable locations; too often the question of whether to decentralize the audi-torium, art gallery and museum, or to organize them into a centre, is being decided on the basis of land costs alone. Existing historic public open spaces are respected but are rarely appreciated to the extent of being duplicated. There are few vantage points where one can step aside to watch the activity of the core, or in maritime cities, to get a panorama of the harbour. Nature is excluded from the core, or at best, is found around its edges. The retail core is an introvert structure, preoccupied with its business and ignoring its surroundings; few efforts are made to relate it to adjoining parts or take advantage of the natural assets of the setting. (One example of this failure is downtown London which ignores the river at the end of its main street, although Maritime cities can be accused of the same indifference.) Lastly, contemporary building, with its anonymous face, is wiping away much of the old character of the central city. Just one example of this: St. John's is the only large city in Canada where you can walk along the main street and at every intersection see the water's edge a few yards away, but this perfect symbol of the city's economy is to be eradicated by a truck access route.

Surely it is now obvious that, because the core is one of the most intensively used parts of the city, there must be comprehensive planning if it is to improve. There must be a unifying power able to set the design "theme" of the core, which will link and draw together its elements, and work out the pattern of expansion. This requires less of the uniform restraint of the zoning by-law, but more emphasis on restraint, and a greater direction of new growth by the municipality.

Acceptance of this degree of intervention by both business and the municipality is the stumbling block, but there are a few hopeful signs: new city halls have involved several municipalities in so heavy an expenditure that they are taking special care to encourage suitable surrounding development. Perhaps Windsor has gone furthest in this direction. It has started pushing a boulevard through from the civic square to the waterfront, which may eventually link up with a riverside park that was recently created at the foot of the main shopping street; this has already become a focal point for an auditorium and a local museum. Too often though, municipal improvements stay within the traditional limits of responsibility. Too often, costly street improvements are unaccompanied by a reassessment of the organization of adjoining land, or by an equivalent concern for the needs of the pedestrian. The closed-street mall is a useful solution to these problems because it causes a minimum of disturbance to the *status quo* of property ownership. In many cities, though, the main shopping street is also a main artery that cannot be closed. If, instead, an adjoining secondary street is successfully converted, this could shift the centre of shopping activity. However, the closed-street mall will not work in all cities; moreover it is a hesitant compromise that does not overcome the inadequacies of the grid street and lot layout. If substantial improvement of the core is to be achieved, the mall, or any other concept, will have to be a component of a master plan that predetermines the redevelopment of each block and is backed by the power to coordinate the use of the constituent properties.

No special case is being made here for any one form of core revitalization. On the contrary, the hope is that every core will continue to typify the city and region it serves, and that core developers can overcome the strong tendency to make it look like every other core. Also, I hope there will be less preoccupation with getting maximum use out of the land and more determination to develop the core as the most colourful and most human part of the city.

LIVING THE GOOD URBAN LIFE

Some parts of a city dictate the type of land and layout they require, or else, they are not influenced too strongly by natural conditions, but residential districts suffer and benefit from the conditions that affect the city as a whole. In Saint John, the ground is so irregular in most parts that it will never permit an even expansion of the city, but will probably ensure that open space, useful or otherwise, will always be plentiful. In extreme contrast, Regina has so flat a site that development in any direction is unimpeded. London is properly called the "Forest City," and for most of its life has had to be more concerned with clearing away natural growth than with replanting. Regina again presents the opposite situation of a city where every tree has been planted by man. Canadian cities also differ in their inheritance of past growth. The railroads chopped up cities like Winnipeg, which grew rapidly during the boom of that era, while Hamilton, which developed as a heavy industrial centre, acquired street after monotonous street of mass-produced housing. A large portion of Saint John was destroyed by fire and rebuilt at a time when it lacked prosperity and, in the opinion of some, architectural taste. Past growth has also bequeathed particular problems to each city. Windsor has an extensive rootlike form of outer development that is very difficult to fill in and rationalize; pre-war Quebec includes numerous rabbit warrens of high density; the prairie cities possess large, old grid subdivisions still only sparsely settled.

These inheritances give shade and emphasis to the contemporary residential growth and problems of each city. But if there are differences, there are also common prospects facing the cities. All should recognize that, as they grow to metropolitan dimensions, the range of types of housing that their populations need and want extends in the direction of higher densities. It is more likely, however, that the experimentation and even the permissive atmosphere necessary to produce the new forms will not appear because of ignorance or prejudice; certainly many examples of multi-family housing reflect a failure to understand the requirements of higher density housing. It must also be recognized that the process by which neighbourhoods are now built is considerably different from that which occurred before the war. If earlier subdivision pushed far beyond actual need, it also tended to create odd-shaped lots and leave vacant land in most neighbourhoods, which later were taken up as parks, or filled in with houses of a different period. The end product, the older districts of the cities today, display diversity and contrast as a result. Post-war subdivision, on the other hand, leaves little to chance and less to future decision. Irregularly shaped land is no longer neglected but is made usable by the bulldozer. The building process is not left to the different tastes of different times but is carried through by one developer on a full-block basis. Open space in a neighbourhood has been reduced to the minimum parkland required by the municipality. Thus, modern subdivision development is more orderly and efficient, but so far, also tends to be more sterile and characterless.

The Good Neighbourhood

When walking through the many different areas in which people live, one realizes that each district, regardless of calibre, has benefited if certain qualities are present and has suffered if they are lacking. These qualities are described below as a basis for the appraisal of the different types of residential areas that follows.

1. *An atmosphere of community.* Some districts are more successful than others in binding individual properties together as a larger whole. With them, the stranger senses, first, that he is entering an area, then, with some embarrassment, that he is an intruder. Here the home life of the resident does not stop at his front yard; it takes in the activity of the street, the local store at the corner, the church and the playground. And over all these things there is a feeling of ownership that only the residents can possess.

This feeling of community is strongest in the older, more densely built up districts of the Eastern cities. There the inside of the house is often worn and dark, and the yard space usually negligible. The street is the communal front yard where children play and men fix their cars, while the rest of the family sits on the porch watching the world go by. Of necessity, such people must participate more in the local community,

and for all its failings and ugliness, there is a homely feeling to the street that better-class apartment and suburban districts rarely possess. A sense of community, perhaps not as intimate as in the poor areas, also prevails in the old areas that were once separate communities. The district in Edmonton that was once the municipality of Strathcona, and the old centre of Sandwich, now part of Windsor, still retain a unity and spirit that sets them apart from the surrounding development. One of the most carefully structured residential districts is Walkerville, also in Windsor, created to house employees of the Hiram Walker distillery which, in layout, reflects the socio-economic differences of the residents.

This sense of community is not simply a product of age; a study of these communities reveals that physical layout, though often natural and unintentional growth over the years, is responsible to a considerable extent. If, therefore, one observes that few recently built districts possess a sense of community, it implies a criticism of their design.

2. *Good neighbourhood design.* Without attempting a detailed discussion of principles and techniques, we may acknowledge the importance, in this period of mass housing programs, of subdivision design that is comprehensive and sufficiently detailed to build a community that is efficient, visually pleasing, suited to the particular needs of its residents, includes the elements needed to make it complete, and has qualities conducive to producing a sense of community. Some of the more important elements of good design are the subject of the next five sections.

3. *Protected isolation.* Perhaps one of the most obvious and universal truths about residential properties and neighbourhoods is that they stay in better condition and have a longer life if they are cut off from surrounding non-residential areas and activities. Houses on a dead-end street are better off than those on a through road; a residential district several blocks wide stays in better repair than a street of houses within a non-

residential area; a pocket of residential blocks located on one side of a main artery keeps in better condition than a district lying between two arteries. There are numerous examples of old, poorly built districts, with no open space or other community amenities, that nevertheless are able to retain their integrity and good condition because they are not subjected to a daily pounding from through traffic, or riddled with non-residential uses. While such conditions of "protected isolation" prevail, the neighbourhood remains healthy, but if, as the area declines, it is opened up to any kind of potential development, the majority of housing is doomed, not to extinction unfortunately, but to continued misery.

4. *Visual contrast.* This rudimentary principle of design must be emphasized because contemporary residential growth is so persistently uniform, in type of house and its siting, in the facade of its streets, in the evenness of its development, in the appearance of its shopping centres, and so on. The forces producing uniformity are so powerful, that if a greater diversity is ever to be achieved, a deliberate effort will have to be made to plan residential areas so that they look different.

5. *A community centre.* The centre about which the neighbourhood clusters is its point of identification: the place where the local community comes to a head. Most community centres in Canada have grown naturally—along a main road and usually at an intersection—but the more successful ones, from a planning viewpoint, are much more than a line of stores. Hamilton has a pleasant example in its Westdale district, where the street system of the neighbourhood radiates from a tight cluster of stores fronting on a short section of the main road made extra wide to handle parking and circulation. A good community centre is not just a retail centre, but the place where most of the institutions serving the community are concentrated, and therefore, it is the main centre of local community activity. Like the neighbourhood as a whole, the centre has identifiable limits and a point of

focus, usually a main intersection. The centre should be sufficiently well organized that people can get to it and move around it easily. Lastly, although it may never be an architectural triumph because it is the product of many hands, it must attract attention and must be a human setting, not one dominated by the automobile or the symbolism of national corporations.

6. *Open space and landscaping.* Open space and landscaping are probably more under-rated than any other feature. The trees and landscaping of a district can mark its boundaries, establish its character, soundproof it, cool it, and create its contrasts in colour and light intensity. A district is improved more by attention to plant and tree growth than by adding to the value of the houses. Much of the luxury atmosphere of Rockcliffe Village in Ottawa, for instance, is achieved by keeping properties very large, and the roads very narrow, so that the whole district is dominated by natural growth. The same result is more consciously achieved in Wildewood, a middle class district in Winnipeg, which is one of the best Canadian examples of the Radburn principle of fronting properties on a common central park and hiding the roads behind the houses. And again, most of the poorest areas on the outskirts of cities, if they had nothing else, enjoy empty lots and open fields. It seems as though the rich and the poor can have good open space, but the middle-income groups, whose housing needs are catered to on a mass-production basis, have to get along without it—for the first 10 years at least.

7. *Subordination of the auto.* When the private automobile is treated as an all-important factor in the layout of a residential area, it becomes objectionable. It confines the neighbourhood design to the few alternatives enabling the car to get right up to the house. It requires most of the land that can reasonably be set aside for public use. It dominates the landscape and introduces an accident hazard precisely where small children most frequently play. It adds significantly to development costs through the installation of pavements, curbing and

dual sidewalks. In short, in the moderate-cost subdivision, the dictates of the automobile can be satisfied only at the expense of other features of residential life.

8. *Minimum standards for family districts.* Cities in Western Canada are generally built at lower densities than those in the East; this includes their central residential districts, which, although old and poorly built, manage to keep gardens and a bit of grass. Largely because of this, these areas never seem to reach the level of nastiness and disrepair that can be found in the cities farther east. At a time when so much private building seems to have the objective of getting as much floor space on a lot as will be tolerated, and with the need to rejuvenate old housing increasing every year, it is necessary to recognize that there are minimum standards for building and rehabilitating family housing areas, which, if undercut, destine a district for premature decline. Setting a maximum gross density is not enough. There must be standards of light, air and privacy for each dwelling unit, and there must be space for recreation and other activities in the neighbourhood.

The Shortcomings of Suburbia

Criticism of post-war suburbs has become almost a national sport, perhaps because in the part of the city least restricted by the dead hand of the past, modern city building has regularly produced the trite and the obvious. One can criticize suburban architecture, which from coast to coast follows the "strawberry-box" school, and one can continue with a detailed critique of the street and neighbourhood. But perhaps the failings of suburban growth can be reduced to two main faults.

First, suburban neighbourhood design is dominated by an indiscriminate desire for space. This desire has produced lots of space for all things and between all things. The typical residential street is dominated by an expanse of road, boulevards, sidewalks and front yards, against which the single-storey house offers insignificant contrast. This low line of houses is occasionally

broken by a square of park, but rarely by more imaginatively planned open space designed to contrast with the street. The shopping centre stands in asphalt isolation from the surrounding development, typically a bald line or "L" of low stores. Other components of the suburbs — school grounds, community centres, and the like—add to the expanse. Largely because of the developer's "scorched earth" policy, there are few trees and bush growth to fill the void; instead, the popular standard of lot layout —front lawn, back lawn, path to front door, driveway to one side of lot, low hedge if any, not much garden, a picture window unobscured—contributes to the visual monotony.

The second failing of suburbia is that it has few of the unifying features that the term "neighbourhood" implies. One area runs into the next, divided only by main roads and roadside development. The methods used to protect and isolate the houses are half-hearted and only partly successful. In Edmonton, much of the new subdivision has service roads and grass boulevards between the main arteries and the residential properties, but in other cities the curvilinear road design alone is usually relied upon to discourage non-local traffic, with few attempts made to add a buffer. Also, in every city there are countless examples of houses fronting on main streets, and shopping centres or gas stations flanking a row of houses. The average suburb is without a centre, too. Rather, development is turned inside out, so its shopping centres are on the edges and oriented toward the road. Besides, the shopping centre fails as a community centre because, being organized strictly to do business, it has no space or time for other community needs; unfortunately, the other community facilities are only occasionally combined with the shops to form a well-rounded neighbourhood centre.

It is perhaps too obvious to conclude that improvement will only come with more detailed neighbourhood planning that relates the essential components to each other, and more than anything else, minimizes the influence of the automobile over neighbour-hood design.

Preserving the Old Districts

In 1941, 40 percent of the dwellings in urban Canada were 31 years old and more, with the greatest volume of residential construction having taken place in the Twenties.[2] Numbers of these dwellings have since been demolished, but it is obvious that in every city there are still large areas with houses that are now forty to sixty years of age. Despite old age, and deficiencies, these districts include centres and pockets of development that have been built up and elaborated on over the years, so that they now are among the more interesting sections of the city, full of character. At the same time, much of the building in these districts is not good housing. Three-storey houses achieving high densities are common, especially in Ontario; many of the smaller houses are of simple frame construction that soon deteriorates if neglected; some were jerry-built; streets of houses are wedged tightly between older developments; much is dull and monotonous.

This old housing is being subjected to pressure and strains created by the post-war growth of the city. Central districts lie in the path of the expanding core, which under speculative development, tends to open up more area than it needs. In a similar fashion, institutions that previously were reasonably compatible with their residential neighbours, want to expand their plant and parking space, while more institutions try to locate cheap central quarters among the old houses. Small industries that never were compatible, grow into bigger industries. The old mansions of a city create a problem because they were designed for the particular use of a past era; obsolete and expensive to maintain, the owners seek to put them to any use that is economical, with the result that they, more than smaller houses, are in danger of being over-used and mis-used. And all residential districts close to the city centre are incapable of handling either the volume of traffic to and from the down-town area, or the automobiles their residents now own.

If the old houses of the central city were destined to disappear in the near future, their present living conditions would be of passing interest, but it is inconceivable that the majority of them will be taken over by non-residential operations or replaced by new housing. In fact, the old districts will probably continue to supply a wide range of accommodation — single rooms, flats, and low-cost family housing; they will remain the customary locale of ethnic settlements; and they will increasingly reflect a tendency, apparent so far only in the larger cities, for some middle-class families to live in the central city.

It is surprising how well this housing is being maintained on the whole. But the owners are labouring under disadvantages that only the municipality can overcome, and yet, the older districts seem to be less protected by municipal policy than are the new suburbs. Zoning is the typical defence, a negative safeguard with legal non-conforming loopholes that have particular significance in this part of the city. Slum clearance projects have removed relatively small amounts of deteriorated housing, and could not cope financially with much larger areas; even if possible, it is debatable whether a large amount of public or subsidized housing is desirable in the central city. What is needed are broad, continuing municipal programs to protect and improve the integrity of the older houses and their neighbourhoods. The appropriate action would have to vary with the particular need of the area: a decision *not* to widen a heavily used road, or to close it; punching a hole in a tight block of houses; the removal of non-conforming uses and houses in bad condition. The appropriate action should also take into consideration the attitude of the residents. In districts displaying local pride, it might be logical, and economical, to give indirect public support (e.g., a supply of house paint at cost) rather than to undertake public works. Lastly, the appropriate action will have to maintain the existing use of housing, or provide for a transition to another use; if the objective is to assist but not eradicate an area of low-cost housing, it will be important to avoid removing or driving away such features as the corner store, cut-rate shops and space to work on the car.

Rehabilitation on a large scale has not yet been faced by Canadian municipalities. Because it involves new expenditures, extra administration, and a degree of community improvement not yet practised, municipalities will likely resist assuming this responsibility, especially if they are already committed to expensive slum clearance projects.

The Good Life at High Densities

The number of apartments built varies a great deal from city to city, but on the whole, they are one of the most distressing products of post-war urban growth. The buildings are usually of a trite, plain design, sited with a minimum of imagination and landscaping. Much of the fault stems from a preoccupation with achieving a maximum density on a lot designed for low-density use, complicated by the need to provide off-street parking space. The end-product is necessarily a structure that hogs a site surfaced in large part with asphalt. When several apartment blocks are built side by side, the effect is so arid one wonders why any man, whose wife is less than very loving, bothers to come home at night.

Up to now, apartment sites have been selected more on the basis of the advantages of the site than in accordance with an overall plan; zoning is the main regulator, but in some cities, this has limited significance in the older districts. Properties that are deep and wide are invariably sought out, and this has resulted in the replacement of many old, but nevertheless handsome mansions, as well as the filling in of irregular parcels that had previously been neglected. (What will happen, one wonders, when the supply of large lots is used up? Will apartment standards drop, rents go up, or will land values go down?) Apartment developers usually ignore areas in need of reconstruction, preferring to capitalize upon the attractiveness of well-landscaped, stable districts. As the apartments appear, that charm disappears, and too often, the stability of the district is weakened. In the new suburbs, apartments are cramped on the land at higher densities than the surrounding houses but with no compensating features: frequently they lie alongside a main road or butt up against a shopping centre. In some instances, one is led to conclude that these buildings are intended to be sound barriers rather than housing. In the older districts, the introduction of more intensive housing brings drastic changes to the existing pattern of development. At present, the allocation of land for high-density housing is still treated as a zoning matter, and the implications of introducing larger buildings and more people into an area are substantially ignored. It is difficult to see how any improvement can be made here without assembling properties in accordance with a plan designed to prepare a district for high-density use. Obviously, to do this, a fair degree of public participation will be necessary.

Post-war building trends substantially ignore the fact that many families, particularly in Eastern and Maritime cities, live in multiple-unit buildings of one kind or another. Row housing, duplexes, and the like, permit a moderate increase in density over the single-family unit, but to compensate for this gain, they require a skillful site layout. Because site layout is a characteristic failing of Canadian building, there are many bad examples of multiple-unit buildings across the country, which probably accounts for the strong resistance to new efforts in this direction. There are some good examples though. A street of row houses in Walkerville (Windsor), some approaching the charm of ivy-covered cottages, comes to mind as an example from the past. Among the recent products, public-housing projects seem to represent the best we can do. The rows are not built too long to be monotonous, the street system and siting of the buildings create a diversified scene, even though they have a straightforward design. The project site is large enough that the buildings are oriented away from the main road, and interior roads are kept to a minimum; instead, foot paths and stretches of grass run between many of the units. Parking space is organized on a communal basis.

Nevertheless, the projects have several common failings. Invariably, they are devoid of trees and shrubbery. Perhaps this is a maintenance economy, but it does much to establish an impersonal, institutional, and in the summer, oven-like atmosphere. Open space is not usually segregated, and inevitably, the children play around the houses. At these densities there is a need for at least one organized recreation area removed from the houses. Perhaps the most striking omission is that there is no outdoor privacy, nor any demarcation to show that certain grounds belong to a particular unit. As a result, each unit is never more than an anonymous part of the institutional whole, bound together by unbroken stretches of open space. Surely each resident should be given the opportunity to enjoy a feeling of possession, if only to encourage his responsibility for maintenance.

WIDENING THE SUPPLY OF OPEN SPACE

No feature of contemporary urban growth suffers more than open space from our insistence on accounting for every scrap of urban land. The error lies, not only in inadequate provision, but also in an over-simplified concept of its form and use. Open space has become a parcel of land set aside for use as a park or a playground, which means it is either equipped for sports, or is a square, or perhaps a triangle of land that is levelled, grassed and cleared of all growth except for a selection of trees, shrubs and neat beds of plants. Somehow, the lessons offered by the past have not been taken to heart, so that with very few exceptions, we must still look at pre-war products to appreciate the wide range of open space it is possible to have.

In a sense, the largest open space is that which offers a panorama. This may be a view of the sea, a river, a section of the city, or one of its features. The City of Edmonton has a well established policy of developing as public boulevards the lands overlooking the broad and deep North Saskatchewan River valley. Regina has no opportunity to look down or up; so early in the century, the creek running through the city was dammed to make a lake, and extensive parkland was planted with trees. This area now forms a very cool setting for the provincial government buildings, with ample space for even socialist government expansion. Other cities have inherited similar bounties, but too often—in the East more than the West, as is my impression—they are not adequately preserved or developed. Fort Howe, in Saint John, for example, affords an aerial view of the city and harbour that Toronto would give its eye teeth to possess, yet it is undeveloped and neglected.

Even more common is a failure to have adequate points of public access to the sea or riverfront, especially in the older parts of the city. The railroads, in this case, are a frequent obstruction. Every city has at least one major developed park that is its pride and joy. To my knowledge, all are inheritances from the past. A surprising number were military holdings originally. Others are ravines and valley lands, unsuitable for intensive development. But clearly, some cities have been more methodical than others in parkland acquisition and preservation. For example, Edmonton has four golf courses, two municipally-owned, close to the centre of town (compare this with the Shaughnessy Golf Course in Vancouver, now close to being subdivided, and the shortage of golf links common to most large cities).

One other form of major open space is a connected system of parks, public grounds, boulevards and paths. A good example is in Calgary, where a boulevard runs along the top of the Bow River valley from a park overlooking the valley, to the grounds of the Jubilee Auditorium about a mile and a half away. With its emphasis on variety, an ability to utilize a wide range of types of land, and an effect of tying a district together, this is one of the most potentially useful concepts of open space.

177

It is not necessary to recite the standard need for local parks, but the all too standardized approach should be pointed up. Too many parks look as though they were surrendered under the subdivision regulations rather than included as a component of the neighbourhood design; their original lack of justification perhaps explains why they remain open stretches of grass, clearly labelled "Park," and set apart from their surroundings. Some of the better subdivisions use space as a visual breather, or to preserve a natural feature, or as a setting for the houses. Again, the Wildewood subdivision in Winnipeg, with a winding park as its central theme, provides an exceptional example. Occasionally, too, one comes across a juxtaposed shopping centre and park hinting at a way to create, within a park setting, a community centre of shops and local public services. An obvious scheme perhaps, but rarely is it attempted.

The attitude of keeping open space in its place also prevents it from finding its way into the downtown core. I am tempted to generalize that one of the main design inadequacies of Canadian cores is caused by an ignorance of the value and versatility of open space. Traditionally, central open space is limited to that which is immediately utilitarian, or else obviously decorative. Most fall into one or more of these categories:

(1) the commemorative square;
(2) a green space set apart for people to rest;
(3) a natural feature that has been preserved;
(4) a setting for a public building;
(5) open space geometrically positioned in the original town plan; or,
(6) an outdoors business activity, most frequently a market.

The more successful open spaces, if measured by intensity of use or aesthetic appeal, are areas that have outlived their original function and have been allowed to convert to a more casual use. Quebec, within the walls, offers several examples: Dufferin Terrace, once a battery site commanding the St. Lawrence, evolved into the main promenade of the city; and the two squares nearby that are focal points of the tourist city, previously functioned as a parade square and a market.

These represent a type of open space that only recently has received adequate recognition—an area of public activity that functions as part of, or is physically related to, the main function of the core. The sidewalk and the public market are the traditional versions, the mall is recent. Such open space does not have standard dimensions or layout; it can use the odd, left-over scraps and corners, the pieces that are most readily surrendered for such a negative use. But if more is to be done in this direction, we must have a better appreciation of the social psychology of open space in the core. We must learn that people like to watch people, to see them act, and to get into the act themselves. The citizen comes to see what is new with his fellow citizens, and to take part in the daily performance of the human comedy, whether as glamour girl or undeniable romeo, superior observer or lonely isolate, humanitarian or proselytizer. However, if the people are to create the drama, the planner-architect must first create the setting, and in North America, we have much to do and to learn in this direction.

The supply as well as the quality of all kinds of open space varies considerably from city to city. The quantity seems to increase from east to west, but it is probably true that every city has a shortage of open space, and in addition, under prevailing conditions of compact development, needs a continuing program of parks acquisition to anticipate future requirements. One can easily become pessimistic about this planning being done. Some cities have never been overly concerned about open space; others have enjoyed such an abundance within the city and without that the need to prepare for the time when the population will outgrow the natural supply is not taken seriously. In fact, all cities are too close to their small town and frontier days to appreciate what they are in the process of losing. Besides, open space is never regarded as

178

the most pressing need. How many cities have even stopped polluting the waterway within their boundaries? The prospect of a wider, more intelligent use of open space is gloomiest of all in the core, for in that part of the city, where land values are highest, what chance is there for a land use that cannot be judged by sound business principles?

THE CITY AS A "COLLECTIVE WORK OF ART"

The previous paragraphs have followed the customary planning approach of looking at each functional part of the city separately. This approach overlooks, or at any rate underemphasizes, a quality of all contemporary urban growth that should be primary cause for concern among planners.

From one viewpoint, this can be described as a lack of visual diversity. The development patterns of Canadian cities have basically two origins. Some cities have grown naturally, coping with land conditions piecemeal, following original trails, and growing at the whim of individual developers. Victoria, St. John's and old Quebec are good examples. But more often, city growth, including parts of these three cities, has adhered to, and up to the post-war period extended, a grid system. Thus, order and regularity prevail in the layout of most cities, but contrast, change and surprise are rare, and what does exist is not a product of forethought, but of chance and gradual evolution. Planning removes the element of chance in urban growth but so far has not often been successful in introducing intentional contrast, and recent building has only consolidated the regularity of urban design.

The growing tendency to urban uniformity may also be expressed as a loss of local and regional differences in urban growth. In terms of founding dates and site conditions, Canadian cities probably represent as wide a range as will be found in any country. It is still possible to see these cities as the products of several generations of people whose historical backgrounds still influence their attitudes, their values and their tongues, who make a living in partic-

ular ways, and who have adapted their way of life and their city to the weather, landform, changing technology and external pressures. But the image of each city is blurring. More and more they are coming to look alike, as new construction reflects national and international patterns. Gas stations are identical in appearance from coast to coast; roadside developments on the outskirts of every city use identical building and advertising techniques; basically the same subdivision and house plans are adopted across the country (because, one is tempted to believe, they are most likely to be accepted by, often were designed by, the federal mortgage authority); office buildings reflect international architectural styles and use module-dominated, standardized building techniques and materials. No doubt, there is something to be said about these tendencies being desirable or unavoidable, but there is much to be said, and little being done, about preserving and encouraging individualism in Canadian cities. It is naturally repugnant to the average resident if he is told his town looks like many others, or has no character: it implies that the same may be said about him, and certainly, there is an interaction between a city and its people that shapes both. It may well be that the universal urban forms that are imported will eventually prove unsatisfactory under local conditions, or worse, will work to obliterate the identity of the residents. The planner, who presumes to rationalize the building process, will be increasingly challenged by this wholesale depersonalization of cities and will be confronted with the aesthetic issue of contemporary urban growth: the issue of whether, with such an impoverished background of conscious urban design in this country, the strong tendency to regularize development can be overcome.

WILL IT BE AN ERA OF PLANNING?

These are the major issues I see cropping up in all metropolitan centres during this period of rapid urbanization. Some are inheritances from the past to be overcome. Some involve foreseeing the long-range needs of the future. All are intensified by the present rapid pace of growth. A sound planning program is essential if we are to cope with these problems; more to the point, more effective planning is needed than now prevails. Will cities face the issues that arise as they grow to metropolitan stature? Fortunately, many of the important decisions still have to be made, but we must admit that one pedestrian mall does not represent a revolution in downtown design, nor one Don Mills (Toronto) a major breakthrough in suburbia.

The planner must recognize that Canadian society has ingrained attitudes and values that are inimical to the truly urban way of life, and particularly, to the planning action it entails. It might be well to summarize these.

First, many older Canadians (including those now in positions of authority) grew up on a farm or in a small town, and if younger adults did not take part in this way of life, many inherited a family background that did. Moreover, a generation ago our largest cities were small enough and sufficiently loose-knit for the countryside to be near at hand. There was land to spare, and people displayed a frontier-like indifference to what was done with it. Roderick Haig-Brown describes how each of the resources of British Columbia was, of necessity, ransacked in order to open up the province.[3] With Canada still a country of wide open spaces, the traditional attitude to the land has been carried over to the urban setting where, as in earlier times, the cost of indifference is not yet fully realized.

Second, this heritage, in combination with a failure to import other than the most straightforward concepts when the cities were laid out, has resulted in Canadians having almost no first-hand knowledge of the more sophisticated and architecturally interesting forms of civic design.

Third, in this country, the social value of land ownership is high, while tenancy and public ownership is held suspect. Private land ownership is so sacrosanct in public opinion and law, that it can question the propriety of planning proposals and defeat

legitimate community objectives. One of the most effective limitations it imposes on planning comes from the extreme stability it gives to property lines, even when they divide the land into parcels unsuitable for modern use.

Fourth, the municipality is expected to provide the necessities of public life, but its responsibility for community pleasures is limited. Compared with European cities, municipal responsibility in Canada for promoting the arts is small and uncertain. Several metropolitan cities do not have a public auditorium. Many art galleries and museums are privately organized operations billeted in old houses. The preservation of local history has only recently started to receive public support. City halls are singularly lacking in outdoor space for civic celebrations.[4] A hard thing to generalize upon certainly, but I am tempted to conclude that, on the municipal level of government, no great responsibility is accepted to propagate the cultural and social pleasures of the people. Nor does the populace, in turn, either expect these pleasures or hold them in very high esteem. The attitude toward city building is that it be concerned with the pioneer tasks of supplying a roof over one's head, an adequate supply of water and other basic necessities, and rarely are there aggressive demonstrations of public dissatisfaction with an urban setting that is visually trite or ugly, or one that is unable to provide the finer, non-essential things of urban life.

Fifth, there is a widespread reluctance to subordinate the ideas and wishes of the individual to any plan designed to meet the needs of the community as a whole. Whether explained as human nature or a continuation of the pioneer spirit, this attitude, in its extreme forms, directly opposes the basic premises of planning. The attitude, "We have no right to tell a person what to do with his land" still prevails, and a fear of flouting public opinion and private interests that hold this view can be read into municipal hesitation to undertake new planning programs.

The criticism of these social values is, not

that they are inherently wrong, but that they are out of date as far as city building is concerned. The period of exploration by rugged individualists is over, and we are confronted in our cities with a version of the world problem created by our growing population: the problem of determining to what extent man must submit to common objectives and common operations in order to survive. We are at the stage of urban growth in Canada that requires, not grudging recognition of the need to plan in the trouble spots, but positive recognition of the value of taking everything into one's ken, of knowing, as you pay Paul, how much you are liable to rob Peter. The alternative is to have, not collapse, but mediocrity and inadequacy in our cities.

The problem is still being resolved, but with it, one nagging question persists: How much time do we have before the city of the next few generations is built? Too often the answer seems to be that it is half built already—in all its archaic splendor.

REFERENCES

[1] This article has been moderately edited. In particular, a short section on urban transportation was deleted. This section simply recognized the automobile problem, supported the growing importance of good public transit systems, and bemoaned the sacrifice of other urban assets to improve roads.

[2] Subcommittee on Housing and Community Planning, Final Report, (Ottawa: Queen's Printer, 1946), page 29.

[3] Haig-Brown, R., *The Living Land* (Toronto: Macmillan, 1961), Chapter 1.

[4] Since this article was written, some municipalities have done more—for example, Victoria and Toronto have excellent civic squares. But I am not convinced the importance of outdoor meeting places is widely recognized yet.

POSTSCRIPT: The decade since the above observations were made has been as significant for the cities of Canada as it has for the life styles of its people. Most of our larger cities lost the last bit of country innocence they had in the early '60's; they have, to some degree, been bought and made over by the international jet set of the development world or by a local emulator. Their new adornments are slicker and more massive than before, and more sophisticated elaborate to appeal to urban dwellers who are told every evening via telstar what is currently stylish in the world's trend-setting capitals.

The major building forces operate downtown. The new scale of core redevelopment is now in block dimensions, though for every block rebuilt in a co-ordinated way, twenty follow the traditional piecemeal practice or lie fallow. Thus, we still have the spectacle of towering office slabs looking down on countless two-storey shops and parking lots. Few people question if this should be.

Thanks perhaps to a now established immigrant population and more young people who have more urban tastes, the ethnic-boutique strain is strong in the central shopping areas, though not always within the high rent central business district. Some of these entrepreneurs are discovering our historic urban relics, refurbishing and exploiting them in a generally palatable style.

Big investment is also in the suburbs. Super shopping centres sport two department stores and a theatre, tied together by an indoor mall, but still alas, surrounded by acres of asphalt. The scale is bigger, but the concept is not much changed really — these complexes certainly aren't part of any community fabric.

EXPO was the urban experience of this decade, pointing out what a joy it is to be a pedestrian. Some of its more obvious lessons, particularly about human design, show up in new bits and pieces of our cities.

The selection of housing offered in Canada has broadened, particularly in the direction of low density multiple housing: the stigma of "Row Housing" has been removed by "Garden Court Apartments." In the bigger development projects there is now a broader residential mix, but the high rises offer the same fixtures with a little more green grass, a sauna bath and perhaps taller towers, spaced to let in the sunlight. There is now some resistance to bulldozer urban renewal projects to the extent that authorities are questioning these wholesale programs. Instead, more importance is placed on rehabilitating the old that still has some life in it. Suburbia remains about the same.

In the field of urban transportation, the decade has brought a disenchantment with the freeway. Now *in* is rapid transit, unfortunately too often promoted as a panacea just as the freeway way not long ago. Technologically, the transportation revolution is still round the corner. Besides, public transportation is not as easily financed by gasoline taxes as freeways and at present the provincial and federal governments are going through the initial agonies of working out shared responsibilities for this new heavy capital investment in urban living.

When one revisits Canadian cities, these changes are obvious; certainly we are introducing a few experiments, though most have been tried elsewhere first. Yet I recently decided that Canadian cities are creatures of mediocrity — and likely will continue to be for two reasons: one our strong tradition of doing the unspectacular in city-building; two, the powerful international development forces, which apply monotonously similar plans, building techniques and sheer advertising gimmicks to their projects all over North America. These forces build the corporate image service stations, office slabs, everyman shopping centres, ticky-tacky suburbia and standard approved freeway designs, which characterize much of Canadian cities today. There may be a rising generation of urban dwellers in Canada who are starting to insist that their city have character and humanity, but compared to these forces of monotony, they seem small and frail participants in the process of city building.

Some Aspects of the Geography of Finance in Canada

Donald Kerr

Urban geographers engaged in diverse tasks have virtually ignored finance as an urban economic activity. It is obvious that financial activity ranks high in some cities and not in others, but little attempt has been made to measure its magnitude and to study its functional organization. A major aim of this article is to select and employ a variety of statistical and non-statistical criteria to describe spatial variations of financial activity in Canada, and thereby, it is hoped, to cast some light on the broader problems of measurement and definition. Particularly in any attempt to define the financial centre, it seems necessary to distinguish between volume of financial activity and financial functions. This article seeks also, in a preliminary way, to show the significance of the hinterland concept in offering explanation for the rise and fall of financial centres; and to make brief reference to their connectivity. It should be noted that financial systems, although basically similar in broad structure in all capitalistic countries, do differ in detail, and therefore conclusions reached may or may not have relevance in the United States and other nations.

MEASUREMENT AND DEFINITION

Finance is varied and complex, and the problems of securing data to measure its magnitude and character spatially are formidable. Little can be gained by comparisons with other economic activities, for there is in finance no over-all total figure of value, and consequently, no possibility of abstracting "value added" (which has proved to be so meaningful in studies of manufacturing). Furthermore, it is unlikely that an aggregate figure, if one were available, would have much meaning; broadly speak-

ing, finance is not part of the productive process as are agriculture and manufacturing, but on the other hand, it is not entirely a service industry.

Unquestionably, the most accessible body of statistical data is that of employment in finance, and indeed, such figures have been used by Harris, Bogue, and others in systems of classification and analysis.[1] But, at least in Canada, other types of statistical data may also be assembled. If the focus is changed from employment to financial institutions, it is possible to reveal a somewhat different geographic pattern. For this reason, assets of financial institutions, according to head office locations, have been tabulated and mapped. Third, the selection of one important financial activity—the movement of cheques through clearing houses—allows yet another regional pattern to be identified. All these criteria have serious limitations, however, and in terms of defining the financial centre, one must pay attention to selected financial functions.

Employment

No record of employment in finance is kept by census areas. However, by making the broad assumption that the labour force in finance is approximately the same as employment in numbers and location, it is possible to measure variations in the number of people engaged in financial activity from one urban centre to another.[2]

Two very obvious characteristics of finance and the labour force can be listed: employment in finance is quite small (only 3.5 percent of the Canadian total in 1961), and it is concentrated to a greater extent than any other economic activity in large cities. Whereas wholesaling, manufacturing,

● Donald Kerr *is Chairman and Professor of Geography, Department of Geography, University of Toronto.*

Reprinted from the *Canadian Geographer,* Vol. IX (4), 1965, pp. 175-192, with permission of the author and the editor.

and retailing, on the basis of employment, are rather widely dispersed throughout the urban hierarchy, finance is very much concentrated in Toronto and Montreal (41 percent) and in all major urban centres having a population over 100,000 (74 percent) (Table 1). It is not surprising that the spatial distribution of financial activity (Figure 1) mirrors that of major urban centres. Toronto and Montreal stand out, the radii of the circles for Winnipeg and Vancouver not reaching half the values of the two eastern cities. Southern Ontario contains the largest number of centres that have an employment in finance over 500, while southern Quebec, in spite of a relatively large population, supports only

Quebec City, Trois Rivières, Sherbrooke, and Chicoutimi outside of Montreal. The pattern on the prairies shows the dominance of Winnipeg; in the maritimes the absence of a large centre of financial activity is noteworthy.

Table 2 provides more detailed information by ranking the cities according to employment in finance; thus Toronto stands first over Montreal, and Calgary outranks Edmonton. By also giving employment data for sub-group 704 (investment companies and security dealers, see reference 2), Table 2 further emphasizes the concentration of finance in Montreal and Toronto, where indeed approximately 60 percent find employment. To single out this category has merit,

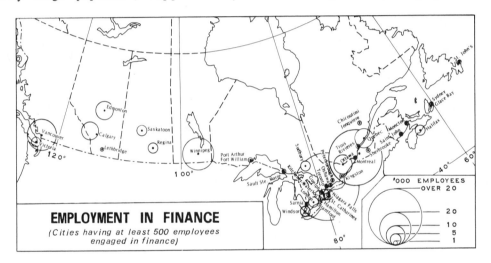

Figure 1

Table 1

DISTRIBUTION OF EMPLOYMENT IN SELECTED ECONOMIC ACTIVITIES AND POPULATION BY MAJOR URBAN AREAS GIVEN AS A PERCENTAGE OF THE CANADIAN TOTAL IN 1961

	Population	Manufacturing	Wholesale trade	Retail trade	Finance
Urban areas with population greater than 1,000,000 (Montreal and Toronto)	21	35	31	26	41
Urban areas with population 100,000 to 1,000,000	24	24	33	29	33
All others (rural and urban areas)	55	41	36	45	26

Source: *Census of Canada 1961*, vols. 1 and 3.

Table 2

LABOUR FORCE IN FINANCE AND CHEQUES CASHED AT CLEARING HOUSES FOR SELECTED CENSUS
METROPOLITAN AREAS (OVER 100,000) IN 1961

Census metropolitan area	Labour force in finance†		Ratio of labour force in finance to total labour force		Labour force in investment companies and security dealers‡		Cheques cashed§	
	No.	Rank	Ratio	Rank	No.	Rank	$000,000	Rank
Calgary	5,566	6	5.0	7	445	5	10,300	5
Edmonton	5,467	7	4.1	11	328	6	6,700	7
Halifax	2,988	12	4.1	10	137	13	2,800	12
Hamilton	4,969	9	3.3	15	213	7	6,000	8
Kitchener-Waterloo	3,152	11	4.8	8		15	1,300	15
London	4,580	10	6.2	2	163‖	10	3,700	11
Montreal	41,984	2	5.2	6	3,649	2	78,600	2
Ottawa	7,480	5	4.4	9	205	9	5,900	9
Quebec (City)	5,127	8	4.0	12	211	8	7,900	6
Regina	2,530	14	5.4	3	113	14	4,900	10
Sudbury	994	16	2.5	16		16	700	—
Toronto	52,338	1	6.6	1	5,703‖	1	109,600	1
Vancouver	15,918	3	5.4	4	1,165	3	17,800	4
Victoria	2,244	15	4.0	13	143	11	2,700	13
Windsor	2,690	13	3.9	14	155	12	2,400	14
Winnipeg	10,252	4	5.3	5	830	4	20,900	3
TOTAL FOR CANADA	228,905		3.5		15,459		293,784	

Source: *Census of Canada 1961*, vol. 3.

†Division 9 in D.B.S., *Standard Industrial Classification Manual* (Ottawa, 1960), p. 42.

‡Sub-group 704 in *Ibid.*

§Source: D.B.S., *Canada Year Book 1962* (Ottawa, 1962), p. 1105.

‖Less than 100.

Table 3

ASSETS OF FINANCIAL INSTITUTIONS (MILLIONS OF DOLLARS) ACCORDING TO HEAD OFFICE LOCATION FOR SELECTED CENTRES
(BASED ON AN AVERAGE 1959-1963)

	Montreal	Toronto	Vancouver	Winnipeg	London	Ottawa	Kitchener-Waterloo	Halifax
Chartered banks	10,300	9,000	nil	nil†	nil	nil‡	nil	nil§
Insurance (life, property, and casualty)	3,500	4,500	50	950	850	850	900	30
Trust and loan companies	2,100	1,900	‖	10	540	nil	100	270
Others (finance companies, investment funds, etc.)	500	2,100	350	440	210	nil	nil	‖
TOTAL	16,400	17,500	400	1,400	1,600	850	1,000	300

Sources: *Financial Post Survey of Industrials* (Toronto, annually); *Report of the Superintendent of Insurance for Canada* (Ottawa, annually).

†Bank of Western Canada has not yet been established.

‡Bank of Canada not included.

§It has been assumed that the executive head office of the Bank of Nova Scotia is in Toronto.

‖Less than 10.

because many of its employees are involved in all aspects of the capital market and are in fact specialists in financial transactions. It is revealing that Vancouver, Winnipeg, and Calgary are the only other cities, besides the aforementioned, in which a substantial number are employed in this category.

In order to refine descriptions of financial activity, it is possible to manipulate statistical data on employment in a number of ways[3]; but in view of the over-all aims of this article, only one very simple exercise is presented, that is, the relating of employment in finance to total employment. The calculation of these ratios gives insight into the relative importance of finance in the total structure of various cities. The arithmetical mean for all urban areas over 100,000 population is 4.4 percent, which, as Table 2 shows, is exceeded only by Toronto, London, Regina, Vancouver, Winnipeg, Montreal, Calgary, Kitchener-Waterloo, and Ottawa, in that order. In fact, on the basis of this ratio, with modifications, Toronto stands fourth among the large cities in North America, exceeded by Dallas, San Francisco, and New York; Montreal is seventh, exceeded by Minneapolis and Atlanta.[4]

Assets of Financial Institutions

A tabulation of assets of all financial institutions according to head office location gives some measure of the geographical concentration of finance in Canada, at least for an average of the five-year period 1959 to 1963[5] (Table 3). Quite arbitrarily, a decision was made to include only private financial institutions (non-government), namely, chartered banks, life, casualty, and fire insurance companies and societies, trust and loan companies, instalment finance and small loan companies, and investment trusts. Montreal and Toronto dominate the financial structure of Canada by accounting for 85 percent of all assets (Figure 2), although Winnipeg, London, Kitchener-Waterloo, and Ottawa rank reasonably high mainly because of the wealth of large insurance companies. Western cities (other than Winnipeg) have spawned very few financial institutions, and even Vancouver, ranking third on the basis of employment, drops to seventh place.

When total financial assets are separated into categories (Table 3), the dominance of Toronto and Montreal is again apparent. Both cities show a concentration of chartered banks;[6] indeed, when compared with American cities, Montreal ranks third behind New York and San Francisco, and Toronto fifth behind Chicago. Insurance companies and trust and loan companies are somewhat more widely dispersed, but mainly through eastern Canada (see also Figures 3 and 4). Nonetheless, the concentrations in Toronto and Montreal are very striking; in all of North America these two cities are

Figure 2

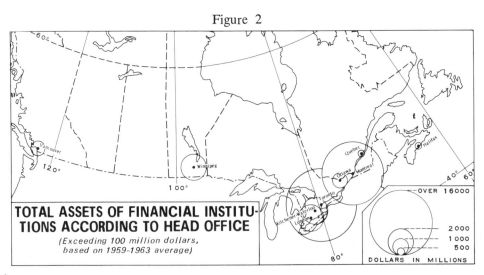

TOTAL ASSETS OF FINANCIAL INSTITUTIONS ACCORDING TO HEAD OFFICE
(Exceeding 100 million dollars, based on 1959-1963 average)

DOLLARS IN MILLIONS

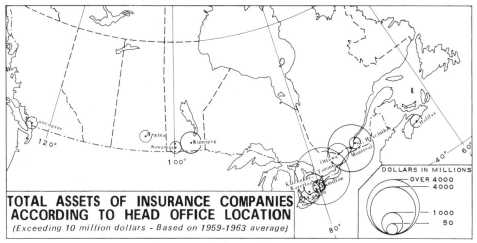

Figure 3

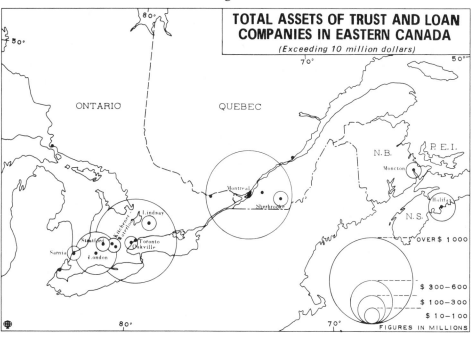

Figure 4

Only Winnipeg in western Canada has assets over 10 million dollars (based on an average 1959-1963). This pattern has important historical ramifications.

exceeded in assets of insurance companies only by New York, Hartford, Boston, and Milwaukee.

Cheques Cashed at Clearing Houses

A record of the value of cheques charged to customer accounts in all bank offices in major clearing house centres gives some insight into the amount of banking activity in various cities,[7] and as such, it is an index of a very common financial transaction. The relevant data for Canadian cities appears in Table 2, from which a ranking has been drawn up (see also Figure 5).

187

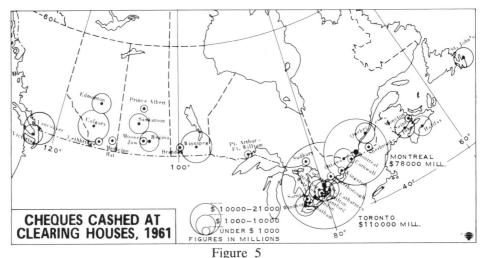

Figure 5

Apart from Montreal and Toronto where almost two-thirds of all cheques are cashed (by value), the pattern reflects quite well nodal points in the distribution of population.

Again, Toronto and Montreal are dominant, having respectively 37.2 percent and 26.7 percent of the Canadian total, or together, almost two-thirds. It is interesting that Winnipeg outranks Vancouver, and that Calgary is ahead of Edmonton. The positions of Quebec and Hamilton are somewhat higher in this system of ranking than in employment.

It may be said that on the basis of these three criteria—employment, assets of companies, and bank clearings—a measure of what might be called the volume of financial activity in Canadian cities has been achieved. Toronto clearly ranks first and Montreal second, a reversal of their rankings in almost all other categories (population, manufacturing, wholesaling, and so forth). It appears that Winnipeg, to a very slight degree, outranks Vancouver, even though the latter leads in population and industrial rankings. Calgary stands fifth, and London and Kitchener-Waterloo show up strongly on the basis of assets.[8]

Clearly there are limitations in the use of these data. Total employment in finance embraces a great variety of occupations, ranging from real estate agents through bank messengers, trust officers, investment analysts, and bank presidents. Certainly, the meaningfulness of an aggregate figure is a problem and must be questioned, but it is always such in a summing of the values of disparate categories.

The problem of emphasizing assets of financial institutions, of course, focuses on the function of the head office and invites the question of whether it is realistic to attribute so much importance to a single location within the system of a large financial company. No simple answer can be made, but clearly this is a major limitation in the use of these aggregates of assets. Even more serious is the fact that accounting procedures between firms are not always uniform and disparities may arise in the tabulation of assets.[9] However, although double counting is a distinct possibility in manufacturing concerns in which physical as well as monetary assets are included, it is less likely to occur in financial institutions.

In many ways bank clearings are more indicative of the amount of economic activity within a city than of a specific financial activity. It can be argued that most of the population is very much involved in the process of writing cheques and that therefore bank clearings measure this energy rather than financial transactions *per se*.

Despite the above and other limitations, these three criteria do portray impressions of the amount of financial activity by urban areas, although they give little, if any, insight into the concentration or dispersal of certain financial functions.

188

SELECTED FINANCIAL FUNCTIONS

Although financial activity is extremely complex, it is possible for the purposes of this article to identify three major functions: the collection and accumulation of funds, their disbursement or investment, and trade in financial instruments.[10] Some of these activities are nation-wide in the sense that they take place in towns or even villages, but unquestionably there is a concentration of substantial transactions in large centres.

It is not difficult to envisage the neighbourhood bank performing the task of collecting, accumulating, disbursing, and lending money within a fairly well-defined trade area. Nor is it unusual that a large insurance company will direct all of its collections through its head office, at which point large sums accumulate to be disbursed through payment or invested in mortgages or securities. The former activity is in a sense ubiquitous, the latter centralized. The third function, trade, is highly concentrated, and its importance in the workings of what might be called the capital market cannot be minimized.[11] Such trade makes provision for liquidity of non-monetary financial assets or, in short, for the shift in ownership of outstanding liabilities. For a number of reasons, largely of historical significance but focusing to a great extent on the notion of economies of scale, this trade goes on only in a few centres. It may be carried on through the facilities of an organized market such as the stock exchange, or in a less formal way such as "over-the-counter" bond sales.

It is indeed unfortunate that the statistical data to measure the regional variations of these different facets of finance are fragmentary or entirely lacking. For this reason, and partly because of the great complexity of financial systems, the commentary which follows deals only with underwriting and trade in stocks and bonds. Although the discussion is of necessity restricted, the significance of the financial centre is underlined, and indeed some progress is made towards its definition.

Underwriting

One important way in which capital can be raised is through the process of underwriting, that is, by making new security issues available for sale by one or more investment bankers. It is not the purpose of this article to review the various aspects of underwriting; a comprehensive and informative description has been provided by Fullerton, especially with reference to Canadian conditions.[12] What can be emphasized, however, is that a new issue is organized by an investment dealer or, more commonly, by a syndicate, which is simply a collection of investment dealers. An advertisement describing the securities to be sold and listing the investment dealer, or dealers, handling the sale appears in a financial newspaper. No record of these sales is made available; consequently it is impossible to compare the amount of underwriting in one city with that in another. All that can be said is that Montreal- and Toronto-based investment firms appear most frequently in advertisements,[13] and it is not unusual for a syndicate to comprise several different firms from both cities.

There are 181 investment firms in Canada, of which fourteen, without any question, have national significance.[14] Of these, seven have head offices in Toronto, five in Montreal, and one each in Winnipeg and London. It is not meant to imply that *all* new issues emanate from the Toronto and Montreal markets, for a significant number of provincial and municipal bonds are floated in local markets, the sales of which are helped along by local laws and prejudices.[15] Also, conditions may make it desirable for governments, or even private companies, to borrow money in New York. Nevertheless, it is not an exaggeration to state that practically all corporate and most governmental issues arise in either the Toronto or the Montreal markets, or both. On the basis of interviews with a number of investment dealers, selected at random, it can be noted that there is at present a somewhat greater volume of underwriting in Toronto than in Montreal.

A number of specialists engage in underwriting. Investment dealers perform the specific tasks of underwriting, but bankers, investment counsellors for insurance companies, corporation lawyers, chartered accountants, geologists, and others help to prepare and support such ventures. The propinquity of these specialists within the financial district of the financial centre is noteworthy. In Canada such concentrations are to be found almost entirely in Montreal and Toronto, where as previously noted 60 percent of the employment in sub-group 704 (investment companies and investment dealers, Table 2, reference 2) is found. It may be added that this concentration of the machinery for underwriting often makes it necessary for a company or a business venture, when contemplating the raising of money through underwriting, to proceed directly to either Toronto or Montreal. There is little possibility that a manufacturing company in Moose Jaw, for example, would be able to arrange for the issue of new securities in Regina, or for that matter in Winnipeg, and would of necessity have to deal with investment bankers in either Toronto or Montreal.

Trade in Bonds and Stocks

It has been indicated that provision must be made for the liquidity of non-monetary financial assets in a capitalistic economy. Through the course of history, markets have grown up in various centres where securities may be traded. These markets form an integral part of the larger capital market and are, in fact, essential to its proper functioning. New York and London contain the largest number of both organized and less formal financial markets.

Some measure of the magnitude of this trade can be gained by tabulating stock market transactions on organized stock exchanges. In Canada, five such exchanges handle on the average 3.5 billion dollars of trade annually (Table 4). Toronto clearly dominates, and in North America stands second to New York; the Montreal exchanges account for over a quarter of all trade, and Vancouver less than one-

twentieth; Calgary and Winnipeg are quite insignificant.

Typically, the stock exchange building occupies a central position within the financial district around which investment dealers and other financial institutions are grouped. Very often the street on which it is located, such as Bay Street in Toronto, St. James in Montreal, and Howe in Vancouver, has become synonymous with the whole financial community and carries a connotation into the hinterland analogous to that of Wall Street in the United States.

Trade in bonds and debentures outranks that of the stock market in value, but in view of the fact that all transactions are "over-the-counter" no record is published. The trading system is rather intricate and is based on the connectivity of investment firms by private telephone lines and teletype services; in fact, some investment firms have as many as ninety direct telephone lines, making it possible to sell and buy bonds very quickly. Banks and some insurance companies are also linked by wires

Table 4

AVERAGE ANNUAL VALUE OF STOCK MARKET TRANSACTIONS ON STOCK EXCHANGES IN CANADA, 1962-64

City	Value in millions of dollars	% of total Canadian
Toronto	2,420	68.8
Montreal†	939	26.7
Vancouver	153	4.3
Calgary	3	0.1
Winnipeg	3	0.1
TOTAL	3,518	100.0

Sources: *Toronto Stock Exchange Year End Review*, 1962-64; *Statistical Summary Montreal and Canadian Stock Exchanges*, 1962-64; *Vancouver Stock Exchange*, 1962-64; *Calgary Stock Exchange Official Yearly Summary*, 1962-64; *Winnipeg Stock Exchange Official Yearly Summary*, 1962-64.

†Figures for the Montreal and Canadian stock exchanges have been combined. The latter is also located in Montreal and has emerged from the old Curb Market.

in the trading system. Most trading goes on within Toronto and/or Montreal and between the two centres. Connections are also maintained with New York, and in Canada with major centres such as Winnipeg and Vancouver.

It is obvious that underwriting and trade in bonds and stocks, among other activities, are highly concentrated in a few places, and indeed help to define the nature of the financial centre. The implication is one of primary functions stimulating the flow of funds not only in the financial centre but also over a wide area which may be regional, national, or even international in scope. The substance of the financial centre is to be seen in the composition of the financial district within the central business district. The buildings comprise not only the offices of investment dealers, headquarters of banks, and the stock exchange, but also a great variety of other establishments including legal firms and chartered accountants specializing in underwriting and investment in general, consulting geologists, investment analysts, and financial newspapers and magazines. In short, employment in finance in a financial district may be distinguished from that in other cities by a retinue of specialists in financial affairs.[17]

CLASSIFICATION

The preceding discussion on measurement criteria suggests that, in any classification of cities based at least in part on finance, functional organization must be recognized. Clearly, employment measures only the volume of activity; thus, employees performing certain services in finance (bank tellers and insurance agents) are found in all urban places regardless of size. It is not difficult to define the trade area of a bank around a village or suburban shopping plaza, or that of an agency of a life insurance company in larger context. These service functions in finance are, therefore, widespread and would find closest correlation with population density and income levels.

Other activities in finance are very much tied up with certain specific workings of the capital market, most of which, as have been noted, take place in only a few centres.

Thus, the financial centre has been defined as that place in which markets are provided for financial trading and in which facilities are available for large-scale borrowing and lending. It can therefore be distinguished from other urban places, regardless of size, in which these activities do not take place. A hierarchy can be envisaged in which, on a world-wide scale, New York and London would rank at the top. The question of which cities among Tokyo, Amsterdam, Paris, Milan, Toronto, San Francisco, and so forth would occupy successively lower strata is challenging, but impossible at present to answer.

With respect to non-financial centres further research is needed to designate subclasses. Does the regional city (for example, Saskatoon) function in any way that is different from the small town (such as North Battleford), with its branch banks and insurance offices? It may be that in this class there is no hierarchy at all but rather a continuum. Also, provision should be made for the city which lacks the trappings of a financial centre but houses the head offices of financial institutions. In Canada London and Kitchener-Waterloo may be identified (Table 2), and in the United States Hartford, Connecticut, stands out.[18] Accordingly, such a city does not function as a financial centre but, on the other hand, the contribution of its financial institutions to the economic structure of the city—providing employment, paying taxes, and demanding office space—is significant.

According to the definition given above, only Montreal, Toronto, Vancouver, Winnipeg, and Calgary qualify as financial centres in Canada. It is rather obvious that the first two stand out as national and even international centres and have first rank; the others are important regional centres and are of second rank. Worthy subjects as well for further research are those cities such as London, Halifax, Quebec, Regina, and Kitchener-Waterloo that do not rank as financial centres but nevertheless carry out important financial activities.

191

THE HINTERLAND CONCEPT

The classification and ranking of Canadian cities according to financial functions and activity raise a number of questions on spatial distribution, some of which have historical relevance. Whereas in most countries of the world one city clearly dominates as financial capital, in Canada two share primary functions. Furthermore, the present growth of Vancouver and Calgary, the slow and long decline of Halifax, and the vigorous growth of Winnipeg just before and during World War I are noteworthy. Obviously, to lay bare the rationale of the changing significance of various centres is beyond the scope of this article; in fact, much is central to fundamental problems in Canadian economic history and historical geography. But it may be fruitful to comment briefly on the hinterland concept and illustrate with reference to Toronto and its relationships with the Canadian Shield.

According to Gras, the emergence of financial centres fulfills the fourth and final stage in the development of a metropolitan economy, especially with reference to the control that the metropolis exerts in financing the hinterland: "from the nature of things the metropolis will develop financial resources to the limit to care for both the extended and the hinterland trade, the intermetropolitan and intra-metropolitan commerce."[19] It follows that not only does the financial centre play a vital role in the development of the hinterland, but in turn that that very development profoundly influences the financial centre. By way of illustration, the development of interrelationships between Toronto and the Canadian Shield can be briefly, if somewhat superficially, examined.[20] Of necessity, discussion will be very limited and will refer only to the financing of mineral resources.

There is little question that the ability of some segments of the financial community to wrest control of mineral developments on the Shield has brought great wealth to the city and laid the basis for vigorous growth of the stock exchange, particularly since the late 1940's. That part of the Shield which has fallen under the dominance of Toronto includes a broad belt from Rouyn and Noranda in western Quebec to the Northwest Territories; the eastern Shield has, to a large extent, and especially in recent years, come under the control of Montreal interests. Quite obviously the tentacles of New York, but no longer those of London, also extend through the whole Shield, but those of Toronto are dominant. It need only be reiterated that the great wealth of mineral resources has proved to be a remarkable attraction to mining companies, investors, speculators, and the like.[21]

The roots of Toronto's conquest of the Shield may be traced to the building of the Ontario Northland Railway in the early years of the twentieth century. Intended to stimulate agricultural developments on the Clay Belt, this railroad directly and indirectly resulted in the discovery and development of mineral resources. Although it was later extended into western Quebec, giving Toronto an advantage over Montreal in trade[22] and accessibility,[23] it has been supplemented increasingly by highway facilities and airline services.

The historical record indicates that Toronto groups were interested in promoting mining properties in British Columbia[24] and on the Shield as early as the 1880's. The formation of two small mining exchanges, later to merge in 1898 to form the Toronto Standard Stock and Mining Exchange, facilitated the promotion of mining ventures which the more conservative Montreal and Toronto Stock Exchanges did not wish to handle.[25] Clearly, some of the techniques, facilities, and energies for financing mineral properties had taken hold in Toronto before the discovery of very rich deposits of silver at Cobalt in 1903, and of gold at Porcupine in 1911 and Kirkland Lake in 1912 had made the Shield so attractive. On the basis of very scanty evidence it appears that similar traditions had not developed to any extent in Montreal; nor did there seem to be much desire to build them up until a later date. Some capital for development came from varied interests in Montreal,

New York, Buffalo, and other cities as well as Toronto, but the mining exchange became a focus of a great deal of speculation, and increasingly mining activities centred on Toronto.

Thus, the financing of the hinterland revolved around the nature of the resources (rich minerals), the initial advantage of transportation links, and the traditions of the financial community, and increasingly in the early twentieth century evolved inter-relationships between the Shield and Toronto that have had a profound influence on subsequent developments. In the 1930's, in an otherwise depressed market, gold issues became buoyant following an increase in the price of gold, and activity increased on the Toronto market. Following the crash of 1929, the Montreal Stock Market remained weak, and the volume of trade on the Toronto exchange in the early 1930's, although relatively small, exceeded that of Montreal for the first time. A merger of the old Standard Stock and Mining Exchange with the Toronto Stock Exchange in 1934 was followed by more stringent regulations in speculation, and as the financing of mining issues became more sophisticated other segments of corporate and governmental financing increased in Toronto at the expense of Montreal. The post-war period brought forth great activity in mining issues (uranium, copper, oil, and so forth), which, when combined with the improvement of trading facilities (high-speed electronic equipment and data processing centre) and the accelerated establishment of branches of Toronto brokerage houses throughout Canada, led to a further acquisition of the market in industrials. In short, it is suggested that the prominence of the Toronto financial market at present may be traced to earlier developments of mining properties on the Shield which, in turn, strengthened the financial community and made Toronto a somewhat more attractive trading centre than Montreal during and, in particular, after the depression.

Such broad assertions are based on scanty evidence, but even so, suggest the relevance of the hinterland concept in any analysis of the growth of financial centres. Only part, however, and a very small part, of the nature and significance of the rivalry of the Montreal and Toronto financial communities comes to light from the previous discussion.

Quite briefly and superficially, the relationship of other cities to their hinterlands can be reviewed. At the time of World War I Winnipeg was on the verge of becoming a metropolitan centre, in a financial sense, but the lack of variety and substance in its hinterland economy, so overwhelmingly dominated by grain, made this leap impossible. Can the decline of Halifax, so well endowed with financial institutions in the nineteenth century, be attributed to the poverty of its hinterland? To what extent may the rise of financial power in Calgary and Vancouver be attributed to their abilities to capture extensive hinterlands?

It may be that, if attention is directed to the relationship of the financial centre to the hinterland—a geographical approach of long standing—while at the same time colleagues in other fields write histories of financial institutions and of the men who directed their destinies, further progress can be made towards understanding the evolution of financial centres in Canada.

SPATIAL INTERACTION AND FINANCIAL CENTRES

Financial centres grew up, if not entirely independent one of the other, at least largely so. As communications improved, however, and as financial capitalism grew more important,[26] interactions between financial centres increased. The questions may now be asked how closely are financial centres linked and, therefore, how inter-dependent are they, and to what extent do they still function as independent entities? Some comments can be made with reference to Montreal and Toronto.

Spatial interaction[27] between Toronto and Montreal in commodities, people, information, and so forth, is the most intense in Canada. There is some evidence to support the view that the interaction of these two financial communities is such that they tend to participate, at least in some activities, as one large financial centre (as defined in this article) and are not necessarily the custodians of two separate financial markets. For example, it is not unusual for a Toronto-based investment firm to arrange the underwriting of new securities for an industrial company that has its head office in Montreal; nor is it so for a Montreal investment firm to handle the financing of a Toronto industry. The syndicates referred to earlier comprise investment dealers in both cities, and it has been emphasized that Montreal and Toronto are intimately connected by all sorts of electronic communications equipment which allow instantaneous trade in bonds or stocks between a desk in one city and that in another. Fullerton comments that with reference to the bond market the rivalry of the two centres is, in some respects, academic.[28] Furthermore, Walter and Williamson state that, among other factors, because there is a large amount of inter-listing of stocks on the Toronto and Montreal exchanges and because the large brokerage houses have memberships on both exchanges, the interaction between the two exchanges is significant.[29] From time to time there is even discussion about the possibility of merging the stock exchanges in Canada to form a national one, an initiative, however, which comes entirely from Toronto circles. Most Montreal interests are vigorously opposed, and the fact that they have moved into a new stock exchange building and have plans for a vigorous publicity campaign suggest that the independence of the two centres and indeed their rivalry have by no means disappeared. Nonetheless, it is not entirely unrealistic to view these two historic centres as being increasingly drawn together and in some spheres of financial activity functioning as one.

CONCLUSION

It is of relevance, geographically, to distinguish between general services in finance and primary financial functions. The former, epitomized best in the workings of the neighbourhood branch bank or local insurance agent, are found in all urban centres; the latter, comprising mainly trade in financial instruments and large-scale borrowing and lending, are highly concentrated, notably in Montreal and Toronto, and to a certain extent in Vancouver, Winnipeg, and Calgary. However, the distinction between these two types of activities is by no means sharp, for such cities as London, Kitchener-Waterloo, Halifax, and Ottawa house the head offices of one or two large financial institutions or a number of small to medium-sized ones. Clearly these cities do not function as financial centres, but they do nevertheless carry out important financial activities. Furthermore, research is needed to discover the significance of divisional or regional offices of financial institutions in the Canadian financial system and thus to define more precisely financial functions in such cities as Winnipeg, Calgary, Vancouver, Halifax, and Quebec City. Particularly, the role of Regina, which has among all cities the third highest ratio of employment in finance to total employment, might be discovered.

To answer why two large financial centres should have emerged in Canada instead of a single dominant one as is the case in nearly all countries would be to engage in a long and complex historical discussion beyond the scope of this article. It is obvious that the status of a financial centre may change through time; noteworthy has been the striking rise of Toronto in recent decades to exceed Montreal in value of trade and in underwriting. At present, considerable evidence suggests that Vancouver may be on the threshold of becoming a primary financial centre, thus to compete more vigorously with its eastern counterparts.

In conclusion, it is apparent that varied and more substantial research is suggested by this article. Investigations into the changing spatial organization of financial institutions from the nineteenth century to the present should be made to define the changing status of cities and to cast light on urban growth processes. In what way does a concentration of financial wealth and power affect the growth of a city? Clearly the whole question of agglomeration theory is very much at issue. What is the threshold at which the grouping of financial institutions affects, and is in fact interrelated to, other decision-making units of the economy such as industrial corporations? Finally, it appears that a vigorous and expanding financial community exerts a profound influence on the structure of the central business district. In fact, the vitality or decline of central areas of the city may, in large measure, be related to the nature and volume of financial activity.

REFERENCES

[1] Harris, C. D., "A Functional Classification of Cities in the United States," *Geog. Rev.,* XXXIII (1943), 86-89; Bogue D. J., *The Structure of the Metropolitan Community* (Ann Arbor, 1950).

[2] Employment in finance is tabulated under two broad headings: (1) financial institutions, and (2) insurance and real estate industries. Five sub-groups have been recognized: no. 702, savings and credit institutions; no. 704, investment companies and security dealers under group (1); no. 731, insurance carriers; no. 735, insurance and real estate agencies; no. 737, real estate operations under group (2).

For a detailed description of the types of employment see Dominion Bureau of Statistics, *Standard Industrial Classification Manual,* no. 12-501 (Ottawa, 1960).

For the purposes of this article, and despite obvious drawbacks, the labour force and employment in finance are considered to be one and the same. All data on the labour force have been taken from *Census of Canada 1961,* vol. 3, no. 2, "Labour Force for Industries by Metropolitan Areas."

[3] One paper can be cited: Morrissett, Irving, "The Economic Structure of American Cities," *Papers and Proc. Regional Sci. Assoc.,* IV (1958), 239-52. See Florence, P. S., *Investment, Location and Size of Plant* (Cambridge, 1948), for a review of statistical procedures to be used in describing the location of industry.

[4] In order that the Canadian figures would be more comparable with those of the United States, employment in real estate was deleted from the total figures for Toronto and Montreal. See Robbins, S. M., and N. E. Terleckyj, *Money Metropolis* (Cambridge, Mass, 1960), p. 277.

[5] *Financial Post Survey of Industries* (McLean Hunter Co., Toronto, annually) and *Report of the Superintendent of Insurance, Department of Finance* (Queen's Printer, Ottawa, annually). From these two publications it was possible to identify all non-governmental financial institutions. The assets for each insurance company, finance company, bank, etc., were compiled for a five-year period (1959-63) and averaged. The resulting value for each company was then inserted in the column of the appropriate city as determined by the location of the head office. Total assets for each major category of finance, and for finance as a whole, were summed for each city.

[6] Branch banking development in Canada from the outset. Amalgamations, mergers, and bankruptcies characterized the period of the late nineteenth and early twentieth centuries, and now only a few very large banks dominate chartered banking in Canada. The merger movement also affected other financial institutions. See Jamieson, A. B., *Chartered Banking in Canada* (Toronto, 1955). At present the federal government, for the first time in several years is considering applications for the formation of three new banks: the Bank of Western Canada, the Bank of British Columbia, and the Laurentide Bank.

[7] D.B.S., *Canada Year Book 1962* (Ottawa, 1962), p. 1105.

[8] Because of the thinness of the statistical data, it was impossible to apply any sophisticated statistical techniques to achieve a formal ranking of cities.

[9] For a useful discussion see Porter, John, *The Vertical Mosaic* (Toronto, 1965), Appendix II, p. 570.

[10] For a discussion of various functions in Canada, see Hood, W. C., *Financing Economic Activity in Canada,* vol. XXIV of Report of Royal Commission on Canada's Economic Prospects (the Gordon Commission) (Ottawa, 1958), and Ashley, C. A., *Corporation Finance* (Toronto, 1949).

[11] Unquestionably the Bank of Canada which controls currency in Canada is an integral part of the market. Other government bodies such as Central Mortgage and Housing Corporation and the Industrial Development Bank play a much less significant role in the workings of the capital market.

[12] Fullerton, D. H., *The Bond Market in Canada* (Toronto, 1962), pp. 138-55.

[13] Each week the *Financial Post* (Toronto) lists the new issues and the participating dealers, and each January a list of the year's underwriting by months appears in an early edition.

[14] A firm having national significance is arbitrarily defined in this paper as one with at least ten branch offices scattered in at least five provinces. For an inventory of all firms see *Membership List of the Investment Dealer's Association of Canada on December 31, 1964* (Toronto, 1965).

[15] Fullerton, *The Bond Market in Canada*, p. 91.

[16] *Ibid.*, pp. 164-66.

[17] For a description of Toronto's financial district see Kerr, D., and J. Spelt, *The Changing Face of Toronto: A study in Urban Geography* (Geog. Branch, Dept. of Mines and Tech. Surveys, mem, no 11, Ottawa, 1965), pp. 153-57. For a description of the money market and New York's financial community see Robbins and Terleckyj, *Money Metropolis*, chap. 2.

[18] Goodwin, W., "The Management Center in the United States," *Geog. Rev.*, LV (Jan. 1965), 12.

[19] Gras, N. S. B., *An Introduction to Economic History* (New York, 1922), p. 243.

[20] Kerr and Spelt, *The Changing Face of Toronto*, pp. 83-85.

[21] For a useful analysis and review see Currie, A. W., *Canadian Economic Development* (Toronto, 1963), pp. 222-43.

[22] Field, F. W., *The Resources and Trade Prospects of Northern Ontario* (Board of Trade pub., Toronto, 1911), p. 73.

[23] The influence of Toronto on Cobalt, the leading mining centre in the first decade of mining developments, may be measured by examining advertising in copies of the *Daily Nugget* of Cobalt, an incomplete set of which is available in the Provincial Archives, Toronto. For example, in advertising by law firms and stockbrokers of Toronto and Montreal origin, the number from the first exceeded that from the second city by a ratio of 3: 1.

[24] Innis, H. A., *Settlement and the Mining Frontier* (Toronto, 1936), p. 281.

[25] "The Stock Market in Canada," *Monthly Review* (Bank of Nova Scotia, Sept. 1960).

[26] Gras, N. S. B., *Business and Capitalism* (New York, 1946), pp. 246-59.

[27] Ullman, E. L., "The Role of Transportation and the Bases for Interaction," in Thomas, W. L. (ed.), *Man's Role in Changing the Face of the Earth* (Chicago, 1956), pp. 862-80.

[28] Fullerton, *The Bond Market in Canada*, p. 90.

[29] Walter, J. E., and J. P. Williamson, "Organized Securities Exchanges in Canada," *J. of Finance*, XV, no. 3 (Sept. 1960), 311, 313.

AGRICULTURE

An Introduction

Less than 10 percent of the total land area of Canada is in farmland, and slightly over one-half of this is in cultivation. The bulk of the improved farmland is in southern Canada where physical and economic factors are amenable, although major outliers do occur in the Lake St. John district of Quebec, the Great Clay Belt of Ontario-Quebec, and the Peace River Country of Alberta-British Columbia.

Given the environmental and economic restraints, Canadian agriculture is diverse, ranging from extensive grain growing in the prairies to the capital intensive tobacco industry of southern Ontario. It is the intent of this section to elaborate on selected characteristics of the Canadian agricultural scene. Nevertheless, some of the diverse problems facing the Canadian agricultural industry are discussed briefly. (For a fuller elaboration of the problems one should consult *Canadian Agriculture in the Seventies,* Report of the Federal Task Force on Agriculture, 1969.) In the introductory paper to this section, I. F. Furniss discusses the reasons why Canadian agricultural productivity has increased by two percent annually over the period 1949-1969. His analysis reveals marked changes in the input mix over the 20-year period. Of the four major input categories—labour, real estate, machinery and equipment, and "other"—labour has decreased by one-half, (from 48 to 24 percent) and "other" (purchased feed and seed, fertilizers, pesticides and other purchased inputs) has doubled from 16 to 32 percent. Real estate has remained almost stationary whereas machinery and equipment increased modestly from 16 to 23 percent.

Although specialized crops such as fruit, vegetables, and tobacco occupy a very small proportion of the farmland in Canada, they nevertheless constitute a significant segment of the Canadian agricultural industry. In his paper, Ralph R. Krueger reviews one of these specialties—the tree fruit industry—on a national basis, and then provides a summary of the physical basis of orcharding, the orchard trends and economic conditions in each of the major orchard provinces. The most significant trends in the industry are decreasing acreages, increasing specialization, and higher productivity. Krueger concludes that marketing is, and will continue to be, the major problem facing Canadian orchardists.

Three aspects of Western agriculture are examined in the concluding papers. Changes in farm operation, organization, and types of farming are operative throughout Canada. Although the kind, rate and magnitude of change may vary regionally and even between farms, certain trends in resource-use adjustment are emerging. Illustrative of some of these trends is R. A. Stutt's paper, in which he summarizes some of the major adjustments that have occurred in farm organization and land use in the prairie area of Saskatchewan from 1951 to 1966.

Stewart Raby's analysis of irrigation development in Alberta reveals an over-extension of irrigation facilities in relation to normal demand, and that many of the present-day problems stem from an inadequate evaluation of physical, economic, and social factors. The amount of land that can be devoted to irrigated cash crops is restricted by market opportunities, and in areas where dryland farming is a reasonable alternative, the irrigation facilities are seriously under-utilized. He concludes that future government subsidization of irrigation development can be jusified only on the basis of broad regional objectives—stability and growth—and that the promotion of irrigation development solely to encourage dense settlement is no longer valid.

Perhaps in no other area of Canada is the juxtaposition of extensive and intensive types of agriculture as apparent as in the southern interior of British Columbia. In the narrow, longitudinal valleys, irrigated fruit and vegetable growing predominate, while in the adjacent upland, extensive livestock ranching is supreme. In the concluding paper to this section, Thomas Weir analyzes the physical basis for and the characteristics of the ranching economy in the southern interior plateau.

Post-War Productivity Gains in Canadian Agriculture

I. F. FURNISS

The objective of this paper is to analyze the changes in overall agricultural productivity during the two decades, 1950 to 1969. It attempts to identify the major sources of productivity gains and to make some short-run predictions. Finally, an attempt is made to draw some inferences for the agricultural community.

The study of agricultural productivity is important because changes in productivity have implications for farmers and farm organizations in terms of income effects, for the suppliers of farm inputs in terms of the demand for factors of production, and for farm product processors in terms of the cost and supply of raw materials for processing. There are important implications, too, for consumers in terms of food costs, and for governments in terms of the development of policy measures.

PRODUCTIVITY GROWTH

The rate of growth in agricultural productivity during the past twenty years is estimated to have been almost two percent a year. This increase was the net effect of a rise of more than two percent a year in farm output and an annual growth in the volume of farm inputs of less than half a percentage point. Since 1960, the rate of growth in farm productivity has increased because of an increased rate of growth in output more than offsetting the increases in inputs, of more than one percent a year, resulting in a net productivity growth rate of two and a half percent a year.

The index of overall agricultural productivity was estimated at 173 (1949=100) for 1969 (Table 1). This was an increase of more than five percent from the level of the 1968 index and was entirely due to a corresponding rise in total farm ouput, since the level of farm inputs was estimated to be unchanged from 1968 to 1969. Much of the increased total farm output in 1969 was due to greater crop output (not shown in the tables). Crop production (the words "output" and "production" are used synonymously in this paper, but, at all times, both terms refer to "production intended for market") rose by almost 10 percent from 1968 to 1969 while livestock and livestock product output increased by less than two percent between the same years.

Projecting the trends in output and inputs would indicate a level of agricultural productivity of 176 (1949=100) by 1972 on the basis of the 1950-69 trend (Projection I, Table 1). This level would still be less than the peak productivity index for 1966 of 180 (1949=100). However, should the more recent trend; that is, the trend from 1960 to 1969, be maintained, then the productivity index could reach almost 184 (1949=100) by 1972, exceeding all previous levels (Projection II, Table 1). On the basis of the longer-term trend, the index of production inputs may be little changed from the 1969 level by 1972. However, if the trend of the

● I. F. FURNISS *is Economist and Assistant Director, Research Division, Economics Branch, Canada Department of Agriculture.*

This paper is based on three articles by the author published in *Canadian Farm Economics* as follows: "Productivity Trends in Canadian Agriculture, 1935 to 1964," Vol. 1(1), 1966, pp. 18-22; "Trends in Agricultural Productivity," Vol. 2(1), 1967, pp. 15-21; and "Agricultural Productivity in Canada: Two Decades of Gains," Vol. 5(5), 1970, pp. 16-27. The article is reprinted with permission of the author and the Canada Department of Agriculture.

past ten years is maintained, then the input index could reach a level of 107 (1949=100) by 1972. The longer-term trend in output would indicate the possibilities of an output level by 1972 only one index point above the 1969 index. However, projecting the trend of the past decade would indicate a level of output nine to ten percent more than the 1969 output and 16 percent more than the 1965-69 average.

The longer-term trend for Canadian farm production indicates that output of livestock and livestock products is rising faster than crop output. The volume of crop production is estimated to have increased by two percent a year since 1950, compared with an increase of almost three percent a year in livestock and livestock production. In the shorter-run, however (that is, since 1960), the rate of growth in crop production has outstripped the annual increases in livestock and livestock products. The former averaged more than five percent growth a year while the latter grouping averaged less than three percent annually since 1960.

Table 1

INDEXES OF OUTPUT, INPUTS AND PRODUCTIVITY, CANADIAN AGRICULTURE, SELECTED PERIODS AND YEARS, 1950 TO 1969 WITH PROJECTIONS TO 1972.

Period or Year	Farm Output	Production Inputs	Productivity Ratio
	1949 = 100		
1950-54	120	95	126
1955-59	127	94	135
1960-64	140.5	96	147
1965	160	99	162
1966	180.5	100	180
1967	158	103	153
1968	168	102	164
1969	177	102	173
1965-69	169	101	166.5
1972 Projections			
I	178	101.5	176
II	196	107	183.5
		percent	
Annual Growth Rates			
1950 to			
1969	+2.3	+ .4	+1.9
1960 to			
1969	+3.6	+1.1	+2.5

Projecting these trends in crop and livestock production would indicate that, by 1972, the crop output index, if on the longer-run trend line, will be at a level of 201 (1949=100). This projected level of output is less than the actual 1969 output but is above the levels for all other years in the last twenty years, except for 1963 and 1966. With the usually more stable growth rate exhibited for livestock and products, output of this grouping in 1972 could lie between index levels of 172 and 174 (1949=100). These predicted output levels would, of course, exceed the actual output levels recorded for any previous year to date, and the increase over the actual 1969 level would amount to between eight and nine percent.

SOURCES OF PRODUCTIVITY GAINS

In a broad sense, productivity gains have occurred in agriculture because the rate of increase in total output has exceeded the rate of increase in production inputs, as they have been measured. However, the existence of this relationship is not sufficient to explain the source of the productivity gains. If, for example, the total input mix were to remain constant over time, then the main source of productivity gains would have to be found in increased efficiency of resource use or even in cyclical improvements in weather conditions. But the input mix in farming has not remained constant over time, so sources of productivity gains are found not only in improvements in production organization but also in the nature of the new inputs and in the substitutions among inputs. Figure 1 illustrates the changing relative importance of the major input groupings since 1949. In the following sections, some of these changes will be discussed in detail.

Changes in the Input Mix

Total *real estate* inputs in Canadian agriculture have been relatively stable during the past two decades, both in terms of volume and share of total inputs. The growth of real estate inputs averaged one percent a year during the period 1950 to 1969 but, since 1960, the average growth rate has been negligible (Table 2). Real estate inputs reached a peak index of 116 (1949=100) in 1967 but declines were recorded successively in each

Figure 1

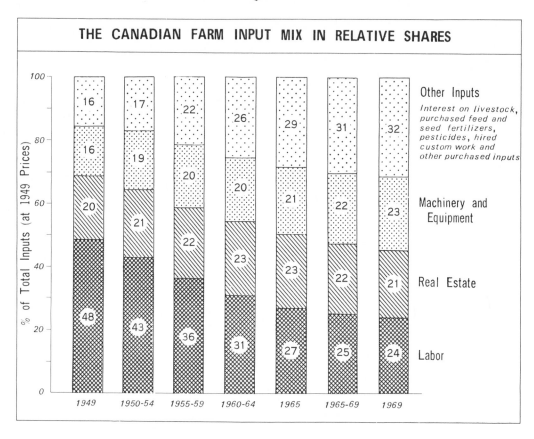

THE CANADIAN FARM INPUT MIX IN RELATIVE SHARES

year since to 1969. The decrease from 1968 to 1969 was almost five percent. All of the items comprising the real estate input category, with the exception of one relatively minor item, contributed to the total decrease between those two years.[2] This situation reflects in large measure the depressed farm product market conditions of 1969, especially in the Prairie Provinces. In 1969, total Canadian farm building construction was reduced from year-earlier levels as were building repairs, all in real terms. (Estimated expenditures on building repairs and new construction declined even in current dollar values.) However, a large part of the longer-term stability in the total real estate input category is, of course, due to the relatively fixed nature of the total land supply for agriculture.

The total land area in farms in Canada was unchanged from 1951 to 1966 and it increased by only about one percent from 1961

to 1966. Improved land area in farms showed a rising trend during these years. In 1966, it was about 12 percent more than in 1951 and about five percent greater than in 1961. More definitive descriptions of the trend in land area in farms since 1966, both total and improved, will have to await the 1971 census. However, in view of the general economic conditions in farming since 1966, it does not seem likely that there would have been any significant increase in the total land area, and the increase in improved land in farms may have leveled off. Present evidence indicates that most of the increased investment since 1966 in farm real estate (in current dollars) is a result of general increases in land values, occasioned in part by nonfarm demand for land, at least up to the end of 1968. On the basis of recorded sales, land values in the Prairie Provinces actually declined (in current dollars) from 1968 to 1969 with the

Table 2

INDEXES OF PRODUCTION INPUTS (FACTORS OF PRODUCTION), CANADIAN AGRICULTURE, SELECTED PERIODS AND YEARS, 1950 TO 1969 WITH PROJECTIONS TO 1972

Period or Year	Real Estate	Labour	Capital Inputs					Total Production Inputs
			Machinery and Equipment	Purchased Feed and Seed	Fertilizer and Limestone	Other	Total Capital	
				1949 = 100				
1950-54	99	85	115	95	107	110	108	95
1955-59	105	70	123	123	116	131	124	94
1960-64	112	61	124	142	174	160	138.5	96
1965	112	55	136	170	243	170	157	99
1966	115	50.5	142.5	181	282	178	167	100
1967	116	52	148	185	311	184	173	103
1968	115	51	151	180	335	185	174	102
1969	109.5	50	153	200	273	188	179	102
1965-69	114	52	146	183	289	181	170	101
1972 Projections								
I	120	40	151	206	321	206	185	101.5
II	115	43	165.5	220.5	392	203.5	199	107
Annual Growth Rates				percent				
1950 to 1969	+1.0	—3.3	+1.5	+4.5	+8.8	+3.5	+3.1	+ .4
1960 to 1969	+ .2	—3.1	+3.1	+5.0	+4.1	+2.6	+4.0	+1.1

greatest decline, more than 11 percent, in Saskatchewan. Such a decline as this reflects purchasers' diminished expectations of the income potential of farmland as well as increased mortgage credit costs.

Labour employed in Canadian agriculture[3] has shown a continually declining trend throughout the 1950's and 1960's (Table 2). Since 1950, the rate of decrease has averaged about three and a third percentage points a year but this rate of decline eased off in the 1960's to just over three percent a year. However, the rapid rate of attrition which characterized the farm labour force for much of this period, particularly since 1966, may be leveling off. Total employment in agriculture in 1969 at 535,000 persons was only about two percent fewer in numbers than in either 1968 or 1966. Preliminary data for 1970 would indicate a possible decline in numbers of five percent from year-earlier levels. Projecting the trend of the past ten years would indicate a total farm labour force of about 465,000 persons by 1972, 13 percent fewer than the actual numbers in 1969.

The rate of net outmigration of labour from farming is explained, in a perfectly competitive, free-enterprise economy, by the desire of workers to shift to occupations yielding the highest labour income which they can command. Thus, when longer-run incomes are depressed in farming, farmers, if they are income maximizers, will seek to move to occupations yielding higher returns. If this theory were to operate fully in practice, then incomes should tend to be equalized between farm and nonfarm workers, for comparable skills and resources employed. There is, however, a large gap between theory and practice and various economic studies bear this out.[4] Rather, the hypothesis is that the national unemployment rate is much more important in explaining the net outmigration of farm labour than is the farm/nonfarm wage ratio. The Canadian unemployment rate during the 1950's averaged slightly more than four percent but, during the 1960's, it averaged about five percent a year. By mid-1970, the unemployment rate was running at almost seven percent (seasonally adjusted) a year, equaling the high unemployment rates of the 1958 to 1961 period. These unemployment rates support the hypothesis that the reduced rate of net outmigration of farm labour in the 1960's can be traced to the generally higher unemployment rates, principally in the nonfarm sector, of those years.

In addition to the overall decline, during the past two decades, in the size of the farm labour force, there have been important structural changes in it. When the farm labour force is classified into its three main structural components (farm operators, unpaid family labour and hired labour), it is found that the greatest relative decrease in numbers has been in the family labour portion (operators plus unpaid family labour) and the least, in the hired labour component.[5] Consequently, there have been important shifts in the relative importance of each group of workers in agriculture. Paid labour in the 1950-54 period comprised 12 percent of the total farm labour force but, by 1965-69, it had increased to 18 percent. While both farm operators and other family help (unpaid) have declined in relative importance, the greatest relative decrease since the 1950's has been in the numbers of unpaid family workers. Thus, in the first half of the 1950's, unpaid family help comprised almost a quarter of the total farm labour force, but, by the last half of the 1950's and the first half of the 1960's, it had decreased to a fifth. However, in the last half of the 1960's, unpaid family help increased again in relative importance to 22 percent of the total farm labour force. This reversal of trend further supports the hypothesis that it is the national unemployment rate which is an important factor in determining the rate of outmigration from the farm sector. A rise in the national unemployment rate which results in a slowdown in the net outmigration of labour from farming most directly affects the youngest portion of the farm labour force, that is, the unpaid family labour plus the younger farm operators.

The rate of growth in *farm machinery* inputs[6] amounted to one and a half percent a year from 1950 to 1969 (Table 2). However, the last decade of the period saw an increase in the growth rate to more than three percent a year—the converse to the rate of decline in the farm labour force during the same years. The increase in all farm machinery inputs from 1968 to 1969 was slightly more than one percent, reflecting mostly an increase in the interest on investment and depreciation portion of the total machinery and equipment inputs. Inputs of fuel, oil and lubricants were estimated to be down by almost one percent between these two years.

Feed and seed (including nursery stock) inputs[7] purchased by farmers from the nonfarm sector have been rising by four and a half percent a year in the past two decades (Table 2). This represented the second highest rate of increase in a major input grouping, exceeded only by the growth rate for fertilizer inputs (if fertilizer inputs are categorized as a "major input grouping"). However, since 1960, the rate of increase in purchases of feed and seed increased to five percent a year, exceeding the growth rate in fertilizer inputs for the same period. Although purchased seed and nursery stock inputs increased relatively more than purchased feed inputs during the two decades since 1950, they still represented only a small proportion of the total inputs in this grouping of purchased feed, seed and nursery stock. For example, in 1950, purchases of seed and nursery stock represented about two percent of that total, but by 1960, seed and nursery stock represented seven percent and, by 1969, almost nine percent.

Of all the groupings of farm inputs, the rate of expansion in *fertilizer and limestone* used by farmers in the past two decades has been the highest, almost nine percent a year.[8] But farmers cannot continuously expand the consumption of one input such as fertilizer at a high rate, especially in the face of declining factor/product price ratios and sales per farm unit. (There are also likely to be technical reasons which will inhibit continued rapid expansion of nitrogenous fertilizer inputs but these will not be dealt with in this article.) Fertilizer and limestone consumption peaked in 1968 at an index of 335 (1949=100) and then dropped by 18 percent in 1969. This sharp decline in fertilizer use in 1969 was a major contribtutor to the reduced average annual rate of growth in this input grouping during the 1960 to 1969 period to four percent.

The greatest absolute and relative expansion in fertilizer consumption has been in the Prairie Provinces. Comparing the first half of the 1960's with the last half, the tonnage of mixed fertilizers and fertilizer materials sold in the Prairies tripled, while the increase for all of Canada was about 70 percent between the same two five-year periods. Fertilizer consumption by Prairie farmers increased, therefore, from about a fifth of total Canadian consumption to more than a third of the total, in one decade. Furthermore, this increased consumption was largely in the form of nitrogeneous fertilizers. For the whole of Canada, sales of fertilizers in total elemental forms doubled from 1960-64 to 1965-69. However, the increase in the nitrogen content of fertilizers in the same period was more than 150 percent. As a percentage of total sales, the nitrogen portion of fertilizers increased from about a quarter of the total in the first half of the 1960's to a third of the total in the last half of the decade.

Finally, that grouping of farm inputs which is designated *Other Inputs*[9] increased by 3.5 percent a year from 1950 to 1969, and by 2.6 percent annually since 1960 (Table 2). Within this broad category of inputs, the most significant increases have occurred in the use of electric power, custom work, pesticides (including herbicides), and livestock services. However, from 1968 to 1969, consumption of a number of inputs in this grouping decreased resulting in an increase of less than two percent for the "other inputs" grouping as a whole. Livestock services, for example, decreased by slightly more than four percent.

Changes in Farm-Produced Inputs

So far in this section, the nature and sources of productivity gains which are attributable to the changes in consumption of specific inputs have been discussed. It is important also to consider the changes which are occurring in the proportion of inputs which arise from the farm sector itself and those which are obtained from the nonfarm sector. Throughout the 1950's and 1960's, the agricultural sector continued to show an increasing dependence upon the nonfarm sector as a source of production inputs. However, the rate of rise in this increasing dependence slackened off in the last half of the 1960's, and these five years became a period of relative stability in this relationship. Should this situation continue, it will mark an important watershed in the structure and source of inputs for agricultural production. From 1960 to 1969, the rate of growth in *purchased* inputs[10] averaged more than two percent a year (Table 3). This was about one-half the rate of growth in the "capital" input grouping but, in addition to including most inputs classified as capital inputs, the purchased input grouping also includes hired labour, and this explains, in part, the lower growth rate for the purchased inputs group as a whole.

Table 3

INDEXES AND RELATIVE SHARES OF PURCHASED AND NON-PURCHASED INPUTS, CANADIAN AGRICULTURE, SELECTED PERIODS AND YEARS, 1950 TO 1969 WITH PROJECTIONS TO 1972

Period or Year	Purchased Inputs	Non-Purchased Inputs	Purchased Inputs	Non-Purchased Inputs
	1949 = 100		percent of total inputs	
1950-54	97	94	39	61
1955-59	110	84.5	44.5	55.5
1960-64	122	80	49	51
1965	134	77	52	48
1966	140	75	53.5	46.5
1967	143	78	53	47
1968	142	78	53	47
1969	143	77.5	53	47
1965-69	140.5	77	53	47
1972 Projections				
I	152	70	57	43
II	157	76	56	44
Annual Growth Rates			per cent	
1950 to 1969	+2.5	−1.3		
1960 to 1969	+2.2	− .3		

Non-purchased inputs, that is, the inputs offarm operators (proprietors) and their families, declined by almost one and a half percent a year during the twenty-year period, 1950-69 (Table 3). However, since 1960, the decline has been less than one-half percent a year. In terms of relative proportions, farm-produced (non-purchased) inputs made up

205

just over 60 percent of total inputs in the 1950-54 period. They declined to about 50 percent in the 1960-64 period (1964 was the watershed year when purchased and non-purchased inputs were each one-half of total inputs) and averaged 47 percent of total inputs during the last half of the 1960's.

Research and Development

One of the major sources of productivity gains implicit in the index of agricultural productivity are expenditures on research and development, mainly by public agencies. These expenditures are not included in the estimates of production inputs, hence, part of the productivity gains shown derive indirectly from this source. In 1967-68, total operational expenditures on agricultural research and development (R and D) were estimated to be $74.7 million, of which governments (principally the federal government provided almost 64 percent.[11] Expenditures by the universities on agricultural R and D in the same year comprised 29 percent of the total, but only seven percent was contributed by the nonfarm sector. Total federal government agricultural R and D expenditures increased during the 1960's. Total current expenditures of the federal government on scientific activities in the fiscal year 1967-68 were estimated at $37.6 million, 43 percent more than the *current* dollar expenditures in 1963-64.[12] Even after discounting this increase for the price inflation of this period (amounting to about 11 percent on the basis of the General Wholesale Price Index), the rise in real expenditures was significant.

Research and development expenditures are, however, only a part of total government expenditures on agriculture and related activities. Although the following figures are not directly comparable with R and D, expenditures for research, education and extension were estimated to total about one-fifth of total government (all levels) expenditures on agriculture in the 1960's.[13]. This proportion was almost unchanged from the proportion found in the early 1930's. According to the same source, total government expendi-

tures on agriculture in Canada amounted to about $20 per capita in the mid-1960's compared with $90 for national defence and $8 for the Canadian Broadcasting Corporation. Thus, total research and development expenditures on agriculture by governments can be deduced to be about $4 per capita or less, roughly equal to one-half the per capita expenditures on the national radio and television broadcasting system.

How do expenditures on agricultural R and D affect productivity? Briefly, research and development for agriculture contributes to improved varieties or strains of crops and livestock, to improved machinery and equipment, to the control and/or eradication of diseases and pests of crops and livestock, to the development of newer and improved processes of food preparation and preservation, to an understanding of the factors affecting the rate of adoption of technology and to the improved management of resources in agriculture.[14] Public investments in agriculture have been justified up to the present time largely on the basis of increased productivity and the high payoff to society, and because of important characteristics of the agricultural industry which distinguish it from other sectors of the economy.

There are two principal distinguishing characteristics involved, one economic and one biological-physical. The economic characteristic, which is partly due to past public land settlement policies, is manifested in the large number of relatively small-size firms (farms) in agriculture (as compared with nonagriculture), none of which is large enough to support an independent research establishment. Nor are these individual firms (farms) able to recoup the cost of independent research through the market pricing system since all producers sell in the same product markets and at basically the same prices. The biological-physical characteristic lies in the fact that agricultural production, for the most part, is subject to vagaries of the weather and to the nature of the biological growth processes which can vary from a relatively short time-period for broilers to three or more years for cattle. Thus, a farmer's production plans are subject to the influence

206

of economic and physical forces largely beyond his individual control.

A further consideration in support of public research in agriculture is "who are the ultimate beneficiaries?"[15] The benefits to farmers of agricultural research have been stated usually in terms of enabling them to increase output and/or save inputs, that is, to increase productivity. But this does not indicate much about the effect of agricultural research which contributes to productivity gains on farm income except that, as noted previously, increased productivity is a *necessary* condition for higher farm incomes. But it is not a *sufficient* condition. Because of the generally inelastic demand for farm products, especially at the farm level, there is little variation, if any, in the amount of farm products purchased by consumers, with a given change in product price, other things being constant. Furthermore, in a situation of rising agricultural productivity (other things constant), the highly competitive nature of the primary agricultural industry, with a large number of relatively small firms producing farm products, unrestricted entry to the industry, etc., tends to depress farm product prices and to reduce total farm revenues.

Consumers of farm products, in the institutional and economic environment described for agriculture, benefit from agricultural productivity gains occasioned by agricultural research and development by lower real costs for food materials. In fact, much of the rise in retail food prices is the result of added costs for services, in the form of convenience packaging, pre-cooking and other complementary services, rather than of any large increases in the costs of the raw materials of food (and fiber). Of course, it is correct to say that farmers gain also from improved productivity as consumers of food products, but only a relatively few farmers are able to reap significant monetary gains by early adoption of new inputs or techniques of production. Farmers, as a group, could retain the full benefits of productivity gains due to output-increasing innovations only if it could be assumed that any increase in output resulting therefrom would be absorbed by the product market at the given (current) product price. The effects are somewhat different for input-saving innovations. In this case, more of the benefits could be retained by producers, at least in the short-run, for a given output level. But the nature of the organization of the primary industry in Canadian agriculture is such that the indivdual farmer would be worse off if he does not adopt new input-reducing and/or output-increasing technology so long as his competitors (other farmers) in the industry do so.

FARM AND NONFARM LABOUR PRODUCTIVITY

In making comparisons of productivity on an interindustry basis, one commonly accepted method is to relate output to a single input such as labour. However, in this case, real output usually consists of the value-added portion (including capital consumption allowances) for each industry in the economy. (It can be described also as "net output.") For agriculture, value added corresponds closely to real net farm income where gross output and nonfarm inputs are valued at implicit prices. (To this is added the investment capital consumption allowances of farm operators.) Although such a measure of productivity purports to relate one input, labour, to the value added in the industry and hence is referred to as "labour productivity," it is still a partial productivity measure. It is a "partial" measure because gains to labour reflect not only changes in the nature of the labour force in terms of skills, but also the changes due to the nature of the productive resources with which labour works, together with the effectiveness with which resources are combined and organized for production.[16]

Real gross domestic product (value added) per man in agriculture has increased at an annual rate of more than 4.5 percent since 1950 and at a slightly faster rate since 1960 (Table 4). These rates of growth in agricultural labour productivity exceed by a considerable margin the growth rates in labour productivity for the rest of the economy. Furthermore, although the growth rate for agriculture accelerated slightly since 1960, there was a deceleration in the nonfarm sector. In this sector, the growth rate in gross domestic product per man averaged almost three percent a year since 1950. However, from 1960 to 1969, the growth rate was just over 2.5 percent a year compared with almost five percent for agriculture.

Table 4

INDEXES OF REAL GROSS DOMESTIC PRODUCT PER MAN EMPLOYED, CANADA, AGRICULTURE AND NON-AGRICULTURAL COMMERCIAL INDUSTRIES, SELECTED PERIODS AND YEARS, 1950 TO 1969 WITH PROJECTIONS TO 1972.

Period or Year	Agriculture	Commercial Non-Agricultural
	1961 = 100	
1950-54	81.5	77
1955-59	102	90
1960-64	124	103
1965	146	112
1966	183	114
1967	144.5	116
1968	156	121
1969	167	121
1965-69	159	117
1972 Projections		
I	179	130
II	191	131
Annual Growth Rates	percent	
1950 to 1969	+4.7	+2.8
1960 to 1969	+4.7	+2.6

Sources: (1) *Aggregate Productivity Trends, 1946-1968,* Cat. No. 14-201 Annual, May 1970, Dominion Bureau of Statistics.
(2) *D.B.S. Daily,* Feb. 12, 1971, Cat. No. 11-001, Dominion Bureau of Statistics.

In the past decade, the growth rate in agricultural labour productivity reflects an annual increase of more than one and a half percent in real gross domestic product and a decline of about three percent each year in employment in farming. In the nonfarm sector, by contrast, output in the form of real gross domestic product has been rising each year by more than twice the agricultural output growth rate. However, employment has been rising also, by three and a third percent each year, with the result that the gain in labour productivity in the nonagricultural sector was just over 2.5 percent a year.

PRODUCTIVITY TRENDS IN CANADIAN AND UNITED STATES AGRICULTURE

Since much of the technology which is applied to Canadian farming arises partly in the United States, a comparisonof the trends in agricultural productivity in the two countries is useful (Figure 2). There are many similarities, too, in Canadian and United States farm output but the larger agricultural base of the latter country tends to dampen annual fluctuations in the volume of total output, while in Canada the size of the western grain crop can cause wide annual fluctuations in total output from year to year. United States farmers compete, too, with Canadian farmers in export markets, and especially the North American market. For this reason, alone, farm productivity in Canada must equal or better United States farm productivity.

The rate of growth in Canadian farm productivity since 1950 has been 1.9 percent

Figure 2

PRODUCTIVITY TRENDS IN CANADIAN AND UNITED STATES AGRICULTURE, THREE-YEAR MOVING AVERAGES, 1949 to 1968 (Indexes 1949 = 100)

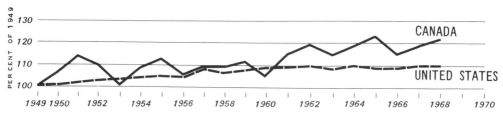

annually compared with 1.3 percent for United States farming.[17] Since 1960, the rate of growth in productivity has been much higher in Canada, about 2.5 percent a year as compared with 0.2 percent in the United States. The trends in the use of farm inputs (resources) since 1950 or 1960 have been almost the same in either country; less than 0.5 percent a year increase over the twenty-year period or about one percent a year increase since 1960. However, the annual growth of United States farm output has been much less than for Canadian agriculture; 1.7 percent a year since 1950 compared with 2.3 percent for Canada or 1.5 percent a year since 1960 compared with 3.6 percent. It is this much higher growth rate in Canadian farm output during these years that has been largely responsible for the greater growth rate in Canadian agricultural productivity compared with that of the United States. The importance in the Canadian farm output pattern of grain production, which lends itself to extensive mechanization and the use of other specialized inputs such as herbicides, appears to be a significant factor contributing to the greater growth rate of Canadian agricultural output as a whole over the particular periods reviewed.

REGIONAL AGRICULTURAL PRODUCTIVITY[18]

Productivity Growth

Since 1946, the rate of agricultural productivity growth in Canada has been highest in Quebec, 3.6 percent a year,[19] followed by Ontario with a growth rate of 2.9 percent (Figure 3). In Quebec, the increase in productivity is a combination of an annual increase of 1.9 percent in farm output and a decline of 1.7 percent in inputs. The Maritime region showed a decrease of 0.1 percent a year in output, but the rate of decrease in inputs was greater, 2.4 percent a year, with a resultant increase in productivity of 2.6 percent a year.

On the Prairies, farm output rose from 1946 at a rate of 1.8 percent a year while inputs declined slightly. Consequently, agricultural productivity in this region increased at 2.0 percent a year. The heavy weight which Prairie agriculture carries in the all-Canada indexes is reflected in the close correspondence between the growth rates for the region and for Canadian agriculture as a whole, 2.4 percent a year since 1946.

British Columbia agriculture showed the highest rate of growth in farm output, 2.6 percent a year. However, production inputs rose by almost one percent annually, with the result that the rate of productivity growth was the lowest for the five regions, or 1.7 percent a year.

Changes in Inputs

The shifts which have occurred since 1946 in the composition of farm inputs help to explain the differences in productivity growth among the five main regions of Canada (Figure 4).

In the Maritimes, real estate inputs declined by 1.5 percent a year. All other regions showed increases, although that for Quebec was small. The highest rate of increase in farm real estate inputs was in British Columbia, an estimated 2.2 percent a year since 1946.

Labour inputs declined in all regions, with the greatest decrease being in Quebec, 4.7 percent a year, followed by the Maritimes with a decrease of 4.2 percent a year. This high rate of decrease in the Quebec farm labour force is the main contributing factor to the overall productivity growth rate in that province.

The annual rate of growth in capital inputs in Quebec was higher than that of Ontario but below the rates for the Prairies and British Columbia. British Columbia had the highest rate of growth in capital inputs, 2.5 percent a year. This is the principal factor contributing to the productivity growth rate shown by this region. (The growth rate in total output was almost exactly the same as the growth rate in capital inputs.) In the Maritimes, there was a small annual decline in capital inputs.

Thus, the Maritime region has shown a decrease in all of the three major factors of production. Farm output also declined

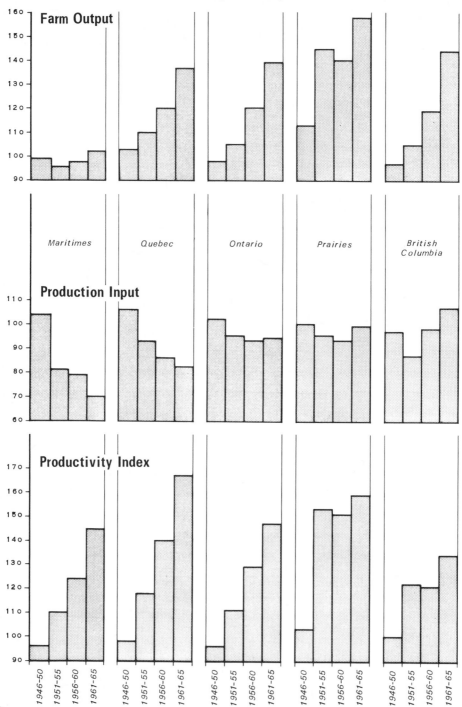

Figure 3
INDEXES OF INPUTS AND OUTPUTS AND PRODUCTIVITY RATIOS
BY REGION 1946-1965
(1949=100)

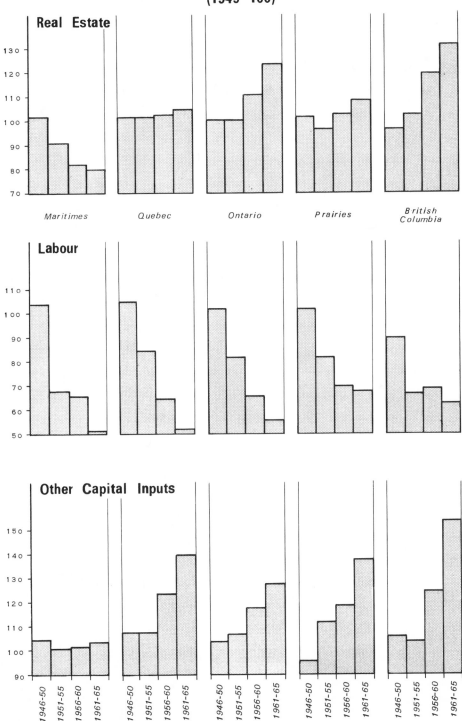

Figure 4

INDEXES OF INPUTS, CANADIAN AGRICULTURE BY REGION
1946-1965
(1949 = 100)

211

fractionally, but since the decrease in all inputs was greater, this region has shown an increase in productivity exceeding that of the Prairies and British Columbia. The case of the Maritimes illustrates the effect that an adjustment in resource use can have on productivity growth when total output remains unchanged.

In Figure 4, the changes in the three major categories of farm inputs are compared by region. Within the capital grouping, however, there have been important differences in the trends of the various items. In the Maritimes, although the major declines in farm inputs were for labour and real estate, decreases from 1946-50 to 1961-65 occurred also for livestock, feed and seed, and fertilizers and limestone. Inputs of machinery rose by about one-quarter and electric power consumption doubled. The Maritime region was the only one of the five areas studied that showed a decrease in fertilizer use. However, 1959 marked the low point in fertilizer consumption in the Maritimes; it has been rising slowly since then.

In Quebec, all major input components, including the various subgroupings of the capital input component, showed increases in the 1961-65 period compared with the five-year period, 1946-50, with the exception, as noted before, of labour. Machinery inputs were up by 70 percent; fertilizer and limestone by 56 percent; and electric power by more than three and a half times. Miscellaneous inputs were up by over a fifth.[20]

One notable difference in the trends between Ontario and Quebec has been a reduction of 6 percent in purchased feed and seed in Ontario compared with an increase of 11 percent in Quebec. Fertilizer and limestone consumption by farmers in Ontario was up by two and a half times. Miscellaneous inputs, including pesticides, rose by 32 percent in Ontario.

On the Prairies, all major input components, with the exception of labour, increased from 1946-50 to 1961-65. Livestock inputs were up by 86 percent, fertilizer by five-fold and electric power by thirteen times. In British Columbia, the notable features of the changes in resource use between 1946-50

and 1961-65 were the increases of 36 percent in real estate and of 45 percent in capital inputs. The increase in these capital inputs was distributed more uniformly among the various items than possibly in the other regions. The overall increase of 45 percent represented increases of 57 percent in machinery; 81 percent in livestock; 35 percent in purchased feed and seed; 76 percent in fertilizer and limestone; and almost three and a half-fold in electric power. Miscellaneous inputs, including pesticides and irrigation water, rose by 15 percent.

SUMMARY AND CONCLUSIONS

During the two decades since 1950, overall agricultural productivity in Canada increased at a rate of almost two percent a year. This is the result of a rise in total output of more than two percent a year and an increase in production inputs of less than half a percent a year. However, in the last decade of the period, productivity gains accelerated to two and a half percent a year. This resulted from both a greater volume of inputs employed and improved efficiency in resource combination. Since 1960, total output has increased by more than three and a half percent a year while total inputs rose by just over one percent a year.

In the years of the 1950's and 1960's, there was a marked decline in the labour input in primary agriculture averaging more than three and a third percent a year since 1950. However, this rapid replacement of labour with capital as an important source of productivity gains may be nearing an end, barring the introduction of any new labour-saving innovations of significant proportions. Since 1966, the decline in the size of the farm labour force has been much less than it was in previous years. Part of the explanation for the reduced rate of outmigration from farming lies, however, in the rising national unemployment rate in recent years. Significant productivity gains can be attributed also to the rapid expansion in the use of such factors of production as chemical fertilizers and pesticides. Much of this increased expansion in fertilizer use occurred in the Prairie Provinces and this is where much of the increased output occurred in the past two decades.

Farm-produced inputs declined in both absolute and relative importance during the years since 1950. However, in the latter part of the period, that is, since 1960, the rate of decline in the volume of non-purchased inputs leveled off. Most of the increase in total inputs during this decade was in the form of purchased inputs and particularly in such items as purchased feed and seed inputs. Since 1960, the growth in volume of purchased feed and seed inputs used has exceeded tthe annual growth rate for fertilizer consumption.

Labour productivity in agriculture in the two decades reviewed has shown a much higher rate of gain than labour productivity in the nonfarm sector. In the last decade of the period, agriculture further increased its lead over the nonfarm sector in this respect. The declining size of the farm labour force, coupled with rising employment of capital inputs, has resulted in much greater relative gains in efficiency in resource combination in agriculture than in the nonfarm sector. In this latter sector, both net output and employment have been rising during the past twenty years.

Much of the research and development which contributes to productivity gains in agriculture is funded in Canada by public agencies. The per capita cost, however, is relatively low, about $4 a year, or roughly one-half the per capita cost of the national radio and television broadcasting system. On the other hand, much of the benefit from agricultural research and development accrues to consumers in general through lower real costs for food. Farmers benefit, too, as consumers but it is largely the early adopters of technical innovations who benefit in terms of higher resource earnings.

The trends established and the sources of productivity gains identified lead to the conclusion that a plateau may have been reached as to the major gains which can be achieved by a further reduction in number of farm workers, *per se*. This pleateau may, of course, be short run in nature. Future aggregate gains, insofar as the productivity of the labour input is concerned, seem likely to lie more with improvements in the quality of the inputs. Following from this and assuming a continuation of the predominance of the individual proprietorship in farming, it seems likely also that there will be no further significant reductions in the total inputs arising from the farm sector, but that the proportion of inputs arising from the nonfarm sector will continue to increase. What form these inputs will take (and whether they will be reflected in the data) is another question. One likely area for expansion would seem to be in the form of managerial services with increased integration between the production, processing and distribution of farm products. This area seems likely to offer greater possibilities for future gains in productivity in Canadian agriculture than does a continuing substitution of capital inputs for labour in the production part of farming.

Gains in productivity in agriculture are a necessary, but not a sufficient condition for improved incomes (and resource earnings). Productivity gains that lead to total output increasing more rapidly than total demand can lead to depressed farm product prices and to reduced total revenues for many commodities. The implication is, therefore, that unless production is market-oriented, increased resource earnings in agriculture cannot be expected in the short run. Furthermore, resource earnings in a perfectly competitive society are basically determined by their opportunity price. The earnings of labour resources are, in turn, determined in part by the qualifications of the labour and the demand for these skills in relation to the supply, and also to the quality (and amount) of the other resources with which labour is combined. The many imperfections which exist, both in product and factor markets, contribute to the disequilibrium, between resource earnings in the farm and nonfarm sectors. One conclusion is, therefore, that the removal of the market imperfections, which include uncertainty, or the development of countervailing powers are necessary conditions for the equalization of resource earnings between agriculture and other sectors of the economy.

REFERENCES

[1] Agricultural productivity, as used in the context of this paper, is defined as the relationship between total farm output, measured in constant prices and total farm inputs, also measured in constant prices. Thus, the relationship between output and input, which is called agricultural productivity, is a measure of output per unit of input. Since inputs are measured in constant prices; that is, the effects of changes in factor and product prices have been removed, the result is a relative measure of the real (constant-price) costs of production. A rising index of agricultural productivity indicates, for example, that the real costs of producing farm products are declining.

[2] Real estate inputs refer to both owned and rented real estate and include interest on investment in land and buildings, depreciation and repairs on buildings, property taxes and fencing maintenance—all valued at 1949 prices.

[3] Labour inputs are the sum of the annual average numbers of "own account workers," "employers," "unpaid family workers" and "paid workers," all valued at hired farm wage rates regardless of class of workers. For purposes of this study, farm operators are the total of "own account workers" and "employers" in agriculture. The data used are annual averages of 12 monthly sample surveys. See: *The Labour Force*, Cat. No. 71-001, Monthly, Dominion Bureau of Statistics, Ottawa.

[4] Heady, E. O. and L. G. Tweeten, *Resource Demand and Structure of the* (United States) *Agricultural Industry*, Iowa State University Press, Ames, 1963, Ch. 9.

[5] *The Labour Force*, Cat. No. 71-001, Monthly, Table 6, Dominion Bureau of Statistics.

[6] Farm machinery inputs include interest on machinery investment, depreciation, repairs, fuel, oil, lubricants and other operating expenses—all valued at 1949 prices.

[7] Feed and seed inputs refer to inputs of feed, seed and nursery stock purchased by farmers from nonfarm sources, valued at 1949 prices.

[8] Unless otherwise stated, the index of fertilizer and limestone refers to total consumption of commercial fertilizers, both fertilizer materials and mixed fertilizers, together with agricultural lime and limestone (including marl). The index largely reflects fertilizer consumption trends. Consumption of *agricultural lime* in Canada in the last half of the 1960's was at about the same level as in the first half of the 1950's. During the intervening years, consumption dropped to as low as a fifth of the consumption in the other years. Consumption of *limestone for agricultural use* has been rising and, in the first half of the 1960's was almost double the amount consumed in the previous five-year period.

[9] "Other inputs," for purposes of this analysis, purports to include, unless otherwise stated interest on investment in livestock and poultry, interest on investment in purchased feeder livestock (purchased through commercial channels, that is, excluding interfarm sales and transfers), livestock services (A.I. fees, breed association fees, veterinary services and supplies), hired custom work by nonfarm operators, electricity, telephone, insurance, pesticides (including herbicides), containers, twins, irrigation water levies and, finally a "miscellaneous" grouping intended to account for tools, small hardware, etc—all valued at 1949 prices.

[10] Purchased inputs include the interest on investment in rented real estate; depreciation on rented buildings; all building repairs (rented or owned); all property taxes (on rented or owned real estate); hired labour; machinery repairs and operating inputs; interest on purchased feeder livestock; purchased feed, seed and nursery stock; fertilizers; limestone; pesticides; hired custom work; and all other purchased (from the nonfarm sector) goods and services: Non-purchased inputs represent the interest on investment in owned real estate, livestock and machinery of the farm operator (proprietor), (together with his labour and that of his family (where applicable). Non-purchased inputs also include the depreciation on owned buildings and machinery. All inputs, purchased and non-purchased, were valued at 1949 prices.

[11] B. N. Smallman, *et al.*, Background Study for the Science Council of Canada, 1970 Special Study No. 10, *Agricultural Science in Canada*, Cat. No. 5521-1/10, Queen's Printer, Ottawa, 1970, Table 1, p. 38.

[12] *Federal Government Expenditures on Scientific Activities*, Cat. No. 13-401, Biennial, Dominion Bureau of Statistics, February 1969, Table 2, p. 25. The figure cited refers to expenditures classified as research and development, scientific data collection and scientific information dissemination. It does not include federal capital investment expenditures for scientific activities.

[13] *Canadian Agriculture in the Seventies*, Report of the Federal Task force on Agriculture, December 1969, Cat. No. A21-15, Queen's Printer, Ottawa, 1970, pp. 276 to 277.

14 An "improved" crop variety is one which yields a greater output (all other resources held constant) than previously grown crop varieties. In livestock production, improvements are associated, for example, with greater output of livestock and livestock products per pound of feed input—all other inputs constant. Likewise, machinery improvements are said to occur when one man one man is enabled to plant and harvest a greater volume of production than previously, or when time lost from machinery breakdowns is reduced, or when an improved machine enables an increased rate of operation so as to take advantage of favourable harvesting weather thus reducing the value of crop losses to weather by more than the additional costs of the higher speed operation.

15 This discussion is based upon several published sources which treat the subject in more detail than is possible here. The references consulted include:

(a) Peterson, W. L., *The Returns to Investment in Agricultural Research in the United States,* Staff Paper P69-5, April 1969, Department of Agricultural Economics, University of Minnesota.

(b) Farmers in the Market Economy, Iowa State University Press, Ames, Iowa, 1964, Ch. 8.

(c) Heady, E. O., *Agricultural Policy Under Economic Development,* Iowa State University press, Iowa, 1962, Ch. 4.

16 A more detailed discussion will be found in Study No. 3 of the Royal Commission on Farm Machinery by C. J. Maule entitled: *Productivity in the Farm Machinery Industry,* Ch. 2, "Labour Productivity—Meaning and Measurement," pp. 5 to 13, published by the Queen's Printer, Cat. No. Z1 - 1966/4-3, Ottawa, 1969. This reference deals with the inherent conceptual and statistical problems in making interindustry and intercounty comparisons of labour productivity.

17 Based on statistical information given in *Agricultural Statistics 1970,* United States Department of Agriculture, Tables 649, 652 and 653 and (previous issues). The indexes given in the cited source were converted arithmetically from 1957-59=100 time base to a 1949=100 time base.

18 For the purpose of this paper the regions are: Maritimes (Prince Edward Island, Nova Scotia, and New Brunswick); Quebec; Ontario; Prairies (Manitoba, Saskatchewan, and Alberta); and British Columbia.

19 An earlier study of regional agricultural productivity for the periods 1944-48 to 1954-58 reported a similar trend. See W. Mackenzie, "Regional Changes in Income, Terms of Trade and Productivity within Canadian Agriculture," *Can. Jour. Agric. Econ.,* Vol. XI (2), 1963, p. 50.

20 "Miscellaneous Capital Inputs," unless otherwise specified, include: farm share of electric power, fruit and vegetable containers, nursery stock, veterinary, twine, and irrigation charges, pesticides, fencing, breeding fees, breed association fees, and other miscellaneous purchased goods and services.

The Geography of the Orchard Industry in Canada

RALPH R. KRUEGER

CANADIAN ORCHARD TRENDS

Over the turn of this century, orchards were planted extensively in British Columbia and Eastern Canada, and reached their greatest extent by 1911, when the Census of Canada recorded almost 16 million apple trees in the country. A decline in orchard plantings began during World War I, and has continued ever since. The Canadian orchard acreage of 297,000 acres in 1921 was reduced to 148,000 acres in 1961, and 143,000 in 1966. Ontario had the largest share of the 1961 acreage, followed in order by British Columbia, Quebec, Nova Scotia and New Brunswick (Figure 1).

Apples

Despite the decline in orchard acreage, apple production in Canada has been constantly increasing. For instance, while apple tree numbers decreased by more than fifty percent between 1931 and 1966, the average annual apple production almost doubled in the same period (Figure 2).

The increased apple production has been coming from a constantly decreasing number of farms. *Census of Canada* data indicate that the number of farms producing apples commercially declined by 43 percent between 1961 and 1966. The average size of apple orchard in 1966 was 10 acres.[1]

The different provinces have not shared equally in the increases in total Canadian apple production since the 1920's (Table 1). From the production period of 1928-32 to that of 1965-69, the Maritimes dropped from 37 percent of the total to only 17 percent. In the same interval, Quebec's share increased five-fold from 5 percent to 26 percent. Ontario's share of the national apple production fell slightly in the late 1930's but rose again until in the 1958-62 and 1965-69 periods it stood at 28 percent. British Colum-

Table 1

AVERAGE ANNUAL APPLE PRODUCTION IN CANADA
AND PERCENTAGE BY PROVINCES, 1928 TO 1969

Production Period	Canada (Millions of Bu.)	Percentage of Canadian Production			
		Maritimes	Quebec	Ontario	British Columbia
1928-32	11.5	37	5	21	32
1938-42	13.7	36	6	17	41
1948-52	14.7	18	13	20	49
1958-62	16.5	16	25	28	31
1965-69	21.4	17	26	28	29

Source: Dominion Bureau of Statistics, Ottawa.

● RALPH R. KRUEGER *is Professor of Geography, Department of Geography, University of Waterloo.*

This article has been modified from Ralph R. Krueger, "The Geography of the Orchard Industry of Canada," *Geographical Bulletin*, Vol 7(1), 1965, pp. 28-71. It was updated by the author in 1971.

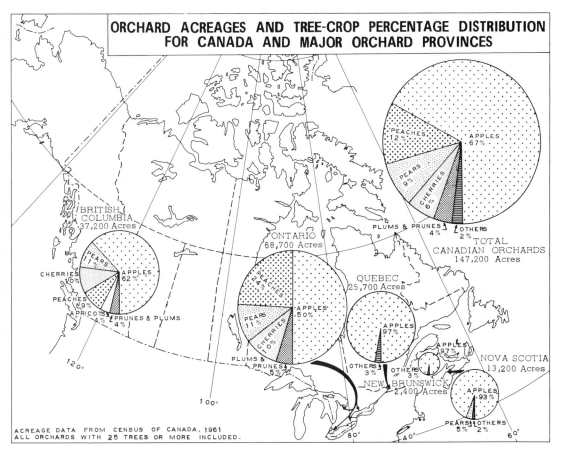

ORCHARD ACREAGES AND TREE-CROP PERCENTAGE DISTRIBUTION FOR CANADA AND MAJOR ORCHARD PROVINCES

Figure 1

bia's share of the national total rose from 37 percent in the 1928-32 period, to 49 percent in the 1948-52 period and then declined to 29 percent in the 1965-69 period.

Canadians, who since their school days have heard of the famous apple orchard regions of Canada, may be surprised to learn that Canada produces only about 2 percent of the world's apples and stands only seventh among the apple producing countries of the world (Table 2). It is also surprising for Canadians to learn that the United Kingdom produces approximately as many apples as Canada. In fact, Canada did not catch up to the United Kingdom until 1966.

Between 1930 and 1960, the total world apple production almost doubled, but the rate of increase levelled off in the 1960's. The great expansion in apple production in the earlier period partially explains Canada's

drop in its apple exports from more than six million bushels in the 1930's to between one and two million in the 1950's. Exports increased slowly in the 1960's, reaching three million bushels by 1968.

Canada's greatest loss of apple market was in the United Kingdom. In the late 1930's, the United Kingdom obtained 40 percent of its apple imports from Canada. By the early 1950's European competition and currency problems reduced Canada's share of the British market to 7 percent. This increased to 10 percent in the mid-1960's but declined to 4 percent by 1968.

In the 1963-66 period, Canada sent 43 percent of its apple exports to the United Kingdom and 35 percent to the United States. In the 1966-69 period, 56 percent went to the United States and only 25 percent to the United Kingdom.

217

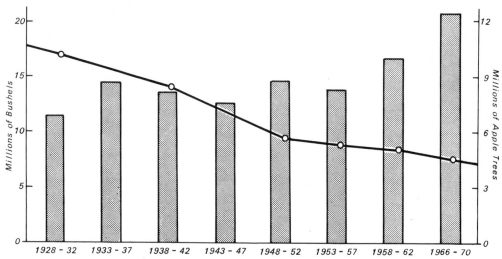

APPLE PRODUCTION AND NUMBER OF TREES IN CANADA, 1928-1970

Each bar represents the average annual production for a five-year period; the line represents the number of apple trees. The raw production data are from Statistics Canada, and the numbers of apple trees are from the Census of Canada. Up to 1941 all orchard trees were enumerated; from 1951 on the only orchards recorded were those with 25 or more trees.

Figure 2

Table 2

AVERAGE ANNUAL APPLE PRODUCTION FOR TWELVE LEADING COUNTRIES

(millions of bushels, dessert and cooking apples, for five-year periods, 1960-69)

Country	1960-64	1965-69
United States	125.3	127.9
Italy	100.1	98.8
West Germany	74.6	69.8
France	39.2	58.7
Japan	46.6	50.9
Argentina	20.2	22.1
Canada	18.9	21.6
United Kingdom	23.8	21.0
Turkey	13.6	20.9
Australia	15.6	17.2
Spain	13.3	15.7

Note: Data from "Deciduous Fruit World Production and Trade Statistics," 1960 to 1969, Foreign Agricultural Service, U.S.D.A., Washington. Note that the data in this table includes dessert and cooking apples only, whereas the table in the original article included cider apples as well.

Soft Fruit

Since 1960, there has been relatively little change in Canadian production of the soft fruits of pears, cherries, and plums and prunes; nor has there been a significant change in the provincial share of production (Table 3). However, both the acreage and production of peaches and apricots have declined since the mid 1950's. The production of apricots was cut by half within a decade. From 1956 to 1966, the peach acreage declined from 19,800 acres to 13,500 acres; and in the 1960's, average annual peach production declined by almost one million bushels.[2] This reduction in apricot and peach production has resulted primarily from serious low temperature injury in British Columbia and urban expansion in the Niagara Fruit Belt of Ontario.

DEVELOPMENT OF REGIONAL SPECIALIZATION

Except in the Prairie Provinces, one of the first farm activities undertaken by the early settlers of Canada, was the planting of a few apple trees to provide the family with much need fresh fruit during the fall and winter

Table 3

AVERAGE ANNUAL SOFT FRUIT PRODUCTION IN CANADA AND PERCENTAGE BY PROVINCE, 1958-62 and 1965-69

Tree Fruit	Canada (bu. '000) 1958-62	Canada (bu. '000) 1965-69	Percentage of Canadian Production Nova Scotia 1958-62	Percentage of Canadian Production Nova Scotia 1965-69	Percentage of Canadian Production Ontario 1958-62	Percentage of Canadian Production Ontario 1965-69	Percentage of Canadian Production British Columbia 1958-62	Percentage of Canadian Production British Columbia 1965-59
Peaches	2,695	1,782	0	0	78	88	22	12
Pears	1,506	1,528	3	4	58	58	39	38
Cherries	634	734	0	0	79	78	21	22
Plums and Prunes	545	446	1	0	58	57	41	43
Apricots	258	112	0	0	0		100	100

Source: Dominion Bureau of Statistics, Ottawa.

months. In the areas with a more favourable climate, orchards were enlarged year by year, and as the trees matured, they provided not only the family needs, but also a welcome cash crop. Thus the farmstead orchard became a part of the mixed farm operation.

Except for the Okanagan, where the agricultural economy changed abruptly from livestock grazing directly to commercial orchards, all of the present orchard regions evolved from these farmstead orchards. As the demands for both quantity and quality of apples rose in domestic markets, and as Canada began exporting apples to European markets, only those who took advantage of better varieties and improved equipment and management methods were able to survive the competition. To maximize returns on his investment and labour, the grower had to become expert at such orchard activities as insect and disease control, pruning, grafting, thinning, tilling, fertilizing, harvesting, storing, and selling.

The choices open to the general farmer in areas with sufficiently moderate winters were either expansion of his orchard into a commercial operation or complete abandonment. The collective decisions made by farmers in any given region led either to orchard specialization or orchard decline. Some farmers have combined commercial orcharding with general crop and livestock farming; others have supplemented orcharding with cash crops such as small fruits and vegetables.

Geographers are interested in not only describing and analyzing man's response to

his environment from place to place, but also in explaining how and why certain regional patterns have developed. As difficult as it may be, it is the task of the geographer to attempt to determine the factors most instrumental in governing the land-use decisions made by groups of people in given areas from time to time. Some of the physical factors that should be considered are often quite obvious, but the reasons for the human response to those physical conditions are usually difficult to ascertain. It often requires a study of historical documents and interviews with people who have lived many years in the region in question. In Canada, most of the orchard specialization has occurred since the turn of the century and there are consequently many people closely associated with the orchard industry who remember why certain decisions were made.

Olmstead states that in the United States the evolution towards regions of orchard specialization was dependent upon (1) advantageous location with respect to market, (2) climate and land-type advantages, and (3) a particular desire to specialize in orchards.[3] By looking at a map of the major orchard areas of Canada, it would appear that the orchard areas around Lake Ontario and in southern Quebec have definite domestic market advantages because of their proximity to the large cities of Toronto and Montreal. Although the closeness to market was an advantage in these two areas, it was the superior orchard land-type and climate combination which influenced farmers to special-

219

ize in orchards. In fact, the Chateauguay-Napierville orchard region immediately south of Montreal is declining in importance because the land is too flat for proper soil and air drainage while other orchard regions, farther removed from Montreal, but with superior slope and soil conditions (for example, the Huntingdon and Missiquoi regions) are continuing to increase their orchard production. Likewise, in Ontario, the way in which the major orchard areas hug the lake regardless of the location of major urban centres, indicates that superior physical conditions were more important to orchardists than proximity to market. Those immediately around Toronto have the double advantage of good orchard growing conditions and closeness to market.

Nova Scotia provides an example of an orchard producing area in which its general location relative to the United Kingdom market was an important factor in farmers' decisions to turn to apple specialization. However, within the province, orchard specialization occurred in the Annapolis Valley because of superior physical conditions.

The Okanagan Valley provides an example of an apple and soft fruit orchard area that had to overcome both physical and market location liabilities. In the case of the Okanagan, real estate promotion and the success of some of the first large orchard enterprises seemed to be the most important factors leading to orchard specialization.

Nowhere in Canada is there evidence that orchard regions evolved because of a particular desire by settlers to specialize in orchards. Within areas with superior land-type and climate conditions, some districts turned to orcharding because of the financial success of one or two commercial orchardists. When one farmer found a crop that proved to be more profitable than any other, it was not long before many of his neighbours had sought his advice and copied his operation.

Orchard specialization in a region leads to special research, service, processing, advertising, storage, shipping, and selling organizations, all of which tends to encourage greater efficiency and even more specialization. It is difficult for a number of small orchard operators or even a few large orchard establishments in other areas to compete with growers within the specialized orchard region. Thus, there is a tendency for more orchards to be concentrated in smaller areas.

Since even the milder areas of Canada approach the poleward limits for the production of the less hardy soft fruits, it is not surprising that physical conditions have been the most important factor in the location of soft fruit orchards. The Niagara Fruit Belt, the Kent-Essex region of southern Ontario, and the South Okanagan Valley are the only areas with mild enough winters to permit successful growing of peaches, sweet cherries and apricots. Peaches and sweet cherries are grown in all three areas. Surprisingly, apricots are produced commercially only in the Okanagan. It is generally conceded that apricots could be grown in the Niagara Fruit Belt, but early unsuccessful experiments with apricots in the region discouraged extensive plantings. Recent successes with apricots in the Niagara Fruit Belt indicate the possibility of the crop becoming more important there in the future.

THE ANNAPOLIS VALLEY

Location and Physical Base

Commercial orchards in Nova Scotia are concentrated almost entirely in the Annapolis Valley, a long narrow valley extending some seventy miles from the Annapolis Basin on the west to the Minas Basin on the east (Figure 3). The width of the valley ranges from about two miles at Annapolis Royal to about eight miles at Wolfville. The edges of of the Annapolis Valley are distinctively marked by two escarpments known as North and South Mountains, which rise to more than 700 feet above sea level. The western two-thirds of the valley is drained by the Annapolis River; the eastern third by the Cornwallis and Avon Rivers as well as several smaller streams. Although the proper geographical name for this area is the Annapolis-Cornwallis Valley, the more common term, Annapolis Valley, is used in this paper for the sake of simplicity.

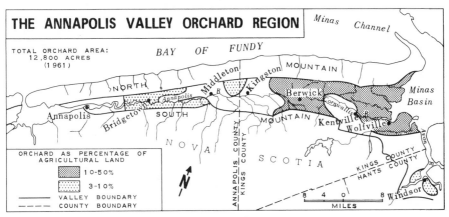

Figure 3

Slightly over half of the soils in the Annapolis Valley have been developed on glacial till. These glacial soils, located primarily on the lower slopes along the sides of the valley, are the most suitable for both orchards and mixed farming. The other residual and water transported soils located on the lowest part of the valley floor are generally too light in texture for orchards and are also less suitable for mixed farm crops. The lighest textured of these soils are used primarily for special crops such as potatoes and other vegetables.

Of the glacial till soils, the well-drained loams are considered to be the best for apple orchards; the clays are too poorly drained and the sands too subject to drought. Fortunately, most of the good orchard soils are found on slopes that afford sufficient air drainage to help protect the fruit crops from frost damage.

The *Soil Survey of the Annapolis Valley Fruit Growing Area* classifies more than 100,000 acres of soil as suitable for apple growing.[4] However, about one-fifth of these soils are marginal for apple growing under the present economic conditions. This still leaves more than 80,000 acres of very good orchard soil in the Annapolis Valley. Since current commercial orchard plantings occupy less than 13,000 acres, it is obvious that there is space in the valley for a six-fold expansion of the orchard industry even if orchards are limited to the very best soils.

More than 80 percent of the orchard acreage is found in Kings County in the eastern end of the valley (Figure 3). This concentration of orchard is likely due to the land-type in the eastern end of the valley, which provides extensive orchard sites with favourable slope and soil conditions.

The casual observer driving along country roads may gain an exaggerated impression of the density of orchards in Kings County because the original orchards were all planted adjacent to the farmsteads. The original farmstead orchards have been expanded into commercial-sized units and thus a large proportion of the orchard area is near the road where it is readily seen.

Because of its geographic location and land-type characteristics, the climate of the Annapolis Valley is very suitable for apple orchards. The Bay of Fundy and Minas Basin moderate the cold continental winds from the north and northwest. North Mountain gives further protection from northerly winds, and the valley slopes provide protection against radiation frost. The moderating effects of the Gulf Stream and the Atlantic Ocean are introduced by winds that enter the valley from the south and southwest.

Thus, for its latitude, the Annapolis Valley has very mild winters. The lowest temperature recorded at Kentville was —24°F. in 1920. Winter temperatures have dropped to —20°F. or lower only three times in thirty years at Kentville and have never hit the —20°F. mark at Annapolis Royal. During

221

the same thirty-year period, winter temperatures have dropped to −12°F. or lower twelve times at Kentville and once at Annapolis Royal. Since −12°F. and −20°F. represent the critical temperatures respectively for dormant buds and wood of peach trees,[5] it would appear that the winters are moderate enough for all the orchard crops grown in the Annapolis except peaches and cherries, which are more susceptible to winter injury. The western end of the valley has winters mild enough for peaches, but in the eastern end there is a chance of peach crop loss due to low winter temperatures in about one out of two years. There is some injury to the new growth on peach and cherry trees almost every winter. The life span of peach and cherry trees is very short. There is approximately a 50 percent mortality rate in young peach and cherry trees before they reach maturity.

An advantage of the winter climate of the Annapolis Valley is that there is a slow change in temperature from fall to winter. This gives the fruit trees an opportunity to harden up in preparation for the low winter temperatures. Consequently, winter damage is a minor problem as far as apple orchards are concerned.

Spring frosts do not generally result in much apple crop damage because the cooling effect of the surrounding water delays blossoming until the danger of frost is past. The Annapolis Valley has as good a frost record as any other major apple producing region in North America.

Valley precipitation of approximately 40 inches, distributed fairly evenly throughout the year, is sufficient for orcharding. The high humidity during the growing season, however, results in disease problems. In apples, scab is a constant threat; in plums, black knot is rampant; and fungus diseases attack the cherries. Rain also splits the cherries if it comes when they are almost ready for harvest.

Unfortunately, the Annapolis Valley lies in a hurricane zone. The frequency of severe hurricanes is low, but when they occur, the severity of damage to fruit crops is very high.

For example, Hurricane Edna in 1955 resulted in an apple crop loss ranging from 30 to 40 percent.

Orchard Trends and Economic Conditions

Until 1939, most of the Annapolis apple production was exported to the United Kingdom. The Annapolis growers specialized in hard-textured apples that could be shipped in barrels at a low cost. The British market was lost during World War II, and after the war, when the British were able to import apples again, the demand had changed to bright red dessert apples such as the Delicious and McIntosh, neatly packed in boxes instead of barrels. The result was that the Okanagan Valley replaced the Annapolis Valley as the major Canadian supplier of apples to the United Kingdom.

With the loss of the British market, the Annapolis orchardists decided that they would have to reduce their acreage and change their varieites of apples. Federal and provincial subsidies were paid to assist growers to pull up trees of unwanted varieties and to replace these with varieties in demand.

In the twenty-year period between 1939 and 1959, there was a reduction of over one million trees, an average of 50,000 per year. Tree removal was more active in the early 1950's, tapering off to about 25,000 trees a year by 1959. In the 1960's new planting began to outnumber tree removals, resulting in a gradual increase to almost 700,000 trees in 1970, with an average annual production of about 3,000,000 bushels. This contrasts with the peak of 7,500,000 trees in 1911, and the record production of 8,288,000 bushels in 1933.[6]

In addition to changing from hard cooking apples to dessert apples, the Annapolis orchardists also turned to packing apples in boxes instead of barrels, and kept the apples in cold storage. As a result, a small part of the United Kingdom market was regained, but marketing still remained a major problem for the Annapolis apple producers. In the late 1960's, only 7 or 8 percent of the apple crop was exported in the fresh state, with 14 or 15 percent being consumed in the province and another 3 or 4 percent going to the other provinces.[7]

As a result of fresh apple marketing problems, Nova Scotia has turned increasingly to the production of processed apple products. In the late 1960's, approximately 70 percent of the apple crop was processed. Thus, the culinary apples are still an important part of the industry. In fact, some growers are again planting cooking varieties.

The Annapolis Valley has never produced significant quantities of soft fruits. In 1966, the total commercial orchard area of approximately 13,000 acres was composed of 93 percent apple, 5 percent pear, and 2 percent peach, cherry and plum. Winter injury and the lack of sufficient summer heat has discouraged production of peaches and sweet cherries. Although the climate is quite suitable for pears and plums, they have never been a major item in the Annapolis orchard industry because of a lack of market.

The results of a questionnaire sent to commercial orchardists in the Annapolis Valley in 1959 showed that 41 percent of the growers had from 1-10 acres of orchard, 42 percent had from 11-30 acres, 13 percent had from 31-50 acres, and 5 percent had more than 50 acres.

As a result of small orchards being removed and most of the new plantings being made by growers who already have sizeable orchards, the average size of orchard in the Annapolis Valley is constantly increasing. Most successful orchardists and horticultural and marketing experts in the valley agree that 20 acres is the minimum size for a full-time commercial orchard operation.

For most of the Annapolis fruit growers, orcharding is only one of a number of farm activities. Orchard activities are usually combined with general mixed farming, with dairy cattle as the most common livestock specialty. Potatoes, other vegetables, and poultry and egg production are becoming increasingly important in the valley, and are often combined with apple growing. Farm records kept by the marketing agencies show that the farmers with large orchards with no other important farm operation have greater yields per tree, better quality fruit, and a higher net income per orchard acre.

Larger orchards with accompanying improvement in varieties and cultural practices have resulted in constantly-rising apple yields. In the 1945-50 period, the average annual yield was only three bushels per bearing tree; in the 1956-60 period it was five bushels and in the 1960-65 period it was six bushels per bearing tree.

The Annapolis Valley has always had a lower per tree cost than other major orchard areas in North America. Land costs are very low, an integrated biological-chemical insect control program has reduced the cost of spray, and labour costs are lower than the Canadian average. With recent changes to more desirable varieties, with increasing size of orchard operations and resulting efficiencies, with rapidly increasing yields and quality, and with modern storage and packing, the Annapolis Valley is again in a position where it can compete favourably with other fruit growing areas for the Canadian and foreign markets.

The greatest disadvantage of the Annapolis Valley is its lack of one cooperative selling agency. Scotian Gold Co-operative Limited handles about half of the total apple production. Most of the individual cooperative companies send the growers' fruit to Scotian Gold for packing and cold storage. Returns are then pooled by variety, grade, and size groups. However, a considerable number of growers refuse to join cooperatives and insist on their right to compete individually for the markets. This is an old problem in the Annapolis. The Royal Commission investigating the Nova Scotia apple industry in 1930 recommended complete cooperative methods and highly centralized control, which, it said, was the universal practice in other parts of the world where the same problems prevailed.[8] Considering the individualistic attitudes of many growers, it is unlikely that Nova Scotia will achieve central selling unless market disasters force the issue.

Figure 4

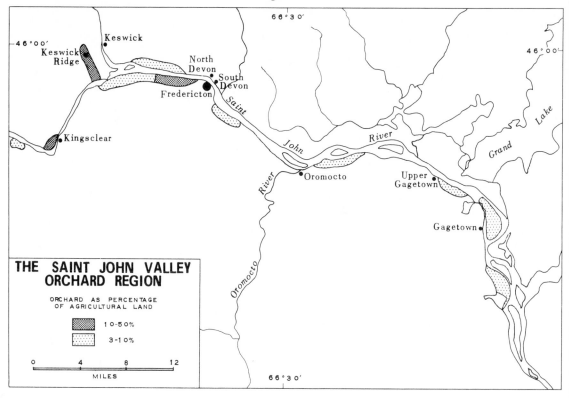

THE SAINT JOHN VALLEY
ORCHARD REGION

ORCHARD AS PERCENTAGE
OF AGRICULTURAL LAND

10-50%

3-10%

0   4   8   12
MILES

SAINT JOHN VALLEY, NEW BRUNSWICK

Location and Physical Base

The orchards of New Brunswick lie almost entirely in the Saint John River Valley. They are most concentrated from a point about ten miles west of Fredericton to Gagetown about thirty miles downstream (Figure 4). In this area, the width of the valley varies from about one mile just west of Fredericton to about three miles farther downstream. The surrounding highland rises approximately 200 feet above the valley floor.

During the retreat of the glaciers, the glacial till that covers the valley slopes was mantled with layers of sand and fine gravel varying in depth from 1 to 10 feet. Most orchards have been planted on these sandy and gravelly loams where the clayey till is

not less than 2 feet but not more than 5 feet from the surface. Such soils provide good drainage, yet are not too liable to drought, and permit good root development.[9]

Although the frost-free season for the Fredericton area is about the same as in the Annapolis Valley, the winters of the Saint John Valley are much more severe. At Fredericton, the temperaure fell to lower than $-12°$F. every winter in the 1950-59 decade, fell to lower than $-20°$F. in five of those ten years, and plunged to $-32°$F. on one occasion. Because of the severe winters there is no commercial production of any tree fruit other than apples, and of these only the hardiest varieties survive.

Orchard Trends and Economic Conditions

New Brunswick farmers began commercial production of apples in quantity in the years following the 1917-18 freeze, which had

224

seriously damaged orchards and reduced the apple crop in Ontario and Quebec. Because their orchards had escaped winter damage, the New Brunswick orchardists felt that they had a good opportunity of capturing the Ontario and Quebec markets. Orchards were planted upstream in the Saint John Valley as far as 25 miles north of Woodstock, and downstream to the river mouth. Orchards were also planted around Oak Bay, along the south shore of the Peticodiac River near Moncton, and even along the shore of Northumberland Strait in the vicinity of Shediac.

By 1921, the orchard acreage had reached almost 8,000 acres. The rate of orchard increase levelled off during the 1930's. Since 1940 the apple acreage in New Brunswick has been increasing very gradually with the average of 150 acres of new orchard being planted annually during the 1940's tapering down to an average of 75 to 100 acres annually since that time.

The provincial orchard census of 1959 recorded 2,500 acres of apple orchard in New Brunswick. This acreage includes orchards of all sizes, many of which are not productive commercial orchards. A realistic eestimate of the commercial acreage is closer to 2,000 acres. From the total of 318 growers, only half are giving their orchards first class care.[10]

The average size of orchard is only 8 acres, which means that there is a large number of very small orchards. There are only a few growers with orchards of more than 20 acres.

The average farm size in the major orchard area is about 200 acres, of which usually less than half is cropland. The farms are laid out in long narrow strips about 200 to 400 yards wide, running back from the river several miles. Only the front end of the farm is cultivated, the back remaining in forest. Most orchards are part of a mixed farming operation. Oats, hay, and pasture occupy by far the largest proportion of the cultivated land. A large number of farmers combine orcharding and mixed farming with lumbering, which is the source of a substantial part of the farmers' income during the winter and spring months. Lumbering sometimes interferes, however, with the proper care of orchards, because the farmers are busy in the forest at times when they should be carrying out orchard activities such as pruning the apple trees.

Apple production in New Brunswick doubled between 1940 and 1969. The average annual production in the 1965-69 period was approximately 500,000 bushels.

From 1920 to 1940, approximately half of New Brunswick's apples went to the Province of Quebec. Since 1940, however, the apple production has been marketed primarily in the home province. In the 1940's sales within New Brunswick ranged from 71 percent to 87 percent of total production. In the 1950's and 1960's, this increased to more than 90 percent. The apples that are shipped out of the province go primarily to Quebec, the other Atlantic Provinces, and England.

Because it does not suffer from the vagaries of an export market, the New Brunswick orchard industry has enjoyed great stability and growers have received higher prices per bushel than their counterparts in Nova Scotia. The New Brunswick Apple Exchange Limited, organized in 1930, has done much to standardize packing and quality, and to stabilize prices of apples, despite the fact that it handles only about 35 percent of the total crop.

QUEBEC

Location and Physical Base

Apple orchards are located throughout the southern portion of Quebec from the Ontario border to a point some sixty miles northeast of Quebec City. Approximately 95 percent of the commercial orchards are found in the southwest corner of the province, within fifty miles of Montreal (Figure 5). Small orchards are scattered across most of the counties south of the St. Lawrence River from the Monteregian Hills region east to a line drawn south from Île aux Coudres. The only significant commercial orchard east of the

Figure 5

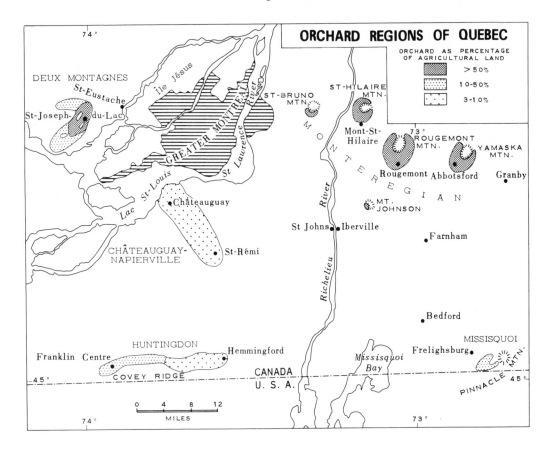

ORCHARD REGIONS OF QUEBEC

ORCHARD AS PERCENTAGE
OF AGRICULTURAL LAND

> 50%
10-50%
3-10%

Monteregian Hills is found on Île d'Orléans and on the Beaupré Coast on the north shore of the St. Lawrence River, opposite Île d'Orléans.

All the important commercial orchard regions are found on land-types that exhibit a considerable degree of slope and well-drained, light-textured soil. The poorest orchard sites are found in the Chateauguay-Napierville region where the land is too flat and the soils too poorly drained for successful apple production.

The climate of the major orchard regions of Quebec is similar to that of the Saint John Valley of New Brunswick, but, of course, more continental in nature.

Tree injury, resulting from low winter temperatures, is the greatest hazard facing Quebec orchardists. Winter temperatures frequently plunge to −25°F. or lower. The low-

est recorded temperatures in the major orchard regions range from −35°F. to −40°F. These low temperatures have resulted in serious orchard losses. In the winter of 1917-18, Quebec lost about half of its bearing trees. In the winter of 1933-34, temperatures, which dropped to between −30°F. and −35°F. in all the major orchard regions, killed 47,000 trees or about 8 percent of the total, and many more trees were severely damaged. The apple crop was reduced from 320,000 barrels in 1933 to 90,000 barrels in 1934.

There was less tree-kill in 1933-34 than in 1917-18 because the Spy and Ben Davis varieties, which were hard hit in 1917-18, had been replaced by more hardy varieties such as McIntosh. In 1933-34, large numbers of the Fameuse variety were killed. These were also replaced with more hardy varieties.

226

Since 1934, despite temperatures that have dropped below −30°F., there has been much less tree-kill because of the predominance of more hardy varieties.

Orchard Trends and Economic Conditions

In the nineteenth century, farmstead apple orchards were common throughout much of southern Quebec. As the population increased particularly in the Montreal area, farmers with favourable orchard sites began growing apples on a commercial scale in order to supply the apple demand of the growing cities. The orchard regions closest to Montreal (Deux Montagnes, Chateauguay-Napierville, and Monteregian Hills) developed most rapidly.

By the early 1930's, Quebec had 11,000 acres of orchard and was producing about 1,300,000 bushels of apples annually for the provincial domestic market. Large quantities of apples were still being imported from Nova Scotia, with smaller quantities coming from New Brunswick, Ontario, and British Columbia.

The 1933-34 freeze resulted in a decline in orchard acreage in the 1930-40 decade, but with good prices for apples, large acreage were planted in the 1940's as farmers discovered that apple growing was the most profitable of all farm enterprises. Because of their proximity to the market, local growers could sell apples at a profit in Montreal at prices that could not be met by competitors from other provinces.

By 1951, Quebec had 30,000 acres of orchard, a three-fold orchard acreage increase in twenty years. By the early 1950's, the Quebec domestic market had been saturated, orchard profits began to drop, and growers had started to neglect marginal orchards. The *Census of Canada, 1966,* showed only 24,000 acres of orchard, a decrease of 6,000 acres in a fifteen-year period. Field observations and interviews indicate that the trend of removing apple orchards in areas with inferior climate and/or soils is continuing. Excluding the small, sub-marginal orchards, a realistic estimate is about 20,000 acres of commercial orchard in the mid-1960's.

Lower profits have forced the inefficient growers out of business, resulting in fewer orchardists who are enlarging their orchard operations to cut the per unit cost of production. A large number of apple growers still run general mixed farms, often with a dairy specialty, but the trend is toward orchard specialization, particularly in the Monteregian Hills region.

Fewer growers and larger orchards have led to increasing higher yields. The average provincial yield per bearing tree in 1956 was 3.4 bushels. By 1970, this had increased to 4.9 bushels per bearing tree. Only in Missiquoi County, with a yield of 6.7 bushels per bearing tree, does the apple yield compare favourably with current apple yields in the Annapolis Valley.

The annual apple production of more than 5,000,000 bushels is sold primarily within Quebec. The average crop during the 1962-66 period was disposed as follows: 74 percent to the fresh market within the province; 14 percent to processing; 6 percent export and the balance to other provinces. There are still many apples shipped to Quebec from outside the province, particularly varieties, such as Delicious, that are not grown extensively in Quebec.

About one-quarter of the apple production is handled by cooperatives. The Monteregian Co-operative is the largest and most progressive in the province. It has found a new outlet for lower quality apples by developing a champagne-type apple cider and has used controlled atmosphere storage to extend the fresh apple market into the middle of the summer. In 1969, Quebec growers put a total of 1.6 million bushels of apples in controlled atmosphere storage.

Quebec's apple production fluctuates greatly from year to year. This creates grave marketing problems in the boom years. For instance, in 1955 the production of apples in Quebec was almost double the previous ten-year average. As a result, the price fell to less than a quarter of the average price for the previous decade. In 1968, Quebec had a record apple crop of 6.6 million bushels, which was 2.5 million more than the poor crop of 1966. However, in 1968, more aggres-

Table 4

QUEBEC ORCHARD STATISTICS BY REGION

Region	Orchard Acreage (1934)	Orchard Acreage (1937)	Average Yield per tree (Bu. 1970)
Deux Montagnes	1,030	2,710	4.9
Monteregian	5,990	7,120	4.9
Missiquoi	1,000	3,430	6.7
Huntingdon	760	3,770	4.5
Chateauguay-Napierville	2,270	1,200	—
Northern Section	—	1,000	—
TOTAL	11,050	19,230	3.4

Note: In order to include only the major commercial orchard areas, the 1961 acreages include only counties with at least 200 acres of orchard. The 1934 acreages are for the same counties and therefore, are directly comparable. The Northern Section includes all counties north and east of the Monteregian Hills.

Sources: 1934 orchard acreage from: "Statistiques des Vergers," in the *Quarante-et-unième Rapport Annuel de la Société de Pomologie et de Culture Fruitière de la Province de Québec, 1934.* 1961 orchard acreage from: *Census of Canada, 1961.* Yield data from Bureau de la statistique du Quebec, section agricole.

sive marketing methods resulted in a considerable increase in the volume of apples exported and used in processed products, thus helping to reduce somewhat the impact on apple prices.

In the 1950's, Quebec's exports of apples were trivial and sporadic. In the 1960's exports expanded to between 300,000 and 400,000 bushels, about 10 percent of the Canadian total.

In the face of stiff competition, it is unlikely that Quebec will be able to capture substantial new markets in Europe, the United States, or other parts of Canada. This means that most of the apple production must be marketed within the province. Assuming that the consumption rate and the amount of imports remain relatively constant, then the future increase in demand for Quebec apples will be dependent upon the province's increasing population. The 1966 orchard acreage could easily provide the volume of apple production required to supply the demand of Quebec's predicted 1980 population because many of the orchards have not yet reached peak production, and orchard management is constantly improving. In light of these conditions, it is encouraging to Quebec orchardists to note that the number of trees under ten years of age has

been rapidly declining since 1951 indicating a levelling off of Quebec's explosive growth in the apple industry.

Some statistics for the major orchard regions of Quebec are provided in Table 4. The Monteregian Hills region is the most important orchard area, annually producing more than one-third of the total provincial apple crop. The highest yields are obtained in the Missiquoi region. The Missiquoi and Huntingdon regions are the most recently developed orchard areas and accounted for over 60 percent of the provincial increase in orchard acreage between 1934 and 1961. Chateuguay-Napierville is the only region in which the orchard acreage declined between 1934 and 1961. Because the land is too flat and the soil too poorly drained, there has been severe winter tree injury and few new orchards have been planted.

ONTARIO

Location and Physical Base

Orchards are found throughout the greater part of southwestern Ontario, along a narrow strip north of Lake Ontario, and in the St. Lawrence and Ottawa Valleys. Within this area, precipitation, length of growing season, summer heat, and hours of sunshine are all adequate for apple growing. In most of the area winter temperatures never reach lower than −30°F. and temperatures of −20°F. are rare. Only in a small area in the interior of southwestern Ontario and in the Ottawa and St. Lawrence Valleys are winter temperatures a major hazard to apple orchards (Figure 6). The major orchard regions are found adjacent to the shores of the Great Lakes where the winter temperatures are more moderate and the cooling effect of the water in spring delays blossoming until most risk of frost is past (Figure 7). All of the orchard sites have some degree of slope, and the soils range from droughty sands to well-drained clay loams. Peaches are grown only on the lighter textured soils.

Apples, comprising one-half of the total orchard acreage, are grown over a widespread area. Peaches, comprising about one-quarter of the orchard acreage, are grown extensively in the Niagara and Kent-Essex regions, the only two regions with mild enough winters and extensive areas of the right kind of soil for that crop. The less important fruit crops of cherries, pears, and plums are grown primarily in the Niagara Fruit Belt.

Because the Kent-Essex region provides the only alternative to Niagara for peach production, it is instructive to compare the peach climates of these two regions. The winters of the Kent-Essex region are similar to those of the Niagara Fruit Belt. The frequency of temperatures of −20°F. and −12°F. (critical temperatures for dormant peach wood and buds respectively) are the same at Leamington as they are at Vineland in the Niagara Fruit Belt. Therefore, there is very little risk of low winter temperature injury to the peach trunk and branches, and a chance of only three years in thirty of some crop loss due to winter frost damage to blossom buds.[11]

Figure 6

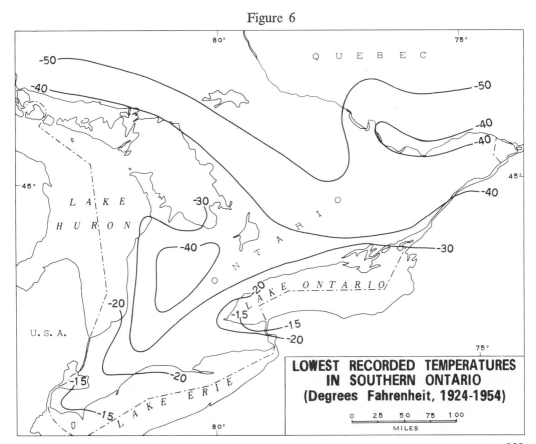

LOWEST RECORDED TEMPERATURES IN SOUTHERN ONTARIO
(Degrees Fahrenheit, 1924-1954)

229

Figure 7

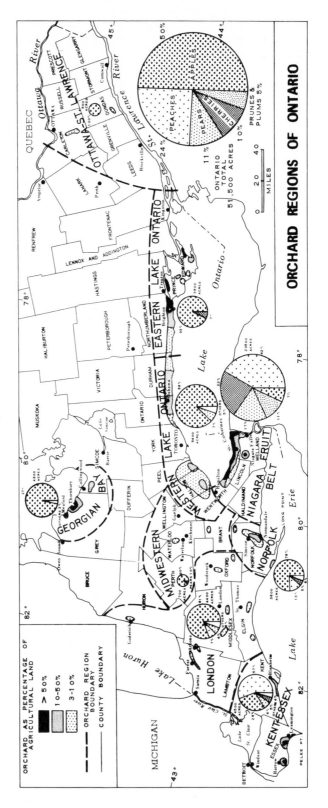

ORCHARD REGIONS OF ONTARIO

In recent years there has been extensive winter injury to the roots of peach trees in the Kent-Essex region. A detailed study conducted by the Agricultural Research Station at Harrow has shown that temperatures just slightly below zero can kill or seriously injure peach trees if there is insufficient snow to help insulate the roots. The damage has been greatest on the lighter textured soils and in orchards with inadequate cover crop. Severe losses of peach trees due to root injury were first noticed in 1958-59 when about 20 percent of all peach trees were killed. There was serious tree-kill and injury again in 1960-61 and 1962-63. A survey carried out by the Agricultural Research Station at Harrow in 1963 showed that 45 percent of the peach trees in the 11-19 year old category had been lost since planting and 20 percent of the 1-4 year old trees had been killed.[12]

As far as spring frost injury to blossoms is concerned, the Niagara Fruit Belt is clearly superior to the Kent-Essex region. The Niagara peach orchards suffer less spring frost damage chiefly because they blossom a week or two later than the Kent-Essex region. The delay in blossoming is sufficient to avoid most spring frosts.

When winter and spring frost damage are considered together the probability of serious crop loss due to low temperatures is twice as great in the Kent-Essex region as in the Niagara Fruit Belt.

Orchard Trends and Economic Conditions

From the early pioneer days, the orchard was an important aspect of the farm operation in much of Southern Ontario. It provided fresh fruit for most of the winter, and as production increased, also provided a cash crop. The farmstead orchard remained important until the early 1930's. In 1931, according to the *Census of Canada*, there were 53,000 farms in the province with orchards of more than 25 trees. Only 2,000 of the orchards were larger than 10 acres.

The trend away from small farmstead orchards began after the severe damage to orchards in the cold winter 1917-18, and

was accelerated by the 1933-34 freeze that killed many trees which had been replanted after the 1917-18 freeze and which were just reaching peak production. From a climate point of view alone, the future of the orchard industry in the mid-1930's looked bleak in all but the mildest parts of Ontario. In addition, apple scab and maggot were becoming an increasing menace. It became impossible to market apples unless the trees were sprayed, but spraying equipment was not economic for an orchard of one acre or smaller. Besides, the spraying had to be done in the spring when the farmer was busy with seeding. Consequently, most farmstead orchards were completely neglected. Some orchards were uprooted, but most of them have remained as derelict orchards ridden with insects and infected with disease. In the less favourable climatic areas, most of the trees have died and have been cut, leaving only stumps or a few scattered trees as a trace of the past.

The total orchard area declined from 181,000 acres in 1921, to 80,000 acres in 1941, and 69,000 acres in 1961. The number of farms with 25 or more fruit trees declined from 53,000 in 1931 to 8,000 in 1961. Only 6,000 of the 1961 number were commercial orchards. In the thirty-year period from 1931 to 1961, the average size of orchard had increased from 2½ to 10 acres. Thus, as farmstead orchards were declining throughout most of the province, some farmers in the more favourable climatic areas were beginning to make orcharding their specialty.

With increasing specialization, yields have been constantly rising. In the early 1930's, the average provincial apple yield was under 2 bushels per bearing tree. By 1961, the average apple yield per tree had increased to over 5 bushels and by 1966, to over 7 bushels, the highest in Eastern Canada.

Apple production was at is peak in Ontario in the early 1920's. According to the Census of Canada, in 1921 there were more than 120,000 acres of apple orchard which yielded over 9 million bushels. By 1945, the acreage had declined to 33,000 acres and production to 2.5 million bushels. Between

Table 5

ACREAGE AND PRODUCTION OF TREE FRUIT CROPS IN ONTARIO, 1945-1970

Crop	Acreage 1945	Acreage 1961	1970	Average Annual Production 1945-50 (millions)	1957-61 (millions)	1966-70 (millions)
Apples	33,000	25,900	28,000	2.5 bu.	4.6 bu.	6.2 bu.
Peaches	15,500	12,000	10,200	55.0 lb.	113.1 lb.	80.2 lb.
Cherries	4,300	5,900	5,300	12.5 lb.	25.5 lb.	28.1 lb.
Pears	5,800	5,000	5,700	18.7 lb.	37.2 lb.	44.7 lb.
Plums and Prunes	4,300	2,700	2,100	16.0 lb.	17.7 lb.	11.8 lb.

Source: Farm Economics and Statistics Branch, Ontario Department of Agriculture and Food, Toronto.

1945 and 1961, the acreage continued to decline, but production increased substantially. Between 1961 and 1970, the acreage began to increase again and the annual average production climbed to over 6 million bushels (Table 5).

The peach acreage and production in Ontario did not begin to decline until the mid 1950's. Total production declined by about 30 percent between the 1957-61 and 1966-70 periods (Table 5). This reduction in acreage and production resulted from a number of factors including winter injury in the Kent-Essex region, urbanization in the Niagara Fruit Belt, cheap imports of canned peaches, and a general cost-price squeeze.

Of the other soft fruits, in the 1960's there was a decrease of 600 acres of both cherries, and plums and prunes, and an increase of 700 acres of pears. In the same period there was a slight increase in production of cherries and pears and a decrease in production of plums and prunes (Table 5).

In the early 1930's Ontario exported between 30 and 40 percent of its apple production. Some of this went to Quebec, but the largest portion went to the United Kingdom. The United Kingdom market was lost for Ontario apple producers during World War I because of a lack of shipping. Immediately after the war the British market for Canadian apples was drastically reduced by currency restrictions and strict quotas. By the time Britain was again able to buy large quantities of Canadian apples, Ontario was unable to compete successfully with Nova Scotia prices and British Columbia varieties, and packaging and selling methods. When Quebec began supplying most of its own demand for apples, Ontario was restricted primarily to its own domestic market. In the 1960's, approximately 50 percent of the apple crop went to the fresh market within the province, 5 percent to the other provinces, and 5 percent was exported.

In the 1960's, between 30 and 50 percent of Ontario's apple production went into processed products. The volume of apples processed is much higher in bumper crop years. Apple juice is the most important processed product, taking up about one-half of the processed apples each year. Controlled atmosphere cold storage has made it possible to sell apples as late as July, thereby considerably extending the fresh apple market season. Of the 1969 apple crop, two million bushels were placed in controlled atmosphere storage.

Despite several marketing organizations, the lack of stable prices is still one of the most serious problems facing Ontario fruit growers. For example, in the fall of 1970, apple prices to Ontario growers was the lowest in several decades.

Table 6 shows the trend in orchard acreage by region. Between 1931 and 1951, there was a significant reduction in orchard acreage in all regions except Niagara. The following regions experienced an orchard reduction of two-thirds or more between 1931 and 1951: Ottawa-St.Lawrence, Eastern

Table 6

ONTARIO ORCHARD ACREAGE TRENDS BY REGION

| Region | Total Orchard Acreage | | | |
	1931	1951	1961	1970
Ottawa-St. Lawrence	6,220	1,760	800	690
Eastern Lake Ontario	16,080	6,900	3,770	6,450
Western Lake Ontario	23,480	15,200	8,290	4,360
Niagara	18,160	29,830	20.530	18,500
Norfolk	5,650	3,650	3,620	2,690
Kent-Essex	6,790	4,680	4,520	4,020
London	24,030	8,190	4,410	2,130
Midwestern	9,650	2,410	660	710
Georgian Bay	13,620	5,320	4,690	5,890

Note: The 1931 and 1951 acreages are from the Census of Canada; the 1961 acreages from *Agricultural Statistics for Ontario,* Ontario Department of Agriculture, 1962; the 1970 acreages from unpublished data from the Ontario Department of Agriculture and Food. The 1931 and 1951 acreages include only those counties with more than 200 acres of orchard in order to make those statistics more comparable with the Ontario Department of Agriculture and Food acreages for 1961 and 1970 which include only commercial orchards.

Lake Ontario, London, Midwestern, and Georgian Bay. Between 1951 and 1961, all of these same regions, except Georgian Bay, which had only a small decline, lost at least half their orchard acreage. Between 1961 and 1970, the London region lost approximately half of its orchard acreage but losses in the Ottawa-St. Lawrence region were small. Eastern Lake Ontario, Midwestern and Georgian Bay experienced small gains in orchard acreage.

Western Lake Ontario, which only lost one-third of its orchard acreage in the 1931-51 period, suffered almost a 50 percent decline in the 1950's and again in the 1960's. As a result, Western Lake Ontario, which had the largest acreage in 1931 (24,000 acres), fell to seventh place (2,000 acres) by 1970. This loss in orchard can be attributed directly and indirectly to the rapid urbanization that occurred between Toronto and Hamilton. Although new orchards have been established some distance back from the lake, and there is considerable good orchard land still available in the region, the impact of the urban shadow is so great that the decline of orcharding is likely to continue.[13]

The trends in orchard acreages have been different in those regions that produce soft tree-fruits as well as apples (Table 6). Norfolk and Kent-Essex suffered major declines

in orchard acreages between 1931 and 1951, but both retained approximately the same acreages during the 1950's. In the 1960's Norfolk suffered a decline of approximately 25 percent and Kent-Essex a decline of 10 percent. These declines are at least partially due to the reduction of peach orchards following winter temperature injury in the early 1960's.

The trend in the Niagara orchard acreages is interesting. Despite a decline in apple growing and some losses due to urbanization, the total orchard acreage increased by almost 12,000 acres between 1931 and 1951, reflecting the rapid expansion of peach growing in the area. However, there was a loss of 9,000 acres between 1951 and 1961, and another loss of 2,000 acres in the 1960's. Thus, it is clear that since the early 1950's new plantings have failed to keep pace with the loss of orchard to urbanization.

Only a small portion of the Niagara Fruit Belt has the light-textured soil (tender fruit soil) required for peaches, sweet cherries and some of the more valuable wine grapes. Two studies have indicated that in half of the tender fruit producing townships, urbanization has proceeded so far that there is no chance of preserving significant acreages of tender fruit soil for fruit production. In the four townships more remote from urban

233

growth, there are some 20,000 acres of tender fruit soil, but Reeds estimate that only half of this can be preserved for fruit growing. Reeds concludes that unless fruit-growing is made more economically viable, either strict zoning or government expropriation will be required to save even 10,000 acres.[14]

If current trends continue, Krueger has forecast the total demise of the Niagara orchard industry before the turn of century.[15]

BRITISH COLUMBIA

Location and Physical Base

More than 90 percent of the orchards in British Columbia are located in the Okanagan Valley, running some 120 miles from the forty-ninth parallel north to Salmon Arm on Shuswap Lake. The only other significant orchard area is found around Creston, just south of Kootenay Lake, where there are approximately 1,200 acres of orchard. Small and generally scattered orchards, comprising less than 4 percent of the provincial acreage, are found in the Vancouver Island-Lower Fraser, Upper Fraser-Thompson, and Grand Forks-Kootenay-Arrow Lake regions (Figure 8).

The major tree crop is apples, comprising about two-thirds of the total orchard acreage. Pears, cherries (primarily sweet), and peaches account for a quarter of the total orchard acreage and apricots, prunes and plums account for the balance. Some pears, cherries, and prunes and plums are grown in all regions; peaches and apricots are limited primarily to the South Okanagan. Although prunes and plums are grouped in the statistics and maps in this paper, the former are by far the more important of the two. A few crabapples are also grown in most regions.

All the orchards in British Columbia, except the few acres on Vancouver Island, are located on the lower slopes and terraces of river valleys or along the sides of lakes. Since most of the interior is climatically marginal for tree fruits, these sloping sites are important in making fruit growing possible in British Columbia. Cold air flows down the slopes and builds up pools of air that can reach tens of feet in depth on the valley floors. It is when the pool of cold air deepens to the height of the tree branches that frost damage ensues. Thus the best orchard site is on a sloping terrace high above a broad valley floor. Gullies cutting across the terrace help even further in draining the cold air away from the orchards.

In all of the orchard regions outside of the Vancouver Island-Lower Fraser Valley, there is periodic tree damage due to extremely cold winter temperatures. Since 1900, most of the orchard regions have suffered severe low-temperature injury to fruit trees in about one out of every seven years.

The winter of 1949-50 was the most devastating of all. In that winter, record low temperatures in most regions killed approximately 20 percent of all trees in the province. While all the tree fruits were damaged to some degree, apples and peaches suffered most with a 30 percent tree-kill rate. The highest percentage of tree-kill occurred in the Upper Fraser-Thompson region where 54 percent of the fruit trees were lost; surprisingly, the best record was at Creston, which did not receive the brunt of the cold wave and consequently had a tree-kill of only 4 percent. Even in the Oliver-Osoyoos area at the southern end of the Okanagan Valley, the tree-kill rate was a high 27 percent.

In November, 1955 an early cold snap plunged temperatures to below 0°F. Because the trees had not had time to properly harden up, this cold wave again killed large numbers of trees, many of which were young trees that had been planted to replace the losses of 1950.[16]

Below-zero temperatures in the winters of 1964-65 and 1968-69 again caused serious tree injury and resulted in substantial crop losses. Peach and apricot trees were hit the hardest with the result that many orchardists are replacing the killed tender fruit trees with more hardy apple trees. In 1965 and 1969 there was virtually no production of peaches and apricots in British Columbia.

Climatic statistics showing the frequency of temperatures critical for peach trees (−12°F. for dormant peach buds, and −20°F. for the tree itself) indicate that the

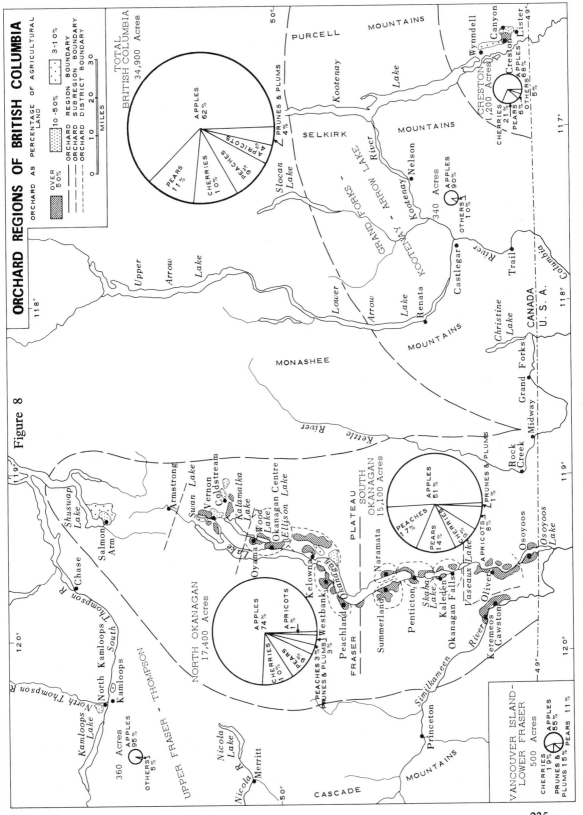

ORCHARD REGIONS OF BRITISH COLUMBIA

Figure 8

ORCHARD AS PERCENTAGE OF AGRICULTURAL LAND

OVER 50% | 10-50% | 3-10%

ORCHARD REGION BOUNDARY
ORCHARD SUBREGION BOUNDARY
ORCHARD DISTRICT BOUNDARY

MILES
0   10   20   30

118°

TOTAL BRITISH COLUMBIA
34,900 Acres

APPLES 62%
PRUNES & PLUMS 4%
APRICOTS 4%
PEACHES 9%
CHERRIES 10%
PEARS 11%

PURCELL   MOUNTAINS

Kootenay   Lake

SELKIRK   MOUNTAINS

Slocan Lake
GRAND FORKS LAKE
KOOTENAY River
Nelson

CRESTON
1,200 Acres
CHERRIES 21%
PEARS 6%
APPLES 68%
OTHERS 5%
Wynndell   Canyon   Lister

340 Acres
APPLES 90%
OTHERS 10%

Castlegar

Renata

Lower Arrow Lake

MOUNTAINS

Christine Lake

CANADA
U.S.A.

Columbia River

Trail

117°
118°

Upper Arrow Lake

Grand Forks

Midway

Kettle River

Rock Creek

50°
49°

119°

MONASHEE

PLATEAU

SOUTH OKANAGAN
15,100 Acres
APPLES 51%
PRUNES & PLUMS 1%
APRICOTS 8%
CHERRIES 9%
PEARS 14%
PEACHES 7%

Armstrong
Swan Lake
Vernon
Coldstream
Kalamalka Lake
Oyama
Wood Lake
Okanagan Centre
Ellison Lake
Kelowna
Westbank
Peachland

FRASER

Summerland   Naramata
Penticton
Shaha Lake
Kaleden
Okanagan Falls
Vaseaux Lake
Keremeos   Oliver   APRICOTS
Cawston
Osoyoos
Osoyoos Lake
Similkameen River

Shuswap Lake
Chase
Salmon Arm

NORTH OKANAGAN
17,400 Acres
APPLES 74%
APRICOTS 1%
PEACHES 3%
PRUNES & PLUMS 3%
PEARS 3%
CHERRIES 10%

Kamloops Lake
North Kamloops
Kamloops
South Thompson R.
North Thompson R.

360 Acres
APPLES 95%
OTHERS 5%

UPPER FRASER-THOMPSON

Nicola Lake
Nicola R.
Merritt

Princeton

CASCADE   MOUNTAINS

VANCOUVER ISLAND-LOWER FRASER
500 Acres
APPLES 55%
CHERRIES 19%
PRUNES & PLUMS 15%
PEARS 11%

120°
119°
50°
49°

235

area around the southern end of Lake Okanagan is the only one where the winter temperatures are as favourable for peach growing as they are in the Niagara Fruit Belt.

Spring frost damage to fruit blossoms is another major hazard in the orchard areas of British Columbia. In districts not adjacent to a lake or with insufficient slope to provide adequate air drainage, there is a risk of blossom damage almost every year. Even in the better districts there is a chance of some spring frost damage in one year out of two. In the Okanagan Valley, production of peaches, apricots and cherries was reduced because of spring frosts in five years during the 1950-60 decade.

The semi-arid climate has made irrigation necessary in the major orchard regions. Although this incurs additional expense, it results in much higher yields than in regions that depend upon natural precipitation for moisture. British Columbia apple tree yields, for instance, are three to four bushels higher than in Eastern Canada. A dry climate also makes disease control easier. Apple scab is not nearly so difficult to control as in Eastern Canada, and brown rot, which is a plague to Ontario peach growers, is relatively unknown in the Okanagan Valley. However, insects thrive in a hot dry climate, and in some years infestation with codling moths and mites reaches epidemic proportions.

The soils on which orchards are planted in British Columbia are formed primarily from glacial tills or glacio-fluvial deposits. The texture of orchard soils ranges from coarse gravel to clay. Because of the low preticipation (ranging from 10 to 20 inches annually), and the resulting reliance on irrigation, which makes control of the amount of moisture possible, good drainage is not as important for orchards in British Columbia as it is in Eastern Canada. For instance, while the best peach soils in Ontario are sands and sandy loams, the best peach soil in British Columbia is a silt loam. In British Columbia the lighter textured soils require more frequent and greater applications of water and fertilizer. There is, also, greater risk of frost damage to the tree roots in the lighter textured soils. On the other hand, the heavier clays restrict root development too much for the tender fruit crops of peaches, apricots, and sweet cherries.

Orchard Trends and Economic Conditions

The Okanagan Valley has not always been the most important orchard region in British Columbia. In fact, up to about 1890, orchards were planted in most of the valleys in southern British Columbia except the Okanagan. However, a superior climate, coupled with spectacular promotion by real estate speculators, led to rapid settlement of the Okanagan Valley in the early 1900's. By 1911, the Okanagan Valley had become the leading orchard region of the province; by 1920, it had over two-thirds of the total fruit trees in British Columbia. Its supremacy as an orchard region in British Columbia has been increasing ever since.

Primarily due to the orchard boom in the Okanagan Valley, the number of fruit trees grew very rapidly in the early part of the twentieth century. By 1920 there was a total of 1,900,600 fruit trees, a number not to be exceeded until 1950, when the orchard survey showed 2,477,900 fruit trees. There was a slight decline in fruit tree numbers as a result of the widespread losses during the winters of 1949-50, 1955-56 and 1968-69.

Because of the small population of British Columbia, the economy of the orchard industry has depended primarily upon sales to the rest of Canada and exports to foreign countries. Export of apples began on a large scale to the United Kingdom in 1900. Prior to World War I, 40 percent of the Province's apples went to the United Kingdom. By providing bright red, carefully packed, high-quality dessert apples, British Columbia was able to maintain a large volume of exports to the United Kingdom during the interwar years. After the interruption of regular trade patterns during World War II, the British market was again opened to large volumes of British Columbia apples. In 1946, British Columbia apples were accepted instead of those from Nova Scotia because of superior quality, packing, and salesmanship. In that year, approximately 2.5 million bushels of apples were exported to the United Kingdom. Because of British currency problems, apple exports were sporadic in the next few years. Finally, in 1954, Britain established an allocation of apple imports from North America. Sixty percent of the allocation was assigned to Canada, and British Columbia filled the largest portion of Canada's share. After World War II, the United States became an important market for British Columbia apples. In the late 1940's, almost 2 million bushels went to the United States annually. During the 1950's and 1960's, British Columbia apple exports to the United States averaged less than a million bushels a year. Although British Columbia has remained the leading Canadian exporter of fresh apples, its share of the total export sales declined from 85 percent in the 1950's to 60-70 percent in the 1960's.

At one time British Columbia had close to 100 percent control of the apple market of all Western Canada. Since 1950, this market control has been declining slightly, due to increased imports and increased sales of Ontario apples in Western Canada. Nevertheless, in the mid-1960's, interprovincial shipments of apples from British Columbia were still eight times as large as from Ontario and constituted 85 percent of total interprovincial sales of apples in Canada. In 1969, almost one-half of the British Columbia fresh apple sales were made in Western Canada.

Of the total British Columbia apple production in the 1960's, on the average, about 35 percent was exported, 30 percent went to other provinces, 15 percent was consumed in the province and the balance was processed. In the latter part of the decade, the proportion exported decreased, while the percentage going to other provinces and processing increased.

The British Columbia fresh soft fruit market is primarily found in Western Canada. The British Columbia control of the soft fruit market is declining in the Prairie Provinces because of the frequency of crop failures (particularly peaches and apricots) resulting from severe winters.[17]

A combination of a number of factors has led to economic difficulties in the orchard industry of British Columbia.

Right from the beginning, the prices of orchard land in the Okanagan Valley were unrealistically high, and they were further inflated by large numbers of Prairie farmers who decided to spend their retirement years operating small orchards. The newcomers were willing to pay high prices for small orchard holdings.

The rapidly growing tourist industry has also forced land prices upward. The warm sunny summers, beautiful bathing beaches, and fishing in the upland lakes, have led to a tourist boom in the Okanagan Valley. New motels, camping parks, entertainment facilities, and cottages are springing up all over the valley. Lakefront lots of less than a fifth of an acre are selling for at least $5,000 in places like Penticton and Kelowna. These prices have increased the market prices for adjacent orchard land.

The net result has been that orchard land prices in the Okanagan Valley range from $2,000 to $3,000 an acre, which is much too high in relation to present orchard profits. The price of fruit is about the same as it was in the 1930's while fruit growing costs have approximately tripled. With a reduced margin, growers have found it imperative to increase the size of their orchard holdings in order to make fruit growing economical. However, many of the growers feel that it is uneconomical to increase an orchard holding at a price of $2,500 an acre. Farm economists believe that $1,000 an acre reflects more closely the present economics of fruit growing.

More than 70 percent of the growers in the Okanagan Valley have less than ten acres of orchard. As a result of small orchards and declining profits, approximately one-half of the orchardists in British Columbia cannot make a living off their orchards. The average grower can expect his orchard investment and labour to provide him with a house, garden, and enough cash to pay depreciation on his equipment. For his family's livelihood he must find employment away from the farm. Many growers with orchards of less than the average size find that they must subsidize the orchard operation from off-farm earnings.

The large tree losses and drastic crop reductions following the freezes of 1949-50 and 1955-56, along with marketing problems resulting from United Kingdom restrictions on imports, led to a clamour from British Columbia orchardists for governmental assistance for the industry. After the 1949-50 freeze, the federal and provincial governments paid out $250,000 to assist orchardists in replacing killed trees. Demands for further assistance following the 1955-56 freeze led to a Royal Commission investigation of the tree-fruit industry of British Columbia.[18]

Instead of outright financial assistance, the Royal Commission Report recommended government controlled long-term, low interest loans. The report stipulated that such loans should be made only to farmers with more than 10 acres of orchard with favourable site, soil, and climate conditions. It added the further condition that loans should be made only to those who had demonstrated managerial capacity better than average in their districts.

Because of a lack of a sizeable provincial domestic market, and because of the stiff competition in the markets in the rest of Canada and in foreign countries, the British Columbia orchard industry learned early that it had to organize its selling activities in order to survive. After several unsuccessful attempts at cooperative selling ventures, B.C. Tree Fruits Limited was established in 1938. From that time on, fruit from all the orchard regions (except the Vancouver Island-Lower Fraser region) has been sold by B.C. Tree Fruits. Returns and expenses of individual fruit crops are pooled. Packing houses or growers are permitted to sell to retailers, roadside stand operators or consumers in their immediate vicinity. A consumer from outside the orchard regions may take home no more than 20 boxes of fruit a year. The selling regulations are strictly enforced.

The cooperative central selling agency has made it possible to maximize sales and returns to the growers. The promotion of British Columbia fruit is very effective. There is strict adherence to quality standards and the packing ensures that the quality of the product is maintained until it reaches the consumer. There is also control over the flow of fruit onto the local domestic market. Special quality and varieties of fruit are saved for foreign markets where the competition is keenest. A cooperative processing plant helps to increase returns from sub-standard quality fruit and helps to reduce the surplus of bumper crop years.

In summary, an effective selling agency has been largely responsible in making it possible for the orchard industry to survive in the face of many serious physical and economic problems. Up to 1970, British Columbia has been able to market successfully almost all of its fruit production. However, with current world supply and demand conditions, it may have difficulty marketing the increased volumes of apples that are projected in the next decade.

PROBLEMS

The physical environment would permit a substantial expansion of the apple orchard industry in the four orchard provinces. The tender fruits (apricots, peaches, and sweet cherries) can be produced successfully on a large scale only in the South Okanagan Valley, the Kent-Essex region and the Niagara Fruit Belt. Climate records and tree-kill and crop loss data indicate that the Niagara Fruit Belt is the best tender fruit region in Canada. It is unfortunate that this region, which has both a superior tender fruit climate and a superior location in relation to market, is being threatened by haphazard urban expansion.

Major problems facing the entire Canadian tree-fruit industry include uncertain export markets, foreign imports, relatively low prices for produce in the face of rising production costs, and organizing the growers in order to have more effective control over long-range production volume and quality of produce reaching the market.

In view of European apple production trends, Canada is likely to have difficulty keeping, let alone expanding, its traditional

European market. The United Kingdom, which in the 1930's produced only 10 million bushels a year, is now producing as many apples as Canada. France has increased its production by almost 20 million bushels in a decade and has begun penetrating European markets, including the United Kingdom. Within the 1960-70 decade, France's share of the British apple market increased from less than 2 percent to approximately 25 percent. When the United Kingdom enters the European Economic Community, France will have an additional advantage in that market. Much of the French increase in production has come from new and renewed orchards that are producing high quality apples of desirable varieties including the Golden Delicious. With some 10,000 acres of apple trees coming into bearing age each year, France will undoubtedly provide stiff competition to Canadian exporters. It is interesting to note that in 1968, France shipped 370,000 bushels of apples to Quebec.[19]

In the late 1960's Canada increased substantially its export of apples to the United States. However, this is an unsure market that fluctuates according to United States production volume. A bumper crop in both countries in the same year would be of serious consequence to the Canadian producers.

There would appear to be considerable potential for expanding Canadian apple exports to countries outside of Europe and the United Statees. Apples cannot be grown in tropical countries, and in southern hemisphere countries. Canada can supply apples fresh from the tree at a time when domestic apples in those countries have been stored for at least six months. South Africa has demonstrated how to take advantage of this hemispheric difference. In 1967-68, South Africa exported more than a half million bushels of Granny Smith apples to Canada and because they were fresh and crisp in the spring and summer, they demanded a premium price.

The annual apple consumption of the average Canadian in 1970 was slightly under one bushel. This consumption is not likely to rise to more than one bushel, because with increased standards of living, Canadians tend to increase their consumption of citrus fruit and other imported out-of-season fruit rather than increase their consumption of domestic fruit. Since Canada exports about 2 or 3 million bushels more a year than it imports, it would appear that the average apple crop of over 21 million bushels is approximately sufficient to meet the demand of some 21 million Canadians.

Studies have indicated that apple production in Canada will increase by at least 25 percent by 1978.[20] Since the domestic market will not be able to absorb this increase, and there is a risk of exports to traditional markets declining, serious marketing problems could beset the apple growing industry. If Canada cannot find new export markets, it would appear to be wise policy to greatly curtail new plantings in the 1970's.

The marketing of the soft fruits presents even more problems than that of apples. These fruit crops are very perishable and therefore must reach the market within a day or two of being harvested. If picked too green, the fruit does not achieve its full flavour; if picked too ripe, it is very easily damaged in shipping. As a result, the bulk of the fresh sales of soft fruit must be made on the adjacent domestic market.

The Ontario soft fruit producers have an advantage in being able to place tree-ripened fruit of high quality on the large adjacent urban markets with low transportation costs. They have not fully capitalized on this advantage because of the lack of an efficient selling agency, inadequate packaging and quality control, and competition from United States imports.

Because of their earlier growing season, United States growers can place end-of-the-season surplus fruit on the Canadian market at low prices. A higher tariff during the harvest season has not been sufficient to discourage imports from growers who have already saturated their own domestic market. In the twenty-year period preceding 1967, the volume of imports of fresh peaches from the United States approximately tripled.

A large proportion of Canadian soft fruit production is processed. In canned fruits,

Canadian growers also face stiff competition. The total imports of canned peaches increased from an average of 9 million pounds in the 1948-52 period to more than 50 million pounds in the 1963-67 period—more than a ten-fold increase in fifteen years. The imports of canned pears doubled during the same period.[21]

The United States has always been a major exporter of canned fruits to Canada. However, by 1967, Australia led all countries as the supplier of canned pears to Canada and was second only to the United States as the supplier of canned peaches.[22] Canadian producers attribute this surge in imports from Australia to subsidies paid to the Australian fruit grower. The United States canned peaches are generally considered to be inferior to the Canadian product, but they are sold at lower prices, and it is the price tag along with aggressive promotion that usually determines the housewife's choice of canned goods in the supermarket. The Canadian canners have done a poor job of promoting the higher quality of the Canadian product. In some cases, the task of achieving a good name for Canadian canned fruits is complicated by United States ownership of canneries in Canada. With the same trade name used on both sides of the border, it is difficult to establish any identity for the Canadian product.

Canadian fruit growers and processors have been pressuring the Canadian Government to either raise the import tariffs on fruit or drastically change the structure of protection from imports. They believe that a system of flexible quotas that considers the total supply and consumption of each fruit in Canada, and then allows for imports to make up the difference, would give adequate protection for both the growers and the processors. However, in view of present trends toward freer world trade, it is highly unlikely that the Canadian Government will be willing or able to heed the requests of the fruit growing and processing industries.

The best hope for the orchard industry's marketing problems is a more aggressive export policy and more orderly marketing procedures with a central selling agency closely associated with the processing industry in each province. Thus far, this has been achieved only by British Columbia.

REFERENCES

[1] Burns, J. R., "The Canadian Apple Industry," *Canadian Farm Economics*, Vol. 4 (5), 1969.

[2] Burns, J. R., "Some Observations on the Long Term Prospects of the Canadian Peach Industry," *Canadian Farm Economics*, Vol. 3 (5), 1968, and Siddiqui, R. M., "Pear Production in Canada," *Canadian Farm Economics*, Vol. 3 (5), 1968.

[3] Olmstead, C. W., "American Orchard and Vineyard Regions," *Economic Geography*, Vol. 32 (3), 1956.

[4] Harlow, L. C., and Whiteside, G. B., *Soil Survey of the Annapolis Valley Fruit Growing Area*, Canada Department of Agriculture, Ottawa, 1943.

[5] Mercier, G. G., and Chapman, L. J., "Peach Climate of Ontario," Report of the Horticultural Experiment Station, Vineland, 1955-56.

[6] Morse, N. H., *An Economic History of the Apple Industry of the Annapolis Valley in Nova Scotia*, Ph.D. thesis, University of Toronto, 1952.

[7] Redmond, L. V., and Gervason, P., "Apple and Pear Tree Surveys: Planting and Removal, Annapolis Valley of Nova Scotia, 1970," D.B.S. and N.S. Department of Agriculture and Marketing, Truro, N.S.

[8] *Report of the Royal Commission Investigating the Apple Industry of the Province of Nova Scotia*, Nova Scotia Department of Public Works and Mines, Halifax, 1930.

[9] Stobbe, P. C., and McKibbin, R. R., *Orchard Soils of the Fredericton-Gagetown Area, New Brunswick*, Canada Department of Agriculture, Ottawa, 1937.

[10] Orchard Census, Department of Agriculture, Fredericton, 1958-59.

[11] Krueger, R. R., *Changing Land-Use Patterns of the Niagara Fruit Belt*, Transactions of the Royal Canadian Institute, Vol. 32, Part II, Toronto, 1959.

[12] Wensley, R. N., Unpublished table giving peach-tree mortality due to low winter temperatures. Canada Department of Agriculture Research Station, Harrow.

[13] Hallman, D., *Orchard Resources in the Toronto-Hamilton Area, 1931-1965,* (unpublished Master's thesis, Department of Geography and Planning, University of Waterloo, 1967).

[14] Reeds, L. G., *Niagara Region Agricultural Research Report: Fruit Belt,* Regional Development Branch, Department of Treasury and Economics, March, 1969.

[15] Krueger, R. R., "Recent Land Use Changes in the Niagara Fruit Belt." Manuscript prepared for Provincial Government study in 1968. Upon going to press, the study has not yet been publicly released.

[16] Krueger, R. R., "The Physical Basis of the Orchard Industry of British Columbia," *Geographical Bulletin,* No. 20, 1963.

[17] *Financial Statements,* B. C. Tree Fruits Ltd., Sun-Rype Products Ltd., October 1969 and October 1970.

[18] MacPhee, E. D., *Report of the Royal Commission on the Tree-Fruit Industry of British Columbia,* Victoria, 1958.

[19] Al-Hashimi, M., *Trends in the Production and Marketing of Apples in Canada, with Special Reference to Ontario,* Farm Economics, Co-operatives and Statistics Branch, Department of Agriculture and Food, Toronto, 1969.

[20] *Ibid.*

[21] Burns, J. R., *op. cit.*

[22] Burns, J. R., *op. cit.* and Siddiqui, R. M., *op. cit.*

Changes In Land Use and Farm Organization In the Prairie Area of Saskatchewan, 1951 to 1966

R. A. STUTT

In the prairie area of Saskatchewan, a number of important and effective changes in land use and farm organization have taken place during the last two decades. These adjustments have served to enhance the inherent physical and economic capabilities of the land resource. As a result it is believed that a larger proportion of farmers are attaining a higher degree of stability in agricultural production, financial security and a level of living comparable to that of urban dwellers.

Despite periodic setbacks because of climatic conditions and marketing problems, there is a tendency to forget or to downgrade the amount of progress by farmers.

The purpose of this article is to show that change has taken place on Saskatchewan prairie farms during the period 1951-1966.

METHOD OF ANALYSIS

The study was done by using Census data for a selected group of rural municipalities[1] representative of poor, medium and good grades of land in the Brown and Dark Brown Soil Zones of Saskatchewan. The statistics were drawn from the 1951, 1961 and 1966 Censuses, supplemented by the series of farm organization studies conducted by the Economics Branch during the 1951-1966 period.

●R. A. STUTT *is an Economist with the Farm Management and Agricultural Adjustment Division, Economics Branch, Canada Department of Agriculture.*

Reprinted from *Canadian Farm Economics,* Vol. 5 (6), 1971, pp. 11-19 with permission of the author and the Canada Department of Agriculture.

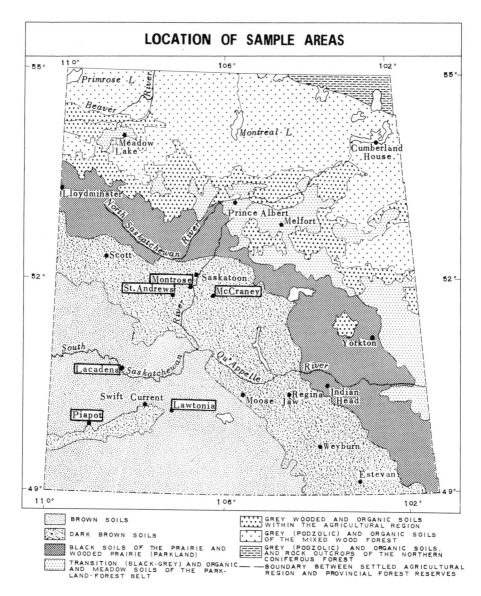

LOCATION OF SAMPLE AREAS

BROWN SOILS	GREY WOODED AND ORGANIC SOILS WITHIN THE AGRICULTURAL REGION
DARK BROWN SOILS	GREY (PODZOLIC) AND ORGANIC SOILS OF THE MIXED WOOD FOREST
BLACK SOILS OF THE PRAIRIE AND WOODED PRAIRIE (PARKLAND)	GREY (PODZOLIC) AND ORGANIC SOILS, AND ROCK OUTCROPS OF THE NORTHERN CONIFEROUS FOREST
TRANSITION (BLACK-GREY) AND ORGANIC AND MEADOW SOILS OF THE PARK-LAND-FOREST BELT	BOUNDARY BETWEEN SETTLED AGRICULTURAL REGION AND PROVINCIAL FOREST RESERVES

Figure 1

Indicators selected deal with land use, size of farms, number of farms, livestock numbers, types of farms, economic farm classes, land tenure, farm population, age of farm operators, farm operator residence and off-farm employment. Some additional data for the period since 1966 from economic studies carried out by the Economics Branch were also used to provide the latest information available at time of writing.

Farms in the rural municipalities chosen for this study are located on land representative of each land class based on their suitability for wheat production which has been considered to be the most profitable land use (2). The rural municipality selected to represent a specific land class had a predominance of the land area in that particular category. For this study, a modification was used whereby the selected rural municipalities

Table 1

TOTAL ACREAGE AND PERCENTAGE OF ACREAGE IN EACH LAND CLASS,
REPRESENTATIVE MUNICIPAL UNITS, PRAIRIE AREA OF SASKATCHEWAN

	Rural Municipality Name	No.	Total Acreage	Land Classes				
				I	II	III	IV	V
Brown Soil Zone				—percent—				
Poor grade	Piapot	110	205,183	74.2	14.8	10.1	0.9	—
Medium grade	Lawtonia	135	206,093	11.1	15.9	55.6	17.4	—
Good grade	Lacadena	228	288,009	20.4	6.5	11.6	20.1	41.4
Dark Brown Soil Zone								
Poor grade	Montrose	315	217,803	59.5	18.3	19.7	2.5	—
Medium grade	McCraney	282	235,086	31.4	27.9	40.7	—	—
Good grade	St. Andrews	287	204,096	6.3	3.7	6.1	14.3	69.6

Source: *An Economic Classification of Land,* Unpublished data, Economics Branch, Canada Department of Agriculture, 1937-1951.

were assembled into three broad grades of land—poor, medium and good. Under this grouping, land rated as Land Classes I and II (submarginal and marginal) was designated as poor; land rated as Land Class III (fair wheat land), as medium; and land in Land Classes IV and V (good and excellent wheat land), as good. The grouping is shown in Table 1. This abbreviated study approach was used in an attempt to portrary the wide variability in soil capability and economic productivity of farming that exists in this are of Saskatchewan. Using this approach in the study, it was thus possible to show the range, scope and effects of changes on each specific grade of land.

Level of Physical Productivity

Evidence of the level of physical productivity of the selected municipal units is provided by information on wheat yields (Table 2). Wheat yields in the 1951-1960 and 1961-1969 periods are considerably above those of previous decades.

Number of Farms and Farm Population

The number of farms in the representative municipal units follows a general downward trend. Between 1951 and 1966, in the Brown Soil Zone, the number decreased by 21.1, 22.5 and 20.5 percent in the poor, medium and good grades of land, respectively. In the Dark Brown Soil Zone, decreases of 13.1, 22.2 and 21.8 percent occurred in the comparable grades of land during the same period.

Farm population also decreased during the same years. This reduction was about 25 percent on all grades of land except in good grades of land in the Dark Brown Soil Zone, where the overall population decrease was 34.5 percent in the 1951-1966 period.

The total percentage decline in population was similar to the total percentage decline in farm numbers for each grade of land, except in the Rural Municipalities of Montrose (no.

Table 2

AVERAGE WHEAT YIELDS IN RURAL MUNICIPALITIES REPRESENTATIVE OF POOR, MEDIUM AND GOOD WHEAT LANDS, SASKATCHEWAN, 1918 TO 1969

	Brown Soil Zone			Dark Brown Soil Zone		
	Grade of Land					
Period	Poor	Medium	Good	Poor	Medium	Good
	—bushels per acre—					
Average						
1918 to 1930	13.0	14.8	14.9	11.3	13.4	18.1
1931 to 1940	7.2	4.9	10.3	8.2	8.9	14.7
1941 to 1950	7.3	11.1	12.7	10.7	13.0	16.4
1951 to 1960	16.3	20.7	20.8	15.3	17.7	23.8
1961 to 1969	15.1	17.8	20.6	16.6	17.9	22.8
1918 to 1969	11.8	13.8	15.7	12.3	14.0	19.0

Source: *Supervisor of Statistics,* Saskatchewan Department of Agriculture.

315) and St. Andrews (no. 287). Here, the population decline was greater. This was probably due to the close proximity of larger urban centers.

This picture suggests the attainment of a favourable trend in farm numbers, population stability, increased farm size and greater income-earning capacity. In the case of the poorer grades of land, most of the farm depopulation occurred during the 1930's and 1940's. In the representative municipal units, the decrease in farm population from 1951 to 1966 was 14.0 and 21.7 percent on poor lands, 23.7 and 23.6 percent on medium lands, and 24.2 and 34.5 percent on good lands in the Brown and Dark Brown Soil Zones, respectively. Changes in farm numbers and farm population have served to facilitate desirable adjustments to resource capacity and farm family needs through a more appropriate ratio of the population to land resources.

Changes in Land Use

Little change took place in the amount of occupied farm land from 1951 to 1966. The total area of improved land increased slightly and also as a percentage of occupied farm land. In 1966 the improved land area was 37.2 and 59.2 percent of occupied farm area for poor lands, 87.0 and 81.8 percent for medium lands, and 75.0 and 93.1 percent for

good lands in the Brown and Dark Brown Soil Zones, respectively. Important changes, however, have taken place in the use of improved land (Table 3).

First, there was a drop in the proportion of improved land under crops,[3] especially in the Brown Soil Zone. This was most apparent on the poorest grade of land. Some of the decrease can be attributed to an increase in summerfallow and a general adoption of a two year summerfallow-crop rotation system on the medium and good lands while a three year summerfallow-crop rotation took place on poorer lands.

Second, there was a reduction in the proportion of improved land used for wheat.[4] The most significant reduction occurred on the poorer grades of land. This was due to a considerable shift and an encouraging increase in the proportion of improved land devoted to grass for pasture or hay purposes on the poorer grades of land. These kinds of land use adjustments have long been advocated as desirable shifts from a "one-crop" type of agricultural economy and as improved conservation practices.

Size of Farm

In terms of total occupied area, farms on the poorest grade of land are considerably larger than on other grades of land but have a smaller proportion of improved land. Con-

Table 3

IMPROVED LAND USE IN REPRESENTATIVE RURAL MUNICIPALITIES IN SASKATCHEWAN, 1951, 1961 AND 1966.

| | *Percentage of total improved land* | | | | | | | | |
| | *Under crops* | | | *Wheat* | | | *Improved pasture and tame hay* | | |
	1951	*1961*	*1966*	*1951*	*1961*	*1966*	*1951*	*1961*	*1966*
Brown Soil Zone									
Poor grade	72.7	60.8	63.3	37.7	29.0	31.5	9.3	12.9	21.6
Medium grade	56.1	53.0	53.0	47.3	45.2	48.8	1.7	2.1	2.6
Good grade	63.1	55.8	59.7	54.6	42.6	50.5	2.5	2.2	3.8
Dark Brown Soil Zone									
Poor grade	70.1	58.3	63.8	39.0	32.3	34.5	8.5	13.2	14.5
Medium grade	53.2	53.5	57.3	46.3	42.9	50.7	4.4	3.7	2.7
Good grade	55.0	54.0	57.5	49.8	34.0	47.3	3.4	1.8	2.3

Source: Census of Canada (Agriculture).

Table 4

LAND USE, RURAL MUNICIPALITIES REPRESENTATIVE OF POOR, MEDIUM AND GOOD
GRADES OF LAND IN SASKATCHEWAN. 1951, 1961 AND 1966.

	Total farm area			Total improved area			Area under crops			Area under wheat			Area under improved pasture and tame hay		
	1951	1961	1966	1951	1961	1966	1951	1961	1966	1951	1961	1966	1951	1961	1966
Brown Soil Zone							acres per farm								
Poor grade	1,244	1,505	1,568	438	524	582	318	318	369	165	152	184	48	67	162
Medium grade	599	665	755	491	560	657	275	297	348	232	253	321	10	12	17
Good grade	769	929	1,050	571	695	788	360	388	470	311	296	398	16	15	30
Dark Brown Soil Zone															
Poor grade	816	1,003	1,052	464	595	623	325	347	398	181	192	216	46	79	90
Medium grade	672	763	861	497	611	704	264	327	404	230	262	357	25	23	19
Good grade	609	803	846	572	759	788	315	410	454	285	258	373	24	14	18

Source: Census of Canada (Agriculture), 1951, 1961, 1966.

versely, more unimproved land is found on the poorer grades of land and thus is available for livestock grazing (Table 4).

Associated with the decrease in the number of farms, largely through a consolidation process, there has been a significant increase in the average size of farm. If reliance can be placed on the above data for the selected municipal units, it can be concluded that farms on the good grade of land have increased in acreage more than farms on poorer land. During the 1951-1966 period, farms on good land increased over one-third in size compared with slightly more than one-quarter for farms on poor and medium grades of land.

These changes, and changes to enterprises more suited to land capability, along with increased use of credit and better management practices, indicate progress toward the aim of attaining appropriate economic commercial farm units.

Livestock on Farms

Other important changes in the area of farm organization and management have been made in livestock enterprises. A comparison of livestock numbers on farms in 1951, 1961 and 1966 shows relatively large increases in cattle, mainly beef type, and a slight increase in the hog population. On the Brown Soil Zone, total cattle per farm on the poor grade of land numbered 93 and 116 in 1961 and 1966, respectively, compared to 35 in 1951. In the case of municipal units

with medium and good grades of land in this soil zone, 1961 numbers were 19 and 26, and 18 and 23 in 1966, compared with 8 and 9, respectively, in 1951.

Similar, though less pronounced, increases took place in the Dark Brown Soil Zone. In all the selected municipal units, cattle numbers doubled from 1951 to 1961. During 1951-1966, cattle numbers doubled on farms in the poor and medium grades of land and increased by two-thirds in the good grades of land. The expansion in cattle numbers has been abetted and supported by the extension and development of Prairie Farm Rehabilitation Administration (PFRA), Agricultural and Rural Development (ARDA) and provincial community pasture programs as well as individual farmer's projects for farm improvement and diversification. All these adjustments point to an appropriate use of lands normally considered to be sub-marginal for wheat and other grain crop use.

Economic Classification of Farms

Commercial farms, defined in the 1966 Census as having a total annual value of agricultural products sales of $2,500 or more, predominate in the prairie area, as throughout Saskatchewan. In the 1966 Census, all census farms, with the exception of institutional farms, were classified into eleven economic classes based upon the gross income received from sales of agricultural products during the previous twelve months (Table

Table 5

DISTRIBUTION OF FARMS, CLASSIFIED BY ECONOMIC CLASS FOR RURAL MUNICIPALITIES REPRESENTATIVE OF POOR, MEDIUM AND GOOD GRADES OF LAND, SASKATCHEWAN, 1966.

Number of farms	Brown Soil Zone			Dark Brown Soil Zone		
	Grade of Land					
	Poor	Medium	Good	Poor	Medium	Good
Census	131	276	416	199	264	262
Commercial	114	258	381	149	243	242
Small scale and Institutional	17	18	35	50	21	20
Total value of products sold	Percent					
$35,000 and over	—	—	1.7	—	1.1	1.5
25,000 to $34,999	3.8	0.7	1.4	—	3.0	5.0
15,000 to 24,999	11.5	12.7	14.2	7.0	17.8	26.7
10,000 to 14,999	17.6	26.8	24.8	11.6	22.3	24.4
7,500 to 9,999	20.6	18.9	18.8	19.6	13.3	13.8
5,000 to 7,499	19.8	18.1	16.3	15.1	20.5	11.8
3,750 to 4,999	5.3	7.6	9.4	8.5	5.3	8.4
2,500 to 3,749	8.4	8.7	5.0	13.1	8.7	0.8
Commercial farms	87.0	93.5	91.6	74.9	92.0	92.4
Total value of products sold						
$1,200 to $2,499	6.1	5.4	7.4	10.6	3.8	4.6
250 to 1,199	3.1	1.1	0.5	9.5	2.3	1.9
50 to 249	3.0	—	0.5	3.5	1.5	1.1
Small scale farms	13.0	6.5	8.4	25.1	8.0	7.6
Institutional farms, etc.	0.8	—	—	1.5	0.4	—

Source: Census of Canada (Agriculture), 1966

5). The classification of census farms into economic classes presents a measure of the productive size of the holdings. This system does not provide a stable long-term measure but only represents the situation with respect to production and prices of agricultural products during the 1965-66 census year.

Although there are a number of commercial farms in the relatively low income categories, the majority had incomes of $7,500 or more. The average annual gross farm income in 1966 (1965 crop year), for all census farms was $9,185, $9,226 and $10,846 for the representative municipal units in the Brown Soil Zone for poor, medium and good grades of land. This compares with $6,430, $10,439 and $12,702 for the respective municipal units and grades of land in the Dark Brown Soil Zone. These figures exceeded the 1961 figures (1960 crop year) by a range of 22 to 64 percent for the respective grades

of land. The average annual gross farm income on the poorest lands was $7,524 compared with $9,819 and $11,565 for the medium and good grades of land, respectively, in 1966. The provincial average was $8,536. Since these data pertain to a relatively recent normal crop year (1965), and since the price factor is constant for all grades of land, these income relationships may be considered representative of an average situation over a period of years. Other information presented in this article indicates the growing importance of income from sources other than wheat, especially on the poorest grades of land. This has helped to improve its income position relative to other grades of land.

Type of Farm

Further evidence of the diversification of farm enterprises is shown in Table 6 for commercial farms in the representative

246

Table 6

COMMERCIAL FARMS CLASSIFIED BY TYPE, REPRESENTATIVE MUNICIPAL UNITS,
SASKATCHEWAN, 1966.

| | Brown Soil Zone | | | Dark Brown Soil Zone | | |
| | | Grade of Land | | | | |
	Poor	Medium	Good	Poor	Medium	Good
			percent			
All commercial farms	100.0	100.0	100.0	100.0	100.0	100.0
Dairy	—	—	—	0.7	—	—
Cattle, hogs, sheep (excluding dairy farms)	46.4	1.9	5.2	18.1	3.3	1.2
Poultry	—	—	0.3	—	0.4	—
Wheat	36.0	95.7	91.1	61.7	92.6	90.5
Small grains (excluding wheat farms)	12.3	1.6	2.6	16.8	2.1	7.9
Field crops, other than small grains	—	—	—	—	—	—
Fruits and Vegetables	—	—	—	—	—	—
Forestry	—	—	—	—	—	—
Miscellaneous specialty	—	—	—	0.7	—	—
Mixed	5.3	0.8	0.8	2.0	1.6	0.4
Livestock combination	4.4	0.4	0.4	0.7	0.4	—
Field crops combination	0.9	—	0.3	—	—	—
Other combination	—	0.4	0.3	1.3	1.2	0.4
Number of commercial farms	114	258	381	149	243	242

Source: Census of Canada, Dominion Bureau of Statistics, 1966.

municipal units. These farms are classified in the 1966 Census according to the dominant source or kind of farm income.

During the last two decades, there has been a considerable change in farm type and sources of income. The farmer, generally had been following a one-crop wheat economy. For instance, almost one half of all commercial farms in the poor land of the Brown Soil Zone were classified as livestock (mainly beef cattle) farms in 1966. Here only about one-third of the farms were classified as wheat farms compared with 62 percent in the poor land of the Dark Brown Soil Zone and more than 90 percent in all other groups. In the prairie area there has been a polarization of farm type or specialization in terms of grain (mainly wheat) and grain-livestock (mainly cattle) farms. At the same time, a number of commercial mixed or small grain (excluding wheat) farmers continued to diversify on the poorer grades of land.

Tenure

Private ownership of land in the past was the main objective of most farmers in this region. It is usual now, however, to find the farm operator and his family operating owned lands and in many cases renting additional land. The practice of renting additional land is a common method used to enlarge the size of farm and increase the volume of business and income. The capital investment in land is thus restricted and permits the use of scarce capital for other inputs.

The group classed as part owner-part tenant made up 47, 38 and 41 percent of all farm operators on poor, medium and good grades of land, respectively, in 1966. They exceeded the number of owners on the poor lands and there were three owner-renters for every four owners on the other two grades of land. The trend to part owner-part tenant has been accelerating since 1951.

Off-farm Employment

A comparison of the 1966 and 1961 Census data indicates farm operators in the prairie area, especially those in the Dark Brown Soil Zone, are devoting an increasing amount of time to, and are obtaining more income from off-farm work. In the Brown

Soil Zone, in 1966 farmers reported 9.5, 25.7 and 28.8 days of off-farm work compared with 11.8, 25.1 and 18.4 days in 1961 for those on poor, medium and good grades of land, respectively. In the Dark Brown Soil Zone, farmers in 1966 reported 36.3, 27 and 40.2 days compared with 16.6, 14.6 and 18.5 days in 1961 for those on poor, medium and good grades of land, respectively.

These data suggest that off-farm work can be an important source of extra income under certain conditions and circumstances. In the prairie area of Saskatchewan, however, this source of income is generally restricted by the relative lack of service or industrial employment opportunities in rural areas and nearby urban centres, compared with other parts of Canada. Off-farm work, is, of course, restricted by on-farm labour requirements depending on the size and type of farm, age of operator, family sources of labour, and the commuting distance between the farm and the job site. In many cases, the opportunity for additional income from off-farm work is enhanced by residence off the farm during all or part of the year. Information from the farm organization studies of the Economics Branch show that off-farm employment was most prevalent among the operators of small-size farms. Between one-quarter and one-half of the operators of half-section (320 acres) farms on poor and medium grades of land had off-farm employment.

Age of Farm Operator

It is often assumed that farmers on the poorer grades of land are older than those on the better grades of land. An examination of data from both the 1961 and 1966 Census does not confirm this belief. Little difference is noted in the distribution of farmers by age groups in the representative rural municipalities. In each municipal unit, about one-half of all farmers were in the 35 to 55 age group.

The proportion of farmers in the younger age groups (up to 35 years) was 17 percent and those in the older age groups (60 years and over) was 20 percent in the representative rural municipalities of each soil zone.

There was little difference in the proportions in each age group between grades of land of each soil zone.

Residence of Farm Operators

The 1966 Census shows the distribution of farm operators according to the number of months of residence on the farm. On all grades of land, at least three-quarters of all farm operators resided on their farms for more than nine months during the year. Off-farm residence was most common on the medium and good grades of land of both soil zones, being 22.1 and 20.1 percent, and 27.6 and 28.6 percent of all operators, respectively. In the case of the poor grade of land, the percentage was 13.0 and 19.1 for the Brown and Dark Brown Soil Zones, respectively.

These data are confirmed by information assembled in the farm organization studies. These studies also indicate that operators of small-size farms (about one-half of the sample) reside in a town or city and thus are in a position to pick up other types of employment and additional income. Most off-farm employment was in service industries.

Income Comparisons

Data on land use, enterprise organization, labour, crop production, cultural practices, and livestock operations are available from farm surveys. These surveys were done on grain and grain cattle farms in the prairie area of Saskatchewan in 1956-1965 by the Economics Branch. The records also provide a financial picture of the capital investment and input-output relationships dealing with sources and amount of income and perquisites, cash operating expenses, depreciation, and measures of financial success.

In general, these data reveal relatively satisfactory returns to the typical farm operator and family for labour and capital after deductions are made for cash operating expenses and depreciation, taking into account a normal yield and farm price situation for the grades of land involved.

Table 7 provides, in summary form, the pertinent data[5] for one section (approxi-

Table 7

AVERAGE LAND USE, FARM CAPITAL, AND NET RETURNS, POOR AND MEDIUM GRADES OF LAND, ONE-SECTION GRAIN FARMS AND TWO-SECTION GRAIN-CATTLE FARMS, BROWN SOIL ZONE AND DARK BROWN SOIL ZONE, SASKATCHEWAN.

SIZE OF FARM—	ONE SECTION GRAIN FARMS								TWO SECTION GRAIN-CATTLE FARMS							
SOIL ZONE—	BROWN SOIL				DARK BROWN SOIL				BROWN SOIL				DARK BROWN SOIL			
GRADE of LAND	POOR		MEDIUM		POOR		MEDIUM		POOR		MEDIUM		POOR		MEDIUM	
	per farm	per improved acre	per farm	per improved acre	per farm	per improved acre	per farm	per improved acre	per farm	per improved acre	per farm	per improved acre	per farm	per improved acre	per farm	per improved acre
LAND USE (acres)																
Total Area	657		688		668		627		1,216		1,285		1,292		1,199	
Improved	553		583		556		598		594		1,016		951		1,049	
Wheat	239		276		255		330		231		418		332		515	
Summer-fallow	262		267		229		234		219		450		328		435	
FARM CAPITAL (dollars)																
Real Estate	21,669	39.18	21,970	37.69	20,561	36.98	37,750	63.13	28,150	47.39	42,190	41.53	36,644	38.53	60,242	57.53
Machinery	7,742	14.00	8,695	14.91	9,710	17.46	13,692	22.89	8,583	14.45	16,612	16.35	14,166	14.90	22,780	21.71
Livestock	647	1.17	967	1.66	946	1.70	365	0.61	10,185	17.15	8,264	8.13	6,823	7.17	8,089	7.71
Total	30,058	54.35	31,632	54.26	31,217	56.14	51,807	86.63	46,918	78.99	67,066	66.01	57,663	60.60	91,111	86.85
NET RETURNS (dollars)																
Return to operator for labour and capital[1]	3,385	6.12	5,682	9.74	4,151	7.47	4,970	8.31	5,807	9.78	10,149	9.99	7,527	7.92	9,164	8.73
Return to operator labour[2]	1,882	3.40	4,100	7.03	2,590	4.66	2,380	3.98	3,461	5.83	6,796	6.69	4,645	4.88	4,608	4.39
Return to operator capital[3]	24	0.04	2,529	4.34	887	1.60	1,054	1.76	2,109	3.55	6,288	6.19	3,734	3.93	4,462	4.25
Study Year	1963		1961		1962		1964		1963		1961		1962		1964	
Study Area	Gull Lake Maple Creek		Swift Current		Asquith Delisle		Davidson		Gull Lake		Swift Current		Asquith Delisle		Davidson	

[1] Crop sales were calculated on the basis of average wheat yields for the 1947-1966 period assuming land use, grades and prices as in the study year.

[2] The allowance for interest on farm capital was calculated at the rate of 5 percent; management allowance for the operator at 2 percent of farm capital; and operator's labour at $2,220, $2,640, $2,760,$2,880 and $3,000 in 1961, 1962, 1963 1964 and 1965 respectively. The operator's labour was based on the current cost of hired farm labour.

Source: *Organizational Characteristics of Grain Farms in the Prairie Provinces*, Pub. No. 69/9, Economics Branch, Canada Department of Agriculture, 1956-1965.

mately 640 acres) grain farms and two-section (approximately 1,280 acres) grain-cattle farms for the grades of land included in the series of farm organization studies during the period.

Returns to labour and capital for grain farms were $6.12 and $9.74 per improved acre on poor and medium grades of land in the Brown Soil Zone. This compares with $7.47, $8.31 and $11.76 on poor, medium and good grades[6] of land in the Dark Brown Soil Zone. Similarly for grain-cattle farms, returns to labour and capital were $9.78 and $9.99 on poor and medium grades of land in the Brown Soil Zone compared with $7.92 and $8.73 in the Dark Brown Soil Zone.

Farmers on medium grade land or on poor grade Dark Brown soil apparently gained little extra return by adding a beef enterprise to their farm. On the other hand, farmers on poor grade Brown Soil Zone land increased their income from $6.12 to $9.78 per improved acre with the addition of an economic beef operation.

Surprisingly, grain farms on the medium quality Brown Soil Zone produced more profit per acre than similar farms on the Dark Brown Soil Zone. Among the reasons for this is that these farms are lower in capital value per acre than farms on the Dark Brown soil, and hence, carry lower interest charges. Since costs of production are lower, farmers on medium grade land on the Brown Soil Zone realized higher returns to labour and investment than farmers on the Dark Brown Soil Zone. This is true for grain farmers and cattle-grain farmers alike.

A study of 50 commercial wheat farms in the Elbow-Hawarden area (loam soils)[7] of the Dark Brown Soil Zone in 1968 indicated returns to the operator and family for labour and capital comparable with returns obtained in previous years. Assuming 1963-1967 average wheat yields and grades and 1968 wheat prices, returns to the operator and family for labour and capital on farms with 480 to 640 improved acres averaged $5,107 or $8.63 per improved acre compared with $10,327 or $9.54 per improved acre for farms with 800 to 1,440 improved acres and

$21,466 or $8.64 per improved acre for farms with 2,000 to 3,200 improved acres. Similar information for grain-cattle farms in the prairie area since 1966 is not available.

SUMMARY AND CONCLUSIONS

This appraisal points to a number of structural factors and production adjustments that have changed the pattern of farming in the prairie area of Saskatchewan in the 1951 to 1966 period. These phases have a meaningful relationship with the broad farmer goals of net income maximization and attainment of a level of living comparable with entrepreneurs outside the agricultural industry. It must not be concluded, however, that the process of adjustment was fully satisfactory or widespread in all cases during the period. The analysis gives a sense of direction to the kinds of adjustments and to the quickening of pace required to meet changing socio-economic and market conditions since 1966.

Important highlights include fewer but larger commercial farm units with concentration on specialization, economies of scale and increased volume of business; a polarization towards two main specialty farm types, i.e. grain (mainly wheat) on the better grades of land and grain cattle on the poorer grades of land, together with some enterprise diversification on mixed farms; a trend to a decreasing proportion of improved land devoted to field cash crops and a slowly increasing proportion used for improved pasture and tame hay; an increasing emphasis on livestock production, mainly beef cattle when it is possible to dovetail the enterprise effectively with an adequate feed and water base and labour supply; a process of farm enlargement effected primarily through the renting of additional land and a part owner-part tenant type of land tenure; and improved labour utilization and total income through increased off-farm employment during the slack farming season. These basic changes have been made possible through increased adoption of new technology and mechanization and improved farm management knowledge and ability.

250

REFERENCES

[1] The selection of representative municipal units was facilitated by their inclusion in a series of change in farm organization studies that were underway from 1956 to 1968. These studies in Saskatchewan represent farms of grain and grain-cattle types subdivided into small, medium and large size categories. Samples of farms are included from light, medium and heavy textured soils of each soil zone.

[2] During the period from 1937 to 1951, the Economics Branch carried out a program of an economic classification of land in Saskatchewan and covered all of the Brown and Dark Brown Soil Zones (39.1 million acres). Each quarter section was rated on a comparative budgetary basis in terms of its wheat-yielding and net income-earning capacity. The classification system included five classes ranging from Land Class 1 (submarginal for wheat production) to Land Class V (excellent wheat land). Coloured land class maps and statistical summaries are available from the Economics Branch.

[3] There was a slight increase in the proportion of improved land under crops in the rural municipalities representing the medium and good grades of land in the Dark Brown Soil Zone.

[4] A very moderate increase in the proportion of improved land and acres per farm devoted to wheat was noted in both municipalities representing a medium grade land.

[5] The data discussed here deal only with one selected size of farms for each product type. Similar information for both smaller and larger farms is found in the two reports cited at the foot of the tables.

[6] Farm capital on grain farms in the Rosetown-Elrose area (good grade of land) averaged $90,742 compared with $31,217 and $51,807 on poor and medium grades of land. The major share of the capital investment was, of course, in real estate, and averaged $124.01 per improved acre in the Rosetown-Elrose area compared with $36,98 and $63.13 in the poor and medium grades of land, respectively.

[7] *Costs and Returns, Commercial Wheat Farms, Prairie Provinces, 1968.* B. Middlemiss and M. Ragush, Economics Branch, Canada Department of Agriculture, 69/12.

Irrigation Development in Alberta

Stewart Raby

Commenting upon "the true arid district which occupies most of the country along the South Saskatchewan," Captain John Palliser, the well-known British explorer who visited the area between 1857 and 1861, noted that "this . . . district . . . can never be of much use to us as a possession."[1] Today almost all of the 990,000 acres under irrigation in Alberta (Table 1) are located in that river basin. This acreage represents about 4 percent of the province's improved arable lands, yet is responsible for nearly 20 percent of its gross farming revenue, including the proceeds from a wide variety of specialty crops which cannot be grown economically in Alberta under dryland conditions.[2]

Since the first major project was undertaken at the turn of the century, a rapid expansion of storage and distribution facilities has resulted in the present allocation to Alberta of over 2.25 million acre-feet of water each year for eventual use in its irrigation projects from the South Saskatchewan River system. Even when allowance is made for the regulatory effects of upstream hydroelectric plants on the Bow River, this total is equal to almost one-half of the river's discharge during a dry year at Saskatoon in Saskatchewan.

In contrast with the scale of existing facilities, both the areas actually irrigated and the quantities of water diverted for that purpose have shown wide variations, depending chiefly upon the climatic conditions experienced in a particular year and area. Table 2 illustrates the range of recent fluctuations in these features in the irrigation districts located in the Oldman River basin of southwestern Alberta. In 1949, a particularly dry year when irrigation assumed widespread popularity, a record 800,000 acres were irrigated. About one-third of this figure is more usual.

Small-scale private schemes were constructed towards the end of the nineteenth century in southern Alberta, and, by 1899, 112 projects were serving nearly 80,000 acres of land.[3] Railway land grants formed the basis for much of the large-scale commercial speculation which continued until the end of World War I. After that time, several community projects were set up as irrigation districts financed by bonds guaranteed by the provincial government. Both federal and provincial agencies have since come to play increasingly active roles in the initiation, remodelling, and management of projects. Past experience has, however, clearly shown that inadequate provision for preliminary surveys of the physical resources, of the economic conditions and trends, and of the social factors involved has resulted in the misdirection of considerable heavy investment from both private and public sources. Unfortunately, post-war policies are exhibiting similar tendencies.

As a framework for discussion of these contentions, the more important irrigation districts and developments can be grouped roughly into four categories.[4] The projects in the Lethbridge area are aligned along the Lethbridge-Taber-Medicine Hat axis, and lie to the south of the Oldman and South Saskatchewan rivers. Proposed extensions to this first group spread southwards to the Milk River and Pakowki Lake, the latter usually an area of internal drainage. The two projects along the lower Bow River extend between the Oldman and Red Deer rivers, while the western districts comprise all those schemes transected by or to the

● Stewart Raby is Assistant Professor, Department of Geography, University of Windsor.

Reprinted from the Canadian Geographer, Vol. IX (1), 1965, pp. 31-40, with permission of the author and the editor.

Figure 1

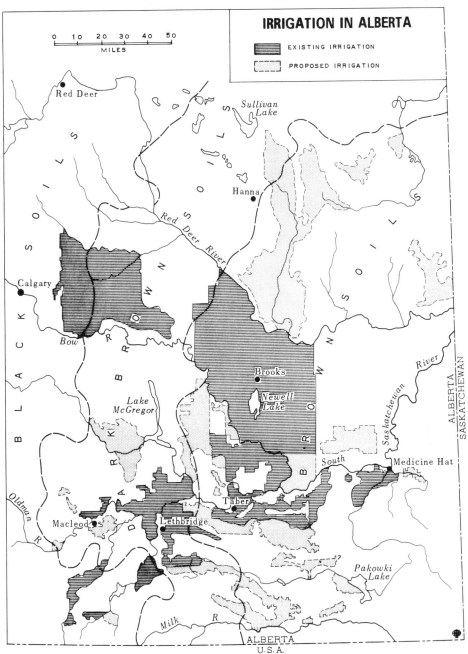

IRRIGATION IN ALBERTA

EXISTING IRRIGATION

PROPOSED IRRIGATION

SOURCE: AFTER S.DUBETZ ET AL. ,GROWING IRRIGATED CROPS IN SOUTHERN ALBERTA
(CANADA, DEPT. OF AGRICULTURE,PUB.NO.1152 ,OTTAWA,1962)

Table 1

MAJOR EXISTING AND POTENTIAL IRRIGATION PROJECTS IN ALBERTA

Source of water	Project and year begun*	Approx. irrigated area (acres) Present	Approx. irrigated area (acres) Ultimate
Bow River	Eastern I.D. (1914)	200,000	281,000
	Western I.D. (1908)	42,000	50,000
	Bow River Development (1918)	123,000	240,000
	Redcliff-Ronalane Project	—	30,000†
St. Mary River	United I.D. (1921)	34,000	34,000
	Mountain View I.D. (1925)	3,700	3,700
Belly River	Leavitt I.D. (1943)	4,600	4,600
	Aetna I.D. (1945)	8,300	8,300
Oldman River	Lethbridge Northern I.D. (1922)	96,100	96,100
	Macleod I.D. (1948)	3,000	3,000
St. Mary, Belly, and Waterton rivers	St. Mary River Development (1901)	204,000	510,000
	Taber I.D. (1915)	40,000	40,000
	Raymond I.D. (1925)	19,000	19,000
	Magrath I.D. (1926)	7,500	7,500
Red Deer and Clearwater rivers	Red Deer River Diversion Project	—	350,000†

Sources: Approximate present acreages from Alberta, Dept. of Agriculture, *Alberta Farm Guide 1963* (Edmonton, 1963), p. 22. Ultimate areas chiefly derived from Canada, Dept. of Agriculture, *Full Development Possibilities in the Saskatchewan River Basin* (Regina, 1952); and from information supplied by the individual districts and developments.

*I.D. = Irrigation District.

†Denotes tentative estimates reported in Gunderson, H., Halmrast Serves Death Notice on Old Redcliff-Ronalane Scheme, *Medicine Hat News*, May 23, 1962; and in Hawkins, S. H., Proposed Red Deer River Diversion Project (unpub. prelim. rept. to govt. of Alberta, Calgary, 1947). Only the Clearwater does not drain to the South Saskatchewan River.

Table 2

DIVERSIONS OF WATER FOR IRRIGATION AND AREAS IRRIGATED IN MAJOR PROJECTS IN THE OLDMAN RIVER BASIN

Year		St. Mary River Development	Lethbridge Northern Irrigation District	United Irrigation District
1957	Diversion*	327,670	114,240	26,950
	Areas irrigated†	169,900	52,110	15,886
1958	Diversion	334,400	86,860	22,380
	Areas irrigated	178,000	52,220	11,032
1959	Diversion	288,850	88,900	25,490
	Areas irrigated	174,000	46,760	17,761
1960	Diversion	371,340	117,380	26,560
	Areas irrigated	195,000	71,310	16,536
1961	Diversion	521,350	165,000	33,840
	Areas irrigated	213,000	73,940	23,095

Sources: Information supplied by the individual districts and development regarding the acreages irrigated, and by P.F.R.A. (Lethbridge) regarding diversion data.

*Diversions in acre-feet.

†Areas irrigated in acres.

west of the Dark Brown/Black soil group boundary. The Western Irrigation District, immediately to the east of Calgary, dominates this latter group. As for the potential schemes, these consist of the Redcliff-Ronalane Project to the northwest of Medicine Hat and the Red Deer Diversion Project which accounts for all of the proposed irrigated area lying to the north of the Red Deer River. Like the eastern part of the Lethbridge group and the projects along the lower Bow River, both of these will be located in the drought-prone Brown soil zone of east-central and southeastern Alberta.

THE LETHBRIDGE AREA

Five active projects centre upon the city of Lethbridge, all of them drawing their water from the Oldman River or its tributaries upon which storage facilities for irrigation supply are currently being extended. The St. Mary River Development has as its nucleus one of the early commercial Canadian Pacific Railway ventures (the Alberta Railway and Irrigation Project), which was sold to the Alberta government in 1946 and is now administered by a Crown corporation. An agreement signed in 1950 provided for the province to make use of water in the development as soon as the federal Prairie Farm Rehabilitation Administration (P.F.R.A.) could make it available. The urgency with which this scheme was pressed forward stems partially from the recommendations published in the Meek Report (1942): "That Canada should construct at an early date the necessary irrigation works to protect by beneficial use its share of the St. Mary and Milk Rivers."[5] The St. Mary is an important tributary of the Oldman and rises in the state of Montana. As in so much of Alberta's irrigation development, and in this case aggravated by the divorce of design and operating agencies, insufficient preliminary survey and research in this area resulted in failure to carry out a detailed classification of irrigable lands, lack of study of potential markets and farm economics, wasteful over-building, and unsatisfactory consultations on a local level.[6]

Ironically, Burchill, in discussing earlier developments, notes that "much of the construction has been admittedly flimsy, and durability has been regularly sacrificed to cheapness,"[7] while similar strictures can be levelled against maintenance standards.

The St. Mary River Development lies between the St. Mary, Oldman, and Milk rivers, and wholesales water to three small irrigation districts which also belong to this first group. These, the Magrath, Raymond, and Taber schemes, were erected as quasi-municipal districts after the passage of the Irrigation Districts Act of 1915. The Lethbridge Northern Irrigation District, which also has its bonds backed by the provincial government, had to be taken over soon after its initiation.[8] It continues to be administered by a trustee of the Alberta government.

Drought intensities and frequencies in this group of projects are such that farmers are not prone to avoid costly and onerous irrigation in the hope that conditions will permit dryland farming. Reservoir storage was comparatively easy to construct, gravity diversion possible, and soils reasonably amenable to irrigation. Colonization problems varied considerably in degree. Skilled Mormon irrigators, a succession of dry years, and the successful introduction of alfalfa on the Alberta Railway and Irrigation Company's lands contrasted with the problems facing the Lethbridge Northern Irrigation District across the Oldman River. There, compulsory reduction of existing dryland farms, "trial leases," and firm supervision coupled with realistic land payments, easy loans, and diversification led eventually to relatively sound development, although without provincial write-offs and advances to cover high construction and operating costs progress at that time would not have been possible. Of particular importance, the commission appointed in 1936 to inquire into various phases of irrigation development in Alberta had maintained that the price of land with water should be cut to an amount which could reasonably be supported by its productivity, and these findings were given legislative backing in 1938. The commission, which set the average price for the land

and water rights combined at approximately one-quarter of the 1926 contract prices, had "been made fully aware that irrigation authorities now agree that the full capital cost of any irrigation project should not be charged up to the lands immediately benefited. The conversion of a non-productive arid area into lands intensively farmed benefits not only the irrigation farmer but also the community, the province and the Dominion as well as many private enterprises such as railways and factories."[9] Costs were much lower in the Taber Irrigation District to the east of the Lethbridge Northern. Experienced irrigators were already settled there when the district commenced operation in 1921. Specialty crops soon came into production to supply local industrial markets which now draw upon surrounding districts as well. This particular project has since been rated a commercial success, a situation unique in the history of irrigation in Alberta. Also, most of the farmers in the district retained their holdings intact both in the change-over period and during the 'thirties.

Although grain crops have maintained an important role throughout the group as a whole, a measure of stability has been attained through gradual diversification into sugar beet, vegetables, fruits, and oil seeds. Livestock have a significant position in residue and by-product consumption, as well as in the provision of manure, a necessity for maintaining the fertility of irrigated land. At present population and income levels, however, it is unlikely that available markets are capable of supporting more than 70,000 acres of sugar beets canning crops, and fresh vegetables in Alberta, and any justification for the expansion of irrigation must therefore be based upon other enterprises with their attendant lower net returns per irrigated acre.[10]

PROJECTS ALONG THE LOWER BOW RIVER

Centring upon Brooks, the Eastern Irrigation District, Canada's largest operating district, was financed and colonized by the C.P.R. Company which accepted its heavy operating losses and eventually turned it over to its water users on particularly advantageous terms.[11] The farming activities developed there are being used as a model for planning in the physically comparable, federally-operated central block of the Bow River Development, also dependent upon the over-appropriated waters of the Bow River for its supplies. The Bow River Development, located between the Bow and Oldman rivers, began as a strictly commercial venture (without the C.P.R.'s incentive of increased freight traffic) under the aegis of the Canada Land and Irrigation Company. The project led a precarious existence, and in 1950 was purchased by the federal government which remained as owner and operator of a substantial portion through P.F.R.A. The western block of the Bow River Development is being actively developed, though with much hardship, by the province through a Crown corporation.

Despite the lack of adequate markets for specialty crops, the Eastern Irrigation District has enjoyed reasonable success. Grain and livestock farms dominate, the latter enterprises making use of the area's rangeland which, owing to aridity, cannot be used for competitive dryland crop enterprises. It is one of the more livestock-oriented districts in Alberta, and mixed farms are surprisingly little developed. Owing to a lack of surplus grazing land resources and to modification of the especially favourable post-war returns from grain crops, there has been a shift towards the feeder aspect of livestock production. Although the strength of livestock farming in the Eastern Irrigation District deserves stress, average labour incomes fall well below those in the Lethbridge group. Alfalfa and tame pasture constitute over one-half of the total irrigated acreage, but wheat makes up most of the remainder. Low charges for irrigated land, experienced operators, and careful organization in the pioneering phase aided a colonization project in the Rolling Hills district in the southeast of the Eastern Irrigation District, though it concentrated upon unspecialized crops. Perhaps most important, it was settled at a time when prices for agricultural products were rising more rapidly than farming costs (in 1939-40).[13]

Regarding the Eastern Irrigation District as a whole, Hargrave comments that:

> . . . as dry range land devoted to ranching, the 1,500,000 acres in and adjacent to this project would provide grazing for 35,000 cattle. It would perhaps support 200 to 3,000 people and both the cattle and the people would have real problems during hard winters and dry summers such as 1961. There are 200,000 acres of irrigated land in this district today and this irrigated area with the million and one-quarter acres of range land now supports 1,400 irrigated farms and makes it possible for 14,000 people to obtain a living in the area.[14]

Unfortunately, the precursor of the Bow River Development suffered cumulative difficulties from low yields and structural deterioration, together with problems of seepage and saline soils. The effects of saline and alkali soils are felt over much of southern Alberta as a whole, though the particularly difficult solonetzic soils, with their hard, impervious sub-soil layer, are concentrated in a belt extending from Medicine Hat 375 miles northwards through Hanna into the North Saskatchewan River basin.[15] The Bow River Development's economy lacked the scale of both the specialty crop and the livestock elements mentioned above, the irrigators producing less of the goods which, unlike grains, are more profitable under their conditions of farming. Farmers with mixed enterprises in the Vauxhall area are, however, tending to reduce their dependence upon markets for grain and to find an outlet for grain and hay through livestock feeding. This procedure allows more satisfactory adjustment to seasonal price changes, though requiring substantial additional expenditure and investment.[16] The original project, now rebuilt by P.F.R.A., is undergoing a programme of rehabilitation and resettlement initiated originally in part to accommodate settlers from dried out sections of Alberta and Saskatchewan. Community pastures have been created in the vacated lands, while P.F.R.A. irrigated community pastures within the Bow River Development itself

are currently being improved in efficiency and carrying capacity.

The irrigation of the western block, now hastily being colonized by the province, was begun soon after the initiation of the St. Mary River Development. Through dry-land adjustment the landholdings were reorganized into large, apparently efficient units utilizing Crown grazing lands for supplementary livestock production,[17] but after only scant revision of pre-war surveys and analyses, the area was subsequently considered capable of sustaining a full-scale irrigation economy. A competent reclassification of land, made after the P.F.R.A. water-supply works had been constructed, reduced the estimated irrigable area by two-and-one-half times to 25,600 acres. As some 60 percent of the land in the western block is Crown-owned, the problems of colonization have also to be surmounted. It is certainly the viewpoint of the farmers involved that regional stability can be achieved both in this project area and in the potential developments without the necessity for taking water to every quarter-section.[18] Unfortunately, the successful integration of irrigated land and rangeland is severely constrained by complexities of land ownership, co-operation, and political conflict.[19]

WESTERN DISTRICTS

The Western, United, and Mountain View irrigation districts experience drought conditions with less intensity and frequency than do those in the area discussed above.[20] Although irrigation facilities were developed on a large scale, the percentage areas actually irrigated are relatively small in most years, as fair yields can be obtained without the need for irrigation. Land productivity is not, therefore, significantly increased by irrigation practices.

In 1944, after thirty years of operation, the western block of the C.P.R. land grant was transferred, together with a cash bonus, to the farmers to become the Western Irrigation District. It is a conspicuous example of project misplacement, and irrigation has made but slow progress there. Initially,

220,000 acres were classified as irrigable, but water allocations made by the Prairie Provinces Water Board now cater only for a more realistic 50,000 acres. The area irrigated attained its maximum in 1937, when water was supplied to some 56,000 acres, and in some recent years irrigation has been absent from this district. In contrast, the area irrigated in the Eastern Irrigation District, for which allocations provide for 220,000 acres, has consistently exceeded 165,000 acres since 1942, and continues its trend to increase.

Livestock attained a greater importance in the United Irrigation District than in the Western, because satisfactory use can be made of hay crops and extensive grazing in the foothills of the Rocky Mountains. Like the Mountain View Irrigation District, it began as an inter-war scheme independent of government, except for the supervisory provisions of the Irrigation Districts Act, but in 1939-40 its farmers failed by almost $200 per farm to cover current operating and living costs and the costs of maintaining farm capital.[21] The province accepted responsibility for the district's debts in 1941 and now manages it through an official trustee.

POTENTIAL PROJECTS

There is strong pressure within the province of Alberta for the construction of two additional schemes. One, the Redcliff-Ronalane Project, is an easterly extension of the Bow River Development whose return-flow could be used in an irrigation and stockwatering project covering 30,000 acres.[22] The second, the Red Deer River Diversion Project, would use both Red Deer and diverted water from more northerly sources in an attempt to aid the economy of a depleted "Special Area" in east-central Alberta at a cost likely to exceed $100 million (excluding operating expenses).[23]

PROSPECT

Irrigation facilities in Alberta have been seriously under-utilized in districts where successful dryland farming proved possible. In the past, it seemed that unless high-return crop production could be developed, irriga-

tion farming suffered heavy competition from the less costly growth of grain on non-irrigated lands. The integration of irrigated fodder crops and pasture with proximate rangeland has offered a partial solution to the dilemma of restricted specialty crop markets and low relative returns from cereal production. Wheat still comprises almost half of the cash crops produced each year on the province's irrigated lands, although by 1961 nearly three-fifths of those lands were under feed crops.

Regarding the expansion of irrigation in the province, if that is to be pursued, livestock production is evidently "the key enterprise in successful and profitable irrigation agriculture."[24] Over twenty years ago, the Meek Report concluded that "the exclusive growing of wheat or any other single crop on irrigated land is neither sound agriculture nor economics, whereas a system of irrigation farming based mainly on feed production and livestock feeding offers, according to evidence submitted to the Committee by successful irrigation farmers, stability and security."[25] Given long-term satisfactory demands for livestock in Canada and abroad, the consolidation and extension of existing projects might be organized with particular concern for feedstuff-stockwater requirements. Small, less spectacular schemes along the design of those initiated throughout the prairies since 1935 by P.F.R.A. cannot be depended upon in all years, although their potential stabilizing effects were well illustrated in 1961 when the severe drought that was experienced throughout most of east-central and southeastern Alberta forced operators to purchase distant feed supplies or sell their stock. Assistance with rail freight and trucking costs had to be provided by the two governments.

All development must necessarily be based upon a realistic appreciation of all the factors involved in what are often multiple-purpose projects, as well as upon an understanding of the scope for alternatives in all decisions regarding both the necessary investment and the uses of the resources involved.[26] Certainly, as far as the water resources themselves are concerned, emphasis is changing from regard for water's function in promoting maximum occupancy of land towards some idea of an optimum among a variety of competing and complementary water- and water-related services.[27] The amount of government subsidy which can be justified for irrigation development has long been a controversial issue in both Canada and the United States. A clarification of the objectives of resource development along the lines of those made by the President's Water Resources Policy Commission is desirable:

Federal water resources policy merely needs strengthening at two points: (a) all private beneficiaries should be required to return the costs incurred to serve them, and (b) social benefits and national interests should be clearly differentiated from those for which reimbursement would be required, and appropriations be made accordingly after full Congressional discussion. The charge of subsidy would then be invalid. Where the public interest is clearly established, public expenditures to promote it cannot properly be regarded as subsidies.[28]

It remains a perplexing political problem as to how far the process of regional depletion, with its high social costs, is to be tolerated by society. In such a situation, regional development to achieve stability and growth emerges as a valid objective which may be attained through provision of irrigation facilities.

REFERENCES

[1] Palliser, J., Journals, *Detailed Reports and Observations Relative to the Exploration of . . . British North America* (London, 1863), p. 246.

[2] Dubetz, S., *et al.*, *Growing Irrigated Crops in Southern Alberta* (Canada, Dept. of Agriculture, pub. no. 1152, Ottawa, 1962), p. 5.

[3] Boan J., *The Economic Significance of Water Requirements Relative to Human Activities and Needs in the Saskatchewan River Basin* (Canada, Dept. of Agriculture, Ottawa, 1961), p. 18.

[4] A similar classification was suggested by Spence, C. C., An Appraisal of Irrigation Development in the Province of Alberta (Paper to the Canadian Agricultural Economics Society, Winnipeg, June 1951, unpublished), 9 pp.

[5] St. Mary and Milk Rivers Water Development Committee, *Report on Further Storage and Irrigation Works Required to Utilize Fully Canada's Share of International Streams in Southern Alberta*, Ottawa, 1942, p. 2.

[6] See *Report of the Irrigation Study Committee to the Government of Alberta* (Calgary, 1958), pp. 8-10.

[7] Burchill, C. S., *The Development of Irrigation in Alberta* (Dominion Economics Division, University of Alberta, Edmonton, 1949), p. 10.

[8] *Ibid.*, pp. 11ff.

[9] *Report of the Commission Appointed in 1936 to Inquire into Various Phases of Irrigation Development in Alberta* (Edmonton, 1937).

[10] See Haase, G., *What is Required in Irrigation Economics?* (Lethbridge Research Station Lethbridge, 1962), 4 pp.; Jorgensen, L. G., *et al.*, "Specialty Crops in Alberta," *Agric. Inst. Rev.*, Nov.-Dec. 1962, pp. 9-10, 46; and Porter, K. D., and B. J. McBain, *Irrigated Specialty Crops in Alberta* (Alberta, Dept. of Agriculture Edmonton 1959), 47 pp.

[11] Dunsmore L. K., "Historical Development of the Eastern Irrigation District," *Econ. Annalist*, June 1950, pp. 55-60 and "Eastern Irrigation District," *The History of the E.I.D.* (Brooks, Alta., 1960), 64 pp.

[12] See Kirk, D. W., *The Bow River Irrigation Project* (Canada, Dept. of Agriculture, Regina. 1955), 63 pp.

[13] Burchill, *Development of Irrigation*, pp. 18-20.

[14] *Western Canada Reclamation* (Western Canada Reclamation Assoc., Redcliff, Alta., June 1963), p. 4.

[15] Alberta, Dept. of Agriculture, *Alberta Farm Guide 1963* (Edmonton, 1963), p. 19.

[16] Porter, K. D., and B. J. McBain, *Vauxhall Farm Management Study, 1956-1959* (Alberta, Dept. of Agriculture, Edmonton, n.d.), 36 pp.

[17] For an examination of this trend in a proposed irrigation district in east-central Alberta, see Darcovitch, W., *Appraisal of Dryland Farming in the Special Areas of Alberta* (Canada, Dept. of Agriculture, Ottawa, 1954), 37 pp.

[18] *Report of the Irrigation Study Committee*, pp. 20-22.

[19] Huffman has indicated various methods of integration and their potentialities for achieving stability and flexibility in production. See Huffman, R. E., *Irrigation Development and Public Water Policy* (New York, 1953), pp. 122-48.

[20] Comparative climatic data are given in Laycock, A. H., *The Climate of the Bow River Project Area: Regional Comparisons and Local Variations* (mimeo, n.d.,), 5 pp.

[21] Spence, C. C., *et al.*, *Farming in the Irrigation Districts of Alberta* (Canada, Dept. of Agriculture, publ. no. 793, Ottawa, 1947), p. 61.

[22] Gunderson, H., "Halmrast Serves Death Notice on Old Radcliffe-Ronalane Scheme," *Medicine Hat News,* May 23, 1962.

[23] Raby, S., "The Red Deer River Diversion Project." *Swansea Geographer,* II, spring (1964), 116-28.

[24] *Western Canada Reclamation.*

[25] St. Mary and Milk Rivers Water Development Committee, *Report*, p. 59. See also, Canada, Dept. of Agriculture, P.F.R.A., *Land Classification of the Bow River Project* (Regina, 1960), 60 pp.

[26] See Raby, S., "Alberta and the Prairie Provinces Water Board," *Can. Geog.,* VIII, 2 (1964), 85-91.

[27] See Fox, I. K., and L. E. Craine, "Organizational Arrangements for Water Development," *Natural Resources Journal*, II, 1 (April 1962).

[28] United States President's Water Resources Policy Commission, Report, vol. I, *A Water Policy for the American People* (Washington, 1950), p. 78.

Range Types and the Distribution of Livestock in the Southern Interior of British Columbia

Thomas R. Weir

The term "range" is commonly used to indicate the area grazed by cattle. In British Columbia, the Forest Service administers 10,000,000 acres of range land, of which some 8,000,000 acres lie in the southern Interior Plateau. In addition, nearly 3,000,000 acres of grassland range are in private ownership. The remainder is leased. Range land is a basic factor in ranching economy and around it is organized the ranch unit. The prime distinction between livestock ranching and farming is dependence on the range in the case of the latter. Knowledge of the range is an important element in understanding livestock distribution patterns.

Ranges may be classified either according to the dominant vegetation associations, or according to the season of use. The latter classification is the one employed by the rancher and the total grazing area falls into four types of range: spring, summer, autumn, and winter range. However, this does not include all lands within the ranching area, as some are cultivated or reserved for hay while some are useless for grazing (Figure 1).

● Thomas R. Weir *is former Professor of Geography, Department of Geography, University of Manitoba.*

Reprinted from *Ranching in the Southern Interior of British Columbia*, Geographical Branch, Department of Mines and Technical Surveys Memoir 4, (Ottawa: Queen's Printer, rev. ed. 1964), pp. 53-78, with permission of the author and Information Canada.

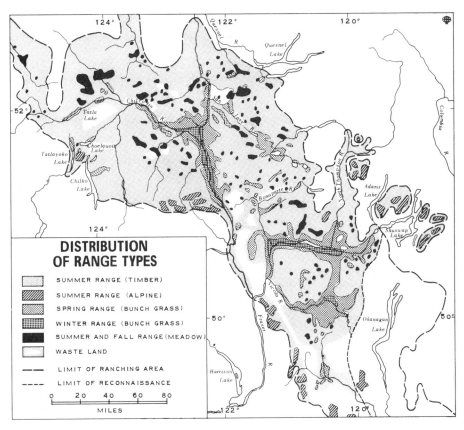

Figure 1

The term "range" implies both physical and human-use elements. The four physical elements — topography, native vegetation, climate, and surface water — determine the use potentialities of a specific range, but it rests with man to decide how and when a given area of potential grazing land shall be used, if at all. The economic factor has gradually assumed such importance in some range areas that competing uses have led to the replacement of grazing. This is particularly true in the Okanagan Valley, along the flood plain of the Thompson River from Kamloops to Chase, and in the vicinity of Ashcroft, where fruit and truck crops are specialties. Recreation in some districts also encroaches on the range.

Range conditions from year to year are by no means static. It would appear that when weather conditions favorable to range growth occur in the north they are accompanied by a moisture deficiency in the south and vice versa. On the average, May 1 is generally the "turn-out" date.

The several elements, both physical and cultural, referred to above, combine to produce four categories or types of range, and the characteristics and distribution of each type are described.

SPRING RANGE

Characteristics

Spring range is grazed immediately after the snow disappears. The vegetation comprising spring range is usually grass that cures on the ground, that is, bunchgrass or, in the Cariboo-Chilcotin region, Kentucky blue-grass. Analyses of the principal species of the grassland ranges show a high nutrient content in the early growth stages marked

261

by a falling off of the phosphorus and protein content when the plant matures. Except in the deep valleys, spring range usually lies on south-facing slopes either along valley sides or on ridges of moraine. Here the first direct rays of the sun produce their maximum warming effect and coarse gravel ridges provide good drainage favoring the growth of "hard" grasses. Usually spring ranges have ample water for stock because minor depressions are filled with run-off at that season of the year.

Spring range is ready for grazing from one to two months before the summer or timber range. When the growth is 5 or 6 inches the range is ready for grazing; in the Thompson Valley this occurs from about April 19 on the lower grasslands to May 13 on the upper grasslands. Early spring range turn-out occurs from early April to the end of May, according to locality and depending upon weather conditions.

Spring range conforms to elevation which, in turn, is related to the length of the growing season. As a result it accords with the zonal distribution of soil and vegetation types. This leads to a further division into early and late spring range, the former conforming to the lower and middle grassland zones, and the latter to upper grasslands (2,850 to 3,000 feet). In the Cariboo-Chilcotin region, because the area below 3,000 feet is small, spring range is restricted to a few localities, but in the Thompson-Nicola region, where low elevations are more extensive, the area in spring range amounts to nearly two-fifths of the region.[1]

Distribution

Conforming to major topographic features, the pattern of spring range follows the major valleys and lowlands (Figure 1) to an upper limit of $\pm$ 3,000 feet. Thus the middle Fraser Valley and the lower Chilcotin Valley, especially at the confluence of the rivers, have extensive areas in spring range. However, where valley walls are steep, as along the Fraser River between Lytton and Big Bar Creek or north of Chimney Creek, the area in spring range is greatly restricted.

Much less extensive, but of great local importance, are certain sections of spring range following some of the minor depressions on the plateau surface. The most conspicuous is the valley of the San Jose River from Lac la Hache to its confluence with the Fraser River. Much of this valley lies at an elevation of 2,700 feet, where open side hills and dry meadows, enclosed by light stands of aspen and fir, constitute limited range for spring grazing. Similar openings and exposed side hills conforming to minor drainage features are typical of the Chimney Creek area between the San Jose and Fraser rivers and in the vicinity of Tatlayoko, Tatla, and Choelquoit lakes in the southwest Chilcotin region. Around a tiny string of lakes east of Canoe Creek and along Bonaparte River, small hard-grass areas are grazed after the snow melts.

In striking contrast to the northern region, the Thompson-Nicola region, characterized by the deeply indented trough of the Thompson River from Chase to Spences Bridge and by the broad basin-like depression of the Nicola River, has the broadest expanse of spring range in the Interior Plateau. The small "island" of early range located in the Princeton basin marks the southern-most limit of this type. Much potential spring range may be reserved for autumn grazing, as discussed later under "fall range."

Carrying Capacity

Carrying capacity, which denotes the ability of a specific section of range to sustain livestock for a definite period of grazing without injury to the plant growth, is highest in the spring range. Variations will occur from locality to locality, but as a general rule it will range from 10 to 30 acres per cow on a 6-month basis.[2] Along the San Jose Valley in the Cariboo, the carrying capacity is 20 to 35, which is typical of most spring range in the upper country. In the Nicola, the carrying capacity[3] is more typically 15 to 20.

Differences in the carrying capacity of the four categories of range vary with the native vegetation and with climatic differences.

Growth is more profuse in the northern region, where spring range is co-extensive with the upper grasslands, than in the south, where precipitation is less. Grazing practices and types of native grasses are important factors in determining carrying capacity.

FALL RANGE

Characteristics

As a rule, areas of bunchgrass allocated to fall grazing are designated as fall range. However, in the Cariboo-Chilcotin region meadows are grazed late into the fall and even early winter. To be consistent with the definition of "range" implying the season of use, meadow land is considered as a division of fall range, although meadow openings are not included in this classification by the British Columbia Dept. of Lands and Forests.

The bunchgrass type of fall range has physical characteristics similar to spring range; the chief difference lies in the season of use. Such ranges have high grazing value as long as snow depths permit cattle to forage for themselves. Fall ranges have the advantage of a second-growth period from approximately August 15 to October 15 under favorable conditions of rainfall and temperature. This is important to their suitability for use, which usually commences in early October and continues into December.

Fall range may be identical with spring range when the same section is grazed a second time in the same year. Ranges having a retarded growth by reason of a northern exposure or edaphic factors are allocated to fall use.

The use of meadows and valley-bottom land for fall grazing is typical of much of the ranch organization in the Cariboo-Chilcotin region. Although the individual meadow may be relatively small, the total area grazed is large.

Distribution

Since spring and fall bunchgrass ranges may be used interchangeably from year to year, or may be the same area grazed twice in a year, they may be considered as identical on a map of range types. Districts deficient in hard-grass range depend heavily on meadow grazing, so that areas of fall range for much of the Cariboo-Chilcotin region are synonymous with meadows (Figure 1). This is particularly true of settled areas remote from major valleys.

Carrying Capacity

The capacity of fall range is similar to that of spring range when the vegetation consists of cured grasses. In areas where spring range is regrazed in the fall, the carrying capacity will be greatly reduced and the land frequently overgrazed.

The capacity of meadows varies enormously with the types of vegetation (browse, sedges, and grasses) and the availability for grazing. The latter will frequently depend on accessibility. The wet meadow is used for fall grazing when a covering of ice makes approach possible. Meadows commonly have a carrying capacity from 10 to 20, but more often 20.

WINTER RANGE

The essential characteristic of winter range is its ability to sustain stock throughout most winters without the necessity of feeding hay. It is among the most valuable of range types and, being relatively scarce, has all passed into private possession.

Characteristics

Winter range is co-extensive with the lower grasslands. The native vegetation, at least in the climax state, is blue-bunch wheat-grass, highly prized for its nutritional value and curing qualities. Snow depths at most are slight—seldom more than a few inches—and frequent thaws often leave the ground bare. The use of winter range, by those fortunate enough to control parts of it, is subject to three disadvantages: shortage of water, low carrying capacity, and in some areas, difficulty of access. The upland swamp meadows, accessible when frozen over, may be grazed in winter, and so may be considered in a sense as winter range.

263

Distribution

Typically, winter range is located on the lower benches and steep slopes of the deeply entrenched main valleys at elevations generally below 2,300 feet. It extends along the middle Fraser River, along the lower Chilcotin River, and in the Thompson Valley from Chase to Ashcroft. There is a small area of winter range along the upper Nicola River between Douglas Lake and Nicola Lake. Possibly the presence of the lakes has a moderating effect on winter temperatures locally. In the case of the deep valleys, föhn winds are the most important of the factors accounting for its availability through the winter.

Carrying Capacity

Owing to low precipitation and a high rate of evaporation, forage is sparse and watering places few. Carrying capacities vary, depending on the previous use of the range, but generally are quite low compared with late spring pasture. Along the north shore of Kamloops Lake the capacity is 40 to 45, the result of a long period of overgrazing. To a varying degree the same is true of most winter range, with the result that its potential under careful management is not known, except that it is less than that of spring range.

SUMMER RANGE

Characteristics

Although sheep and cattle use areas having similar physical characteristics for spring and fall grazing, a sharp division in this respect applies to summer range. Cattle, with some exceptions, graze under timber during the summer, but sheep (except where they are kept as farm flocks) are usually driven to alpine and sub-alpine park ranges. A few cattle make use of alpine range, but these are of minor importance.

Summer range occupies the upland surface of the plateau at altitudes corresponding to the forested zones (above 3,200 feet). About 8 million acres of forest, mostly in the lower forest zone, are grazed annually by most of the cattle in the ranching area for a period of 3.5 months. Pine grass is the chief forage associated with stands of lodgepole pine, and leguminous forage and forbs grow well under aspen and fir.

Readiness for grazing on summer (timbered) ranges averages 7 weeks later than on the lower grasslands, commencing usually about mid-June to early July. As summer range is almost entirely on Crown land, the period of grazing is established by the provincial government. Although varying with districts, use of summer range is permitted for a period beginning usually between June 1 and July 1 and continuing to the end of September or early October.

Alpine summer range, with its lush vegetation of grasses and forbs, is ideal for sheep. Frosts render the forage unpalatable some time in September, reducing the season of use to 3 months or less. Grazing dates, fixed by the government in accordance with readiness for use and carrying capacity, usually run from July 1 to September 30. It would seem that the trend is for the grazing period on alpine ranges to become shorter each year.

The grazing of meadows scattered through tracts of timber constitutes an important element in the use of summer range. Dry meadows are usually grazed in the summer whereas wet meadows are more accessible in the late fall. Cattle will eat pine-grass (*Calamagrostis rubescens*) but, as a rule, congregate around the openings where the more palatable forage such as pea-vine, vetch and meadow grasses are available.

Water is seldom a problem on summer range. Cool temperatures and timber cover reduce the evaporation rate, and the glacially disturbed surface results in numerous swales and poorly defined stream channels, all of which retain water for most, if not all, of the summer.

Distribution

Much more than half of the grazing area comprises summer range under timber. Some of it, however, is not available to cattle. No serious attempt has been made to determine the extent of the unused or unusable area, but estimates suggest about

50 percent. It would appear from Figure 1 that the Cariboo-Chilcotin region has unlimited amounts of summer range, but much is of poor quality. The timbered ranges south of the Nicola basin are superior. The meadow type of summer range covers large tracts in the Chilcotin and Cariboo regions, but becomes much less extensive in the Thompson-Nicola region.

Alpine ranges have a peripheral pattern conforming to the mountainous transition zone. In general, where mountains rise from 5,800 to 7,000 feet, and where ridges tend to be rounded, good summer ranges are found. As sheep gain and hold their weight well on the lush feed of mountain meadows, these ranges are preferred for sheep, although cattle are found on an increasing number of the alpine ranges used.

Carrying Capacity

The variation in carrying capacity among different types of summer range and in various districts throughout the Interior Plateau is very wide. Vast areas consisting mainly of lodgepole pine have the lowest capacity, frequently 45 to 50. Fir and aspen park have a much richer variety and higher yield of herbaceous plants. Where openings occur in light stands of aspen or fir, such palatable herbs as pea-vine, vetch, wild timothy, and June-grass provide excellent feed and a high carying capacity, frequently 20. The range in the Meadow Lake district northwest of Clinton and the Aspen Grove district south of the Nicola Valley is of such quality. The "swamp meadow", consisting of sedges and some grasses found in vast areas of the Chilcotin-Cariboo region, has a carrying capacity of 20 to 40. A balance between spring, summer and fall range, together with sufficient hay-land to supply the herd during the winter, is highly important to an economic ranch unit. About three-quarters of the ranches appear to have achieved this balance.[4] The chief problem appears to be in production of winter feed, especially in the Cariboo-Chilcotin region, whereas in the Nicola region limited summer range is a restrictive factor.

MARGINAL AND NON-RANGE LANDS

High and Rough Areas

There are areas within the plateau that are sufficiently rugged to be classed as mountains. Most conspicuous are Marble and Pavilion mountains lying east of the Fraser River between the lower Thompson River and Big Bar Creek. Steep, rocky slopes, together with a scarcity of forage imposed by a short growing season, have produced an island of waste within the ranching area. Low mountains project into the Chilcotin district from the vicinity of Groundhog Mountains on the south to restrict the usefulness of the range in this area. Again, along the northwest border of the Chilcotin region, topography becomes rough and elevations high, precluding the growth of desirable vegetation. Meadow openings give way to muskeg in the region to the north and west of Chezacut. Areas higher than 5,000-5,500 feet are usually unsuited for range except in the case of alpine grassland.

A large segment of timberland between the Thompson Valley and Guichon Creek extends to altitudes too high for the growth of suitable forage. The same may be said of the divide separating the Okanagan Valley from the Nicola basin. The Princeton basin is an island of grassland surrounded by barren stands of dense lodgepole pine at high elevations.

Areas of Fallen Timber

An undetermined but very large part of the timbered range is inaccessible to cattle because of fallen trees occasioned by fire or destructive insects. Such areas have at present no grazing value. Their location is on a scale too small to delineate in general terms, but it may be said that much of the summer range in the Chilcotin region and considerable amounts in the Thompson-Nicola region are thus rendered useless for grazing. In almost every locality the ranchers complain of "windfalls" reducing the carrying capacity of the Crown ranges.

265

Steep-Walled Valleys

The steep-walled valley is so extensive in its distribution as to merit separate mention. Parts of the middle Fraser River, the lower Chilcotin River, the Thompson River south of Spences Bridge, and the lower Nicola River beyond Canford are so steep sided as to make access virtually impossible to cattle. These canyon-like valleys constitute barriers to the movement of stock and serve only to mark district boundaries.

Dry Areas

Some ranges are too dry during part of the year for stock, particularly cattle. Although these are not entirely waste as far as grazing is concerned, their value as range is greatly reduced. Frequently, dry conditions are found in recently burned-over areas, such as the "Devil's Garden" north of upper Big Bar Creek. In the west part of the Chilcotin region, between the Chilanko and Chilko rivers, the range is very dry. Large areas, now in lodgepole pine reproduction, have been burned so frequently that the soil is despoiled of humus and the range cluttered with charred debris.

Range Problems

Owing to variations in the physical qualities and in the use of the range, a variety of problems have emerged. Seasonal variation in rainfall gives rise to drought conditions with the result that springs and lakes used for stock-raising dry up; forage is light and growth delayed. This, in turn, results in reduced grazing capacity and over-grazing. With the latter, non-forage grasses, weeds and shrubs replace the desirable forage species. Controlled grazing and seeding are two measures taken to revegetate the range. The main problem of timbered ranges is one of accessibility, largely owing to deadfall conditions produced by fire and bark beetles (*Dendroctonus*). Another problem is that of encroachment by timber or grassland range, a matter frequently taken in hand by ranchers who burn back the timber periodically. To this list may be added poisonous plants such as low larkspur (*Delphinium bicolor*) and death camas (*Zygadenus venenosus*), the latter a problem only to sheep, and poison hemlock (*Cicuta douglasii*) and arrowgrass (*Triglochin maritima*) frequenting low-lying sites. Timber milk vetch makes its apearance in the upper grassland and lower wooded ranges and in the latter produces a form of paralysis followed by emaciation, although it is readily treated with thiamine hydrochloride. Finally, infestations of the paralysis tick and the grasshopper add to range problems.

Range Improvements

The greatly increased logging operations in the forested uplands of the plateau have brought about a great increase in the number of access roads leading back from main roads into the summer range. These have opened up large areas of previously ungrazed timber range. Extensive seeding programs including abandoned trails, fireguards and burns by the Grazing Division of the British Columbia Dept. of Lands and Forests have done much to improve the range and reduce erosion. Helicopters are frequently used.

Other range improvements include, in order of importance, the building of drift fences, stock-trails, cattle guards, and stock water holes.

The Canada Range Experimental Farm at Tranquille is a federal station responsible for range research especially "enclosure" studies such as that at Becher Prairie in the Chilcotin region.

RANGE CATTLE

Numbers

Based on statistics for 1948 supplied by the British Columbia Beef Cattle Growers' Association, the ranching area of the southern Interior Plateau, exclusive of the Okanagan-International Boundary districts, had 98,000 range cattle. These were divided nearly equally between the Cariboo-Chilcotin-Lillooet districts in the north (49,300) and the Ashcroft-Thompson-Nicola-Princeton districts to the south (48,700). (The northern districts are included in the term Carriboo-Chilcotin, and the southern districts in Thompson-Nicola. See regional boundary, Figure 2).

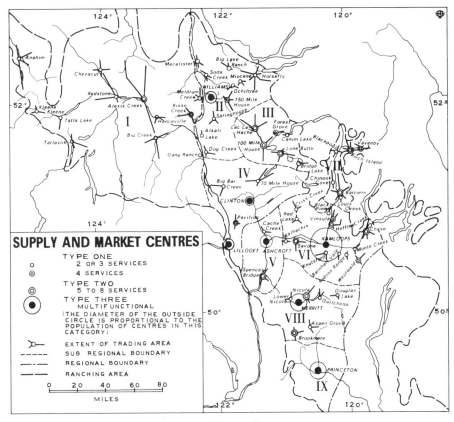

Figure 2

The number of cattle for which grazing permits were issued within the southern Interior Plateau by the Grazing Division, British Columbia Dept. of Lands and Forests, for 1948 was 77,174, approximately 15 percent less than the total, since some are turned out without permit and some are kept entirely on owned land. It is estimated[5] that 25 to 30 percent more cattle are grazed in the southern Interior Plateau than grazing permits issued. In 1960 a total of 96,500 head were permitted to use Crown range, an increase of 25 percent over 1948. The distribution of permitted range cattle by districts, together with the number of permits issued is shown in Table 1, indicating that no appreciable change had taken place in the north-south ratio by 1960.

Dependence upon Crown range had increased considerably, part of the reason being that there was more range available,

Table 1

CATTLE PERMITTED TO GRAZE ON CROWN RANGE BY DISTRICT, 1960

	Cattle	Permits
Cariboo-Chilcotin Region		
Chilcotin	27,665	226
Cariboo-Lillooet	21,778	334
Thomson-Nicola Region		
Kamloops	15,403	243
Ashcroft	8,049	51
Nicola	21,106	72
Okanagan-International Boundary Region		
Princeton	2,517	40
Okanagan-Internat. Bdy.	15,260	228

Source: British Columbia Dept. Lands and Forests, Grazing Div.

267

owing to the opening of otherwise inaccessible tracts through logging roads. If the Okanagan-International Boundary district with 15,200 head were added, the total for the southern Interior Plateau under permit would be 111,700.

Comparative data from the British Columbia Beef Cattle Growers' Association for 1960 is not readily obtainable except that the number reported for the southern division (Thompson-Nicola region) was 65,517, an increase of one-third over the 1948 figure for this district. If a comparable rate of increase is assumed for the northern district the figure would approximate 65,000 thus making a total for the ranching area of 130,000.

Using the 1956 census data, numbers of cattle reported for the northern division (Cariboo-Chilcotin-Lillooet), excepting dairy cattle, were 71,801 compared to 70,571 for the southern division (Thompson-Nicola). A comparison between the three major classes of livestock in the southern Interior Plateau is provided in the table below.

The census indicates that cattle numbers in the 5-year inter-census period have increased nearly one-third. Contrary to popular thinking, numbers of sheep show an increase in the Kamloops-Nicola area. However, horses continue to show a decline in both regions as ranch operations become mechanized.

The number of cattle varies from year to year according to the market, the severity of the winter, and the hay crop. When prices are attractive, cattle move to market in large numbers, thus decreasing the total on the ranges. Heavy snowfall frequently exhausts the hay supply and unfavourable weather

at calving time may result in a heavy loss in the calf crop. If wild hay land becomes inaccessible because of an unusually wet summer or the yield is low because of drought, the rancher is unable to cut sufficient feed to winter his total herd. Therefore, he is forced to sell in the fall or lose considerable numbers from starvation during the winter. The former course is usually followed, resulting in a temporarily large movement to market. Thus, both climate and economic factors cause the cattle population to fluctuate considerably from year to year.

Distribution

The map of cattle distribution (Figure 3) represents an average based on the carrying capacities of the ranges used. It reveals not only a lack of uniform distribution but also certain major concentrations within the ranching area. Three major clusters, separated by areas of lesser density, can readily be recognized. The northern one is centred at the confluence of the Fraser and Chilcotin rivers and fingers outward in three directions, following the lines of drainage. Detached from the major centres, several minor clusters are also apparent. These serve to mark out the smaller ranching communities such as Anahim Lake, Chezacut, and Big Creek in the Chilcotin region, and 150 Mile House, Lac la Hache, and Clinton in the Cariboo region. The central cluster follows the Thompson Valley with minor off-shoots along tributary stream courses. The third major concentration is located in the south, mainly in the Nicola basin, with an outlier in Princeton basin. The small clusters to the southwest mark the use of

Table 2

NUMBERS OF LIVESTOCK 1951 AND 1956 BY MAJOR REGIONS

Major Divisions	Cattle		Sheep		Horses	
	1951	1956	1951	1956	1951	1956
Cariboo-Chilcotin (northern)	57,463	71,801	4,350	3,752	6,161	5,708
Kamloops-Nicola (southern) (not including Princeton area)	49,819	70,571	18,181	23,204	3,269	2,728

Source: Dominion Bureau of Statistics.

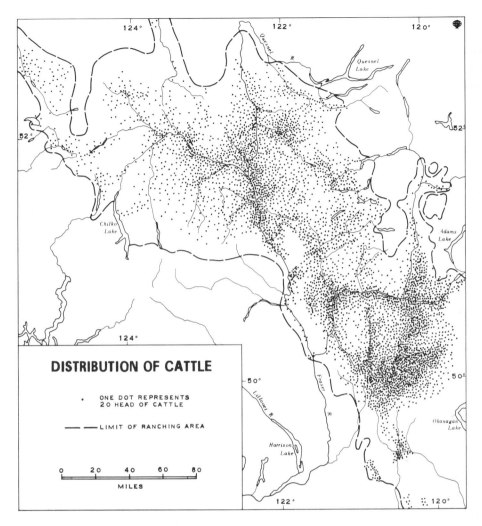

Figure 3

summer alpine range by cattle. Elsewhere, cattle are thinly scattered over wide areas, according to the carrying capacity of the range.

Although no means has been devised of measuring objectively the relative significance of each factor in its effect on the distribution of livestock, it is possible to ascertain the function of each in producing conditions more or less favourable to cattle raising. Four major factors, besides several minor ones, may be cited as fundamental in accounting for the pattern of cattle distribution; (1) native vegetation, (2) topography, (3) drainage, (4) soil. The relative

importance of each varies greatly in different areas, but as a general rule vegetation is the most important single factor because it supplies the raw material from which beef is produced.

Relation to Native Vegetation

The three areas of major concentration of beef cattle are coincident with the most extensive areas of spring and fall range. Carrying capacities, owing to nutritious bunchgrass, are high, ranging from 10 to 20 under careful management, although overgrazing has reduced the capacity of many ranges to about 40. At the same time, these

269

are areas having access to winter range, which reduces the necessity of winter feeding and greatly increases the carrying capacity of the area as a whole. The distribution of native vegetation and range types corresponds so closely with the pattern of cattle distribution that it may safely be said that the vegetation factor is dominant.

Even concentrations of secondary importance in the distribution pattern are readily related to the quality of the range. Such districts as 150 Mile House, Lac la Hache, and the Princeton basin have some hardgrass range with carrying capacities equal to the ranges of Thompson and Nicola valleys.

In the Chilcotin region especially, and to some extent in the Cariboo region, minor concentrations reflect the presence of numerous meadows scattered throughout the timber. Ranching communities such as Big Creek, Chezacut, Anahim Lake, Tatlayoko Lake, Big Lake, and Beaver Valley stand out in the distribution pattern because of dependence on meadow grazing.

Vegetation unsuited to grazing also explains thin spots in the pattern. Timbered ranges, chiefly lodgepole pine, are co-extensive with a sparse cattle population found in large tracts of land throughout the ranching area. Some blank spots on the map, such as that surrounding Princeton, carry such poor forage that grazing is not economically possible. Higher elevations encourage an increase in heavy timber growth and are detrimental to grazing.

Relation to Topography

Although the vegetation factor is dominant, it operates in conjunction with other factors, especially topography. Cattle do better on level to rolling land and will graze rough country only under necessity since steep hillsides and rough land induce foot sores. Consequently, cattle, particularly the heavier animals, left to themselves, refuse to travel long distances in search of forage. Moreover, breeding is adversely affected by rough range. For these reasons, cattlemen avoid rugged topography. This fact is reflected in a thin distribution pattern

around the margins of the ranching area, described as a "transition zone." Areas of rough topography are reflected in the distribution map. Pavilion and Marble mountains stand out as ungrazed areas; the Anahim Lake nucleus is separated from Chezacut by high elevations and rough to mountainous terrain; the steep-walled valleys of the Fraser, Thompson, and lower Nicola rivers, converging on Lytton and Spences Bridge, make good bunchgrass land inaccessible.

However, favorable topography is reflected in three major zones of cattle concentration. In the vicinity of the forks of the Fraser and the Chilcotin rivers, the valleys open and terraces rise in tiers for hundreds of feet above the ribbon-like streams below. These bench lands are usually accessible. Similar topography is typical of the Thompson River above Spences Bridge, continues upstream as far as Chase, and may be traced along the North Thompson River to Heffley Creek.

The third major concentration occupies the broad drainage basin of the Nicola River, where rolling terrain and moderate altitudes combine to favor dense cattle concentrations.

Relation to Surface Drainage

Figure 3 indicates elongations corresponding closely with drainage features, particularly shallow stream valleys on the upland surface, thus showing a direct relationship between surface waters and cattle concentrations. Along minor stream courses, the river bottom-land or the flood plain has a high water-table, frequently giving rise to an interrupted series of meadows containing palatable forage. Then, too, hay land is largely synonymous in the Cariboo-Chilcotin region with meadow openings and irrigable bottom land, which must be considered as an important element in sustaining cattle during 4 months of the year.

Two elongated clusters trend south from the Thompson Valley to the Nicola region. The eastern one, an old glacial stream channel now nearly abandoned except for several lakes, has excellent spring range associated with it. The Guichon Creek valley to the west, principally in the vicinity of Mamit Lake, has an extensive flood plain and a high water-table favourable to abundant hay crops.

Relation to Soil

The three areas of major concentrations have fertile, Brown, Dark Brown and Black (Chernozem) soils corresponding to vertical vegetation zones. Particularly in the Kamloops-Nicola region, wherever topography is suitable, these soils yield a large hay crop under irrigation and greatly extend the cattle capacity of these areas. On the upland surface, Grey Wooded soils predominate. Cultivated areas are thereby greatly restricted, usually to immature soils of flood plains.

Soil, in the sense of the regolith, may exercise a strong negative effect on cattle distribution. Many areas of the upland are strewn with coarse gravelly moraines, which in some localities are so unfavorable that only lodgepole pine can thrive. It was observed that a large segment of the west Chilcotin region between Chilanko and Chilko rivers appears ungrazed. A traverse through this sector between Redstone and Tatla Lake revealed a regolith so stony that herbaceous vegetation was extremely sparse and natural gravels abound.

Some areas, usually too small to appear significant on a map of cattle distribution, have been rendered useless for grazing by destructive fires. Under dry conditions, particularly in midsummer, fires destroy the humus in the soil, making it unproductive except for lodgepole pine. This tree type nearly always appears after burning, no matter how severe the fire. Recent burns in large parts of the Chilcotin region and throughout the timbered ranges generally have turned summer grazing land into waste.

Regional Distribution

The cattle within the reconnaissance area in 1948 were distributed over 567 holdings, of which about 147 had fewer than 25 head. Accepting 25 head as a minimum for an economic unit, there were 420 commercial cattle ranches within the area under study. The remainder are either uneconomic units or depend upon other activities as the major source of income. The numbers and relative importance of commercial ranch units are given, by major regions, in Table 3.

From the data in Table 3, several comparisons for the major regions may be drawn. The Cariboo-Chilcotin has about the same number of cattle as the region to the south and exceeds the latter in numbers of commercial ranches. There are more submarginal and marginal units in the Cariboo-Chilcotin region, but the two regions are about equal in the larger units.

Table 3

REGIONAL DISTRIBUTION OF RANCH UNITS BY CATTLE NUMBERS, 1948*

Region**	Ranch units by numbers grazed						Total ranch units	Total commercial units	Number of Cattle	Average cattle per unit
	Under 25	25-99	100-499	500-999	1,000-5,000	10,000				
Cariboo-Chilcotin	122	127	79	12	7	—	347	225	49,332	142
Kamloops-Nicola	25	112	66	11	5	1	220	195	48,737	221
Total	147	239	145	23	12	1	567	420	98,069	170

*Data are not available for determining exactly the number of non-commercial units (under 25 head) but on the basis of membership in the B.C. Beef Cattle Growers' Association, and of cattle grazing permits, the above figures are given as estimates.

**The boundary between the Cariboo-Chilcotin region and the Kamloops-Nicola region lies north of the Thompson Valley (*see* Figure 2).

Breeds

The predominant breed of cattle in the ranching area is the Hereford. Of 100 ranches, two-thirds reported Herefords and the remainder indicated some cross-breeding with Shorthorns. The resulting strain is believed, by some, to provide more milk (passed on to the calf in the form of increased weight) and a larger-quartered steer. Most, however, consider the Hereford to possess the desired qualities of quick growth to maturity, together with a smaller frame, thus yielding smaller cuts of beef in accord with the present market demand. Scattered herds of Aberdeen-Angus may also be found. There is no indication of regional difference in respect to cattle breeds.

Until the mid-1950's a typical ranch herd consisted of the breeding stock, yearlings, and two-year-olds. In isolated districts of the Cariboo-Chilcotin region even three-year-olds were occasionally marketed off the range, although by the close of the war, the market demanded smaller cuts of beef. With revolutionary changes in marketing occurring about 1957, occasioned by the demand for grain-finished rather than grass-finished beef, most of the ranches, especially those in the Thompson-Nicola region, now sell calves and yearlings to feed lots outside the region. A typical herd now consists of the breeding stock, calves and "long-year-lings" ready to be grain-finished.

It is customary to run one bull for every 20 to 30 breedings cows. The law requires no less than 1 to every 30 head in community breeding pastures. Bulls placed on Crown land must be registered pure-bred animals. As a result, it may be generally stated that all bulls are pure-bred throughout the ranching area. They are acquired by purchase at cattle sales at Kamloops and Williams Lake, and some are imported from Alberta. Small ranchers frequently acquire bulls passed on from large herds at reduced prices. Several large ranches throughout the area maintain a small pure-bred herd to furnish the requisite breeding stock. Breeding cows are not usually registered stock. They serve in the herd for 6 to 12 years and are then marketed.

Out of one hundred ranches sampled, only five raised pure-bred cattle exclusively. This percentage may be considered fairly representative of the area. Three raised Hereford stock, the remaining two, Shorthorn. Herds on ranches specializing in pure-bred cattle are small, because greater care and supervision are required. The total of the five ranches was only 547 head. There appears to be no noticeable areal pattern in the raising of registered livestock; such ranches are scattered from the Nicola region to the Chilcotin region.

The pure-bred stock ranch must be completely fenced to prevent contact with non-registered herds. Accordingly, the land is owned or leased with little or no dependence on Crown summer range. It is necessary to depend entirely on cultivated hay supplemented by grain, and the winter feeding period must be extended by one or two months. Registered animals are disposed of at the Kamloops or Williams Lake sales, or locally by private sale.

The Interior Plateau does not possess any particular advantages for the raising of pure-bred cattle. It is not a grain-raising area, and the kinds of feed required have to be either imported or grown under irrigation at increased cost. Moreover, the long feeding period adds considerably to the cost of production.

SHEEP

Numbers

Sheep are minor in importance compared to cattle in the southern Interior Plateau although there are many more in the south than in the northern region. According to the number of 1960 livestock permits issued, there were 111,000 cattle compared to 18,779 sheep grazing Crown ranges including the Okanagan-International Boundary district. The latter figure, however, does not include the considerable number of farm flocks that are kept entirely on the ranchers' owned land. The 1956 census indicates a total sheep population of 52,387 divided as follows: Cariboo-Chilcotin-Lillooet 3,752; Nicola-Salmon Arm[6]-North Thompson

272

31,306; Okanagan-International Boundary 17,329. From this it is apparent that (1) the number of sheep in the northern region, lacking in alpine range, is small compared to the south, (2) that only 35 percent of the total sheep population of the Interior Plateau including the Okanagan-International Boundary district is grazed on Crown range. Assuming that all commercial sheep ranches use Crown range there are approximately 30 such units in this category. Included are a few that combine sheep and cattle on a large scale. As a rule ranchers do not mix their livestock. A considerable number, however, keep farm flocks.

There are several reasons why the ranchers of this area have not turned more to sheep-raising. Coyotes continue to be a menace. The only protection against the coyote is herding and even then the utmost vigilance is needed to outwit this predator. Although confined to the Cariboo region, where sheep are sparsely represented, timber wolves have been brought under control by predator hunters. The cougar, however, frequents the summer mountain ranges bordering the ranching area and makes the presence of a herder indispensable to sheep protection. The grizzly and the black bear are frequently troublesome, but the bear and cougar, unlike the coyote, do not hunt in packs and can be tracked down with less difficulty.

The second problem is that of finding suitable men as herders; this has in recent years been the chief reason for many ranchers turning from sheep to cattle. The isolated, lonely life of the herder, confined to the hills for three months, together with an absence of prestige associated with the occupation, appear to be reasons why young men are not readily available for this class of employment. Most herders employed on the range today learned the art in Europe.

The third factor is an economic one. The need for hiring a packer as well as a herder throughout the summer results in a greater labor cost than in cattle-ranching, where the practice generally is to turn the cattle out unattended to graze where they will for the summer. Labor costs in recent years have not been offset sufficiently by the price of mutton to justify the widespread raising of sheep. Then, too, the market for sheep prior to the war was little short of chaotic, and many ranchers disposed of their herds as soon as they could change over to cattle.

A fourth problem pertains to marketing. In many cases the rancher is to blame for the low prices of domestic lamb. Poorly finished animals are included with those in prime condition, which depresses the price. Dumping also occurs. A lack of cooperative marketing is a further disadvantage.

The pattern in Figure 4 represents the distribution of sheep on summer (alpine) ranges during three months of the year. As in the case of cattle, sheep must be provided with spring and fall range and winter feed on lots near the home ranch. The significance of the distribution is the importance of summer range, which is complementary to cattle range and involves the seasonal movement of livestock known as transhumance. One rancher in the Tranquille range north of Kamloops trucks 2,400 sheep with lambs over 150 miles to the alpine pastures of Groundhog Mountains.

Two elements of the pattern are readily apparent: (1) the singular absence of sheep in the Cariboo-Chilcotin region, and (2) the peripheral distribution in the southern part of the ranching area.

Only in the French Bar Stock Range of the southeast Chilcotin region are sheep to be found in large numbers, and in this case the home ranch is within the Thompson-Kamloops sub-region. Elsewhere in the Cariboo-Chilcotin, sheep are found only as farm flocks numbering from a few to 40 or 50, and aggregating approximately 3,500.

The most obvious explanation for the comparative void in the northern part of the ranching area lies in the number of predators and the deficiency of alpine range. Although coyotes are a problem in the southern part of the area also, they move in greater numbers in the Cariboo-Chilcotin region. Added to this is the shortage of suitable feed for summer grazing on the alpine borders of the northern region. However, alpine ranges are reported to exist

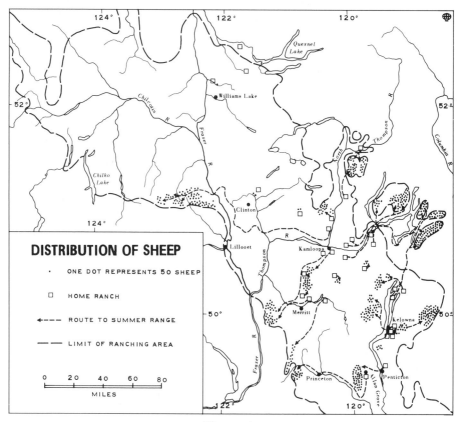

Figure 4

along the margins of the Cariboo Mountains. With the exceptions noted above, sheep ranching is confined to the southern part of the Interior Plateau. The pattern of distribution is mainly peripheral and at the same time concentrated in a number of individual groupings.

This distribution depends on the character of summer range best suited to raising sheep. During the nursing of lambs, sheep require a lush, succulent forage with forbs predominating. Though they will eat grasses and sedges in great quantities, they prefer them to be green; they also do well in ranges where forage runs to browse types. Unlike cattle, they are capable of grazing in areas where flies and insects abound. Moreover, they enjoy a relatively cool summer temperature. They can endure travel for long distances by easy stages and actually put on weight. They are capable of going without water for days and even weeks on a diet of

lush vegetation. They can negotiate steep, rough, and rocky terrain without undue difficulty, but cannot be readily herded in timber, where they are easy prey to predators. All factors considered, especially the character of the vegetation, alpine ranges provide the solution to desirable summer ranges in a semi-arid country such as the Interior Plateau. Accordingly, the summer distribution pattern is related to the distribution of alpine meadows, found principally around the margins of the plateau. As relief increases towards the south, there are extensive areas within the plateau having subalpine characteristics that lend themselves to sheep grazing.

The relegation of sheep to the higher elevations of the ranching area is primarily a matter of "best" land use, the range allotments being made by the provincial Dept. of Lands and Forests. The age-old antipathy of cattlemen to sheep, based on the fact that sheep graze so closely that they kill the grass cover, is still a consideration although of dwindling importance, as sheep ranchers find it pays to transport sheep to alpine ranges by truck. Accordingly, cattle and sheep ranges are kept separate, and where one merges into the other a drift fence is erected. Such cases occur in the Swakum Mountain Range and along that part of Meadow Creek lying to the east of Guichon Creek. Now that summer ranges are being increasingly used by cattle, competition for range is looming.

Exceptions to the use of alpine and sub-alpine ranges for sheep may be found where browse forms a high percentage of the forage and where vetch, pea-vine, and other legumes abound. Such sections occur in the vicinity of Criss Creek, 59 Mile Creek north of Clinton, Meadow Creek, and the Douglas Plateau lying east of Stump Lake. The total number of sheep on these areas is small, approximately 2,000 ewes.

The concentration of sheep in patches is related to altitude; only those parts of the range from 5,800 to 7,000 feet are characteristically alpine.

Regional Distribution

The distribution of commercial sheep ranches shows a preference for the North Thompson Valley and south to the Nicola district. Of the 26 units grazing in 1960, more than 100 head were on Crown range under permit, none were located in the Chilcotin district although one used range there, two were located in the Cariboo-Lillooet district, 15 in the Thompson-Nicola district and 8 in the Okanagan district. These accounted for 17,351 sheep (Table 5 and Figure 4).

The distribution of sheep according to the 1956 Census is shown in Table 4.

The various pastures and ranges comprising the sheep unit are similar to those of the cattle ranch, with the exception of the summer range.

The principal sheep breed in the interior are the Rambouillet or the Rambouillet cross. This breed is desired for its flocking instinct, a valuable trait in a country of predatory animals and unfenced summer ranges and where large bands are economically desirable. The Rambouillet is a dual-purpose breed—large, hardy, and a good forager. The Suffolk cross provides a better meat strain and gives a hardy, good-rustling, quick-maturing animal. About 84 percent of the sheep are bred for mutton, the remainder for wool.

Table 4

REGIONAL DISTRIBUTION OF SHEEP
IN THE INTERIOR PLATEAU

District	Number of sheep
Chilcotin	1,313
Cariboo	1,359
Lillooet	1,080
North Thompson-Salmon Arm	12,168
Nicola-Kamloops	19,138
Okanagan	15,837

Some cross the Suffolk with the Hampshire, the latter being desirable for wool. This cross gives a large, fast-maturing animal, hardy and well able to rustle. The Romney, Corriedale, and Southdown are used to a limited extent for breeding purposes.

The size of sheep ranches in the interior varies from a few hundred head to 3,800. Units, depending on sheep primarily, may be arranged according to size as in Table 5, where 1948 is compared with 1960 based on permits to graze on Crown range.

Table 5

SIZE OF RANCH UNITS ACCORDING TO
NUMBER OF BREEDING EWES

Number of breeding ewes	Number of units	
	1948	1960
100 to 200	6	6
201 to 400	6	6
401 to 600	2	3
601 to 1,000	7	6
1,001 to 1,500	3	3
1,501 to 2,000	2	0
2,001 to 3,000	1	1
3,400	1	1
3,800	1	0
	29	26

Commercial sheep ranches have declined in number since 1948. Factors of location appear to be similar to those governing cattle units, especially the need for spring-fall range. Proximity to alpine summer range appears to have little bearing on the location of the ranch. Therefore drives of many miles are necessary as indicated in Figure 4.

HORSES

Although mechanization in crop production is increasing in the ranching area of British Columbia, the horse continues to hold an important place in ranch operations. Draught animals are less in demand, but the necessity for the "cow pony" in rounding up and herding livestock has sustained the demand for riding animals.

Numbers

Although grazing permits are issued for horses, these do not provide so close a check on the total numbers as in the case of range cattle, as many ranchers prefer to keep their horses entirely on their own land. The total number of horse-grazing permits issued for the study area in 1960 was 3,130. These were unevenly divided between the Cariboo-Chilcotin region (2,746) and the Thompson-Nicola region (384). Figures from the 1956 census (see chart p. 316) show the contrast between numbers of horses in the two regions, although perhaps not as accurately as the permits issued.

Table 6

FARM POPULATION AND NUMBERS OF CATTLE, SHEEP AND HORSES
IN THE RANCHING AREA OF BRITISH COLUMBIA

Census Subdivision	Farm Population			Cattle		
	1951	1961	% Change	1951	1961	% Change
North Thompson	1,095	851	−22	7,347	10,883	+48
Nicola	2,096	1,699	−19	44,634	62,803	+41
Chilcotin South	241	188	−22	16,377	20,726	+27
Lillooet East	992	953	− 4	19,616	22,955	+17
Bridge-Lillooet	579	338	−43	2,995	4,897	+64
Chilcotin North	772	749	− 3	9,720	13,357	+37
Cariboo	1,446	1,111	−23	10,863	14,960	+38
Totals	7,212	5,889	−18.3	111,552	150,581	+34

Census Subdivision	Sheep			Horses		
	1951	1961	% Change	1951	1961	% Change
North Thompson	3,763	5,299	+41	808	469	−42
Nicola	14,418	18,020	+25	2,461	1,850	−25
Chilcotin South	—	17	—	906	1,136	+25
Lillooet East	894	1,446	+62	2,086	1,468	−30
Bridge-Lillooet	67	703	+964	566	359	−37
Chilcotin North	672	981	+46	1,338	1,454	+ 9
Cariboo	2,717	3,596	+32	1,265	965	−24
Totals	22,531	30,062	+33	9,430	7,701	−18

Source: *Census of Canada 1951* and *1961*.

Cariboo-Chilcotin Region 5,708

Chilcotin	2,013
Cariboo	1,547
Lillooet	2,148

Thompson-Nicola Region 2,728

| North Thompson | 554 |
| Nicola | 2,174 |

Distribution

Using the ratio of horses to cattle, the relative importance of horses in the Cariboo-Chilcotin region is much greater than in the Thompson-Nicola region. On the basis of the 1960 census about 73 percent of the horses for the whole area are found in the Cariboo-Chilcotin region. The ratio of saddle horses to work horses is 1:2 for the Cariboo-Chilcotin and 1:1.2 for the Thompson-Nicola region, indicating the declining importance of the work horse in the latter region, where machines are rapidly replacing it. These data are significant as indicating the pioneer type of organization in the northern area, as shown by a greater dependence on draught animals. The Thompson-Nicola region, with a higher degree of mechanization, has less use for the work horse. Greater dependence on swamp hay in the Cariboo-Chilcotin region calls for a greater number of horses, for the scattered nature and roughness of wild hay meadows makes the horse more practicable than the machine in many instances. Then, too, pressure on the range is greater in the south. Today, the importance of the horse is almost entirely related to riding and herding.

A continuous problem in range maintenance exists in the numbers of wild horses, especially in remote areas of the Chilcotin and in the mountainous area east of the Fraser River from Dog Creek to Pavilion. The government at one time tried to cope with this problem by offering a bounty for their extinction. However, at present the government exercises full control through regulated round-ups and shooting of wild horses.

REFERENCES

[1] For elevations consult National Topographic Series, Scale 1:500,000, sheets 92SE, 92NE, 93SE, 93SW, 92NW.

[2] For the sake of brevity carrying capacities are expressed in the number of acres capable of sustaining a cow for 6 months. The sheep-cow carrying capacity ratio is 4:1.

[3] Estimates of carrying capacity were obtained from cover maps of the Grazing Division, British Columbia Dept. of Lands and Forests, Victoria, B.C. Some of these were made 30 years ago when the subjective element entered more largely into the estimate than at present. Lack of uniformity of data, as well as of coverage, is a problem in discussing carrying capacity.

[4] Acton, B. K. and Woodward, E. D., *Cattle Ranching in the Interior of British Columbia, 1958-59.* Econ. Division, Canada Dept. of Agriculture, (Ottawa: Queen's Printer, 1961).

[5] Personal communication.

[6] Salmon Arm census division is included for sheep because of the large numbers using alpine ranges in the vicinity.

RESOURCES AND PROBLEMS

An Introduction

Historically, Canada's economic development has been based largely upon the exploitation of its vast natural resource base. And, while the economy has become more diverse over the years, our dependence upon the export of raw and partially processed resources has not sharply diminished. Fish, farm, forest, and mineral products account for three-quarters of the value of export commodities.

Despite the importance of natural resources to our national prosperity, there is a shortfall in resources inventory; many of the old development problems persist; and new problems have loomed over the horizon. Some of these problems are considered in the articles that follow.

One of the most emotionally argued of these development issues in Canada is the magnitude of foreign investment and ownership of Canadian resources and industry and, concomitantly, the real or potential influence that this can have on public policy, political, cultural and economic independence, employment and so forth. Since the bulk of foreign investment in Canada originates in the United States, the critical and suspicious thrust has been directed to the south, although equally penetrating criticism has been levelled at the government of Canada for not reacting more decisively on this broad issue. One of the earliest and most persistent critics of the magnitude of foreign investment and its impact on Canadian independence was Walter L. Gordon, a former Minister of Finance in the federal government, whose viewpoint was established in *A Choice for Canadians* (Toronto: McLellan and Stewart, 1966). It was largely as a result of Gordon's public blandishments that the federal government established a task force to examine and report on the extent of foreign investment in Canadian industry. Their report, *Foreign Ownership and the Structure of Canadian Industry,* (Privy Council, 1968), clearly outlined the serious degree of foreign ownership and investment in Canada.

Further analysis of the extent of foreign dominance was furnished by Kari Levitt in *Silent Surrender: The Multinational Corporation in Canada* (Toronto: Macmillan, 1970). In September 1970, a group of Canadian businessmen, active and former politicians, and concerned citizens founded the potentially influential and cohesive Committee for an Independent Canada. And in June 1971, the Canadian Association of Geographers, at their annual meeting at the University of Waterloo, sponsored a symposium on this theme.

Less than three months after the Waterloo symposium, the economic dependence of Canada on the United States was dramatically underscored when the latter imposed a temporary 10 percent surcharge on all imports. What the long-term Canadian reaction will be to this economic "insult" is as yet unclear, but it may persuade Canadian manufacturers and trade promotion agencies to mount a vigorous marketing campaign in other lucrative world markets particularly in the European countries.

It is an understatement to suggest that interest is running high on this issue, often to the point of making it difficult to evaluate the real extent and impact of foreign investment. Two papers on this theme are included here. The first contains a brief summary of foreign investment in Canada. The second, by D. Michael Ray, examines the effect of foreign investment in the Canadian manufacturing industry on industrial location.

The second group of papers focus on resource development in the Canadian north and in coastal British Columbia. In the first paper, Dubnie and Buck summarize the influence of federal incentive policies on selected facets of northern development. Davies, in contrast, examines the potential impact of resource development, particularly oil and gas, on northern ecology. In the third paper, Walter Hardwick synthesizes the historical evolution of the forest industry in British Columbia.

Water is the theme of the third group of papers. Canada, like many other countries sharing a common land boundary, has been involved with the United States in the development and control of international waters. Derrick Sewell discusses the history of negotiations between Canada and the United States, which resulted in the ratification of the Columbia River Treaty to develop the upper reaches of this river. The problem of achieving equitable mutual benefits and some lessons that could be applied to future negotiations for comprehensive joint development programs are discussed.

For a number of years the question of water diversion and export to the United States has been discussed and diverse enabling plans have been drafted and promoted for this purpose. The Canadian reaction to such proposals has been largely negative with the majority of the opposition coalescing around the fact that the extent, quality, and amount of water is not yet well known. In his paper, Arleigh Laycock summarizes the major water diversion proposals for Canadian waters and examines the "pros and cons" of water export to the United States.

The federal government's role in resources management relates specifically to the broad provisions of the British North America Act that essentially give provincial governments control over the management and development of these resources. However, there are clearly established cases where federal "rights" prevail, such as in the case of international waters. But in most cases, the rights to jurisdiction are not at all clear with respect to environmenal management (see Dale Gibson, "Constitutional Jurisdiction over Environmental Management in Canada," Government of Canada, Constitutional Study, 1 June 1970, mimeo.)

Despite the complexity of constitutional interpretation, the federal government has long been involved in resources management and their direct or indirect participation is likely to increase as new and more widespread environmental issues arise. One of the areas in which the federal government has recently joined with the provinces is in water management. The analysis of federal involvement in environmental quality issues with detailed reference to the Canada Water Act, is the subject of E. Roy Tinney's paper.

Foreign Investment in Canada

Bank of Nova Scotia

Judged in its broadest context, foreign investment has played a major role in the economic development of Canada. The use of foreign capital and in many cases the related introduction of foreign know-how and technology have provided a higher level of Canadian output and incomes than would have been possible if this country had relied solely on domestic sources of savings and invesment. Yet foreign investment comes in a variety of forms, which in turn carry varying implications and raise differing sorts of questions. Many of the issues involved are quite complex, and all the more so because they extend beyond purely economic bounds. In the past few years, a growing volume of study and research on the subject has been proceeding; this review attempts to outline some of this material, focussing on some of the problems as well as the benefits which arise out of foreign investment.

SOME HISTORICAL PERSPECTIVE

It is no coincidence that the periods of fastest Canadian growth have also been marked by heavy inflows of foreign capital. For Canada's part, there has been a clear need for a high rate of capital formation to provide not only an infrastructure for the economy—for example, British financing for the railways before the First World War—but also the means of developing the country's massive natural resources and a manufacturing base. To the outsider, Canada has quite plainly offered investment opportunities which appeared to be equalled in few other places. Both these closely related factors have shaped the overall course of foreign investment, and between 1926 (the first year of reliable estimates) and the end of 1970, Canada's gross long-term investment liabilities to non-residents grew from $6 billions to almost $44 billions.

The combination of investment requirements and investment opportunities over these years has produced not only fluctuations in capital flows but also a considerable change of pattern. Since the 1920's there has been a steady expansion of U.S. direct investment in controlled companies in Canada, partly to develop new-found resources and partly as a means of getting around the Canadian tariff, while the two marked surges of post-war capital investment in Canada—in the mid-1950's and mid-1960's—were facilitated in large degree by direct participation of foreign companies as well as by bond and equity financing abroad (portfolio investment). By the end of 1970, the book value of total foreign direct investment in Canada had increased to over $25 billions (of which nearly four-fifths was owned by U.S. residents), portfolio investments had risen to about $16 billions, and other foreign holdings of real estate, mortgages etc., to about $2.7 billions. (Figure 1).

The determinants of investment flows are, of course, subject to continual change. Within the last 10 years, the growth of the Canadian economy and the significant move towards a greater liberalization of world trade have improved the market environment for manufacturing companies in particular and have begun to lessen the inducements for operating small-scale replicas of U.S. firms behind tthe Canadian tariff. Of major importance, too, have been the attempts to rationalize production and trade flows in certain industries—most notably the automobile industry, where the Auto Pact brought a massive round of new investment in more specialized and efficient Canadian productive

● This article is reprinted from the *Monthly Review, Bank of Nova Scotia*, April-May 1971. 4 pp. with permission of the Bank of Nova Scotia.

281

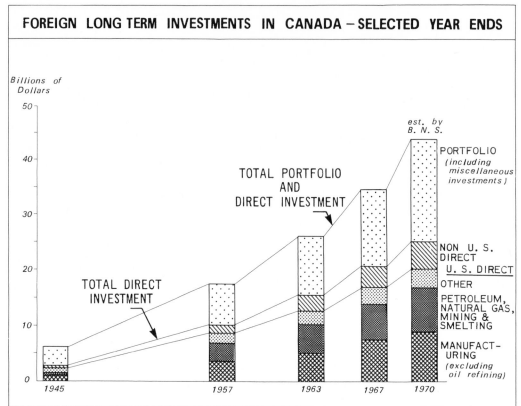

FOREIGN LONG TERM INVESTMENTS IN CANADA – SELECTED YEAR ENDS

Billions of Dollars

est. by B. N. S.

PORTFOLIO (including miscellaneous investments)

TOTAL PORTFOLIO AND DIRECT INVESTMENT

NON U. S. DIRECT

U. S. DIRECT

OTHER

PETROLEUM, NATURAL GAS, MINING & SMELTING

TOTAL DIRECT INVESTMENT

MANUFACT- URING (excluding oil refining)

1945   1957   1963   1967   1970

SOURCE: BANK OF NOVA SCOTIA MONTHLY REVIEW, APRIL-MAY, 1971

Figure 1

facilities as well as an expansion of automotive trade.

While the pattern of investment opportunithus has been changing in recent years, Canada's requirements for foreign capital have also been subject to new considerations. Since 1966, Canada's balance of payments on current account has undergone a major improvement, swinging from an average annual deficit of $1.1 billions in 1965 and 1966 to a surplus last year of slightly more than that amount. While due account has to be taken of the specially favourable factors that have been affecting this country's trade position—including the sluggish course of business investment—there is not doubt that over the recent period the shortfall in Canadian savings has been nothing like as large as in much of its past history. On a "net" basis—

taking account of the entire flow of both short-term and long-term capital into and out of Canada — greater scope is thus being opened for Canadian investments abroad, especially by larger Canadian companies capable of extending their productive activities on a wider international basis.[2] One of the most striking developments of the past ten years, indeed, has been the rising prominence of the so-called multinational company with operations spread out in a goodly number of countries. Extension of such companies into other parts of the world has, on the one hand, introduced a new and significant dimension to the questions relating to foreign investment in Canada. But it has likewise opened up new horizons for Canadians in their own approach to corporate growth and to relevant economic policies.

MAJOR TYPES OF FOREIGN INVESTMENT

Approximately one-third of Canada's foreign long-term liabilities consists of *portfolio investment*. This type of indebtedness arises when non-residents purchase long-term bonds of Canadian governments and corporations or make minority investments in Canadian equity shares. In the case of foreign bond issues, the major source of portfolio investment, the initiative typically is taken by the Canadian borrower, and reflects a desire on his part to take advantage of the lowest interest rates and broader capital markets of other countries. In recent years the total borrowing requirements of Canadian provinces and provincial agencies have increased persistently; and, to avoid undue drawings upon domestic capital markets as well as possibly to secure lower interest costs, provinces have borrowed substantial amounts in the New York market, and some also in Europe. Over the past four years, in addition, there has been a sizeable volume of corporate borrowing abroad, the most notable being the spaced-out drawings of close to $500 millions for financing of the huge Churchill Falls power project.

All told, new issues of Canadian securities in foreign markets averaged $1.4 billions per year over the period from 1963 to 1970. In 1968 and 1969 they ran to nearly $2 billions a year, and even after allowing for retirements net issues were about $1.6 billions. In the past year, however, the marked easing of borrowing conditions in Canada, combined with a government request to avoid foreign financing if possible, has brought a sharp reduction in the volume of new foreign issues, and there may well be less reliance on this kind of financing in the years ahead.

Despite the substantial volume of foreign portfolio financing over the past four years, the flow of foreign *direct investment* has remained the most dynamic element in the growth of Canada's gross long-term investment liabilities. At the end of 1970 estimated total direct investment in Canada at somewhat over $25 billions, accounted for roughly 60% of total foreign long-term investment. This kind of investment, involving foreign control over physical (as opposed to financial) assets, has been the major source of concern among many of the critics of foreign investment in Canada. Unlike portfolio funds, direct investment involves the import of a "package" which may include capital

Table 1

SOURCES OF FUNDS USED BY AFFILIATES OF U.S. COMPANIES IN CANADA
(U.S. $ millions)

	1963	1964	1965	1967	1968
Internally-generated funds					
Retained earnings*	336	497	373	444	529
Depreciation and depletion allowances	552	623	681	800	864
Sub-total	888	1,120	1,054	1,244	1,393
External funds					
Funds from U.S.	168	156	551	242	127
Funds obtained outside U.S.**	241	307	497	423	539
Other sources and adjustments	29	88	75	138	53
Sub-total	438	551	1,123	803	719
Total Funds Used	1,326	1,671	2,177	2,047	2,112

Note: *Retained earnings run around half of total net income of the companies concerned. **Mainly funds secured in Canada, including funds from issues of securities, loans from financial institutions, and trade credits.

Source: Survey of Current Business, Nov. 1970, U.S. Department of Commerce. Based on reports from large U.S. companies accounting for the major share of foreign investment in the manufacturing, petroleum and mining industries. Data available only for the years shown.

but also encompasses such factors as technology, risk-taking ability, management know-how and market access. Capital will usually be part of the direct investment package and, at times, access to capital is itself the main element in the whole process; but direct investment can also proceed without any cross-border movement of capital at all, as occurs when foreign corporations increase their control over assets through an investment in kind or a reinvestment of subsidiary earnings. As may be seen in the accompanying table, funds from the United States actually supplied a relatively small proportion of the total financing needs of U.S.-controlled companies in Canada between 1963 and 1968. Over 60% of the funds used for the expansion of these firms in the indicated years was, indeed, "internally generated" by the affiliates themselves in the form of retained earnings and depreciation and depletion allowances. Of the "external" supplies of financing coming other than from the United States the most important have been through direct borrowings in Canada, either by security issues, by bank loans, or by short-term trade credits.

DIRECT INVESTMENT—
ECONOMIC COSTS AND BENEFITS

Direct investment essentially involves moves by a parent corporation to apply its special advantages in the form of technology, finance or market connections to factors present in a foreign country, such as plentiful natural resources, lower wage levels or a protected local market. The decision to extend production across national boundaries proceeds from the calculation that the corporation's unique advantages will more than offset disadvantages relative to local firms more familiar with the local environment and all its legal, political and social overtones; it also implies a judgment that the corporation could not achieve results as good or better (from its point of view) through such alternative routes as the sale or leasing of production and patent rights, licensing or joint venture arrangements. For the recipient or host

economy, the decision by a foreign corporation to produce within its borders will in most instances contribute to a higher level of capital formation and employment. Even where there is no local shortage of savings, the operation of foreign subsidiaries provides stimulating influences not only through the continued channelling of advanced technical know-how and strong risk-taking capacity in the companies directly concerned but also through the secondary effects on domestic sales and incomes and through the diffusion of technical and managerial skills to local companies.

Foreign direct investment in Canada's resource-based industries has generally fitted into this classic mould. In the forest and mineral developments of the 1920's and 1930's, and in the great postwar expansion in oil and gas, iron ore and other resources as well, direct long-term market outlets, ready access to needed large amounts of capital and advanced technical know-how and experience all had a part to play. It should be clear also that these developments in turn contributed greatly to the strengthening and deepening of the whole Canadian economy through this period, extending added stimulus to the growth of local population and markets and thereby inducing enlargement of other service and secondary industries.

On occasion, excessive bunching of such developments has tended to accentuate upward pressures on Canadian costs and prices and thus to raise problems for less vigorous sectors of the economy. Concern on such grounds is clearly evident in some current discussions of Canadian policies with respect to the development and movement of oil and gas from Alaska and the Canadian Arctic. Yet in being alert to such potential problems it would be foolish to ignore the undeniable long-term benefits of efficient production in lines where Canada does have recognizable comparative advantages.

Foreign direct investment has also, of course, played a major role in the development of Canadian secondary industries, beginning again as early as the 1920's. Such investment, however, has in large part been of a "defensive" type, undertaken not to increase the return available but to forestall loss of markets or of profits in trying to sell over the Canadian tariff (and in the 1930's into Commonwealth preferential countries as well.

While tariff barriers thus were a crucial factor in attracting foreign manufacturing subsidiaries into Canada, the tariff levels in turn have hampered the effective performance of such companies and of Canadian manufacturing in general. Since it took a tariff in the first instance to induce the foreign companies to establish production facilities in Canada, it should not be too surprising that the subsidiary companies would tend to be less efficient than their parents. But because of the nature of the U.S. production and marketing system, and the heavy reliance on brand names and advertising, Canada in general attracted not isolated U.S. producers but virtually the whole range of major companies in particular industries—so establishing the pattern of too many plants producing too many product items with unduly short production runs and unduly high production costs. Studies in recent years have widely documented the nature of these problems, and have examined them also in relation to Canada's anti-combines legislation, and other aspects of the Canadian industrial environment. While some subsidiary companies in fact have been able to achieve high standards of performance, in general it has been found that their operations tend to parallel the shortcomings of domestically-controlled firms. The inherent advantages of the parent companies, thus, have not been as fully transmitted and diffused through the Canadian economy as might be desired; and consumers have not seen all the benefits that could result in terms of price or product quality. Yet even with these quite evident limitations, which the Canadian government has been seeking to remedy (as in the Auto Pact), there is little doubt that, in general, direct foreign investment has made a substantial contribution to the productivity of Canadian factors of production and, thereby, to the pace of overall Canadian economic growth.

The financial cost to Canada for the use of direct investment lies in the profits accruing to non-residents. High profits themselves need not be indicative of an excessive rate of return and may simply be a reflection of efficient operations. It is only where a firm is being subsidized by being permitted to earn monopoly profits that the host country need be concerned. In any case, available data for much of the 1960's show that the before-tax profits of foreign-owned firms in Canada, as a percentage of equity, averaged only slightly higher than those of domestic firms. Canada, moreover, participates in the profits of foreign-owned firms through taxation. The real financial cost of direct investment, therefore, declines as Canadian corporate and withholding taxes rise and as the authorities improve their surveillance of interfirm pricing (in particular, the prices set on raw material exports to the parent). The after-tax rate of return on U.S. direct investment in Canada, averaging about 8% for 1967-69, has been substantially below the average return on all U.S. overseas investment, and in fact in most countries would be considered a reasonable payment merely for the services of long-term debt funds.

As noted in Table 1 showing the financing of U.S. subsidiary companies in Canada, roughly half the profits of these companies are reinvested in Canada—a significantly higher proportion than is true of such firms in other countries. While the percentage has remained relatively unchanged over the years, the actual amounts of dividends paid out have of course increased in line with the greater absolute amount of direct investment in Canada; between 1950 and 1970, for example, total outward transfers have increased from $366 millions to $780 millions. Some concern has been expressed about the implications of these rising outflows for the long-term health of Canada's balance of payments. In fact, however, foreign direct investments have contributed both to growth in export earnings and to displacement of imports, and there is no evidence to suggest that these trade effects have in any way been insufficient to cover the growing volume of dividend payments. This is not to deny that the presence of foreign-controlled companies in Canada increases the risk of occasional instability in Canada's foreign exchange position. But such instability arises essentially from unfavourable patterns of events or misdirected policy developments, and foreign-controlled corporations are only one (though an important one) of the channels for destablizing movements of funds.

One final cost of direct investment, the degree of which is difficult to gauge, is its effect in reducing the incentives to improve skills and technology in Canada. The ease of importing such foreign capabilities (and the pull also which the United States has on talented and educated Canadians) has undoubtedly hampered the local development of entrepreneurial talents and discouraged domestic research and development. But while such impediments may be significant, it is still open to Canadians to seek to improve their capabilities in these respects.

BROAD POLICY ISSUES

Much of the expressed concern about foreign investment in Canada has focussed on the large size of many of the foreign corporations that have established subsidiary operations in Canada, on their dominant position in many important industries, and on potential conflicts of interest in the exercise of their decision-making power.

In fact, as the Corporations and Labour Unions Returns Act returns show, Canadian industries run the gamut from close to 100% control by non-residents (as in petroleum refining) to a near-zero element in many services and utilities. Successive governments have decided that certain "key sectors" of the economy—such as press and communications, air and rail transportation, financial institutions and, most recently, the uranium industry—should be subject to direct limitations on the extent of non-resident participation. Beyond this, various public regulatory authorities establish and administer a wide range of operational standards for vital industries, whether they be domestically or foreign controlled; these include such matters as forest management standards, pro-

vincial mineral regulations and conservation quotas for oil and gas. In important industries, also, operational conditions are much influenced by cross-border agreements worked out by the U.S. and Canadian governments. In all of these respects, there is a continuing challenge to find ways of best serving the national interest.

A more direct, but still rather general, aspect of government policies related to foreign investment lies in the so-called problem of extra-territoriality—i.e. in the extension of foreign laws and regulations to subsidiary operations in Canada. Most prominent in this respect have been some of the U.S. balance-of-payments guidelines of recent years, U.S. anti-trust laws, and the U.S. regulations on trade with communist countries. Essentially what is involved here is a matter of conflicting political jurisdictions, and such problems are becoming of wide international interest with the spreading operations of multi-national corporations. Thus far, however, the only approach open to the Canadian government is to be ready to introduce new national laws or administrative steps in instances where this is needed to countervail against the intrusion of U.S. law.

Probably one of the most significant direct problems arising out of the operation of major U.S.-controlled companies in Canada relates to the size of many of these companies and to their predominant position in important and (very often) the fastest-growing segments of Canadian industry. Though there is little if any evidence to support suggestions of undue political influence, especially in any combined way, the relative size of the companies involved does frequently add a complicating dimension to the pursuance of national objectives, and in some cases, too, there may well be difficulties in reconciling the specific interests of a subsidiary in Canada with the broad international interests of a multi-national organization. In general, however, these problems have been far from unmanageable, and for the most part they have represented just one aspect of the broader challenge to maximize the benefits, while also as much as possible limiting the costs, of foreign investment.

On this broad question of the operation of the performance of foreign-controlled companies, much study has been pursued in recent years especially by Professor A. E. Safarian of the University of Toronto. From his work Safarian concludes that "where behaviour that has been defined as undesirable by public authorities does appear, it can often be related more closely to aspects of the national industrial climate, and not location of ownership, that constitute the main impediment to good economic performance."

On one other source of concern—that is, the predominant position of large non-resident companies in fast-growing sections of the economy—the constructive response surely lies not in choking off the benefits to be secured through efficient expansion of such companies but in fostering more competitive market conditions in Canada and in seeking to strengthen the capacity of Canadian enterprises to participate in new areas of growth and development. Studies made by the Economic Council of Canada have pointed up the extent to which this country has lagged the United States in levels of educational attainment and managerial training as well as in effective local application of technological research. In these and other aspects of the question, the major challenge is to set realistic and balanced policy objectives, and to develop practical means of working towards them.

FOOTNOTES

1 See especially *Task Force Report on Foreign Ownership and the Structure of Canadian Industry,* prepared for Privy Council Office, January 1968; A. E. Safarian "The Performance of Foreign-owned Firms in Canada," for Canadian-American Committee, May 1969; Sidney E. Rolfe "The International Corporation," for Chamber of Commerce, Spring 1969; and C. P. Kindleberger *American Business Abroad,* Yale University Press, 1969.

2 Along with the (approx.) $44 billions of long-term foreign investment into Canada at the end of 1970, other items mainly short-term in character raised total Canadian external liabilities to about $49 billions. At the same time, Canadian assets in foreign countries totalled about $21 billions (the major components being over $11 billions of long-term investment, close to $5 billions of official exchange reserves, and substantial private holdings of short-term investments).

The Location of United States Manufacturing Subsidiaries in Canada

D. Michael Ray

United States residents control more of the manufacturing, the petroleum and natural gas, and other mining and smelting industries than do Canadian residents; the United States has more capital invested in Canada than in any other foreign country.[1] By 1963, 46 percent of manufacturing industry in Canada was controlled by residents of the United States compared with 40 percent by Canadian residents. The remaining 14 percent was controlled by residents of other countries, principally the United Kingdom. These figures represent a substantial increase in foreign control since 1926, the first year for which official data are available, when 65 percent of manufacturing industry was controlled by Canadian residents.

Substantial foreign investment poses potential conflicts between two of the most important institutions of the contemporary world, the multi-national corporation and the nation-state. About 60 percent of foreign long-term investment in Canada is direct investment which constitutes a "package" of product, technology, management, and market access as well as capital. The vehicle for direct investment is the establishment of foreign subsidiaries by multi-national corporations. The subsidiary attempting to implement decisions must deal with the management of the parent company instead of with a multitude of relatively powerless shareholders. If the parent company insists that the subsidiary operate according to the methods and even the laws of the parent's home country, the political and legal foundations of the subsidiary's host-state are weakened. The most frequent problems of this "extra-terri-toriality" in Canada have concerned United States laws on freedom to export, anti-trust law and policy, and balance of payments policy. Other problems of foreign direct-investment that have been studied include its impact on the Canadian capital market, on the size-structure of industrial corporations, and on the Canadian balance of payments.[2] Much less attention has been paid to the impact of foreign investment on industrial location, regional development, and regional disparities.[3]

This study is restricted to an examination of the influence of foreign investment in manufacturing industry on industrial location. The relative concentration of the United States-controlled industry in southwestern Ontario is documented and it is shown that differences in the location of Canadian and American-controlled manufacturing employment are not simply a consequence of the differences in industry-type in which foreign capital is invested. The distribution of the United States subsidiaries is explained in terms of the *economic-shadow* concept which is an extension of the gravity or interactance concept.

THE REGIONAL CONCENTRATION OF UNITED STATES-CONTROLLED MANUFACTURING

Manufacturing employment in Canada stood at one-and-a-quarter million in 1961 or 19 percent of the total labour force. Over one million of the manufacturing employees were

● D. Michael Ray *is Professor of Geography, State University of New York at Buffalo.*

Reprinted from *Economic Geography*, Vol. 47(3), 1971, pp. 389-400 with permission of the author and the editor.

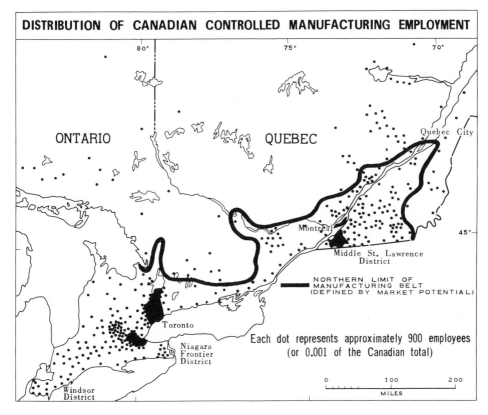

DISTRIBUTION OF CANADIAN CONTROLLED MANUFACTURING EMPLOYMENT

ONTARIO

QUEBEC

Quebec City

Montreal

Middle St. Lawrence
District

NORTHERN LIMIT OF
MANUFACTURING BELT
(DEFINED BY MARKET POTENTIAL)

Toronto

Niagara
Frontier
District

Each dot represents approximately 900 employees
(or 0.001 of the Canadian total)

Windsor
District

0   100   200
MILES

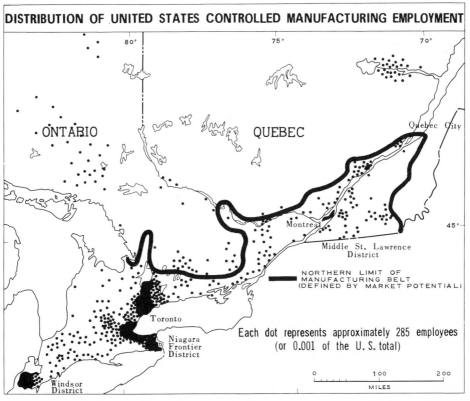

DISTRIBUTION OF UNITED STATES CONTROLLED MANUFACTURING EMPLOYMENT

ONTARIO

QUEBEC

Quebec City

Montreal

Middle St. Lawrence
District

NORTHERN LIMIT OF
MANUFACTURING BELT
(DEFINED BY MARKET POTENTIAL)

Toronto

Niagara
Frontier
District

Each dot represents approximately 285 employees
(or 0.001 of the U.S. total)

Windsor
District

0   100   200
MILES

in Ontario and Quebec, which had 63 percent of the total population of Canada but 81 percent of the workers in manufacturing. This concentration of manufacturing employment constitutes a manufacturing belt stretching from Windsor to Quebec City. The manufacturing belt comprises three manufacturing districts: the Niagara Frontier District which includes Toronto, the Windsor Border District, and the Middle St. Lawrence District which is centered on Montreal (Figure 1).[4] Both the Niagara Frontier District and the Windsor Border District extend across the International Boundary into the United States. Only the Middle St. Lawrence District, located almost entirely in the Province of Quebec, has no extension into the United States.

United States-controlled establishments and employment make a much greater contribution to the two contiguous manufacturing districts than to the Middle St. Lawrence District. Of the 1,618 United States-controlled subsidaries in Canada, 1,132 are in the Toronto-southwestern Ontario region.[5] The only cities outside these two contiguous manufacturing districts with large concentrations of United States-controlled subsidiaries are Montreal with 187, Vancouver with 50, and Winnipeg with 25.

Manufacturing employment in United States-controlled establishments shows a corresponding concentration in the two contiguous manufacturing districts (see Figure 1). This concentration contributes to the disparity in manufacturing activity between the manufacturing heartland and the Western and Atlantic Provinces. In 1961, 83 percent of the American-controlled employment was located within 400 miles of Toronto compared with 70 percent of the Canadian.

SOURCE OF CAPITAL AND INDUSTRY TYPE

Differences in the location of manufacturing classified by country of control may reflect, in part, differences in the type of industry in which Canadian, United States, and United Kingdom capital are invested. United States investment is concentrated in fast-growth industries including the manufacture of rubber products, transportation equipment, and electrical products; it is very low in the slower-growth industries such as leather goods, knitting mills, clothing, wood, furniture, and printing industries.[6] The influences of sources of capital and type of industry on industrial location can be discerned only by disaggregating county manufacturing employment by both type of industry and country of control. Accordingly, manufacturing employment in three industry groups that have substantial employment in both domestic and foreign-controlled establishments (the paper and allied industries, the metal industries, and the machinery industries), as well as in two broader industry divisions (primary manufactures and ubiquitous industry), has been cross-tabulated by country of establishment control.[7] Additional tabulations have been prepared for foreign-establishments of corporations having a book value over $25 million.

Factor analysis and multiple regression analysis have been applied using the 229 census counties of Canada as the observation units. Factor analysis groups the variables with similar geographic patterns into sets such that each set expresses an independent or different spatial pattern. Regression analysis is used to test the effectiveness of a set of independent variables to describe the location of employment in the Canadian, United States, and United Kingdom-controlled machinery industries, respectively. Both analyses provide strong evidence of locational differences when industry-type is further classified by country of control and suggest the importance of source of capital as a factor in industrial location.

The results of the factor analysis are given in Table 1. The first three factors in the analysis groups the variables by country of control into a United States, and a United Kingdom, and a Canadian manufacturing factor. The only industry categories that do not associate closely with these three country-of-control factors are the United States paper and allied industries, the ubiquitous-type industry for all three countries of control, and the U.S. and Canadian machinery

Table 1

MANUFACTURING EMPLOYMENT FACTORS

Variable	Factor Loadings					
	I US	II UK	III Canadian	IV Ubiquitous	V US Paper	VI Machinery
Total manufacturing employment	.03	—.01	—.01	.52	.00	.43
Industry mix employment growth 1949-59	.33	.27	.20	.23	.20	.06
Regional share employment growth 1949-59	.08	—.03	—.13	.01	.03	—.06
Regional share employment growth 1961-64	—.12	—.07	—.10	—.14	—.02	.01
Employment in all United States-controlled manufacturing establishments						
Total	.59	—.04	—.08	.19	.43	.20
Primary manufacturing	.66	.01	—.02	—.06	.56	—.02
Paper and allied industries	.05	.12	.06	—.04	.95	—.02
Metal industries	.92	—.05	—.00	—.04	—.14	—.03
Machinery industries	.00	—.04	—.06	.09	.02	—.57
Ubiquitous industries	.01	.00	—.08	.40	—.01	.02
Employment in large United States-controlled manufacturing establishments						
Total	.73	—.01	—.02	.03	.53	.06
Primary manufacturing	.72	.02	.01	—.09	.58	—.06
Paper and allied industries	.10	.09	.04	—.06	.95	—.04
Metal industries	.94	—.05	—.02	—.06	—.10	—.05
Machinery industries	.01	—.00	.03	—.02	—.00	—.71
Ubiquitous industries	—.01	—.05	—.01	.73	—.01	.12
Employment in all United Kingdom-controlled manufacturing establishments						
Total	—.00	.88	.06	.07	—.02	.05
Primary manufacturing	.01	.92	—.02	.01	.06	—.05
Paper and allied industries	.02	.88	.00	.04	—.00	—.07
Metal industries	—.01	.37	—.04	—.04	.09	.02
Machinery industries	—.01	.08	—.04	.10	—.05	—.03
Ubiquitous industries	.01	—.01	.05	.59	—.05	—.05
Employment in large Canadian-controlled manufacturing establishments						
Total	—.00	—.04	.87	—.01	—.03	.24
Primary manufacturing	.03	—.05	.93	.02	.03	—.12
Paper and allied industries	.07	—.02	.88	.02	.05	—.12
Metal industries	—.01	.06	.23	—.25	—.06	.63
Machinery industries	—.03	—.04	—.05	.24	.00	.33
Ubiquitous industries	—.06	—.02	.32	.49	.04	—.11

Source: Data tabulations prepared by author at the Canada Dominion Bureau of Statistics from the Joint Foreign-Owned Capital Study of 1961. See reference 1 for definition of ownership and reference 7 for definition of industry types. Large establishments are defined as those belonging to enterprises with a book value of over 25 million dollars.

industries and the Canadian metal industry, which form industry-specific factors.[8] Capital size-class of enterprise has no apparent locational effect and no distinctive grouping of the employment variables for large corporations occurs.

The independent variables used in the regression analysis of the machinery industries are market potential, proportion of population classified as rural farm, and distances from Toronto, Halifax, and New York.[9] For the three countries of control distinguished, only employment in the United Kingdom-controlled establishments does not show a statistically significant decline with distance from Toronto. But the decline for employment in the United States-controlled establishments from Toronto is almost ten times higher than for the Canadian-controlled.[10] By contrast, employment in United Kingdom controlled establishments is the only one to show a significant decline with distance from Halifax (the coefficient is —.019).

The elements of the spatial patterns for the three employment distributions in the machinery industries also have a different order of importance. The most important element is market potential for the Canadian-controlled, distance from Toronto for the United States-controlled, and distance from Halifax for the United Kingdom-controlled. The second most important element is distance from Toronto for the Canadian-controlled, distance from New York for the United States-controlled, and market potential for the United Kingdom-controlled. The

Figure 2

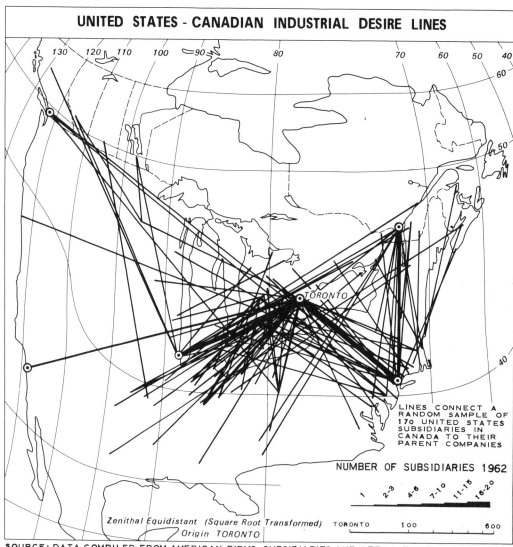

UNITED STATES - CANADIAN INDUSTRIAL DESIRE LINES

LINES CONNECT A RANDOM SAMPLE OF 170 UNITED STATES SUBSIDIARIES IN CANADA TO THEIR PARENT COMPANIES

NUMBER OF SUBSIDIARIES 1962

1   2-3   4-6   7-10   11-15   16-20

Zenithal Equidistant (Square Root Transformed)
Origin TORONTO

TORONTO   100   600

SOURCE: DATA COMPILED FROM AMERICAN FIRMS, SUBSIDIARIES AND AFFILIATES IN CANADA, WASHINGTON, D.C., U.S. DEPT. OF COMMERCE, BUREAU OF INTERNAL COMMERCE, 1962

proportions of the inter-county variation in employment explained by these few, simple elements are very small, but statistically significant (the correlation coefficients are .40 for the Canadian, .30 for the United States, and .16 for the United Kingdom-controlled establishments).

These results support the evidence of the factor analysis that differences in the location of manufacturing employment are not merely a reflection of the different industry groups in which United States capital is concentrated. Instead, these differences represent the influence of external factors on the location of United States-controlled subsidiaries.

INDUSTRIAL INTERACTANCE AND ECONOMIC SHADOW

The influence of external factors on the location of United States-controlled subsidiaries is illustrated by the industrial desire-lines in Figure 2, which connect, by straight lines, a random sample of manufacturing subsidiaries in Canada with their American head office.[11] An examination of this map and of the complete data[12] suggests that the size and distance of United States metropolitan centres affect the number of Canadian subsidiaries which they control. The desire-lines from New York and Chicago are dominant. New York controls 307 Canadian subsidiaries, whereas Boston, much smaller but a little closer to Canada, has 48. Chicago and Los Angeles each have about the same number of manufacturing establishments, yet Chicago controls 197 Canadian subsidiaries compared with the more distant Los Angeles, which controls only 45. These relationships can be expressed by an interactance model in which the number of manufacturing subsidiaries in a Canadian city that is controlled by an United States metropolitan area is proportional to the number of manufacturing establishments in that metropolitan area, and inversely proportional to its distance from that metropolitan area.[13]

Because of the contribution of United States subsidiaries to manufacturing in Canada, there are two important corollaries to the interactance model: first, regional economic development in Canada will tend to reflect the economic health of adjacent regions of the United States; and second, the Canadian regions most likely to acquire large numbers of United States subsidiaries are those closest to the American manufacturing belt.

Two additional elements in the location of United States subsidiaries are evident from examination of the maps showing the location of United States subsidiaries, Figures 3 to 6. The stronger of these is *sectoral affinity,* which is defined as the tendency of subsidiaries to locate in the geographic sector that links the parent company with Toronto, the point of highest market potential. Consequently, most desire lines appear to radiate from Toronto (Figure 2). Toronto provides the optimal market location for American subsidiaries and few subsidiaries locate beyond it. Industrial interactance between a Canadian city and a United States city is severely restricted wherever Toronto becames *an intervening opportunity,* eclipsing sectoral affinity.

Compare Montreal with Toronto. Of the 210 Standard Metropolitan Statistical Areas (SMSAs) in the United States with Canadian subsidiaries, only eight control more subsidiaries in Montreal than in Toronto. Seven of these are located on the Atlantic seaboard: Brockton, Fitchburg, New London, Reading, Patterson, Atlantic City, and Greensboro. The eighth is Portland, Oregon. These eight control 16 subsidiaries in Montreal out of 187 United States subsidiaries in that city, and are dwarfed by the New York SMSA which controls 74 subsidiaries in Montreal. In the cases of New York and several of the eight, Toronto is farther than Montreal and does not constitute an intervening opportunity.

The third element that is apparent from the maps is *spatial momentum* or *sectoral penetration.* This element defines a hypothesis which states that the distance that a parent company penetrates into Canada to locate a branch plant is directly proportional to the distance of the parent company from Canada. The Detroit manufacturer, for instance, can evade the Canadian tariff barrier

293

and the prejudice against foreign products by locating a branch plant across the Detroit River. The marginal benefits of locating closer to the centre of the Canadian market may not compensate for losing the convenience of operating the subsidiary close to the parent company.[14] Consequently, 34 of Detroit's 87 Canadian subsidiaries are located in Windsor where they comprise more than half the total of United States subsidiaries (Figure 5). Eight of Seattle's 11 Canadian subsidiaries are in Vancouver. Detroit controls only 20 subsidiaries in Toronto and Seattle, none. By contrast, Los Angeles has half of its subsidiaries in Toronto, but none in Windsor (Figure 3). In general, the proportion of an American SMSA's Canadian subsidiaries located in Toronto increases with increasing distance from the Canadian Border.

The three elements, industrial interactance, sectoral affinity, and spatial momentum to-

Figure 3

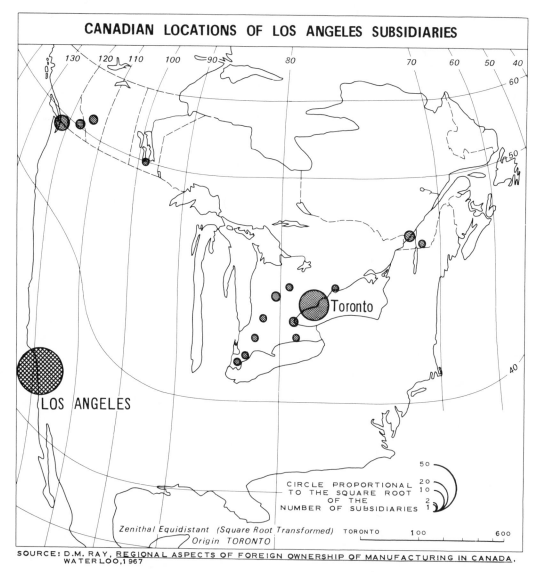

CANADIAN LOCATIONS OF LOS ANGELES SUBSIDIARIES

CIRCLE PROPORTIONAL TO THE SQUARE ROOT OF THE NUMBER OF SUBSIDIARIES

Zenithal Equidistant (Square Root Transformed) Origin TORONTO

SOURCE: D.M. RAY, REGIONAL ASPECTS OF FOREIGN OWNERSHIP OF MANUFACTURING IN CANADA, WATERLOO, 1967

gether comprise the economic shadow concept. Economic shadow defines the relative inability of regions to supplement locally owned and controlled industry with externally controlled branch plants. The foreign component of economic shadow is particularly important for Canada; it could be argued that the same concept applies to the location of manufacturing branch plants of Canadian-controlled enterprises within Canada.

STATISTICAL TESTS OF ECONOMIC SHADOW

The economic shadow concept is tested using both the number of subsidiaries and the total number of manufacturing employees in each Canadian location controlled from head-offices in each United States metropolitan area. The basic interactance components of the economic shadow model are measured by the market potential at the

Figure 4

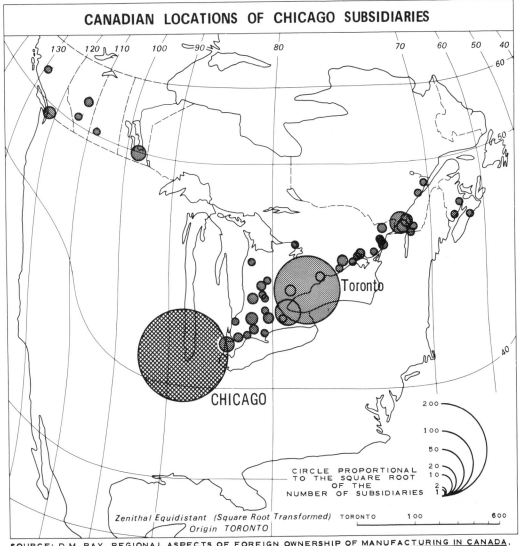

CANADIAN LOCATIONS OF CHICAGO SUBSIDIARIES

CIRCLE PROPORTIONAL TO THE SQUARE ROOT OF THE NUMBER OF SUBSIDIARIES

Zenithal Equidistant (Square Root Transformed)
Origin TORONTO

SOURCE: D.M. RAY, REGIONAL ASPECTS OF FOREIGN OWNERSHIP OF MANUFACTURING IN CANADA, WATERLOO, 1967

295

Canadian centre, the number of manufacturing firms at the United States location with more than 50 employees, and the air distance between the Canadian and United States locations.

The measurement of affinity and momentum is based on an X, Y coordinate system centred on the United States location, with the Y axis passing through Toronto and the X axis perpendicular to it as shown in Figure 7. The X and Y values are scaled by setting the United States-Toronto distance equal to unity regardless of the actual distance involved. The coordinates of each Canadian city with subsidiaries of parent firms at each United States centre are computed. The coordinate system assumes that both affinity and momentum can be described in terms of the *relative* location of Toronto to United States and Canadian cities.

The economic shadow model explains two-thirds of the intercounty variation in the

Figure 5

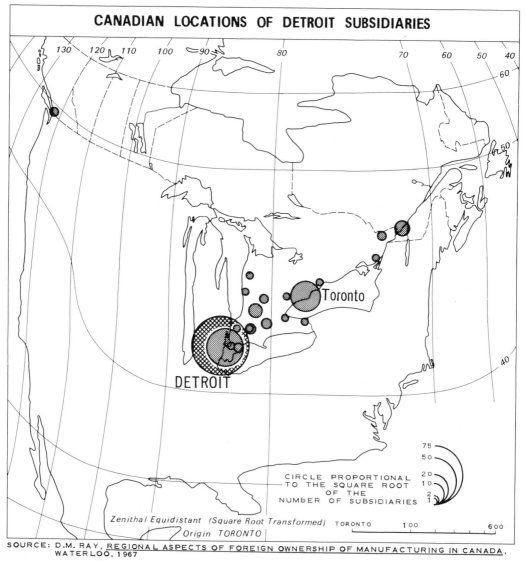

CANADIAN LOCATIONS OF DETROIT SUBSIDIARIES

CIRCLE PROPORTIONAL TO THE SQUARE ROOT OF THE NUMBER OF SUBSIDIARIES

Zenithal Equidistant (Square Root Transformed)
Origin TORONTO

SOURCE: D.M. RAY, REGIONAL ASPECTS OF FOREIGN OWNERSHIP OF MANUFACTURING IN CANADA, WATERLOO, 1967

location of United States-controlled establishments. Unfortunately, a complex form of the model is needed to reach this level of explanation and only the X term (of all the X, Y coordinate values employed) is stable enough to be statistically significant (Table 2). The general form of the shadow is as expected and the number of establishments locating beyond Toronto is fewer than would be predicted if the effects of sectoral affinity and spatial momentum were ignored.

The economic shadow elements of sectoral affinity and spatial momentum are not important in explaining the distribution of United States-controlled employment in Canada or in any of the regions tested. The model thus reduces to the simple interactance elements. The amount of manufacturing employment in any Canadian census division controlled from a United States metropolitan centre is proportional to the number of manufacturing firms at the United

Figure 6

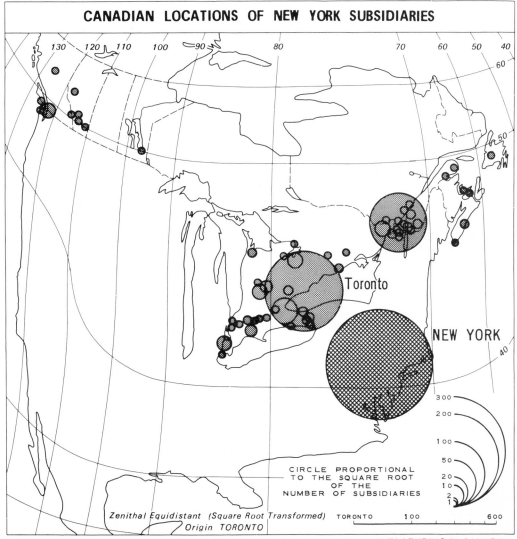

CANADIAN LOCATIONS OF NEW YORK SUBSIDIARIES

CIRCLE PROPORTIONAL TO THE SQUARE ROOT OF THE NUMBER OF SUBSIDIARIES

Zenithal Equidistant (Square Root Transformed) Origin TORONTO

SOURCE: D.M. RAY, REGIONAL ASPECTS OF FOREIGN OWNERSHIP OF MANUFACTURING IN CANADA, WATERLOO, 1967

Figure 7

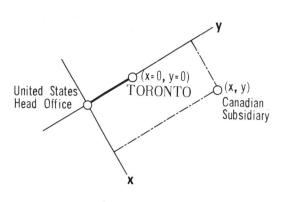

Table 2

ECONOMIC SHADOW MODEL FOR UNITED STATES-CONTROLLED SUBSIDIARIES

Variable	t Value	Variable	t Value
1. P log W	6.01**	6. log D log P	—1.94
2. (logP)/D	1.86	7. X	—3.23**
3. W log P	3.31**	8. X	—1.42
4. W log D	—2.77**	9. XY	— .92
5. P log D	—3.26**	10. Y	—1.36

P = market potential at Canadian location.

W = number of firms at United States location with more than 50 employees.

D = air distance from Canadian to United States location.

X, Y variables are based on coordinate system. Number of observations is 105.

Coefficient of multiple correlation is .82.

* A t value of 1.98 is significant at the .05 level.

** A t value of 2.62 is significant at the .01 level.

States metropolitan centre W, and the market potential at the Canadian centre P, and is inversely proportional to their distance apart D (see Table 3).

The effect of distance in reducing the amount of employment, termed distance decay, shows important regional variations. Distance decay is very strong in Quebec and the Atlantic Provinces, weak in Ontario, and takes a modified form in the Prairie Provinces and British Columbia. In the western periphery, the distance decay effect is incorporated in the shadow variable X which measures distance from the axis linking the

United States centre to Toronto. Generally, employment decreases westward from Toronto more than can be accounted for by differences in market potential (Table 3).

When the model is tested for Canada, the economic characteristics combine in a simple model comprising tthe interactance variables, P, W, and D, to explain the geographic variations in manufacturing employment controlled from each United States metropolitan area. Toronto appears to be much less effective as an intervening opportunity for the location of all manufacturing employment than for subsidiaries.

Nevertheless, the economic shadow models verify the importance of two external factors, the size of United States manufacturing centres, and their distance from Canadian centres; each of these factors exerts a major influence on the distribution of United States-controlled employment.

CONCLUSION

The location of manufacturing activity in Canada is not related solely to internal economic factors. United States subsidiaries comprise almost half of the manufacturing industry in Canada and the number and location of these subsidiaries is significantly related to two external factors, the size and distance from Canada of United States metropolitan centres. Foreign investment has created a set of externally-related development axes, which are a spatial expression of the interactance process, and along which the development impulses are proportional in intensity to the market potential of the Canadian centre and the size of the United States centre, and inversely proportional to their distance apart. The development impulses have focussed on Toronto, the point of highest market potential in Canada, and the two strongest axes are the Toronto-New York and the Toronto-Chicago axes. Indeed, in the case of the location of United States-controlled establishments, though not employment, the attraction of Toronto is greater than would be expected on the basis of market potential alone; and an economic shadow process can be identified in which the num-

Table 3

ECONOMIC SHADOW MODEL FOR
UNITED STATES-CONTROLLED EMPLOYMENT

Region	Number of Observations	Coefficient of Multiple Correlation	Variable	t value
Quebec and Atlantic Provinces	69	.65	$\log D$	—1.8
			$(\log W)/D$	—4.2**
			$(\log P)/D$	4.4**
			$\log P$	—6.1**
Ontario	190	.48	$\log D$	—4.9*
			$\log P \ \log W$	6.1**
			XY	1.4
Western Provinces	42	.60	$\log P \ \log W$	3.8**
			X	2.9
Canada	301	.48	$\log D$	—5.44**
			$W \ \log P$	5.03**
			$\log P \ \log W$	2.59**

Note: The rate at which industrial intractance decays with distance is given by the regression co-
efficients of the D variable. The value is .86 for Quebec and the Atlantic Provinces, .46 for
Ontario, and .35 for Canada.
 * Indicates that the regression coefficient is significant at the .05 level.
** Indicates significance at the .05 level.

ber of subsidiaries in a Canadian city reflects its location relative to Toronto, as well
as the elements of the interactance process.

The location of United States-controlled
manufacturing subsidiaries, and, to some
degree, the geography of regional development in Canada, have been shaped by a process of industrial interactance with United
States metropolitan centres. It is unlikely
that this process of industrial interactance is
restricted either to Canada or to foreign investment, and other case studies are needed
to strengthen the link that this process provides between location theory and regional
development theory. Nevertheless, the process
is particularly important in the case of United
States-controlled subsidiaries in Canada because they represent a level of foreign investment unique among the developed countries of the world.

REFERENCES

[1] An enterprise is considered to be foreign controlled by the Dominion Bureau of Statistics
if 50 percent or more of its voting stock is known to be held in one country outside Canada,
or if effective control is held with less than 50 percent of the voting stock.

[2] The problems are examined in Watkins, M. H., et al. *Foreign Ownership and the Structure of Canadian Industry*. Ottawa: Privy Council 1968; Aitken, H. C., *American Capital
and Canadian Resources*. Cambridge: Harvard University Press, 1961; and Johnson, H. C.,
The Canadian Quandary; Economic Problems and Policies. Toronto: McGraw Hill, 1963.

[3] Efforts to examine this problem include: Ray, D. M., "Market Potential and Economic
Shadow," Research Paper 101. Chicago: Department of Geography, University of Chicago Press, 1965; Ray, D. M., *Dimensions of Canadian Regionalism*. Geographical Paper No.
44. Ottawa: Canada Department of Energy, Mines and Resources (forthcoming); and Brewis,
T. N., *Regional Economic Policies in Canada*. Toronto: Macmillan Co. of Canada, 1969.

[4] White, C. L., Foscue, E. J., and McKnight, T. L., *Regional Geography of Anglo-America.* Englewood Cliffs: Prentice-Hall, 1964, pp. 32-54; and Brewis, T. N., *Regional Economic Policies in Canada.* Toronto: Macmillan Co. of Canada, 1969, pp. 15-18.

[5] Tabulations were made from U.S. Department of Commerce, Bureau of Internal Commerce. *American Farms, Subsidiaries and Affiliates in Canada,* 1962.

[6] Canada. Dominion Bureau of Statistics., *International Investment Position: 1929-1954.* Ottawa: Queen's Printer, 1958, and Canada., Dominion Bureau of Statistics, *The Canadian Balance of International Payments, 1961 and 1962, and International Investment Position.* Ottawa: Dominion Bureau of Statistics, 1964, pp. 95-97.

[7] The industry groups are defined in the *Standard Industrial Classification Manual.* Ottawa: Dominion Bureau of Statistics, 1960. Primary industries are defined in this study as SIC categories 101, 103, 105, 107, 111, 112, 123, 124, 125, 133, 147, 151, 251, 271, 295, 341, 351, 357, 372, and 373. Ubiquitous industries include 128, 129, 131, 139, 141, 145, 254, 273, 286, 287, 288, and 289.

[8] A second factor analysis using sixteen of these variables together with a wide range of demographic, cultural, and economic characteristics gives much the same results. See Ray, D. M., "The Spatial Structure of Economic and Cultural Differences: A Factorial Ecology of Canada," *Papers, The Regional Science Association,* 23, 1969, pp. 3-23.

[9] Harris, C. D., "The Market as a Factor in the Location of Industry in the United States," *Annals of the Association of American Geographers,* Vol. 44 (December 1954), pp. 315-348.

[10] All variables are transformed into logarithms and the distance decay rates are indicated by the regression coefficients, —.024 for the Canadian and —.228 for the United States. Both coefficients are significant at the .01 level.

[11] Distances from Toronto on this map equal the square root of the real-earth distance.

[12] Day, D. M., *Dimensions of Canadian Regionalism.*

[13] See. Olsson, G., *Distance and Human Interaction.* Philadelphia: Regional Science Research Institute, 1965; MacKay, J. R., "The Interactance Hypothesis and Boundaries in Canada," *Canadian Geographer,* Vol. 11, 1958, pp. 1-8; Ray, D. M., "Market Potential and Economic Shadow," Research Paper 101. Chicago: Department of Geography, University of Chicago, 1965; and Wolfe, R. I., "Economic Development," in *Canada: A Geographical Interpretation,* Edited by J. Warkentin. Toronto: Methuen and Co., 1968.

[14] Wolfe finds that spatial momentum occurs in the recreation travel behavior of American tourists in Ontario. See, Wolfe, R. I., *Parameters of Recreation Travel Behavior in Ontario.* Toronto: Ontario Department of Highways, 1966. He later compares this momentum in recreation travel behavior to sectoral penetration in plant location (see Wolfe, R. I., "Economic Development," pp. 226-227).

Factors Affecting Mineral Development in Northern Canada

AMIL DUBNIE AND W. KEITH BUCK

Despite a slow but steady improvement in facilities that strongly influence development, the basic problem of distance from markets will always remain in Canada's North. Location relative to major markets becomes less of an obstacle in conditions of scarcity, but this condition does not exist for northern products. It is recognized that an adequate supply of virtually all metals and petroleum products exists throughout the world and that this condition will remain for some time to come. However, there are some problems related to northern development that can be reduced through appropriate action by government and private bodies. Specific areas of action are in the improvement of transport facilities, the availability of low-cost power, improved communications, availability and stability of labour, and the provision of incentives that make northern industrial activities more attractive economically than they might otherwise be.

TRANSPORTATION

After many years of effort and expenditure of large amounts of money, transportation, so important to reduction of economic disadvantage imposed by distance, remains the most formidable obstacle to northern development. In recent years there has been improvement in all forms of northern transportation, but the greatest advance has been made in overland forms.

Railways

Highlighting recent development has been the construction of the 432-mile Great Slave Lake Railway from Roma, Alberta, to Great Slave Lake, including a 55-mile spur to the Pine Point mine (Figure 1). Construction of the $86 million railway line began in February, 1962, and it was not long before work was ahead of schedule.

Before the railway project was completed, over 10,000 carloads of materials were used, in itself contributing to considerable work elsewhere in Canada. Of the eight major steel bridges and thirty-three smaller steel and timber structures along the route, the largest was the 2,000 ft. steel structure which spans the Meikle River at Mile 73. The Meikle Valley was the only major obstacle which had to be traversed. The general terrain was a series of low plateaus between a number of shallow but wide river valleys.

Construction of the railway was greatly aided by its proximity to the Mackenzie Highway. For example, in one instance a 46-ton locomotive was transported by truck over the highway for 143 miles farther north to allow for simultaneous construction of the Pine Point spur. General ease of access to the site of railway construction from the railway played a large part in facilitating communication and movement of supplies, thereby keeping the construction of the project running smoothly.

● AMIL DUBNIE *is Mining Engineer, Mines Branch, and* W. KEITH BUCK *is Chief, Mineral Resources Division, Department of Energy, Mines and Resources, Ottawa.*

Excerpt modified from "Progress of Mineral Development in Northern Canada," *Polar Record,* Vol. 12 (81), 1965, pp. 683-702, with permission of the authors and the editor.

Figure 1

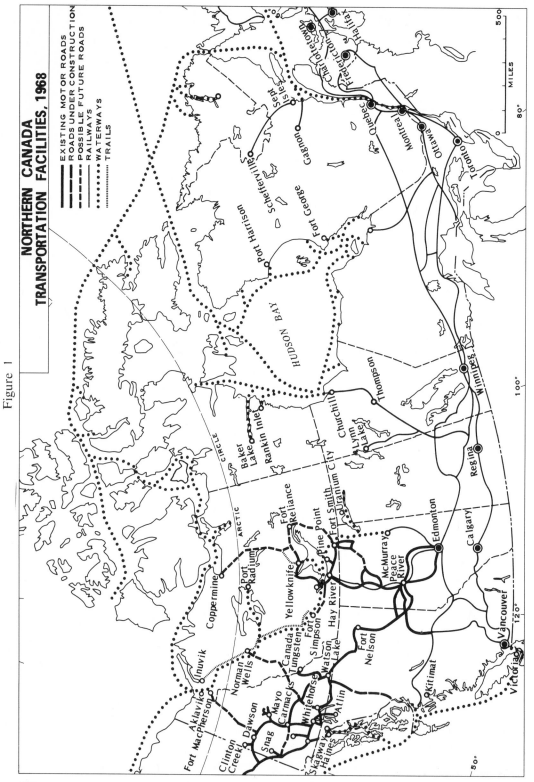

NORTHERN CANADA
TRANSPORTATION FACILITIES, 1968

EXISTING MOTOR ROADS
ROADS UNDER CONSTRUCTION
POSSIBLE FUTURE ROADS
RAILWAYS
WATERWAYS
TRAILS

The railway was not scheduled for completion until December, 1966, but by October, 1962, shipments of grain moved south from Manning, Alberta, at Mile 55. The influence of the railway upon other than the mineral industry is illustrated by the 2,000 carloads of southbound revenue shipments, comprised mostly of grain and lumber, which moved southward by the fall of 1964.

Construction of this railway into the Northwest Territories brings two important considerations into focus. First, the selected location along an established communication route points to rail transport as a superior form of overland transport for bulk shipments of minerals and other products. It may be argued that the southbound freight could be transported over the highway, but if so, would the costs be sufficiently low to make the northern products competitive? The decision to build the railway virtually alongside a good existing highway indicates that the highway is unsuitable to meet the ultimate needs of northern development.

Second, construction of the railway has moved the "jumping-off" point for northern exploration and development 400 miles farther north. If we consider northern exploration supplies and equipment starting from Great Slave Lake versus starting at Peace River, the saving to the mineral industry is readily apparent. Assumed that transportation costs by road will be 4.50 cents a ton-mile against 2.25 cents a ton-mile by railway—the costs at least over the 400-mile distance are immediately halved, a significant advantage.

On minor programs, the total cost of freighting will not be significant as compared with, say, costs for salaries of personnel. On a large-scale program of the kind that involves large total expenditure in the north, freighting of supplies becomes a major cost. One would expect that the prospects of rail service for southern shipments of base metal ores and concentrates may significantly affect the multitude of decisions that result in activity in the Canadian north as opposed to activity elsewhere.

The White Pass and Yukon Railway in Yukon Territory continues to operate much as before. The company that operates it continues to give detailed attention to the combined road-rail-water transport problem, viewing it as a unified service to the areas served. The company has concentrated on containerized cargo movement, which greatly aids trans-shipment, but despite the efficiency of the existing overall transport system, freight volume along the railway is still far below capacity.

With the full cooperation of the White Pass and Yukon Route, a survey of a rail route to reach the Snake River iron deposit has been made by the Canadian National Railway. If the iron deposit is developed, and a standard-gauge railway is built to service it, one would expect part of the route to follow the existing White Pass and Yukon Railway right-of-way.

Some of the prospects north of Great Slave Lake have been under investigation for several years, indicating that results must be sufficiently encouraging to justify sustained attention. If evaluation leads to development of base-metal deposits for production, it appears likely that the Great Slave Lake Railway would be extended, or, depending upon the specific location of properties under development, perhaps railways from northern Manitoba can be extended into the Northwest Territories. In Mackenzie District, it appears logical to expect a railway eventually to follow the Mackenzie River to the Arctic Ocean. Unless some more economical form of bulk transport than railways can be developed, an eventual network of railways can be envisaged in the north.

During 1970 and continuing in 1971, research is being conducted into the feasibility of transporting natural gas and petroleum by pipeline along the Mackenzie River. These activities have been sparked by discovery of petroleum in Alaska and of natural gas in the Arctic Islands.

Sober reflection upon the past and future northern railway developments lead to the thought that little is known about construction and operation of railways in the "barrens." We can expect low temperatures, constant, strong winds, and drifting, compacting snow in winter. One wonders whether research could profitably be pursued now to determine whether standard methods of railway design will be adequate to cope with the environment, or whether modification to design, location, and construction techniques might be profitably employed.

Roads

Transport by road from established jumping-off points into northern areas continues to be aided by government programs of resource and development road construction. Typical examples are to be seen in the Yellowknife and Watson Lake areas.

In 1959, the Mackenzie Highway was extended from Hay River around the west side of Great Slave Lake to Yellowknife. An icebridge and ferry arrangement was provided for traversing the Mackenzie River near Fort Providence. Completion of this important road made it possible to truck supplies directly from Edmonton to Yellowknife except during brief periods of break-up and freeze-up.

Private enterprise has used the opportunity presented by an all-weather road from Edmonton to extend trucking over a specially prepared cross-country route to Tundra Gold Mines, some 185 miles north-east of Yellowknife. In February, 1962, the barrens were for the first time penetrated by truck transport reaching Tundra from Yellowknife. Until the summer of 1964, the private carrier delivered 5,040 tons of freight, part of which was comprised of over 400,000 gallons of fuel oil. The rate by truck from Yellowknife to Tundra was reduced to $65 a ton against $90 a ton by Bristol aircraft and $300 a ton by small, single-engined aircraft. The trucking rate directly from Edmonton to Tundra was $110 a ton for the 1,125 mile haul.

An important "Roads to Resources" project, which was completed in the 1960's, was the road from Watson Lake, Yukon Territory, northwest towards Ross River. From this resource road a development road was pushed on to the Flat River tungsten mine of Canada Tungsten Mining Corporation Limited that has, after some delay owing to the collapse of the tungsten markets, begun production by the open-pit mining method. Operations continued on a seasonal basis in 1970.

A wider use of overland transport for northern servicing into areas previously considered inaccessible to overland transport was indicated by the planned construction during 1965 of a winter road along the Mackenzie River almost to the Arctic. The winter road was planned to follow the telephone right-of-way of Canadian National Telecommunications from Fort Providence to Inuvik. The major users of the road were expected to be oil companies who would independently negotiate a rate with the road builder, Calgary Exploration Services. The road was expected to require seventy-six river crossings and to traverse a number of minor obstacles. The estimated cost of rebuilding the 1,000-mile road was reported to be about $300,000 each year. The freighting season was expected to last five months.

An unexpected use of tracked trucks in the Arctic was the 500-mile traverse from Resolute to Cape Isacson in 1963, conducted for scientific purposes by the Polar Continental Shelf Project. It resulted in operational costs that were considerably lower than by sea or air. The successful completion of the traverse substantially proved the feasibility of using tracked trucks in the Canadian Arctic.

Water

There have been few changes in methods of water transport in northern Canadian waters. Further development of overland transport systems is expected to detract from the Mackenzie water transport system. In the eastern Arctic, the water-transport season, for purposes of insurance, does not start until July in Hudson Strait. During 1970, the shipping season out of Churchill, Manitoba, opened on July 26, and closed on October 26.

Despite the short shipping season, the economic advantage of using water over air transport, is well illustrated from practical experience. One company recently landed 320 tons by freight in Deception Bay, Ungava, by sea transport from Montreal at a cost of $50 a ton. The cost for the equivalent haul by air transport would have been $600 a ton. This relatively low cost of transport by water into the eastern Arctic is a favourable factor, which may advance development of the iron and other mineral deposits in Ungava and elsewhere.

Water transport from Baffin Island has been considered as a means of making production economic at the property of Baffinland Iron Mines Limited. One proposal is to haul by 50- to 60-mile railway from the mine site to Milne Inlet, then during the summer by large water transports to stockpiles on Rypeö, Greenland, some 870 nautical miles away. From Rypeö, a year-round haul of about 2,200 miles would be necessary to reach Rotterdam or Philadelphia.

Air

Air transport continues to be a major means of transporting personnel and freight throughout the north. Unfortunately, the volume of high value freight is insufficient in many cases to allow for utilization of large aircraft and thus reduce the cost to a point that is competitive with other forms of transport. It is worth noting, however, that aircraft of about DC-4 size are now more frequently used than formerly. One company recently found that where an adequate airstrip could be provided for DC-4 aircraft on Precambrian or Cordilleran terrain, costs of air freighting could be brought down to 23 cents a ton-mile. Much lower rates than these have been achieved with similar aircraft on a regular run between Edmonton and Uranium City.

Finding constantly widening use in exploration are helicopters, which are highly adaptable for this type of service. They are unusually useful for mapping from the air where the amount of rock outcrop represents 40 to 50 percent of the terrain, and for dropping off ground-parties for more detailed examination of specific features. Use of chartered helicopters for geological mapping at a scale of 8 miles to the inch has been pioneered by the Geological Survey of Canada and has spread to private interests. Mapping to larger scale has been a logical development, and recent costs of mapping to a scale of one inch to the mile have been estimated at $24 to $30 a square mile depending upon distance from the base.

The Ground Effect Machine (GEM) * appears to offer some promise as a form of transport that might complement railways in the northern environment. Its economic range of operations is estimated to be over distances between 50 and 700 miles. Unfortunately, the GEM has not been tested over northern terrain, despite the interest that has been aroused following demonstrations along the St. Lawrence. It appears to be worth trying out in the northern environment because of the high percentage of water, muskeg and tundra that occurs there. The vehicle can be constructed to large sizes with proportional decrease in ton-mile freighting costs. Such data as have been developed for operation of GEM's under northern conditions result exclusively from design criteria. Operation under field conditions appears necessary to assess fully the vehicles' potential.

POWER DEVELOPMENT

The power resources of Yukon Territory are known to be very large, and of the Northwest Territories to be adequate, for foreseeable needs. Efficient uses of available power resources in the northern areas

---

*Hovercraft. The GEM was tested in the Canadian Arctic in snow and ice conditions during 1966 under the joint sponsorship of the Canadian and British Governments. Suitability of design concepts for the test conditions were fully verified; however, the economics relating to industrial application were not evaluated.

is extremely important, owing to under-development of fossil fuel resources in the west and unfavourable fuel transport costs in the east. Studies undertaken to determine whether nuclear power plants could serve the north have disclosed that costs of nuclear power would be high, owing to the small size of plants.

The Government of Canada has recognized the problems, and as early as 1948, established the Northern Canada Power Commission to develop and manage public power plants. Privately owned plants remained under their original owners and were not affected in any way.

As of 1969, Yukon Territory had an installed generating capacity of 122,000 kw in hydroelectric plants and 35,000 kw in thermal plants. Northwest Territories had an installed capacity of 224,000 kw in hydroelectric plants and 64,000 kw in thermal plants.

Fortunately, many of the undeveloped power sites in the territories are well located to serve future developments in the mineral industry. Sites for substantial blocks of power exist around Great Slave Lake and along the Yukon River. In Ungava, known power sites are south of lat. 60°N, but still within reach of potential mineral developments.

COMMUNICATIONS

Northern mineral development is greatly expedited by continuing improvements in communications that have developed in recent years. Much of the Yukon is within reach of the Canadian telephone service by means of facilities owned by Canadian National Telecommunications. The same company also provides Yellowknife with equivalent service and will ultimately expand service up the Mackenzie Valley.

For much of the north, however, high-frequency radio will probably be the major means of communication. This is particularly true in areas of low density traffic where a microwave or other system would be uneconomic.

LABOUR

In recent years there has been little change in the problems associated with maintaining personnel in the north; the practice of importing skilled labour from southern areas to work the northern mines and oil rigs continues. One author recently suggested that skilled workers from the south be given additional income-tax exemptions to enhance northern service.

Meanwhile, government departments are striving to provide gainful employment for the native population, whose birth-rate exceeds, by a wide margin, the Canadian average. Some success has been achieved in training local labour, and private companies have sometimes taken advantage of the assistance available towards this end. Ultimately, the comprehensive educational opportunities available to Eskimos and Indians will produce an adequate number of highly trained individuals.

When one examines labour turnover records on northern mineral enterprises, one is impressed with the high turnover of workers from the south. This high turnover is indeed costly to employers; they must further provide housing and other amenities virtually to city standards in order to retain labour. For highly skilled occupations, which require advanced technical knowledge, there is no question about filling vacancies from local sources; it just cannot be done. For many other unskilled and semi-skilled occupations, would it not be advantageous to recruit and train local personnel? Advice and assistance is available from the government, but this has not been fully exploited.

INCENTIVES FOR DEVELOPMENT

The federal government is aware of the problems that face producers of minerals and has taken steps to make these ventures more attractive as an incentive to development. At the national level, a three year tax-exempt period will continue to the end of 1973. The present 33⅓ percent depletion allowance will continue to the end of 1976, when it will be replaced by another schedule. Provision for write-off of exploration expenses, accelerated write-off of development expenditures and other encouragements have been provided to assist development of mineral deposits. Additional advantages are available for those who would discover, explore, and develop minerals in the north.

A Prospectors Assistance Program, which came into effect in 1962, is undergoing constant improvement. It offers the serious prospector direct financial assistance towards his costs of transportation into and out of his area, and towards his expenses while prospecting. Virtually all the funds available to the program are committed, each year, illustrating the high level of participation.

A sum of $100,000 is provided each year to the Territorial Governments for tote trails in accordance with the terms of the Federal-Territorial Financial Agreements. Tote trails are minimum-standard roads that provide seasonal or year-round access for the purpose of moving in supplies and equipment for exploration, development, or small-scale production. The bulk of the tote-trail expenditures have been applied towards access into mineral locations.

In the 1971-72 Estimates, the Department of Indian Affairs and Northern Development has provided $160,000 for airstrip construction in the Northwest Territories and the Yukon.

The revised Canada Mining Regulations and the revised Canada Oil and Gas Land Regulations have been in effect since 1961. These regulations, and the Yukon Quartz Mining Act, provide a reasonably attractive climate for mineral exploration and development; however, the legislation is now under review.

Governmental support towards northern development is evident in the activities of Panarctic Oils Ltd., the industry-government company that is exploring for petroleum in Canada's Arctic Islands. This company committed $30 million in 1969 towards the drilling of 19 exploration wells. By the end of 1969 three wells had been drilled on Melville Island and one of these was a major natural gas discovery.

In closing, it may be worth while to recapitulate and enlarge upon the facilities provided for the Pine Point deposit in order to illustrate the extent of concrete government action designed to stimulate northern mineral development. The Great Slave Lake Railway was financed in large part by the federal government. The federal government invested over $9 million in the Taltson River hydroelectric power project and spent about $2 million on a 53-mile all-weather road from the Mackenzie Highway into the Pine Point mine site. For a total expenditure of close to $90 million, in addition to freight charges, a sum not exceeding $20 million will be repaid by the company over a period of ten years. The large government contribution to this development is related to an annual production value of about $22 million, and permanent employment of 200 persons.

Private and governmental activity in the north, so adequately demonstrated by the Pine Point development, by the large acreage under active exploration for petroleum and natural gas, and by sustained activity in exploration for metals, leads to the conclusion that the north is on the verge of a major surge in mineral development.

Ecology of Resource Development in the Canadian Arctic

G. S. DAVIES

It has been said that the Twentieth Century belongs to Canada. Implicit in this statement is the assumption that Canada will accelerate the rate of development of its vast supply of natural resources, particularly in the North, thereby assuming a prominent place in the world economic community. Development has been traditionally slow in the North, however, because of high investment and production costs, lack of available manpower, difficulties associated with establishing transportation routes, and the problems associated with building on permafrost. In recognition of these problems the federal government has developed various incentive programs for potential developers, which include tax exempt status for certain time periods, depletion allowances, etc. However, in spite of these efforts northern development has proceeded until recently at a relatively slow pace.

The discovery in 1968 of large oil and gas reserves on the North Slope of Alaska and the prospect for similar discoveries in the Canadian Arctic have combined to dramatically shift the attitudes of the business community toward northern development. Men and equipment are being moved into the North at a rate that must be reminiscent for oldtimers of the excitement of gold-rush days. Towns like Inuvik and Yellowknife, formerly unknown to most southern Canadians, are now on the map, even to the extent of being mentioned in daily weather reports. While the interest of the business community is focused on northern development, the public is becoming aware of the potentially irreversible ecological damage that could result from permitting certain types of developmental activity to proceed un-

checked. Despite this concern it is expected that the mineral extractive industries along with the oil and gas companies will accelerate their pace of resource development with the result that the Arctic environment will be subjected to an unprecedented intensity and range of man-induced disturbances.[1]

The purpose of this paper is (1) to examine the ecological implications of resource development activities in the Canadian Arctic and (2) to study how private and public institutions have incorporated environmental considerations into their plans for Arctic development. Because of the broad range of developmental activities and the lack of comprehensive data, this paper will consider only the impact of oil- and gas-related activities.

THE FRAGILE NATURE OF ARCTIC ECOSYSTEMS

Arctic ecosystems can be considered fragile for several reasons. First, in the North, as in other harsh environments, there is a reduced number of plant and animal species. Conventional ecological wisdom tells us that in simple systems such as these the elimination of a single species may have a pronounced impact on the entire system, whereas in complex systems, such as those found in temperate deciduous forests, the elimination of a single species is likely to have little effect. The explanation for this argument is usually found in the notion that there is no other species that will fill the ecological role of the eliminated form in a simple community, whereas in complex communities there are usually other organisms that share or could assume the role of the missing

● G. S. DAVIES *is Assistant Professor, Department of Man-Environment Studies, University of Waterloo.*

species. In other words, complex communities have many pathways through which energy is channelled, and the disruption of one of these (by removal of a species) would not be significant, whereas in simple communities few such options are available. Since most of man's activities tend to simplify natural environments, this fact could be of some concern as Arctic ecosystems are simple and intrinsically unstable.

Second, species grow slowly in the Arctic and the recovery time for a population disturbed by unfavourable environmental changes could be significantly longer than for disturbed populations in the south. Available evidence indicates that extended periods of time are required for Arctic vegetation to recover from the disruptive effects caused by forest fires, tracked vehicles, etc. The destruction of vegetation is a particularly serious problem because of the dependency of large grazing mammals such as caribou, muskox and polar bear, on this resource.[2]

Finally, there are problems associated with building on permanently frozen ground. The distribution of permafrost in any region is the result of a dynamic balance between the climatic history, vegetational history, recent climate, present vegetation, substratum, topography, and animal activity, all interacting.[3] Thus, a disturbance that changes this equilibrium may give rise to a host of problems such as severe frost heaving, subsidence caused by melting permafrost, landslides and so on.[4]

AN OVERVIEW OF THE OIL PROBLEM

Since the discovery of the Prudhoe Bay oil reserves in 1968, one of the chief problems that the United States government has had to deal with is the movement of the oil from the production fields to the refineries, in a way that reduces environmental damage to a minimum. The major proposed oil routes for both oil and gas lines are shown in Figure 1.

The most controversial of the three oil routes is the Trans Alaska Pipeline System (TAPS), which plans for an 800-mile pipeline, 48 inches in diameter, from Prudhoe Bay to the ice-free port of Valdez, Alaska. At this southern terminal the oil would be transferred to tankers and carried to Seattle. Environmentalists in the United States have pointed out two major difficulties with this route. First is the fact that to maintain a flow of 2 million barrels per day it would be necessary to maintain an oil temperature between 150 to 180°F. To bury such a hot pipeline in permafrost could induce thawing, and in areas of high ground-ice the resultant subsidence could place undue stress on the pipeline possibly leading to a rupture. The second problem is that the southern portion of the route is characterized by extreme volcanic activity. Of concern to Canadian environmentalists are the hazards involved in transporting the oil from Valdez to Seattle. There is little doubt that the shipping lanes between these two ports are among the most dangerous in the world.[5] The problem is compounded when one considers that 2,000 ships, 7 percent of the world's totals, are involved in collisions every year, and an increasing proportion of these are oil tankers.

The most promising alternative to the TAPS proposal, according to the United States Department of the Interior, is shipment via Arctic waters using ice-breaking tankers like the *Manhattan*. This proposal has been studied extensively and for the present at least, has been found to be uneconomic.[6]

So far, the Department of the Interior has expressed little interest in a proposal to route the oil through Canada via the Mackenzie Valley since this would ". . . serve mainly to shift the location of ecological problems rather than cure them."[7]

David Anderson, a Canadian Member of Parliament, has noted that a primary reason for the Americans favouring the TAPS proposal is based on the notion that the route is entirely within the control of the United States. Anderson points out that the seagoing part of the route is in international waters and that obviously the use of such waters may be endangered by submarine attack. He notes also that the United States and Canada have, by joint agreement, assigned much of

Figure 1

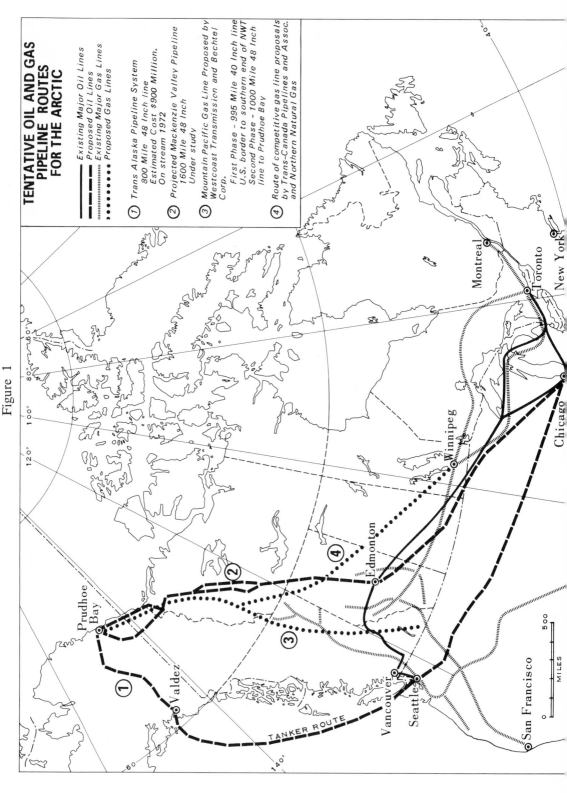

TENTATIVE OIL AND GAS
PIPELINE ROUTES
FOR THE ARCTIC

Existing Major Oil Lines
Proposed Oil Lines
Existing Major Gas Lines
Proposed Gas Lines

① Trans Alaska Pipeline System
800 Mile 48 Inch line
Estimated Cost $900 Million.
On stream 1972

② Projected Mackenzie Valley Pipeline
1600 Mile 48 Inch
Under study

③ Mountain Pacific Gas Line Proposed by
Westcoast Transmission and Bechtel
Corp.

First Phase – 995 Mile 40 Inch line
U.S. border to southern end of NWT
Second Phase – 1000 Mile 48 Inch
line to Prudhoe Bay

④ Route of competitive gas line proposals
by Trans-Canada Pipelines and Assoc.
and Northern Natural Gas

these waters to Canada for anti-submarine defence. "It is curious logic," notes Anderson, "to suggest that a sea route defended by Canadian naval units, is somehow wholly within the control of the United States."[8]

In spite of the lukewarm approach of the United States government to routing American oil through Canadian territory, Ottawa has responded to the possibility of northern pipeline development with a great deal of vigour and has mounted several programs that deal with the environmental impact of such pipelines on Arctic ecosystems. It is possible to justify this response when it appears that a gas pipeline will soon be constructed from Prudhoe Bay to Edmonton and, with the exception of heat-induced problems, the environmental problems of large-diameter gaslines are similar to those of oil lines. Implicit, of course, is the expectation that significant oil finds will be made in Canadian territory in which case the possibility of an oil pipeline in the Mackenzie Valley will assume some real significance.

THE ENVIRONMENTAL IMPACT OF OIL AND GAS RELATED ACTIVITY

Estimates indicate that the potential reserves of the Canadian Arctic are 50—100 billion barrels of oil and 300—500 trillion cubic feet of gas.[9] Of the total area held in lease by oil companies (about 628,000 square miles), three-quarters is on land and the remainder lies offshore. To develop these reserves, the oil and gas industry proceeds through a series of sequential exploratory and developmental processes each of which tends to have a greater impact on the ecology. The sequential processes are outlined in the following sections.

Geological Survey

Geological reconnaissance is the first step of the process. Because bedrock is widely exposed in the Arctic, most of this exploratory work can be done by light aircraft, with the result that there is little environmental disturbance.

Geophysical Survey

The second step is geophysical reconnaissance, which involves moving men and equipment over the terrain either by helicopter, or more economically by tracked vehicles. In the past, much of the geophysical work has been done without consideration of, or possibly, awareness of potential environmental implications. The result has been widespread disturbance and destruction of surface vegetation leading to subsidence in areas where ground-ice content is high. It is now known that much of the impact of tracked vehicles can be reduced by restricting their operation to wintertime, but for economic reasons, the industry is interested in improving its ability to work in summer. To reduce the impact of tracked vehicles a number of modifications and alternatives have been introduced. These include: (1) structural design changes to reduce pressure points, (2) modification of tire tread size, (3) the development of air-supported craft (hovercraft), and (4) the planting of "nurse" crops to reduce thaw and to allow stabilization to occur. The Imperial Oil Company has estimated that geological and geophysical surveys will eventually cover distances of about 150,000 in the Western Arctic. It is clear, therefore, that the potential impact on environmental quality could be far-reaching in magnitude.

Exploratory Wells

The third phase is one of exploratory or wildcat drilling. Estimates indicate that 700 to 1,400 exploratory and 6,000 to 12,000 developmental wells will be drilled to develop Canadian sources of petroleum. Exploratory wells usually occupy a site of 3 to 4 acres and, if insufficient amounts of gravel are used for insulation purposes the impact of these operations could result in the thawing of permafrost and degradation of the surface vegetation.

Developmental Wells

Developmental wells have the same ecological impact as exploratory wells but will usually require a great deal of permanent road construction. Experience in Northern Canada suggests one mile of road will be required for each developmental well. On this basis it has been estimated that development of oil in the Canadian Arctic will require the construction of 4,000 miles of access road and 4,000 miles of trunk road. If suitable quantities of gravel are available it seems likely that these roads can be built with a minimum of ecological disturbance. Unfortunately, the most accessible supplies are in the Mackenzie River and in thawed lakes, removal of which could have a considerable impact on the spawning grounds of the northern fisheries. There are, however, other large supplies of gravel in adjacent mountain ranges but using traditional criteria they would be considered uneconomic. This would be a high but necessary price to pay for maintenance of environmental quality and flexibility.

Pipelines

Feeder Pipelines—Once oil or gas are tapped, the major problem is to transport the material to the southern markets, and the most feasible method is by large-diameter feeder pipelines. Although public attention is focused on the environmental impact of large trunk lines, it is important to note that approximately 8,000 miles of small-diameter feeder pipelines will have to be constructed to deliver the oil to the trunk lines, then to market or to marine terminals. While the impact of hot oil flowing through these small-diameter pipes may be considerably less than that of large-diameter pipes, their cumulative effect could be substantial.

Main trunk pipelines—Let us now consider the environmental impact of large-diameter hot-oil pipelines, remembering that the problems caused by gas lines differ only with respect to the thermal disturbance induced by the high temperature of the oil flowing through the pipe.

(a) Impact during construction: Many ecologists consider that the activities asso-ciated with the construction of pipelines may have more significant impact on environmental quality than the long-term operational effects. The problems caused by construction include: siltation and erosion, creation of barriers to fish migration, physical damage to spawning areas, entrapment of fish, and disturbance of wildlife.

Improper operation of heavy equipment at stream crossings can give rise to severe siltation and erosion problems which, if unchecked, could cause the destruction of considerable wildlife habitat. The greatest effects, however, would be on fish populations, particularly if construction is allowed when there are eggs or fry in the gravel. Faulty construction of haulage roads adjacent to streams or bridges might constitute another source of debris that could be pushed into the water, resulting in barriers to fish migration. Movement of equipment or detonation of explosives in streams would be a significant source of fish kill, depending upon when and where these activities were allowed. Removal of gravel from stream beds might result in portions of the river system being cut off, and fish might become trapped in temporary ponds where falling water levels could result in high mortality rates. Construction activities might also disturb nesting wildfowl populations, which could result in emigration of birds from the area coupled with reduced nesting success. Many of these problems could be minimized by controlling the timing of construction activities to avoid concentrations of wildlife at vulnerable periods in their life histories.

(b) Impact during operation: After construction is complete there will be a number of problems associated with the operation of northern pipelines, but the two most widely discussed appear to be that of thermal impact (in the case of oil) and the possibility that the above-ground portions of a pipeline might interfere with big game migrations, especially caribou.

Lachenbruch of the United States Geological Survey has calculated that "4-foot pipeline buried 6 feet in permafrost and heated to 176°F will thaw a cylindrical region 20 to

30 feet in diameter in a few years in typical permafrost materials. At the end of the second decade of operation, typical thawing depths would be 40 to 50 feet near the southern limit of permafrost and 35 to 40 feet—where permafrost is colder."[10] In situations where permafrost sediments have excess ice and low permeability when thawed, free water could be generated very quickly and the resulting thawed material might persist as a semiliquid slurry. If the permeability is very low and the excess ice content is moderate, ". . . thawing rates could be sufficient to maintain this state for decades."[11] According to Lachenbruch the thawed cylinder would flow like a viscous river and tend to seek a level. If these slurries occupied distances of several miles it would be possible for the uphill end of the pipe to ". . . be lying at the bottom of a slumping trench tens of feet deep, while at the downhill end, millions of cubic feet of mud (containing the pipe) could be extruded over the surface."[12] This process could be self-perpetuating.

It is estimated by Lachenbruch that small changes in soil properties occurring over lateral distances of 10 feet could cause differential settlement and result in undue stress upon the pipeline. These differences would be insidious in that they would be difficult to detect even if test holes were bored frequently along the route. Lack of knowledge concerning the distribution of ground-ice plus related problems caused by extensive areas of thermokarst lakes in the Mackenzie Valley will be major engineering problems encountered in any routing of a pipeline.

A break in the line would result in a spill of several hundreds of thousands of barrels of oil. The result of such a spill would be difficult to assess. Experimental work has shown that mosses and lichens are extremely susceptible to oil damage and their recovery time would probably be a matter of decades. Should it prove impossible to retain the oil on land or should the break occur near water, the damage inflicted on wildlife and fish populations could be catastrophic. The nature of an oil spill is the more serious when it is considered that natural biode-gradation processes are directly temperature dependent and therefore extremely slow in Arctic ecosystems. The persistence of oil fractions in the aquatic environment could have long-term effects on the recovery of certain populations.

In areas of high-ice-content soils a pipeline will normally rest on gravel berms at the surface or be elevated on piles. Fear has been expressed by some wildlife biologists and hunters, that this will form a barrier to caribou migrations. However, evidence so far indicates that caribou seem to be reasonably aggressive, and are not measurably affected by industrial activity or road building.[13] Contrary to prevailing opinion they appear not to be disturbed by the presence of seismic lines but rather use these as trails. It is likely that the disruption caused by the pipeline to wildlife populations during the operational phase will be relatively small compared to the disruption by natural phenomena such as drought and forest fire.

The Human Dimension

Of concern to federal and local government is the impact of proposed pipeline developments on the native peoples. Many of the northerners, Indians, Eskimos and Métis, live in small and remote villages along the northern part of the proposed pipeline routes. Like most communities involved in cultural transition the young are uneasy about modern industrial development, partly because they fear they will be left behind by the new technology, whereas the older people fear social change, which will deemphasize old values.

Lee Hurd of the Northwest Project Study Group has pointed out that jobs for native residents won't be a quick byproduct of oil and gas development because most of these positions will be highly specialized and temporary. He estimates that about 150 people will be needed for operation and maintenance of a pipeline however, and these will likely be native northerners.[14]

It is estimated that about 15,000 (as of 1 June 1971 the population of the Northwest Territories was estimated to be 38,700) oil and gas workers will be attracted to the north and while most of these will be transient, they will have a significant impact on the ecology of the area because of the increased hunting pressure they will exert on fish and wildlife resources. Another problem concerned with direct human impact on the northern ecology is that of disposing of considerable solid and liquid wastes in remote camps.

INSTITUTIONAL RESPONSE TO ENVIRONMENTAL PROBLEMS

Government

The federal government of Canada has made an immediate and dramatic response to the environmental problems of northern development. This response was formalized in August, 1970, when the Hon. J. J. Greene, Minister of Energy, Mines and Resources, and the Hon. J. J. Chrétien, Minister of Indian Affairs and Northern Development, jointly announced the release of a set of pipeline guidelines designed specifically for northern operation. Prepared by an interdepartmental task force, the guidelines established the corridor concept within which an oil and gas pipeline and related facilities would be located.

A number of agencies from three different ministries (Departments of the Environment, Energy, Mines and Resources, and Indian Affairs and Northern Development) are involved in gathering baseline information to ensure that any pipeline would be ecologically acceptable and technologically feasible.

The environmental projects mounted by these agencies include studies on the following: fish distribution and migration routes; location of spawning beds of commercially important species; location of available gravel supplies; bottom fauna studies; caribou migration routes; location for wildlife of sensitive areas, impact studies of industrial activity on surface vegetation; hydrological studies, and protection studies concerned with the reseeding of areas disturbed by construction.

The federal government began the preparation of the northern pipeline regulations

shortly after the Prudhoe Bay discovery was announced. Most of the environmental studies listed above, however, were initiated in the summer of 1971 and are planned to continue until the end of summer, 1972.

Industry

The response of the Canadian oil and gas industry to the serious environmental problems created by northern exploration and development has been impressive. In part, this response may have been influenced by the rapid response of their American counterparts following the passage of the National Environmental Policy Act in the United States. In addition to mounting a number of oil and gas pipeline feasibility studies in the Canadian North, industry is also sponsoring a broad range of studies to measure the impact of pipelines on the northern environment. Some of these are described below.

a) *Mackenzie Valley Pipeline Research Ltd.*—MVPRL was formed in 1969 to investigate the practicability of constructing a large-diameter oil pipeline from Prudhoe Bay to Edmonton. Their program includes an experimental test loop at Inuvik and a series of soils investigations.

The test loop was designed to provide information on the thermal impacts of hot oil flowing through both elevated and buried (in gravel) segments of pipe. The performance of several types of insulation is also being evaluated. A thermal simulation model was developed and is used to predict the thermal response of permafrost to the pipeline.

By the end of the first summer's operation (1970) thaw under the elevated portion of the test loop was negligible and no subsidence was observed. The berm section of the pipe settled about 4 inches, but the site engineer attributes this to the melting of snow and ice trapped under the gravel during construction.[15]

The insulation was removed from a 400-foot section of pipe during October, 1970, and as of April, 1971, the thaw front had advanced 7 feet below grade. The uninsulated segment of pipe has settled 3 feet, while the insulated portions have settled about one foot.

The soil studies conducted by MVPRL are expected to provide information which will be useful in selecting pipeline routes, insulating materials, and support systems for construction in permafrost.

The test loop is an impressive facility, but it has been criticized on the following grounds: (1) too few of the geotechnical properties of soil are being measured and (2) insufficient baseline data (sheer, ice content, etc.) are being obtained *away* from the site, and it will be difficult to determine the magnitude of natural variation, let alone the impact of the pipeline on the environment. Concern has also been expressed that the test loop will not be in operation long enough to enable critical interpretation of results.[16]

b) *Northwest Project Study Group* — NPSG, representing a consortium of 3 gas and oil companies and 3 pipeline companies, has been engaged since 1961 in feasibility studies concerned with solving the problems associated with building a large-diameter gas line from the North Slope of Alaska across Canada to the U.S. border near Emerson, Manitoba. It has a technical research program, which includes the construction and maintenance of the Arctic Test Facility at San Sault Rapids in the Mackenzie Valley.

This facility consists of two separate loops of pipes varying in diameter. The "cold loop" uses compressed air chilled to zero degrees to simulate the flow of natural gas. The "cycling loop" tests the effects of alternately cycling hot and cold air, which simulates the loss of a refrigeration unit due to a failure at a main line compressor station.

The Test Facility has operated for several months and will continue to provide environmental information, such as the stability of a gas line in permafrost, the stability of different types of foundation for above-ground construction, and the impact on surface vegetation.

Other projects conducted by the Study Group include soil temperature studies along the proposed pipeline routes, terrain sensitivity studies, and ecological studies, which are designed to enable route selection to be made with minimum environmental effects.

The ecological studies include surveys of fish, bird and wildlife, but top priority has been given to the study of the caribou. There are 5 major caribou herds in the Canadian Arctic and 4 of these have been studied intensively, but unfortunately not the Porcupine herd. This herd migrates north in the spring to the Arctic coast for calving, and in the fall returns south for winter forage. The proposed pipeline routes pass directly across the migratory routes of the herd and it is feared that if the pipe is above ground it may present a barrier, or at least a significant diversion, to their movement. The Old Crow Indians have traditionally hunted these animals and any threat to the survival of the caribou is essentially a threat to the economy of the tribe.[17]

c) *Alberta Gas Trunk Line Co. Ltd.* — Alberta Gas has formed an Environmental Protection Board whose responsibilities will include a study of the potential impact of a natural gas pipeline on the Arctic environment. The five member Board is an unusual organization for a commercial enterprise because it is independent of the company and acts only in an advisory capacity. The environmental research studies are all done by outside consultants although usually in consultation with a member of the Board. The program involves fish, wildlife and vegetation studies.

Alberta Gas has designs for test facilities in Alaska and at Norman Wells. The studies at Norman Wells will be designed to answer questions concerned with thermal problems at pipelines in an area where temperatures could be both above and below freezing.

d) *The Mountain Pacific Project* — This group, sponsored chiefly by Westcoast Transmission and Canadian Bechtel Ltd. is also concerned with the impact of northern pipeline construction and operation. It has made provisions for initiating research programs in the near future.

PROSPECTS FOR THE FUTURE

It seems likely that the pace of resource development will continue to accelerate in the future. Seismic activity increases, the number of exploratory oil and gas wells grows daily, and there is a renewed interest on the part of the mineral extractive industries in northern development. The continuing improvement of the transport system in the North, as evidenced by the construction of the Dempster and Mackenzie Valley highways, will not only stimulate new development but will also bring increasing numbers of tourists anxious to avail themselves of the wilderness.

All of these activities will result in a greater impact upon fragile Arctic ecosystems.

Although the response of government and industry to the problems posed by the pipeline projects has been gratifying, it is too early to determine how effectively environmental information will be incorporated into the decision-making process. At present, no mechanism exists to ensure that all potential developers will respond in a like fashion. In the United States, the congress[18] has directed all federal agencies to furnish evidence that environmental considerations have been taken into account when a federal project is being funded, planned, and designed. The evidence is presented as a draft Statement of Environmental Impact, which includes a description and justification for the proposed action and an evaluation of its environmental impact. The content of environmental statements and how these should be circulated and reviewed is described in detail in the guidelines prepared by the Council on Environmental Quality.[19]

A similar approach might well be adopted in Canada whereby the federal government would undertake to establish formal ecological impact guidelines. The responsibility for conducting and financing ecological impact studies relative to these guidelines should be the responsibility of the developer (government or private industry). Moreover, it is recommended that all studies of this type be systems oriented since this type of approach has, so far, yielded the most significant information.

It is also suggested that minimum time periods be stipulated on all ecological impact studies. Frequently, experimentation or research periods of one, two, or even three years do not yield sufficient data on which responsible decisions can be made.

Finally, another problem is that of conflicting jurisdictional authority between the Department of the Environment and the Department of Indian Affairs and Northern Development. It is not yet clear which department is responsible for the solution of environmental problems north of the 60th parallel.

REFERENCES

[1] MacKay, J. R., "Disturbances to the Tundra and Forest Tundra Environment of the Western Arctic," *Canadian Geotechnical Journal*, Vol. 7 (4), 1970, pp. 420-432.

[2] Bliss, L. C., "Oil and the Ecology of the Arctic," An address delivered to a symposium on *The Tundra Environment*.

[3] A good account of the importance of permafrost in Arctic communities is found in Pruitt, W. O., "Some Aspects of the Interrelationships of Permafrost and Tundra Biotic Communities" in W. Fuller (ed.), *Proceedings of the Conference on Productivity and Conservation in Northern Circumpolar Lands*, Edmonton, Alberta, 1969, pp. 33-41.

[4] The problems of building in permafrost are discussed by Ferrians, O. J., Kachadoorian, R. and G. W. Greene in "Permafrost and Related Engineering Problems in Alaska," *Geological Survey Paper 678*, United States Government Printing Office, Washington, 1969, pp. 1-37.

[5] From the B. C. Pilots Handbook, Vol. I.

[6] A useful discussion of the alternatives to the TAPS proposal is found in the Draft Environmental Impact Statement of the Trans-Alaska Pipeline System prepared by the U.S. Department of the Interior. Copies of all environmental impact statements are made available to the public by the Council of Environmental Quality through its publication *102 Monitor*.

[7] *Ibid.,* p. 1.

[8] From the brief of David Anderson, M.P. at the Hearings of the Department of the Interior on the Department's Draft Environmental Impact Statement On The Trans-Alaska Pipeline. Washington, D.C. Tuesday, Feb. 16, 1971.

[9] Much of the statistical information related to oil and gas development is taken from a paper presented by Russell A. Hemstock of Imperial Oil to the 20th Alaska Science Conference, University of Alaska, Aug. 24-27, 1969.

[10] Lachenbruch, A. H., "Some Estimates of the Thermal Effects of a Heated Pipeline in Permafrost," *Geological Survey Circular 632,* Washington, D.C., 1970, p. 1.

[11] *Ibid.,* p. 2.

[12] *Ibid.*

[13] Hill, R. M., "Petroleum Pipelines and the Arctic Environment," *North,* May-June, 1971, p. 3.

[14] This information was in an article in *The Albertan* (Edmonton), summer, 1971.

[15] Personal communication, R. Guthrie, Inuvik Test Site, Inuvik, N.W.T.

[16] Personal communication, J. Ross MacKay, Department of Geography, University of British Columbia.

[17] Naysmith, J. K., *Canada North — Man and the Land,* Information Canada, 1971, p. 21. Based on 8 years data, 1962-69, the average annual value of caribou harvested by the Old Crow Indians is $50,000.

[18] The authority is Section 102(2) (c) of the National Environmental Policy Act of 1969.

[19] Council on Environmental Quality, "Guidelines for Statement on Proposed Federal Actions Affecting the Environment," *Federal Register,* V 36(79), 1970.

The Forest Industry in Coastal British Columbia 1870-1970

Walter G. Hardwick

The forest industry situated on the British Columbia coast is one of the largest and most highly integrated to be found anywhere in the world. Well known for the production of lumber, timber and plywood, the manufacture of pulp and newsprint has grown rapidly in recent years. Much of the output is exported, and locational marginality in terms of the world market has made location and intensity of forest activity highly sensitive to minor changes in international markets.

Within this context the structure of commercial forestry shifted during the last century from a few isolated and independent units extracting trees and cutting lumber to large production units, in places agglomerations of complementary mills, which *in toto* comprised a complex interdependent system reaching out to distant domestic and overseas markets. In the process the logging "shows" and converting plants have become larger, fewer in number, and more stable in pattern during the 1960's than in any previous decade. Further, the pattern of individual location, when viewed in the context of each corporation, conformed more closely to classic models of location. This trend took place during a time in which the output of the industry grew in magnitude, and the incidence of vertical and horizontal integration within large corporations expanded.[1]

During the formative years, loggers' "cut-and-run" strategy, creaming the best and most easily accessible timber, contributed to a rapidly changing pattern of log extraction.

Similarly, small, poorly capitalized lumber mills opened and closed quite rapidly, reflecting the vacillations of the timber market and the managerial acumen of first generation timbermen. By 1960 a policy of sustained-yield tree farming with government-authorized annual allowable cuts had established territories for logging shows. In the same period, the full utilization of the log, for lumber, pulpwood chips, veneer, resin, fibreboard and so forth contributed to the grouping of well-financed, centrally managed converting plants bringing about both scale and agglomeration economies.

These goals had been enunciated by corporate management as early as 1956, when a Royal Commission enquired into the industry. The last fifteen years have seen their articulation on the landscape.

In this paper the changing distribution of logging camps and log converting plants is sketched as the industry evolved through three periods: pioneer, liquidation, and sustained yield.

PIONEER PERIOD: 1860-1920

Logging and lumbering have been a central part of the Pacific Northwest economy for just over a century. The growth of the forest industry in British Columbia in the early years was slower than in the American Northwest, reflecting the peripheral location of British Columbia in terms of both Canadian and world markets, and the domina-

● WALTER G. HARDWICK *is Associate Professor, Department of Geography, University of British Columbia.*

Reprinted from *Occasional Papers in Geography* (B.C. Div. Can. Assoc. of Geographers), No. 2, June 1961, pp. 1-7, with permission of the author and editor. The article was revised and updated by the author in 1971.

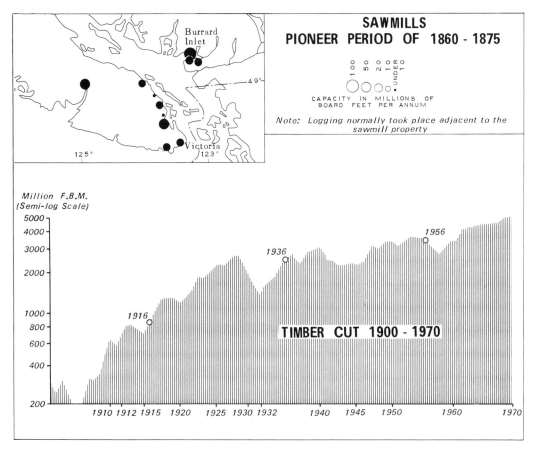

SAWMILLS
PIONEER PERIOD OF 1860 - 1875

CAPACITY IN MILLIONS OF
BOARD FEET PER ANNUM

Note: Logging normally took place adjacent to the sawmill property

Million F.B.M.
(Semi-log Scale)

TIMBER CUT 1900 - 1970

Figure 1

tion of the lumber trade on the Pacific Coast by San Francisco brokers. However, the quantity of timber cut increased rapidly after 1886 with a corresponding increase in the number of logging and sawmill sites. The expansion, which peaked in 1912 (Figure 1), was associated with the coming of the railroad to the west coast cities, the settlement of the wood scarce prairie, and Vancouver's initial 100,000 people.

Logging was restricted in its areal extent to the lower mainland and to the southeastern coastal plain of Vancouver Island in the pioneer period. By the turn of the century, camps had spread to all areas that had access to the sheltered waters of Georgia and Johnstone Straits. Victoria, the original Hudson's Bay Company settlement and colonial capital, never did become a sawmilling centre of any consequence, but by the late 1860's, Alberni, Mill Bay, Chemainus and Moodyville (North Vancouver) had all achieved a certain notoriety.

From the outset, the volume of timber cut far exceeded the modest local needs. Thus, a gold rush in New South Wales, commercial growth in Shanghai, and mining expansion in Peru became the markets for British Columbia's forest products. The sawmills, dependent on overseas shipments, only needed a location where resources were readily available, where year-round water power was abundant, and where docking of sailing ships was easily accomplished.

The initial construction of large sawmills at Vancouver has often been attributed to the arrival of the Canadian Pacific Railway in 1886. This is not quite true. The railway construction and the ensuing urban development certainly helped maintain the sawmills

319

in the Vancouver area; it did not bring them. Burrard Inlet and New Westminster were sawmilling centres, like Chemainus and Nanaimo, in the pre-railway era. The mills had been established to cut the excellent stand of Douglas fir found on the relatively flat lands of Burrard Peninsula and adjacent areas, now occupied by Greater Vancouver.

Prior to World War I, logging spread rapidly northwestward on Vancouver Island and the mainland coast, while lumber milling remained rather concentrated in its areal extent.

After exhaustion of the local supply around Vancouver, development of these up-coast properties started where "capitalist speculators" from the American Midwest and Eastern Canada gained control of large forest tracts along the coast. Large steam tugboats, built to tow sailing ships from the sheltered waters east of Vancouver Island to the Straits of Juan de Fuca and the open Pacific, gradually became deployed in towing large log booms along the coast. Thus a division of labour gradually developed between the up-coast logger and the sawmill operator. Only the occurrence of the resource and its accessibility to tugboats plying sheltered waters limited the areal spread of loggers.

FOREST LIQUIDATION: 1920-1945

Overseas markets expanded rapidly after World War I as Canadians took control of domestic and overseas marketing of their forest products. The opening of the Panama Canal and the formation of Canadian shipping companies to carry lumber provided opportunities for lumber sales unknown prior to the war. Paralleling the new market opportunities, entrepreneurial, technical and government policy changes permitted output to meet the new level of demand. The desire of capitalists who had invested in timber leases in the 1890's and 1900's to exploit their timber holdings for sizeable capital gains brought about increased investment in logging and mills. The development and wide use of logging railways permitted loggers to plunge away from the coast into the rich timbered valleys. A single camp strategically located on the coast or lake could clear logs from a large hinterland. Liquidation of the forest for quick economic return became the unstated policy of government and capitalist alike.

Some logging operations became as large by 1925 (Figure 2) as they ever had been or would be in the post-World War II period.

A characteristic of the logging and milling pattern in this period was the functional forest region, a region self-sufficient in mill inputs and log outputs. The most important region focussed on the city of Vancouver and included most of the camps on the slopes of the basin between the Vancouver Island Ranges and the Coast Ranges north of the city. Several secondary forest regions could be identified as well. The Alberni area and Quatsino Sound on the exposed coast of Vancouver Island were two regions cut off from the general circulation of resources by rough water of the Pacific. The sawmills located on southeast Vancouver Island were each closely linked with adjacent logging camps and maintained quite independent units well established in earlier years. The largest of these was at Chemainus. The Powell River Company's location, eighty miles north of Vancouver, gave this mill an advantage over the Vancouver mills in terms of log-towing costs from northeastern Vancouver Island and Kingcome and Knight Inlets. Thus, the Powell River Company was the focus of a region that stretched to the north of that dominated by Vancouver.

Penetration into the Queen Charlotte Islands for logs first took place in this period, most of the output going to Ocean Falls and Powell River pulp mills. This areal expansion for large-scale forest exploitation had been made possible by the development of the Davis raft, an iceberg-like mass of logs held together by intertwining cables. Since transportation of logs in sheltered waters by flat boom was much more economical than by raft, Davis rafts were restricted to use in exposed areas.

Into the 1920's, logging shows and sawmills usually had different ownership. The goals and strategies of the companies differed. During the 1930's, some corporate

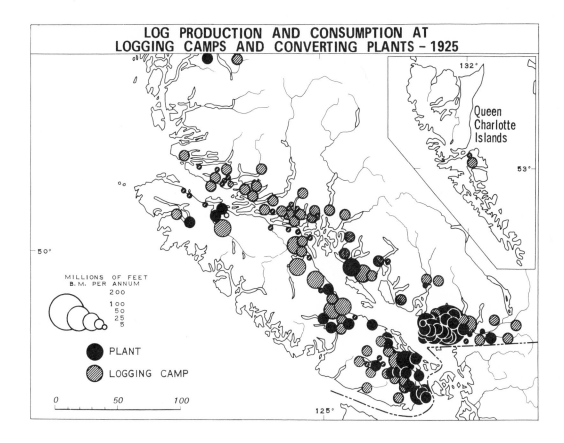

LOG PRODUCTION AND CONSUMPTION AT LOGGING CAMPS AND CONVERTING PLANTS – 1925

Figure 2

links were forged between mills and camps. For example, the Comox Logging and Railway Company, the largest logging operator on Vancouver Island, and Fraser Mills, at New Westminster, the largest mill of its day, fell under common ownership.

SUSTAINED YIELD: 1945-1970

By the 1950's, a new plateau had been reached in the volume of timber cut on the B.C. coast. The gross pattern of forest sites was strikingly similar to that of previous decades but volumes had changed. Peripheral camps showed a marked increase in production, while more centrally located camps showed a decrease. Both were related to the post-war policy of sustained yield, of both the forest industry and the provincial government, and to a change in transportation technology. Of course, the overall market for forest products had expanded to absorb the increased output.

The slash and burn practice of the era of forest liquidation was now abandoned. Sawmills, pulp mills and plywood mills, all involving large capital investment, became conscious of the need for a perpetual source of raw materials. Mergers of timber-rich firms with millers became commonplace. Sustained yield meant tree farms, and tree farms meant more permanent locations for logging operations. Thus, by 1960 (Figure 3), what may be termed the long-term pattern of production camps and consuming mills had evolved.

The interesting change in transportation technology was the development of the self-dumping log barge that replaced the Davis raft. These barges, up to 15,000 tons DWT, could be loaded quickly and dumped even

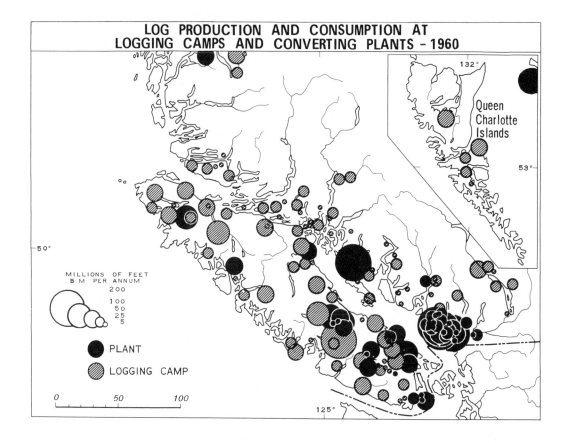

LOG PRODUCTION AND CONSUMPTION AT
LOGGING CAMPS AND CONVERTING PLANTS – 1960

Queen
Charlotte
Islands

MILLIONS OF FEET
B M PER ANNUM
200
100
50
25
5

● PLANT

◍ LOGGING CAMP

0   50   100

Figure 3

more rapidly. Large tugs could tow the loaded barges at six to eight knots—much faster than the two knots attainable with the flat boom. For the first time, barges permitted economical marketing of logs from isolated ports on Vancouver Island's west coast and the north coast to the sawmill and pulp mill centres. In addition, the barges permitted efficient and rapid movement of lumbergrade logs which had previously been left standing or used for pulp to market in the Vancouver sawmills.

Along with sustained yield, corporate integration increased rapidly. Most of the coastal log output has become either cut by or destined for the mills of six major integrated companies. There, the raw material is used more thoroughly than at any time in the past for lumber, plywood or pulp and paper. The waste, pulpwood chips and hog fuel (sawdust), are forwarded by barge to other

mills to be used as raw materials in other manufacturing processes or for fuel for steam generation.

The functional regions noted in 1940 had changed in form by 1960. As a result of corporate integration, sustained yield, and changes in the transportation technology, much of the coast is, in a sense, one functional region. No longer does one mill take the whole output of one logging camp. Now specific species are cut for specialized mills. Metropolitan Vancouver mills and the integrated plants around Georgia Strait now draw on all the coastal forests for at least some grades of the logs.

In 1960, when the first analysis of the coastal forest industry was completed,[2] forecasts were made for the future. These included increased log production within the pattern already in evidence. A later increase in the importance of "grouped mills" centred

322

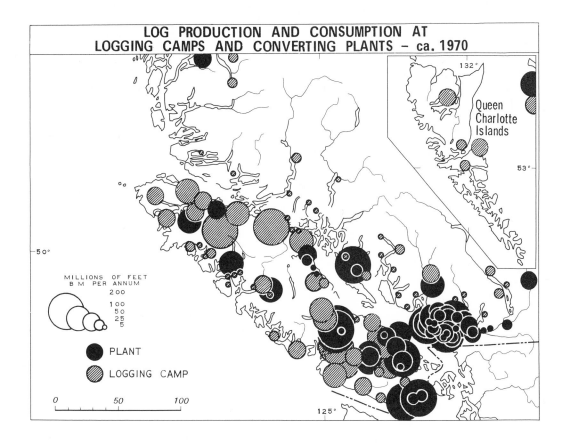

LOG PRODUCTION AND CONSUMPTION AT LOGGING CAMPS AND CONVERTING PLANTS – ca. 1970

Figure 4

upon pulp mills was foreseen. The relative importance of the Vancouver region was expected to decline, even though the forces of inertia were clearly recognized.[3]

The map of the industry for 1970 (Figure 4) illustrates that these forecasts were near the mark. Each sustained-yield unit has stabilized output near the limit of annual allowable cut. Portable camps have been replaced by resource towns and permanent logging roads fan out through the unit as the tree farmers sequentially harvest and plant. Grouped mills have expanded, specifically at Port Alberni and in the Tahsis area. In the north, at places like Kitimat, new grouped mill operations have been established using heretofore inaccessible timber. These operations, not tied to heavy investment in existing plants, are obviously adopting the grouped mill policy. The integration of the mills around Georgia Strait has continued with

marine transportation providing low cost links. Even though the size of the Vancouver market and the large capital investment in plants there have permitted mills to persist to date, the forecast of eventual relocation of sawmills near pulp mills still holds.

In theoretical terms the weight loss analysis of Alfred Weber has application to location of production. When logs were cut into lumber alone, over half the volume was discarded as slabs, bark, sawdust and so forth. Mills were usually built near the raw material source. Now that full utilization of the log is corporate and public policy, the grouped mills adjacent to market or export docks are most appropriate locations. The profitable performance of each unit within the large integrated corporations, a policy enunciated by top management, forces shifts in location that more clearly reflect classic models of industrial location. The logic be-

323

hind industrial location is lost when only aggregate spatial analysis is undertaken.

SUMMARY

It is just over one hundred years since Captain Stamp started the large sawmill at Alberni, a mill that lasted but a few years. It is just over fifty years since the period of speculation ended and companies turned to the serious business of logging and marketing lumber, pulp and paper. It is barely a generation since government and industry agreed on a policy of sustained-yield management. It is but ten years since full utilization became common. The existence and prosperity of the coastal forest industry was and continues to be related to world-wide demand for forest products. On the other hand, locational structure of the industry within the region is a function of the distribution of forest resources, their more intensive use, more efficient transport and corporate organization.

REFERENCES

[1] W. G. Hardwick, "The Persistence of Vancouver as the Focus for Wood Processing in Coastal B.C.," *Canadian Geographer,* IX:2 (1965), pp. 40-44.

[2] W. G. Hardwick, *Geography of Forest Industry of Coastal British Columbia,* Vancouver (1964), 98 pp.

[3] W. G. Hardwick, "Port Alberni: An Integrated Forest Industry in the Pacific Northwest," R. S. Thoman and D. Patton, editors, *Focus on Geographic Activity.* McGraw-Hill, 1964, pp. 60-66.

The Columbia River Treaty: Some Lessons and Implications

W. R. Derrick Sewell

On September 16, 1964, Canada and the United States exchanged notes for final ratification of the Columbia River Treaty. Not only was it one of the most important agreements ever concluded by the two countries, but it was also a major milestone in the history of international river development in North America. In contrast with the Boundary Waters Treaty of 1909 which is cast largely in terms of compensation for losses resulting 'in one country from development in the other, the Columbia River Treaty aims at cooperative development for mutual benefit. Although the arrangement was intended to deal only with the Columbia River, it seems certain that it will be regarded as a precedent in the development of other international rivers in North America. Moreover, some of its principles may be applied in the development of international rivers elsewhere.

THE TREATY PROVISIONS

Briefly, the treaty calls for the provision of 15.5 million acre feet of storage at three projects—at High Arrow, Mica, and Duncan Lake (Figure 1)—in the headwaters of the Columbia River in Canada.[1] Details of the Columbia and of these projects are set out

● W. R. Derrick Sewell *is Associate Professor and holds a joint appointment in the Departments of Geography and Economics, University of Victoria.*

Reprinted from the *Canadian Geographer,* Vol. X (3), 1966, pp. 145-156, with permission of the author and the editor.

MAJOR PROJECTS PROPOSED AND CONSTRUCTED ON THE COLUMBIA RIVER AND MAJOR TRIBUTARIES

Figure 1

in Tables 1 and 2. The storage will be used for two principal purposes: to firm up flows in the United States part of the Columbia River basin during the winter period when natural flows are low, and to reduce flood losses downstream. Canada and the United States will share the benefits resulting from this storage. At the time the treaty was negotiated the power benefits were estimated to be 2.6 million kilowatts of capacity and 13 billion kilowatt hours of energy,[2] and the flood control benefits to be worth about $126 million, representing the losses that would have occurred otherwise. Under the agreement signed in 1964, Canada elected to sell her share of the increased power to the United States for the next thirty years, for which she was paid a lump sum of $273.3 million (Canadian); she will also be paid $69.6 million (Canadian) as her share of flood control benefits.[3]

In addition to the three storage projects to be built in Canada, the United States has the option to build the Libby project on the Kootenay River in Montana. Creating a storage reservoir extending several miles into Canada, it would permit increased generation in both the United States and Canada (550,000 kilowatts in the former and 200,000 kilowatts in the latter) but the increases would not be shared. Both countries would also be provided with flood control benefits, but these too would be enjoyed where they occurred rather than through formal sharing arrangements. The treaty further permits either country to divert water for domestic purposes at any time, and Canada to divert up to 1.5 million acre feet per annum from the Kootenay River into the Columbia River for power purposes after twenty years.

The negotiators seemed well satisfied with the arrangements. Canada, on the one hand, received enough money from the United States to pay for the construction of the three storage projects, and to make a substantial contribution towards the cost of installing generating facilities at the Mica project, the power from which will not have to be shared with the United States. Canadian officials estimated that this arrangement would reduce the cost of power from Mica to 1.5 million kilowatt hours.[4] They also pointed out that the financial arrangements would help to reduce the Canadian external trade deficit with the United States,[5] and that the construction of the projects would provide employment for up to 3,000 men directly and to thousands of others indirectly for as many as fifteen years.[6] The United States officials, on the other hand, seemed to feel they had also got a bargain, for they would be able to forestall an anticipated power shortage in the Pacific northwest and to postpone an increase in power rates in the region. Both parties felt that the treaty would stand out as an example of large-scale international co-operation, hopefully to be imitated elsewhere.

Table 1

COLUMBIA RIVER: GENERAL AND PHYSICAL CHARACTERISTICS

Data	Canada	United States
Source	Columbia Lake	
Mouth		Astoria, Ore.
Length (mi.)	480	740
Drainage area (sq. mi.)	39,500	219,500
Total fall (ft.)	1,360	1,290
Average annual runoff (million acre ft.)	73	107

Table 2

TREATY PROJECTS

Data	Arrow Lakes	Duncan Lake	Mica Creek
Location	5 mi. upstream from Castlegar	Outlet of Duncan Lake	90 mi. upstream from Revelstoke
Drainage area (sq. mi.)	14,000	925	8,220
Average flow (c.f.s.)	39,800	3,600	20,700
Max. recorded flow (c.f.s.)	220,000	21,400	112,000
Min. recorded flow (c.f.s.)	4,800	268	2,140
Dam height (ft.)	190	120	645
Live storage (million acre ft.)	7.1	6.4	7 (stage 1: storage only) 12 (stage 2: with at-site generator)

The agreement, however, was not easily reached, requiring twenty years of studies, investigations, negotiations, and bitter parliamentary debate before the final ratification took place. Furthermore, the treaty itself was severely criticized particularly in Canada[7] and to some extent in the United States. Some engineers claimed that the scheme was technically sub-optimal in the sense that it would not make the best use of the available head and flow of the river,[8] and certain economists suggested that it was economically sub-optimal, pointing out that there might have been cheaper ways to obtain the benefits claimed for the treaty scheme.[9] Some Canadians believed that Canada was "sold down the river,"[10] just as a few Americans felt that Canada "got a steal." There are still a few murmurs of discontent even now. Whether there may have been an economically efficient scheme and whether one country got a better bargain than the other will no doubt provide subject matter for numerous academic diatribes as well as a topic for political debate for some years to come. Interesting as these questions may be, it is not the purpose of this paper to dwell upon them. Instead the focus is upon two rather more important questions: first, what can be learned from the Columbia River experience that will be of value in contemplating the development of other international rivers in North America and elsewhere; and second, what are the implications of this experience for water resource development in Canada as a whole?

A COMPREHENSIVE APPROACH TO DEVELOPMENT

Until the negotiation of the treaty, the two countries had proceeded independently of each other in the development of the river. Such an approach was typical of that followed in the development of international rivers in most parts of the world, for generally it is not until a crisis occurs, such as a water shortage or a conflict in water use, that countries sharing a river basin get together to consider possibilities of joint development.[11] In the case of the Columbia River, the spark to international action came from proposals to build storage projects, which would be located in the United States portion of the basin, but which would back up water into Canada. After referring several of these proposals to the International Joint Commission, the two governments decided, in 1944, to request the commission to study the opportunities for development in the upper part of the basin, especially in Canada, and to make recommendations. Although it is not at all certain that a goal of joint development was envisioned at that time, the door was opened for a comprehensive approach to the management of the Columbia River.

Underlying Concepts

Comprehensive river basin development has come to be regarded as an ideal approach to water management in many parts of the world.[12] The river basin is regarded as an organic entity, of which the development of any one part affects other parts, either beneficially or adversely. Similarly, it is felt that development for one purpose affects possibilities of use for other purposes. Efficient management, therefore, requires a careful analysis of the potential over-all effects of each project on other projects and of each water use on other water uses in order to reveal a scheme which will provide the greatest over-all net gains. Determination of this optimum scheme is the purpose of comprenhensive planning.[13]

Theoretically, maximum advantage can be derived from the use of the river only if the operations of the various facilities are carefully co-ordinated or integrated. This goal is most easily accomplished through some kind of central administration.

Problems of Application

Although the idea of the comprehensive approach is widely accepted, several factors have militated against its application to the development of international rivers. Important among these factors is intense competition for the waters of the basin, as in the case of the Indus River and the Jordan

River. Other difficulties include conflicting political ideologies and fears on the part of one country that its participation may be of greater advantage to the other partners than to itself—factors which appear to have inhibited development of such rivers as the Amur and the Tigris-Euphrates. The Columbia River, however, seemed to offer an ideal opportunity for the adoption of a comprehensive approach. Here were two countries with a long history of cooperation; they had similar political ideologies; both had had considerable experience in river development; there was plenty of capital available for investment in water resource projects; both were growing economically and expanding their power demands; and interest in co-operative development was growing. All these factors were present at a time when the technical advantages of comprehensive development were being lauded by various scientific advisory groups.[14] Thus, the stage seemed to be set for an integrated approach to the development of the river which would embrace both parts of the basin.

The IJC Reference

Following the request of the two governments in 1944, the International Joint Commission (IJC) set up the International Columbia River Engineering Board (ICREB) to study the Columbia River and its major tributaries, to identify possibilities for developing the river, and to determine a scheme which would make maximum use of the available head and flow. The board used the river basin as a basin planning unit, and assumed that the international boundary did not exist. It tried to trace the effects of development of one project on other projects in the basin, and considered several possible purposes of development, particularly hydroelectric power generation and flood control, but also irrigation and navigation.

The ICREB report,[15] presented in 1959, outlined three alternative maximum schemes of development. Each would provide about 17 million kilowatts of firm power, would use about 50 million acre feet of storage,

and would cost about $4 billion to develop (that is, in addition to present investment in the Columbia River scheme). The report noted that there were possibilities for developing some 23 million acre feet of storage in Canada, and that such storage was especially attractive in view of the problems of developing storage in the United States portion of the basin. However, because the stored water would have value to Canada only if power could be generated from it, and because opportunities for the development of head in the Canadian portion were quite limited, some inducement would have to be given by the United States to Canada to develop the storage sites in her portion of the basin.

Although various incentives were suggested, including compensation for lands flooded by reservoirs, the real value of the storage was in the additional power that could be generated downstream and the reductions in flood losses that the upstream storage made possible. Thus arose the concept of downstream benefit sharing, a new concept in international river development in North America, and one which, theoretically, should have motivated the two countries to adopt a comprehensive approach in search of an arrangement that would bring them the maximum net gain from developing the river. The beginning of the search for a formula for sharing and apportioning downstream benefits, however, seems to mark the point where cooperation declined and where a narrower approach to the development of the river began to be taken. It was then that the reality of the international boundary became apparent, and that various institutional factors began to limit the practical range of choice from which the negotiators could select a combination of projects.

Determination and Apportionment of Downstream Benefits

Because the ICREB report provided no advice on the determination and apportionment of downstream benefits,[16] the two governments asked the IJC to consider the matter and to offer recommendations. This was indeed a formidable assignment, particularly when there was no precedent to draw on and when there were strong biases towards particular projects. Although there was a conscious attempt to apply economic principles in the development of a formula for this purpose, several factors appear to have prevented their wholehearted adoption. For example, there may have been a desire to tie in as far as possible methods of analysis already in use by United States federal agencies. These methods have several weaknesses, and have been criticized for their avoidance of economic principles.[17] Had other, more soundly based methods been recommended by the IJC, certain projects that had already been built in the Columbia River basin might, if re-evaluated, have shown up in a bad light, and a number of United States federal agencies might thus have been placed in an embarrassing position. Another factor may have been the knowledge that there was considerable pressure for the development of particular projects, such as Libby, which would be shown to be highly infeasible if rigorous economic analysis were applied; this situation may well account for the vague wording of the principles recommended by the commission.

The recommended principles, which were set out in a report presented to the two governments in December, 1959,[18] seem to have satisfied very few people. Economists in particular have been severely critical of them, noting logical inconsistencies and the wide latitude for interpretation.[19] In any event, the principles seem to have been consistently avoided in the actual negotiations. For example, if the first general principle—that projects should be added in order of relative economic merit—had been followed, the Libby project would not have qualified, for there were better alternatives available.[20] Also, although the second general principle recommended that in the selection of projects the advantage of a project to each country should be compared with that of the independent alternatives in each country, Krutilla, among others, has suggested that there were superior alternatives in both countries to some of the projects actually

included in the treaty scheme.[21] Power principle no. 6 called for downstream power benefits to be shared equally between the two countries, even though such an arrangement had no particular logic; in any case this principle was not strictly adhered to in the negotiations, the United States receiving a somewhat greater share of the increased downstream power output resulting from Canadian storage.[22] The principles were in fact so vaguely worded and extensively qualified that there was considerable latitude for interpretation. The negotiators took maximum advantage of this latitude and applied the principles only "to the extent that it [was] practicable and feasible to do so."[23]

Institutional Inhibitions to Economic Optimization

As noted earlier, a basic goal of comprehensive planning is the identification of the combination of projects that would produce the greatest net benefits. Although there is frequent mention of comprehensive planning in various documents relating to the negotiation of the treaty it does not seem that the negotiators had this particular goal in mind. They appear to have been content to pursue a more limited objective, that of finding a scheme on which the two parties could agree, and in which the benefits exceeded the costs.[24] Even if the negotiators had aimed at the more ambitious goal, there were severe institutional inhibitions to to its being attained.

For example, several factors militated against consideration of the full range of possible projects. First, there was a "prior commitment" on the United States side of the border. The Americans were anxious to ensure that the Libby project would be part of an international agreement, particularly because it had reached an advanced stage of design and because it had been a major element in previous major plans drawn up by the Corps of Engineers for the Columbia River basin. Thus, even though it appeared that there were cheaper projects in Canada, the Americans placed such severe

329

restraints on their adoption that the Canadian alternatives were dropped from consideration. In the case of the Dorr-Bull River scheme, for instance, the Americans demanded assurance of the same amount of flood control as would be provided by the Libby project, and compensation for the power that would have been produced by that project. Such demands were so onerous as to make the Canadian project economically infeasible. They also reflected the fact that a completely comprehensive approach was no longer being sought, that the IJC principles were not being followed, and that decisions were being made on non-economic grounds.

Second, the bargaining became so tough that once a point had been won there was little chance that further negotiation could take place even though the latter might lead to a more advantageous position for both sides. Thus, although advances in technology, power developments in British Columbia, and changes in power policies on both sides of the border opened up a whole new range of possibilities between 1961 and 1964, neither side seemed willing to consider them in formal discussions for fear of losing what had been gained in previous negotiations.

Ideally, the Columbia River scheme should have been compared with all the possible alternative ways of providing the benefits claimed for it. In Canada, there were major barriers to such comparisons, many of them stemming from the fact that the federal government had the responsibility for negotiating the treaty, but did not have access to information on possibilities of developing the Peace River or on costs of developing the Hat Creek coal deposits. Data on the Peace River were jealously guarded by the British Columbia government, and data on the Hat Creek coal deposits were the property of a private power concern, British Columbia Electric Company. Also, although the federal government and the British Columbia government had been undertaking studies of the Fraser River,[25] development was not contemplated by either government at that

time, despite the well-known threat of a major flood; consequently, studies were not being pushed to the stage where realistic estimates of cost of Fraser River power were available for comparison with the Columbia River scheme. These factors may partly explain why the Canadian federal government concentrated its attention on variations of a Columbia River scheme rather than comparing costs of such schemes with costs of other means of providing power and flood control in British Columbia.[26] No doubt there were other reasons too.

A comparison of costs of developing the Peace River with those of developing the Columbia River was undertaken at a subsequent stage by the British Columbia Energy Board. Although no attempt was made in this study to estimate costs of developing power on other rivers such as the Fraser or Homathko, or of developing thermal power, the report did show that the costs of developing the Peace and Columbia were closely competitive.[27] This conclusion, together with the Energy Board's view that there would be substantial advantages to exports of electric power, strengthened the provincial government's hand in its feud with the federal government over the development of the Columbia River. The end results were the "two river policy" of the provincial government, and the reversal of the federal government's traditional policy of non-export of electric power.[28] There was some doubt whether the former was a good thing for British Columbia, but the latter was clearly a good thing for Canada.

There were also difficulties on the United States side of the border in appraising the relative merits of the proposed Columbia River Treaty scheme. These stemmed in part from the changing policies of the federal government with respect to federal participation in power development. Under the Eisenhower administration a policy of "no new starts" brought to a halt the development of storage projects, and thus made Canadian storage seem especially attractive. When the Kennedy administration took over, power policies changed radically, and there was once again federal support for water resources development. In addition, there was active interest in the possibilities of a California-Pacific northwest inter-tie, as well as in federal participation in nuclear power development in Hanford, Washington. These changes of policy, of course, came after the signing of the treaty and too late to influence the outcome. Each of them might have resulted in the development of less Canadian storage but in an economically more efficient Columbia River scheme. As was suggested earlier, however, the die was cast and neither government was willing to alter its position.

The United States Senate quickly endorsed the Columbia River Treaty in 1961, but it took three years for Canada to do so, largely because of the criticisms that were levelled against the treaty in Canada, and the unwillingness of the British Columbia government to agree to the return to Canada of the Canadian share of downstream power benefits. On January 22, 1964, the two countries signed a protocol agreement which was intended to clear up the vagueness in certain of the provisions of the treaty. No major departures from the original agreement were made, however, except for the provision that Canada could sell her share of the downstream benefits to the United States for a period of thirty years.

LESSONS AND IMPLICATIONS

What can be learned from the Columbia River experience that will be useful in dealing with other rivers shared by the United States and Canada, and perhaps international rivers elsewhere as well? What implications does this experience have for the development of other rivers in Canada?

First, the Columbia River case served to highlight the problem of federal-provincial relations in the development of international rivers. The British North America Act gives to the federal government the responsibility for protecting the national interest. Interpretation of this responsibility led to the enactment of the International Rivers Improvement Act in 1955 that makes it necessary for federal approval to be obtained for the construction of any works on international rivers in Canada. In addition, the federal government has responsibility for negotiating treaties with other countries. The provinces, however, have title to the resources within their boundaries and so have the right to say how and if the river shall be developed. This arrangement places the federal government in an extremely awkward position: it can make treaties with the United States for the development of an international river, but it lacks the power to carry them out unless the province agrees.

The Columbia River case emphasized the need not only for obtaining provincial approval but also for obtaining it in formal terms. Much of the delay in ratifying the Columbia River Treaty can be traced to the failure of the federal government to obtain approval of the proposed arrangements from the British Columbia government.[29] The treaty was signed in 1961, but provincial sanction was not obtained until July, 1963, and final ratification did not come until September, 1964. The lesson is that, if the federal government does not obtain formal approval from a province *before* engaging in negotiations with the United States, Canada may run the risk of international embarrassment: that of having a treaty which it is unable to honour.

It has been suggested that the federal government was far too hasty in signing the treaty with the United States. With the advantage of hindsight this assessment may well be correct. But the signing did have at least one major advantage: it forced public attention to be focused on the proposed arrangement, attention that up to that time had been lacking. It also brought to a head the need for a definition of the role that the federal government ought to play in the development of rivers in Canada. Should it continue its role as that of a fact finder and a policeman, or should it actively participate in development? If the latter, should it do so as a partner or as an entity in its own right? The feud between British Columbia and Ottawa seems to have made it clear that the federal government's role is likely to be confined mainly to that of a fact finder and a policeman, with participation in development being confined to sharing the financial burden when the province is unable to find the money itself and asks for federal aid.

Second, in more general terms, the Columbia River experience shows that the comprehensive approach, although appealing in concept, is extremely difficult to put into practice. It seems to be relatively easy to apply at the data collection and preliminary planning stage where the only commitment is the determination of possibilites for development, but once the time comes when decisions have to be made about sharing costs and benefits of development, and when priorities need to be assigned, cooperation is much more difficult to obtain. At this stage the "national interest" (or the "provincial interest") tends to dominate the decision-making, backed up by such factors as "prior commitments" and pressures from various interest groups.

The chances of a sustained comprehensive approach in planning and development are further diminished when there is nothing to give the basin internal cohesion. In the case of the Columbia River, because the developed state of the lower basin contrasts vividly with the underdeveloped state of the upper basin, the activities in the two portions compete with rather than complement each other. In addition, the demand for the services of the Columbia River in Canada comes mainly from outside the basin. Moreover, because these services can be derived from a variety of alternative sources such as power from the Peace or Fraser rivers, there is no particular dependence on the Columbia. This situation contrasts with that in the Lower Mekong River in southeast Asia, where people in the four countries sharing the basin have few alternative sources of supply for the services which the Mekong could provide, and where the river does provide a means of fostering internal cohesion.[30] This may well be an important reason why the application of a comprehensive approach has succeeded (so far at least) in the planning of the Mekong

River, whereas it has failed or has been only partially successful in most other international river basins where it has been attempted.

The Columbia River experience also emphasizes that economic considerations tend to play a subordinate role in the final decision-making. The importance of taking such considerations into account is generally recognized by those in charge of the planning, but various factors inhibit the necessary evaluations—past decisions, prior commitments to particular projects, the engineer's scepticism of the value of economic studies, the lack of analytical tools, and the lack of trained personnel to carry out the studies. The general result is that economic studies occupy a rather minor place in over-all investigations, and are often delayed until most of the decisions have been made or are not undertaken at all. Millions of dollars were spent on engineering studies, most of them using basically the same information; yet only a few thousand dollars were spent on economic studies, even though investments of several hundred million dollars were being contemplated.[31]

Finally, it may be asked what influence the Columbia River experience has had on water management elsewhere in Canada. It is probably a little early for any clear answer to be given, but it does seem that the federal government is much more sensitive than previously to the manner in which the provinces jealously guard the stewardship of the water resources within their boundaries. At the same time the federal authority is keenly aware that a number of serious national water problems are emerging— such as water pollution, flood losses, and interprovincial conflicts over water use— and that there is a need for national leadership to overcome these problems. New institutional devices will be required to enable the federal government to exercise this leadership without treading on provincial water management toes.

Despite the criticisms of the Columbia River studies by economists and others, there has been very little change in the approach to planning the development of

Canada's major rivers. The traditional approach prevails, the major emphasis being placed on taking inventory of the possibilities for development, and little or no attention being paid to economic or social aspects. For example, in the planning associated with the St. John, Yukon, Saskatchewan-Nelson, and Fraser rivers, only a relatively narrow range of possible purposes of development is being perceived; little attempt is being made to review alternative ways of furnishing the benefits claimed for the proposed schemes; and no investigations are being undertaken to determine what institutional problems need to be solved before the schemes can get under way.

The slow progress that has been made is reflective of several factors. First, there has been no crisis to trigger a change in the approach to water management. Abundance of resources and availability of alternatives have permitted a leisurely inventorying of possibilities for development and have not encouraged a careful appraisal of the needs for development. Until now there has been no urgency to weigh the value of one water use against other water uses. It is clear, however, that the day is fast approaching when decisions between competing alternative uses will have to be made on certain rivers—for example, fish preservation and other uses on the Fraser River. Second, there is a scarcity of social scientists skilled in undertaking studies of non-engineering aspects of water resource development. At present there are not more than half a dozen top-flight water resource economists in Canada, whereas many United States agencies each employ several times that number. Unfortunately there are at present no university programmes in Canada of the type required to provide the necessary training, although some progress has recently been made at a few universities. Many of those who go south of the border to obtain their training never return to Canada to practise their craft. It is true that many Americans come across the border and help to balance the trade in brains, but unfortunately few of them are social scientists with an interest in water resources. Clearly, expanded university programmes will help to provide the personnel to do the studies. Just as important, however, will be modifications in the present institutional framework to ensure that the studies are in fact undertaken.[32] The latter is perhaps the greatest challenge to Canada's water resources future.

REFERENCES

[1] For a complete description of the provisions of the treaty, see Canada, Dept. of External Affairs, *The Columbia River Treaty: Protocol and Related Documents* (Ottawa, 1964), pp. 58-81.

[2] Subsequent studies, using different sets of assumptions about stream flows and power demands, have shown that power benefits would be considerably greater than these estimates. Capacity benefits would be 5 to 7 percent greater and energy benefits 14 to 18 percent greater. (*Ibid.,* p. 132)

[3] In United States funds these figures were $253.9 million and $64.4 million respectively. Canadian officials used the United States figures to convince Canadians that they had got a bargain, whereas American officials used the Canadian figures to convince their citizens that they too had got a bargain.

[4] Under independent development, the estimated cost of power from Mica would be about 4 million kilowatt hours (Dept. of External Affairs, *Columbia River Treaty: Protocol and Related Documents,* p. 134).

[5] In 1963 this was about $319 million.

[6] Dept. of External Affairs, *Columbia River Treaty: Protocol and Related Documents,* p. 135.

[7] The principal debates took place in the hearings of the Standing Committee on External Affairs. See Standing Committee on External Affairs, "Columbia River Treaty and Protocol," *Minutes of Proceedings and Evidence* (Ottawa, 1964).

[8] See, for example, Ripley, James A., "The Columbia River Scandal," *Eng. and Contract Record,* April 1964, pp. 45-60.

[9] See, for example, Krutilla, J. V., *The Columbia River Treaty: An International Evaluation* (Resources for the Future, Inc., reprint no. 42); and Higgins, L., "Columbia River Treaty: A Critical Review," *Inter. J.,* XVI (1961), pp. 399-404.

[10] The most vociferous critic was General A. G. L. McNaughton, the architect of the early studies of possibilities of development. See, for example, McNaughton, A. G. L., "The Proposed Columbia River Treaty," *Inter. J.,* XVIII (1962-63), pp. 148-65.

[11] Glos, Ernest, *International Rivers: A Policy-Oriented Perspective* (Singapore, 1961), p. 11.

[12] Origins of the concepts underlying comprehensive river basin planning are described in White, Gilbert F., "A Perspective of River Basin Development," *Law and Contemporary Problems,* XXII (1957).

[13] Steps involved in the determination of an optimum scheme are described in Weber, Eugene W., and Maynard M. Hufschmidt, "River Basin Planning in the United States," *United States Papers for United Nations Conference on Science and Technology,* vol. II (New York, 1963), pp. 299-312.

[14] See, for example, United Nations, Dept. of Econ. and Social Affairs, *Integrated River Basin Development* (New York, 1958).

[15] International Joint Commission, International Columbia River Engineering Board, *Water Resources of the Columbia River Basin* (Ottawa, 1959).

[16] The board had been specifically instructed by the IJC not to consider these in its studies.

[17] There is now a vast literature on the subject. For a thorough discussion of United States agency practices, see Eckstein, Otto, *Water Resource Development: The Economics of Project Evaluation* (Cambridge, Mass., 1958).

[18] International Joint Commission, *Report on Principles for Determining and Apportioning Benefits from Co-operative Use of Storage of Waters and Electrical Interconnection within the Columbia River System* (Ottawa, Dec. 1959).

[19] See, for example, Krutilla, *Columbia River Treaty,* and Higgins, "Columbia River Treaty."

[20] See, for example, Krutilla, *ibid.,* p. 13.

[21] *Ibid.*

[22] The regulation provided by upstream storage in Canada will make possible an increase in firm power output. The Americans, however, were willing to share only part of this increase, claiming that a large block of the secondary power which would be firmed up is already available for most of the time. This secondary power is now sold as firm power in the United States, and the Americans insisted that it be netted out of the increase in firm power resulting from Canadian storage.

[23] This qualification was added to several of the IJC's recommended principles. See Dept. of External Affairs, *Columbia River Treaty: Protocol and Related Documents,* p. 41.

[24] The costs were defined in financial rather than in broad economic terms.

[25] See Fraser River Board, *Preliminary Report on Flood Control and Hydro-Electric Power in the Fraser River Basin* (Victoria, B.C., 1958).

[26] Most of the cost estimates in the federal government's presentation to the External Affairs Committee, for example, relate only to the Columbia River. See Canada, Dept. of External Affairs, *The Columbia River Treaty and Protocol: A Presentation* (Ottawa, April 1964), p. 37.

[27] Both schemes could deliver power to the Vancouver area for about 4 million kilowatt hours. If the flood control benefits attached to the Columbia River scheme were included the cost of the latter would be somewhat less than the costs of the Peace River scheme. Moreover, if account were taken of the fact that the Columbia River scheme would provide more power than would the Peace River scheme, over-all costs for the latter would be higher than for the former because of higher costs of providing the additional increment to Peace River power. British Columbia Energy Board, *Report on the Columbia and Peace Power Projects* (Victoria, B.C., July 31, 1961).

[28] The actual trigger to reversing the traditional policy of non-export of electric power, however, was the expropriation of British Columbia Electric Company by the British Columbia government in 1961 which effectively removed the market for Columbia River power in the province. The expropriated utility was instructed to proceed with the development of the Peace River, thus pre-empting the British Columbia power market for years to come.

[29] British Columbia government officials were present during the negotiation of the treaty and appeared to be in agreement with its terms. Once it was signed, however, the provincial

334

government politely told the federal government that it was not prepared to proceed with the development required by the treaty.

[30] See White, Gilbert F., "The Mekong River Plan," *Sci. Am.*, CCVIII, no. 4 (April, 1963), pp. 49-59.

[31] The lack of economic studies is typical in Canadian water resource development. In the case of the Fraser River, for example, the federal and British Columbia governments spent over $4 million on studies and investigations over a fourteen-year period. Less than $30,000, however, was spent on economic studies, and then only in the last few months of the Fraser River Board's investigations. Yet the board recommended a scheme estimated to cost over $400 million. See Sewell, W. R. Derrick, *Water Management and Floods in the Fraser River Basin* (Univ. of Chicago, Dept. of Geog. Res. Ser., no. 100, June 1965).

[62] Modifications might include: social scientists on advisory committees and in water-planning agencies; specification in federal-provincial cost-sharing agreements that all applications for funds be accompanied by a benefit-cost analysis of the project.

Interbasin Water Transfer – The International Dimension

Arleigh H. Laycock

A number of interbasin transfers involving more than one country have been developed and many others have been proposed.[1] Some of the proposals are spectacular in concept and as many as three countries are involved—e.g., in proposals that Congo Basin waters should be diverted northward toward the Sahara Desert and in schemes in which Alaskan and Northern Canadian waters should be diverted southward into the United States and Mexico. Most of the actual developments are more modest and local and involve diversion from international streams, usually at the expense of neighbouring countries. The diversions from the Great Lakes Basin at Chicago, from the Saskatchewan Basin in the St. Mary Diversion in Montana, from the St. John Basin in the Aroostook Diversion in Maine, from the Colorado Basin in trans-mountain diversions in Colorado, Utah, New Mexico and Wyoming, and from the Indus Basin in diversions to the Ganges Basin in Punjab are among the better examples. Most have been made possible by the promulgation of convenient doctrines by the more powerful country to the effect that the upstream nation might do as it wished with water originating in that nation—e.g., the "Harmon Doctrine"[2] of the United States, or by precedents set before an international boundary had been defined—e.g., the boundary of India and Pakistan. A smaller number of developments involve transfer into an international basin for use downstream by the diverting country. The Long Lac and Ogoki Diversions from the Hudson Bay Drainage into Lake Superior are illustrations of this.

● Arleigh H. Laycock *is Professor of Geography, Department of Geography, University of Alberta and served as President of the American Water Resources Association, 1970-71.*

Reprinted from the *Water Resources Bulletin* (Journal of the American Water Resources Association, Urbana, Illinois), Vol. 7, No. 5, October, 1971, with permission of the author and the American Water Resources Association.

I know of no significant interbasin transfer from one nation to another for use primarily by the recipient. Some minor transfers are marginally in this category—e.g. from New Brunswick to serve Calais and Milltown in Maine, but all points involved are in the basin of the St. Croix River and Estuary. Sweetgrass, Montana buys water from Coutts in Alberta but both are within the Missouri Basin.[3] These are largely for convenience and economy, and legal precedents for larger transfers are not being established.

A large number of international transfers are technically possible and a number have been proposed in North America and elsewhere. The lack of development has been attributed to constraints associated with the presence of the boundary. Roy Tinney and Frank Quinn[4] have noted that there are few interbasin transfers across state boundaries in the United States yet there are a number within states. In Canada too, most transfers are within provinces rather than between provinces—and with the completion of projects now being developed, interbasin transfer will be significantly greater in volume in Canada than in the United States—perhaps partly because the United States have five times as many states as we have provinces. In both countries, however, there are growing inhibitions concerning such transfer. The resistance of basin of origin people is strengthened by the more general opposition by "Ecologists" to such disruption of natural flow patterns. Frank Quinn[5] has noted this alliance.

My focus will be upon potential transfers from Canada to the United States and I'll try to explain why development prospects are currently dim, what is needed if this picture is to change and how the presence of a border could become an advantage for development rather than a handicap. Let me begin by reviewing some of the major Canadian-American transfer proposals and some Canadian and American reactions to them— partly to explain why they have not gone past the proposal stage.[6]

The North American Water and Power Alliance (N.A.W.A.P.A.) scheme of the Ralph M. Parsons Co., of Los Angeles is probably the best known of the various proposals, largely because of company promotion and the political support given to it in the United States.[7] It would involve the transfer of over 100 million acre feet of water from northwestern Canada and adjoining Alaska southeastward into the dried parts of Canada, the United States and Mexico. The benefits to the United States would be major but those to Canada would be smaller and less well defined and would be greatly outweighed by damages and hazards. There is no reference to the sale of water or water rights: only to the costs of transfer. There is no good reason why this valuable resource should be considered a free good—free for the taking from another nation. It was assumed that the multiplier effect of engineering development expenditures would be a major benefit—but this is doubtful if most of the power resource of the region is to be allocated to pumping water southward and much of the valley bottomland is to be flooded to provide storage, which would be released according to American demands. Less than 4 percent of British Columbia is arable and some of this is in valleys along proposed routes. A part of the power would be recovered at higher prices in the southern parts of the system in the United States but this would contribute to reduced market prices for water, not to the development of the source region. The irrigation benefits in Canada would be illusory because we have much cheaper water near at hand for these purposes. Why should we contribute to the costs of an overriding system that could make irrigation and other development more costly to us? The navigation benefits would be of doubtful value because many of the canal routes lead from nowhere to nowhere and are frozen much of the year. The building of major dams in areas prone to faulting might result in severe flooding and the reservoirs would make rail and road re-location expensive and difficult. The major objection in Canada would be to the development of a supra-national authority that would develop and manage our resources according to its needs—which would not necessarily be ours. Moreover, there would be no way for Canada to be released from delivery commitments whatever future need it might have for this water for its own development. Such a loss of sovereignty would not be acceptable and some Canadians believe that any long-term export arrangement would link our economies more closely than would be in Canada's best interest. Ecological and other objections concerning the consequences of diversion in the North are of growing significance. Some of the greatest objections are economic. In this and other grandiose schemes, it is assumed that water will have to be obtained from distant northern areas of Canada because southern water rights will be reserved for Canadian development. It is also assumed that transfer facilities must be large from the start because later reconstruction to larger scales would be costly. The very high costs and financial carrying charges in these schemes would make the water so expensive that there would be little or no market for the large potential deliveries. Alternative water developments including desalination, weather modification, re-use, re-allocation and local surplus diversion would be precluded despite their possible cost advantages. In summary, the proposal contains few advantages and many disadvantages for Canada and the negative attitudes engendered are widely applied to all export proposals. Many authorities in the United States equate import with this scheme. They accordingly discount Canada as a water source and, upon learning of Canadian opposition to it, are inclined to focus upon other alternatives within the United States. Numerous questions have also been raised concerning the provision of water to arid areas at less than cost.[8]

Other proposals for the diversion of Canadian water southward include one by Tom Kierans of Sudbury, Ontario.[9] It would involve the diversion of water now flowing "unused" into James Bay, southward via a G.R.A.N.D. (Great Replenishment and Northern Development) Canal into the Great Lakes for use by both Canada and the United States. The purpose would be partly one of supplementing Great Lakes supplies to dilute the partly polluted water present.

It is now generally agreed that dilution is no solution to Great Lakes eutrophication problems and water level control by varying inflow can be difficult.

Knut Magnusson[10] has proposed tapping the Peace River in Alberta and diverting much of its flow across the Prairies into the Missouri Basin. The physical possibilities are promising but again there were few benefits assigned to Canada.

Ed Kuiper,[11] of the University of Manitoba, has proposed that water from the northeastern parts of the Prairie Provinces should be exported via the Manitoba Lowlands to the Interior Plains of the United States. This water would be sold at prices that would be competitive in the United State and provide a fair return to Canada—possibly 40 dollars per acre foot.

Roy Tinney[12] proposed a wider gathering system in the Canadian North and a wider distribution system in the United States—including the Texas High Plains and the Southwest. Both Kuiper[13] and Tinney, and Tinney and Quinn[14] now doubt that markets would be available at the prices that would be necessary.

Lewis Gordy Smith[15] has proposed that water from the Liard and Mackenzie Rivers should move southward via the Rocky Mountain Trench to the Centennial Valley in Montana. From there it would flow in several directions to the drier areas of Western United States.

Some additional benefits for Canada were suggested by Smith but the assumption that engineering development is virtually enough in itself appears to be present in this and most other proposals. Kuiper has gone farthest in suggesting that the water should be sold at a profit above transfer and opportunity costs. Few of the proposals stress or are adaptable to complementary or phased development. Little mention is made of compensation for "environmental" damages. The water to be transported is generally assumed to be surplus to long-term Canadian needs and specific supplies are noted. There is an assumption of relatively early need for water in the United States and the thought that a greater present American need might be for assured back-up supply is not expressed.

It should not be surprising that some of the proposals would contain very few advantages and some very real hazards for Canada. Counter proposals that would be highly advantageous to Canada and hazardous to the United States might be made. It would appear that water is a major resource of growing value and arrangements for transfer that might have been acceptable relatively few years ago are no longer adequate. It would appear also that relatively fixed terms negotiated at an early date might be advantageous to the United States and flexible terms negotiated as late as possible would be better for Canada. In practice, any terms that are obviously unfair at any stage to either country would be bad because of the hostility and bad faith engendered. Continuing negotiation in good faith and for mutual benefit is essential if any long-term export arrangement is to be successful.

The major schemes, proposals and promotions are widely equated with Canadian water export. This is unfortunate because there are better prospects for smaller-scale and phased developments where diversion might serve Canadian as well as export interests. Some of these complementary transfer developments may be feasible now or in the near future and a lead time for research, planning, discussion and negotiation should probably start now.

In the Great Lakes Basin, Chicago is currently permitted by the United States Supreme Court to divert as much as 3,200 cubic feet per second to carry treated wastes southward via the Illinois River to the Mississippi.[16] This has helped to keep the eutrophication problem in Lake Michigan at lower levels than otherwise might have been present. William Ackerman, Chief of the Illinois State Water Survey, and others, suggest that this diversion should be expanded for Chicago and extended to serve other centres on the western and southern shores of Lake Michigan and a similar diversion southward to the Ohio River should be provided for all of the centres on and near Lake Erie from Detroit to Buffalo.[17] About 25,000 cubic feet per second flow would be needed but the Great Lakes pollution problem would be greatly reduced—to the benefit of both Canada and the United States. Replacement supplies would be needed if objections regarding hydropower losses, navigation impairment, and low lake levels harmful to recreational and other uses are to be overcome. Such supplies can be obtained from streams north of Lake Superior now flowing into Hudson Bay—as we now obtain flow in the Ogoki and Long Lac diversions for Ontario Hydro purposes. We should sell this water at a good price to cover the costs of transfer, the "opportunity costs" (value of benefits lost in Canada), an allowance for other consequences of diversion so that we may have environmental improvement rather than despoliation of the area of origin—*and* possibly a profit.

Additional water might be used by most if not all of the Great Lakes states to supplement stream, reservoir and other supplies outside of the Great Lakes Basin for urban, industrial, supplemental irrigation, recreational, wildlife, aesthetic and other purposes. Integrated networks of water supply systems might meet many needs better than numerous separate local systems. Additional benefits in the United States might include a use of the supplemented Mississippi flow by Texas. The present opposition by Louisiana and other states to diversion by Texas from the Mississippi might be lessened if extra flow is available and the "natural" flow or a better regime can be assured. The selection of a point of diversion might also be easier.

In the Prairies, we might have earlier and larger-scale transfers from the Athabasca and Peace River Basins into the drier plains if extra flow is exported. The Saskatchewan-Nelson Basin Board[18] and Alberta P.R.I.M.E.[19] diversion proposals for Canada could be extended and the Prairies could benefit from the economics of scale and earlier project development as well as from the sale of water—again with opportunity costs, environmental costs and a profit added. Later diversions from farther north to

338

meet export needs are possible but there is little reason why the more available surpluses from the Peace and Athabasca basins should not be sold first. Water rights to streamflow from specific rivers need never be allocated specifically and permanently for export. The natural spillway patterns in the Prairies make phased, sequential development a very reasonable possibility; thus the earlier deliveries need not be extremely expensive. Later deliveries from farther north could gain from improving technology and economies of scale.

Also in the Prairies, we suffer from an imbalance in freight traffic flow—much more goes out than comes in. We might ease this situation in the future by using products pipelines in which water would be used to carry goods in solution or suspension, possibly in plasticized pellets.[20] If, for example, a pipeline is built to carry Saskatchewan potash to Chicago, why shouldn't we sell the water as well as the product carried? The volumes may be small but the extra revenue might make such a project feasible. Later pipelines carrying agricultural, forest and mineral products including many in processed form to Los Angeles and other water deficient areas might greatly help our economy. The water export involved could be significant, partly because of the export precedents that might be set.

In British Columbia, the proposed Shushwap diversion into the Okanagan as replacement supply for irrigation withdrawals would actually involve a type of export. Canada is fully entitled to consume as much Columbia Basin water as it wishes within the Basin, thus replacement supplies from purely Canadian basins will increase downstream flow in the United States for power, irrigation and other needs. Additional flow is needed in the Okanagan but denying that export is taking place merely cuts off a potential source of sales revenues that might pay many of the costs of diversion, cover opportunity costs and enable us to improve the environment of the source areas. Other southward diversions are possible. British Columbia alone has as large a stream flow as the Western two-thirds of the United States and the ex-

port of a small part of it could provide large long-term revenues without net impairment of environments. The technology of pipeline and tunnel construction is improving rapidly and major valley flooding is not essential for water transfer. We should keep our alternatives open for future development.

We have indicated earlier that there isn't a market at present for major scheme water supplies. There may be in the future, possibly fifty or more years from now. If we are interested in exporting water we might start with some of the smaller projects mentioned, but even better opportunities may be present. The United States has more than enough water in most regions to meet its consumptive use requirements for many decades, perhaps centuries. There are major shortages or impending shortages in West Texas, Arizona, Southern Nevada, Southern California and in other areas but large surpluses relative to current demands are present farther north and east in the Columbia, Missouri and Mississippi Basins. Transfers are physically possible but alternatives to transfer need further study first. Institutional constraints upon transfer are growing rapidly. The streams with surpluses in the West are being reserved for future use in their basins and transfers are becoming less and less likely. If replacement supplies could be guaranteed for future use in basins with current surpluses, much of this water might become available for transfer—with due allowance for opportunity costs and environmental improvement costs in the areas of origin. Such guarantees might be provided by Canada and its Provinces through the sale of options on given amounts of water to be delivered to border points at some time in the future— possibly twenty to fifty years away. The needs probably are not as urgent in most areas as are sometimes suggested because alternative solutions such as re-allocation still have cost advantages. With option agreements, actual deliveries could be small, but the assurance of having adequate supplies in the future could be a major benefit.

In the United States option agreements could facilitate the planning and development of numerous diversions, and the mining of ground water, in anticipation that replacement or alternative supplies would later be available. Significant volumes of water that are surplus to present demands in the Columbia, Missouri and other basins might be diverted to water short areas because reservations for future development could be based in part upon Canadian supplies. In the drier areas, if extra supplies are available, or are in prospect of becoming available, many of the litigation and speculation costs now developing might be avoided. Integrated water supply networks might be developed from Canada southward with guaranteed base flow from Canada and more efficient development of local supplies en route. Other alternative supplies—e.g. from desalination or weather modification might be phased into these networks when feasible.

Trade-off agreements could be facilitated —for example, the upper tributaries of the Missouri and Columbia might have flow reallocations to upstream and other basin use if downstream needs could be served from Canadian sources. Similarly, some of the proposed diversions from the Upper Colorado Basin might be avoided in favour of downstream use if substitute supplies can be found. For example, the diversions proposed, in Type I studies and Wyoming State plans, from the Green River eastward to the North Platte and Powder River Basins might be avoided in favour of Lower Colorado Basin use. Present waters supplied to Nebraska by compact might be used in Eastern Wyoming if substitute supplies could be obtained from the Missouri River—with replacement supplies from Canada when needed.

With additional credible replacement supply guarantees, it is quite possible that a part of the resistance in Northern California to water transfer southward might be reduced and a more complete sequence of diversions from Northwestern California and Southwestern Oregon might be developed. Columbia River water, including supplemental supplies from Canada in the long run, might be needed in sequence but not until after local surpluses had been employed. Offshore submerged aqueducts taking water southward from Northwestern California, Southwestern Oregon and eventually the Columbia look very promising.

With replacement supply guarantees, a fuller use of local surpluses would be possible in the short run. Integrated supply networks would be more realistic because lesser parts of the flow would need to be allocated to future development in each basin. Various regional benefits in flood control and other improvements plus opportunity costs are assumed. Better use of ground water storage in the network might be anticipated.

If water export is to be arranged, Canada must be confident that it has surpluses available. For consumptive uses, there should no longer be any question of this. We know that Canada has approximately $2\frac{1}{2}$ to 3 billion acre feet of streamflow per year.[21] This is approximately twice as much as conterminous United States and close to twenty times the per capita supply of the 48 states. We currently consume only 0.1% of this supply and it is extremely unlikely that we will consume ten times this amount or 1% fifty years from now. The physical potential for irrigation use is limited by climatic, topographic and other factors to relatively low levels. If irrigation in the Prairies did expand to 20 million acres or about fifteen times our present development, the consumptive use would not greatly exceed 20 million acre feet because much of this would be supplemental irrigation where average annual deficits are only 5 to 10 inches.[2] Other consumptive uses are small. Most of our cities have a greater net water yield to our rivers than would have occurred in the pre-urban environment.

Non-consumptive withdrawal uses are contributing to significant quality changes in some areas, particularly in our more populated and industrialized areas near the United States border—adding to the problems of international rivers and lakes to which the United States is contributing more than its share.

On-stream uses for hydro power production, recreation, fish and wildlife needs and navigation cannot be ignored. For example, hydro power production in Canada is now two-thirds as large as in the United States and it is still growing rapidly. We have hundreds of thousands of natural lakes with an area much larger than the total areas of Nevada plus California and recreational development is very significant—in part for our major tourist industry. Rapidly growing numbers of Americans have very positive and sometimes possessive attitudes concerning this resource. Some would be delighted if we had no other development. Our wildlife is important for recreational and other reasons—e.g., most of the migratory waterfowl of North America nest in Canada. American support of this industry through "Ducks Unlimited" and other agencies is significant but close analysis might show that we are paying heavily in less intensive land use than we might otherwise have and in grain consumption so that hunting in the United States might flourish. Navigation on the Athabasca and Mackenzie Rivers is still much cheaper than other means of northern transport and the effect of taking a few feet of depth off the top of the latter has been well illustrated in the past few years, while Williston Reservoir on the Peace River has been filled. In most cases, however, these non-consumptive uses would not be greatly affected by water export or there would be compensating benefits with diversion and water control.

We do have large amounts of water and we might reasonably decide, if given adequate sales incentives, to allocate a small part of our supply for export. One percent of our supply would be twice the flow of the Colorado River. The gross revenue from the sale of this amount might well be 500 million to 1 billion dollars per year and option agreements might add to this amount. If we were apprehesive about selling this amount, we might work out arrangements by which replacement supplies would be provided sequentially by Alaska—if and when they are needed.

An issue of rapidly growing importance is that of the "new ecology" relating to water storage and diversion. According to some people, water development of any kind is bad—it "wrecks" the natural environment and "destroys" wildlife. It is undoubtedly true that single use oriented development of

water resources for hydro power production, irrigation, navigation or whatever *may* result in significant losses to other users and to wildlife. We have numerous illustrations of this, but I do not subscribe to the other single use extreme of widely making the preservation of wildlife and the natural environment the exclusive objective of river basin management. It is interesting that opposition to water export by ecologists in Canada is now being strongly reinforced by American ecologists who would like to see most of Canada preserved for wildlife purposes.

Surely, multi-purpose development is possible in which watershed management and water control can contribute to environmental improvement for both wildlife and man. Most environments can be more productive of fish and game with water diversion and control, oriented in part to their needs, than without it. This is not the sole objective of many ecologists who believe that there are values inherent in a natural system no matter how destructive nature may be. It can be argued that many aesthetic and other features of the environment can also be enhanced by judicious development. Some of us believe that the works of man can be attractive and striking and incidentally useful additions to our environments. Numerous illustrations of each point may be cited, but the increasingly organized and militant opposition of ecologists to diversion will be extremely hard to overcome because much of the public is being conditioned to their extreme views and simplistic solutions. Proponents of diversion are partly to blame because most of the proposals contain too little reference to multi-purpose regional development and consideration of environmental improvement with water diversion and control. The public image of both should improve and the implausibility of the claim that there is no such thing as a water surplus anywhere and anytime should become apparent.

This discussion has been focussed upon transfer from purely Canadian river basins to the United States. The special problem of international river basins has been noted in part—i.e., that there is a marked imbalance in development, with the United States having virtually all of the diversion from these basins and providing a disproportionate share of the pollution to the rivers and lakes of these basins. Pressures were placed upon Canada not to transfer water from and within the Columbia Basin and it is recognized that Canada could undertake other diversions to its advantage if the Harmon Doctrine approach is to be used. Many of the growing issues might be resolved in part *if* Canadian water is to be exported to meet growing American needs. Transfers from international basins could involve benefit sharing proportional to the volume of contributions and cost sharing proportional to the demands of each country.

The United States has a greater obligation than Canada to reduce pollution in the Great Lakes because of its greater contributions to the problem. In time, Ontario centres might transport treated wastes by pipeline to below the outlet of Lake Ontario to help to relieve the problem.

In summary, the N.A.W.A.P.A. and most other schemes involving international water transfers have inadequate provisions for the needs of the areas of origin. Opposition in Canada is similar to that in the Pacific Northwest of the United States to diversions to the Southwest, but it has been accentuated because it appears that one nation is attempting to take advantage of another. This has resulted in negative nationalistic responses, which may grow and make future co-operation difficult. The usefulness of the concepts in stimulating thought and discussion may be more than overcome. It might be suggested that non-compensation for environmental and opportunity costs within both the United States and Canada is becoming obsolete. An additional payment or profit is needed between nations because just breaking even economically is not enough if significant advantage appears to be gained by the receiver. The environmental and opportunity costs must be generous because the changes made tend to be lasting and the use alternatives are reduced. Moreover, when another country is involved, the opportunity cost pay-

ments must be more lasting than the re-adjustment period for individuals because the economy of the source country may be handicapped for much longer. Numerous compensating benefits directly from the sale of water and indirectly from the multiplier effects of development upon the economy, greater water control, lower unit costs for water and use of diverted water by Southern Canada en route to the United States can be anticipated and allowances must be made for them in negotiations.

The presence of the international boundary could be an advantage for water transfer. It is possible that Americans might be more likely to accept the concept of payment for imported water than that of payment, in excess of delivery costs, for American water. This does the least violence to the great mass of institutions that have been developed around water distribution in Western United States. Tenure security in water rights might be maintained yet tenure flexibility might be provided.[23] The use of replacement supply options for Canadian water might well be the cheapest way to achieve greater efficiencies in the use of water within the United States because changing or modifying legal and other institutions could be very costly. The political choice of taking options on Canadian water is a much easier one than the forced re-allocation of regional surpluses or of local water rights. If some Canadian water can be considered as a back-up supply permitting easier short-run re-allocation within the United States, and a long-run reserve to supplement the American supply, two very important needs will be served— probably at less cost, on a more massive scale and with greater ease and flexibility of attainment than for any of the alternatives. It is a development alternative that should be considered very seriously, not to the exclusion of others because there isn't any single best way to solve all water shortage problems, but as one that should be opened and kept open for use. Having it in reserve may be very important in your planning and development. At the very least, the anticipation of this development option being available, by some people in your area and in other areas, will help to reduce speculation and litigation in your resource development, and it might help in providing confidence in long term investment—and these are useful services. At the most, it may be the most important of the alternatives available. I urge you to study it. If you want Canadian water, you should anticipate that it will be costly and that both countries should benefit from any transfers. Without mutual benefit there will be virtually no international dimension in interbasin transfer.

REFERENCES

1 Whetstone, George A., *Interbasin Diversion of Water: An Annotated Bibliography,* Texas Tech. University, Water Resources Center, Lubbock, Texas, 1970, 323 p., and "The Role of Water Exportation in National Resources Planning," a presentation to the Engineering Institute of Canada, Sept. 1970, 12 p.

2 Pope, H. W., "The Legal Aspects of Water as it Affects Provincial and International Boundaries," in *Proceedings of a Water Resources and Conservation Seminar,* sponsored by Central Alberta Chamber of Commerce, Red Deer, Alberta, 1967, pp. 13-27.

3 Robinson, G., Letter to Professor I. MacIver, from Secretary-Treasurer, Village of Coutts, Alberta, 25 August, 1970.

4 Tinney, E. Roy and Frank J. Quinn, "Canadian Waters and Arid Lands," in W. G. McGinnies and B. J. Goldman (eds.) *Arid Lands in Perspective,* Univ. of Arizona Press, Tucson, pp. 412-415.

5 Quinn, Frank J., "Area-of-Origin Protectionism in Western Waters," (Unpublished Ph.D. thesis, Department of Geography, University of Washington, 1970).

6 See, Laycock, A., "Canadian Water for Texas?" *Water Resources Bulletin,* Vol 6 (4), 1970, pp. 542-549; Bingham, Jay R., *A Review of Inter-Regional and International Water Transfer Proposals,* Western States Water Council, Salt Lake City, 1969, 13 p.; Alberta Dept. of Agriculture, *Water Diversion Proposals of North America,* Water Resources Division, Development Planning Branch, Edmonton, 1968, 49 p.; Tinney, E. Roy, (ed.), *Proceedings the Seminar on the Continental Use of Arctic-Flowing Waters,* Pullman, Washington, 1 April — 20 May, 1968, 176 p.; and Clark, Chapin D., "Northwest-Southwest Water Diversion-Plans and Issues," *Williamette Law Journal,* Vol. 3(4), 1965, pp. 215-262.

7 Parsons, Ralph M. Inc., *N.A.W.A.P.A. — Water for the Next One-Hundred Years,* Los Angeles, 1964; United States Senate, *Western Water Development,* Special Subcommittee on Western Water Development of the Committee on Public Works, 89th Congress, 2nd Session, US. Govt. Printing Office, Washington, D.C., 1966, 62 p.; and Dolman, Claude E., (ed.), *Water Resources of Canada,* Proceedings of a Symposia presented to the Royal Society of Canada, Sherbrooke, Quebec, 1966, Univ. of Toronto Press, 1967, 251 p.

8 Carlson Jack W., "Water Resources Investments and National Goals," Third Western Interstate Conference, Fort Collins, Colorado, 26 August 1969, 18 p. (Multilith).

9 Kierans, T. W., "The Grand Canal Concept," The Engineering Journal, Vol. ( ), 1965, pp.

10 From 1967-1969 Magnusson, now deceased, distributed mimeographed brochures on this topic.

11 Kuiper, E., "Canadian Water Export," *Engineering Journal,* Vol. 49(7), 1966, pp. 13-18.

12 Tinney, E. Roy, "N.A.W.A.P.A. — Engineering Aspects," Bulletin of the Atomic Scientist," Vol 23(7), 1967, pp. 21-27.

13 Kuiper, Edward, "Feasibility of Water Export," Proceedings American Society of Civil Engineers, Vol. 94(HY 4), 1968, pp. 873-891.

14 Tinney, E. Roy, *Proceedings of the Seminar on Continental Use of Arctic-Flowing Waters,* and Tinney, E. Roy and Frank J. Quinn, "Canadian Waters and Arid Lands", in *Arid Lands in Perspective.*

15 Smith, Lewis Gordy, *Western States Water Augmentation Concept,* Federation of Rocky Mountain States Inc., Denver, 1968, 46 p.; and "Toward a National Water Plan," *Irrigation Age,* 1969, 24 p.

16 Maris, Albert B., *Report of Special Master in the Supreme Court of the United States,* The Legal Intelligencer, Philadelphia, 1966, 574 p.

17 Ackerman, William C., "Water Transfers — Possible De-eutrophication of the Great Lakes," National Research Council, Annual Meeting, 1968, Washington, D.C.

[18] Saskatchewan — Nelson Basin Board, *Annual Report Year Ending March 31*, Regina, Saskatchewan, 1969, 22 p.

[19] Bailey, R. E., "P.R.I.M.E. — Alberta's Blueprint for Water Development," in *Water Balance in North America*, American Water Resources Association, Urbana, 1970, pp. 227-234.

[20] Jensen, E. J., and H. S. Ellis, "Pipelines," *Scientific American*, Vol. 216(1), 1967, pp. 62-72.

[21] Canadian National Committee for the International Hydrological Decade 1969, Preliminary Maps, *Hydrological Atlas of Canada*, Ottawa, 1969.

[22] Laycock, Arleigh H., "Water Deficiency and Surplus Patterns in the Prairie Provinces," in Prairie Provinces Water Board Report No. 13, Regina, 1967, 185 p.

[23] Kelso, Maurice M., "Competition for Water in an Expanding Economy: Policies and Institutions," Paper presented to the American Society of Civil Engineers, 1 Nov. 1967, Sacramento (ditto)

The Federal Government and Environmental Quality

E. Roy Tinney and J. G. Michael Parkes

PART I: THE LEGAL BASIS OF FEDERAL INVOLVEMENT

The basis of federal involvement in environmental quality issues is established basically in the British North America Act.

Canada, in common with virtually all federal countries, faces problems of divided jurisdiction in the management of her resources.[1] The inherent difficulty in Canadian resources management is that it straddles the two groups of constitutional powers, federal and provincial, created by the B.N.A. Act. The distribution of rights and responsibilities under this Act is complex, particularly with regard to air, water and land management. One important aspect of this distribution of powers is that of proprietary rights.

Proprietary Rights—Provincial

By virtue of Section 109 of the B.N.A. Act, the provinces are entitled to substantial proprietary rights. The Act states that with certain exceptions, "all lands, mines, minerals and royalties" that were owned before Confederation "shall belong to the provinces." One exception to this section involved public

● E. Roy Tinney *is Director-General, Policy and Planning Directorate, Department of the Environment, Ottawa, and* J. G. Michael Parkes *is Geographer and Water Resources Planner, Inland Waters Branch, Planning Division, Department of the Environment.*

This paper is an amalgamation of two papers: (1) E. Roy Tinney, "The Canada Water Act — A Vehicle for Action," *Optimum*, Vol. 2 (1), 1971, pp. 35-48; and (2) E. Roy Tinney and J. G. Michael Parkes, "Enhancing the Quality of the Environment: Current Federal Legislation and Programs," *Habitat*, Vol. XIII (5 and 6), 1970, pp. 14-24. In the latter case, Dr. Tinney is the principal author. The former paper is reproduced with permission of the author and *OPTIMUM*, Bureau of Management Consulting, Government of Canada. Portions of the latter paper are reproduced with the permission of the authors and the Central Mortgage and Housing Corporation.

344

lands in the Canadian west. The proprietary rights to these lands were held by the federal government in the Prairie provinces until 1930. In that year, an amendment to the constitution transferred these rights to those provinces, so that in effect all provinces have approximately the same proprietary rights with regard to resources.

It is important to note that only "land" and "minerals" are mentioned in Section 109. No mention is made of "water" or to other natural resources. This is because the law has never recognized ownership of such commodities as fish, wildlife or water while they remain in their natural state. In the case of water, because of its "fugitive nature," it must be captured and reduced to a possession, such as a pail or a reservoir, before it may be owned. However, by making ownership of public lands provincial, Section 109 has given usufruct, or rights of use, of the water and its contents to the province for its own exploitation.

Proprietary Rights—Federal

Federal proprietary rights in resources are nevertheless significant. Section 108 of the B.N.A. Act states that "The public works and property of each province . . . shall be the property of Canada" (at the time of Confederation in 1867). These include "Canals, with lands and water power connected therewith; public harbours; rivers and lake improvement; and . . . lands set aside for general public purposes."

In addition there are several other ways by which the federal Crown may acquire property rights over various aspects of resource management. Section 117 of the B.N.A. Act provides for the rights of Canada to assume any lands or property, if required for the defence of the country. Also, the federal government has recognized expropriation powers in connection with other federal activities; in addition, the Crown may purchase land like any other citizen. It is important to note that the federal government has full proprietary rights in the Yukon and Northwest Territories equivalent to those held by the provinces.

Legislative Rights—Provincial

The right to legislate in respect of resources and the power to make laws concerning resource rights is also divided between the federal and provincial governments. Ownership of minerals and lands means automatic provincial power to legislate in respect to their use. Federal rights to legislate are derived from the B.N.A. Act insofar as certain aspects of resources may be considered of significant national interest. For example, the federal government controls strategically vital minerals such as uranium.

In the B.N.A. Act, a number of sections give the province legislative rights over resources. Under Section 92, the provinces may pass laws relating to the management and sale of the public lands belonging to the province, property and civil rights in the province and "Generally all matters of a merely local or private nature in the province." Further jurisdiction is found in clauses relating to local works and undertakings, municipal institutions and agriculture. Taken with their proprietary rights, provincial power would present a formidable force in dealing with resource management problems in the environment, if it were not for certain impediments.

Provincial powers are tempered by a number of relevant factors. First, no provincial statute may encroach on rights beyond provincial boundaries. Secondly, all federally incorporated companies are immune from provincial law concerning "all matters which are a vital part of their operations." Thirdly, because of unequal distribution of wealth, many provincial governments cannot finance all the operations they have the power to carry out.

Legislative Rights—Federal

The federal government derives considerable power from its legislative rights under the B.N.A. Act in the field of resource management. Under the Act, the federal government has exclusive jurisdiction in coastal and inland fisheries, navigation and shipping (in both inland waters and marine international waters), and the criminal law.

345

This considerable power means, for instance, that provincial constructions such as bridges, power dams, flood control projects or other resource management schemes, must conform to federal navigation regulations. Similarly in fisheries, the federal government controls the regulatory aspects of fishing such as conservation, anti-pollution measures and fishing rights. The provinces handle proprietary and marketing aspects of fishing only, unless additional federal power has been delegated to it.

Under certain international treaties, such as the Boundary Waters Treaty of 1909, the federal government has responsibilities with regard to pollution of boundary waters. The full extent of this Treaty has not been realized, and without provincial co-operation in this area, meaningful results against large scale pollution problems are difficult to obtain.

Since the federal government is responsible for legislation pertaining to the Criminal Code, any contravention of prohibitions established under the Code are criminal offences. Pollution could be made an offence under the Code. Furthermore, if another federal act designates certain prohibitions regarding the pollution of our water, violations of these prohibitions become criminal offences even though not under the Criminal Code.

In addition to these powers, the federal government may pass laws "from time to time" respecting agriculture. In the past this has provided the authority for direct federal planning and development of farm water supplies, land drainage and irrigation works. The limit of this power has not been tested, but there are implications for pollution control.

Finally, a very general provision in the B.N.A. Act enables the federal government to make laws for "the Peace, Order and Good Government of Canada" in matters not exclusively assigned to the provincial legislatures. There has been considerable debate as to the extent of the powers created in this preface to the specific outline of parliamentary powers. Some legal scholars have predicated that if a subject matter of legisla-

tion has great national significance such as pollution, there would indeed be jurisdiction for utilizing such a general power for initiating legislation. This step was taken in drafting the Canada Water Act, although the Act itself is based on a number of other constitutional positions.

In summary, there is a divided jurisdiction over resources management in Canada. The federal government has clear responsibility in fisheries and navigation; provincial proprietary rights extend over all lands, minerals and royalties. Both parties share responsibilities in agriculture, as well as other aspects of resource management. However, it is not just a case of dividing up resource responsibilities and pursuing different programs piecemeal. To be effective and efficient, joint federal-provincial consultation, negotiations and co-operation are essential. In the past this has sometimes meant interminable wrangling, bickering and little action. Yet the urgency of the environmental pollution problem, its crisis significance for all Canadians, and its effect on the very quality of our way of living provide the catalyst for the co-operation missing in the past.

The remainder of this paper is divided into two parts. In Part II the origins and operative mechanisms of the Canada Water Act are discussed; in Part III other legislative measures through which the federal government is involved in resources management are briefly summarized.

PART II: The CANADA WATER ACT— A VEHICLE FOR ACTION

The inadequacies of Canadian water resource development and utilization were, and still are, abundantly clear. Many single purpose projects have been built; dams have been constructed that contribute to grave ecological damages in downstream provinces; the implementation of plans have had to be aborted because local interests and concerns had been overlooked; Canadians have degraded the quality of certain waters flowing into the United States; the U.S. has polluted Canadian waters; mercury and oil slicks emerge from coast to coast; algae blooms have become commonplace; masses of herring have turned red, some salmon runs continue to dwindle, and one shellfish bed after another must be closed. Under these circumstances, the federal government felt it had to equip itself with the legislative authority to meet fully its responsibilities to remedy these deficiences. Recognizing on the one hand the divided jurisdiction in water matters and, on the other hand, the very real gains in efficiency and effectiveness by a comprehensive multi-purpose approach to water development and utilization, the federal government set out to build a legislative platform for joint federal-provincial undertakings. The Canada Water Act creates this basic platform.

The Canada Water Act

This Act has its origins in two major national meetings—*the Resources for Tomorrow Conference* in 1961 and the *Pollution of our Environment Conference* of 1966. It was conceived in 1967 and drafted mainly in the spring and summer of 1969. Its concepts were presented to the Canadian public in a policy paper in late August of 1969 and it was formally introduced to the House of Commons on November 5 of that same year. After many days of debate in both the House of Commons and the Senate and nearly forty public hearings it was passed by Parliament on June 26, 1970.

The Canada Water Act contains six principal parts, which together form a process for joint federal-provincial undertakings:

Consultation—Joint examination of problems to reach accord on those areas and subjects requiring joint action and on those most appropriately undertaken by each government alone

Comprehensive water management—Joint comprehensive planning of river or lake basins, regions or coastal strips and joint implementation of those plans

Water quality management—Joint designation of water quality areas, formation or designation of a management agency, water quality planning, implementation of those plans, a prohibition against the deposits of wastes except as prescribed by the regulations and the assessment of fees and charges

Nutrient control — A prohibition against the import or manufacture in Canada of cleaning agents or water conditioners containing more than the prescribed content of nutrients

Inspection and enforcement — Provision for inspectors to examine effluents, cleaning agents and water conditioners; fines for contravention of the prohibitions; plus the power to close the offending plants if necessary

Public information — A program to inform the public of the state of water resource development in Canada, to increase the understanding by the public of the problems of environmental degradation, and to use public leaders in an advisory capacity.

These are the key features of the Canada Water Act. How do they work or how is it intended they will work?

Consultation

In the past, governments have generally consulted each other on water matters only when one jurisdiction acted as a constraint on the other or for the purpose of obtaining funds. Occasionally governments have proceeded with inadequate consultations despite the clear overlap of responsibility, thus incurring acrimony on both sides. Then too, some problems have been shuttled back and forth without resolution. It was apparent that if cooperation, so essential in this field of divided jurisdiction, was to succeed, consultation had to be recognized as a key initial step and the appropriate arrangements for such consultation had to be created. Section 3 of the Canada Water Act makes such special provision for consultation. To this end, committees of officials from the federal and provincial governments in question are now meeting to decide which problems need joint action, which could be more appropriately handled by one government alone, and to determine priorities for joint action. These ten proposed consultative committees,

347

most of which are now in operation, consist of six or eight officials, half from each government, drawn from departments having major water resource programs. They come together at least semi-annually and their discussions will hopefully bring order to the pace and to the sequence of water resource development across Canada.

Comprehensive water resource management

Whenever a river or lake basin, region or coastal strip is chosen by the two governments through consultation for joint planning, joint basin boards are set up under a formal agreement to supervise broadly based planning committees. These committees select a study director and organize working groups drawn from the agencies concerned.

The exact form of this institutional arrangement is not described in the Canada Water Act. It is primarily the subject matter of agreements between the federal and provincial governments. Experimenting will continue with different arrangements until the best arrangement becomes evident.

It is the federal hope that more and more members will appear on these boards and committees from local governments and the private sector and that a significant portion of the work will be contracted to expert groups outside of government.

The comprehensive planning process itself is a complex, highly technical matter involving substantial inputs from the sociological, physical and biological sciences and from economics and engineering. Particularly striking is the necessity for much more sophisticated techniques to determine the perception and attitudes of the people affected and to involve them directly in the planning process. There is also an awareness of the great concern for ecological effects and the increased value that society is placing on environmental quality generally.

In describing this planning process, the Act calls for the formulation of "comprehensive water resource management plans including detailed estimates of the cost of the implementation of those plans and of revenues and other benefits likely to be realized from the implementation thereof based upon examina-

tion of the full range of reasonable alternatives and taking into account views expressed at public hearings and otherwise by persons likely to be affected by implementation of the plans."

Most of the basin studies are of three or four years duration involving hundreds of man-years of effort to develop alternative plans that would result in the best development of the area in question. The Canada Water Acts refers to this objective in several places. In the preamble it speaks of "planning with respect to (water) resources and for their conservation, development and utilization *to ensure the optimum use for the benefit of all Canadians."*

The optimum use must, of course, be measured against material objectives such as income growth, income redistribution, environmental quality, and social well-being. For the first objective water resource targets can be expressed at two levels. Firstly, they can be depicted in terms of increased outputs of goods and services or more efficient production of irrigated agriculture, increased or more efficient production of industrial output or increased or more efficient output of recreational services. The specific targets can be determined by translating the first step into specific needs for water resources. In the context of the examples used above, the targets would then become water supply for industrial use and water and related land requirements for recreation. These targets would have quantity, quality, space and time dimensions. In a similar manner water resource targets to achieve broad environmental quality objects can be defined. Examples of such targets with resource requirements are as follows:

(a) socially optimum water quality standards;

(b) preservation of scenic views;

(c) preservation of the eco-system;

(d) protection against erosion.

Finally targets to meet social well-being objectives would include:

(a) the increase in per capita income of certain disadvantaged groups or individuals;

(b) reduction of flood or drought risks to stabilize regional incomes and increase community security;

(c) water supply free from health hazards; or

(d) more equitable distribution of urban and rural population distributions.

Implementation and cost-sharing

The implementation of plans developed jointly requires another formal agreement between governments and perhaps the creation of an implementation board which could well be the joint planning board.

The cost-sharing of these two phases—planning and implementation—is flexible and negotiable in the Act, but the pattern is emerging of 50-50 cost-sharing for planning. The current view is that cost-sharing of implementation, however, should be determined in each case by the distribution of benefits and the degree of responsibility. It is perhaps worth noting here that the Canada Water Act is confined to those interjurisdictional water bodies in which there is a significant national interest and to those waters under the exclusive legislative jurisdiction of Parliament, such as the waters in the Northern Territories. It is not the intention of the federal government to become involved in minor river basins.

Choice of strategies

It is frequently stated that we need only a set of national stream quality standards with severe penalties against those who violate them, i.e., we simply need to crack down on polluters. In a totalitarian system which gives high priority to environmental concerns (which itself is an unlikely happenstance) such an approach would probably succeed, but in a democratic, free-enterprise society the system of stream standards and fines alone has been a dismal failure. We need more sophisticated measures to regulate the system at points where effective influences can be made. If we attack environmental quality management only at the end, that is by standards on receiving waters or even effluent standards, we come face to face with those who want to preserve their income and employment. Such a head-on confrontation between environmental concerns and immediate livelihood concerns has traditionally resulted in decisions in favour of jobs and profits.

The public perception of environmental degradation in Canada is developing rapidly, but there always exists the danger that with increasing polemics amongst the other interests the subject may lose its glamour and attractions. It is imperative that there is a clear understanding of why environmental disruption occurs, what its costs and consequences are, and what alternatives exist for improvement of the situation. It would be unfortunate indeed if the opportunities for innovation provided by the current public interest were lost without putting new measures, mechanisms and institutions in place.

During periods of high public concern some "wins" can be made by environmentalists but a more successful and enduring approach is to influence the system at many points. The trick is to identify the "pressure points" and to learn the techniques of applying such pressure. The range of choices of mechanisms and policy instruments which can be employed is indicated by the list on page 350 but, of course, even this long list is not exhaustive.

It is with these concepts in mind that the water quality management section of the Canada Water Act was drafted. Under the Act, if the Consultative Committees agree that the restoration of quality or the preservation of it in a particular basin or coastal strip is of concern to both governments, then that basin or coastal strip can be designated as a water quality management area under the Act. After that step a variety of actions can follow. A new federal or provincial corporation can be established to carry out the water quality management of that area or an existing federal or existing provincial corporation could be named. The intent is to identify a relevant governmental corporation as the responsible agency for that designated area on behalf of both governments. This agency

349

System element	Some control instruments or mechanisms for influencing the total system
1. Final demand sector goods and services (consumer desires)	Alter quantity and nature of the goods and services desired by: (a) public education to influence public attitudes toward consumption; (b) taxes on those goods and services that are large contributors of pollution; (c) restrictions on use of certain goods and services (use of DDT).
2. Production of goods and services	Controls on manufacture of goods, e.g. restriction on the formulation of detergents; incentives to change designs and processes, e.g. loans or subsidies to achieve a non-polluting car; incentives to use by-products and re-cycle, e.g. effluent discharge fees; research on how to produce less-polluting goods and create less-polluting manufacturing processes; assignments of liability for pollution damage; requirements for ecological tests of new products.
3. Storage, collection and transportation of wastes and residues	Creation of waste management agencies; consolidation of regional governments; research on better waste handling methods; economic incentives to create control collection systems.
4. Waste treatment and transformation	Effluent discharge standards and prohibitions; effluent discharge fees; research on waste treatment and by-products; financial incentives, e.g. government loans for construction of treatment plants.
5. Government apparatus	Revision of the constitution to permit a more rational division and sharing of responsibilities and powers between governments; consolidation of environmental agencies into manageable, sufficiently inclusive departments; explicit statement of over-all environmental quality objectives; formulation of a total environmental strategy; reallocation of fiscal and human resources to higher priority aspects of the environment defined by the strategy; problem-shed planning, e.g. comprehensive river basin planning; research on management systems; new improved legislation; judicial machinery.
6. Direct public influence	Encouragement of anti-pollution pressure groups; dissemination of information on a wide range of environmental matters; use of private sector in research and policy advisory roles; education to create an "environmental ethic" in young Canadians.
7. Receptors (people, flora, fauna and inanimate objects affected)	Research on effects of environmental quality on humans — physically, mentally and spiritually; research on quality criteria to preserve the biosphere; research on economic impacts; research on the impact of a degraded environment on flora and fauna and on man-made structures, e.g. effects of dust falls, corrosive air emissions; research on restoration of diminishing species.

would be required to draw up a plan following the guidelines set forth under the joint agreement which set up the agency or assigned to it this particular responsibility. The plan would include the following activities:

Assessments of the present and future wastes to be disposed of in the basin and those likely to be disposed of in the water body of the area;

Determination of wastes that the water bodies can transport or absorb while still maintaining a level of biological activity, with a comfortable margin of safety, that is essential to maintain the quality of that water body in perpetuity;

An analysis of the present and potential uses of the river;

Optimum receiving water quality levels with due regard to the three items above to achieve the maximum net stream of socio-economic benefits from the water bodies of the area;

Effluent standards required to maintain the optimum receiving water quality levels;

The time sequence and levels of waste treatment;

The new treatment facilities that would be required and those that the agency itself could construct to achieve economies of scale;

A financial plan including loans required and user charges and effluent discharge fees to be assessed to make the corporation self-sustaining in the designated area and to induce less production of waste.

Under these plans optimal water quality for the receiving waters would be determined in the following way: each major basin would be analyzed for demographic and economic growth and a study of public attitude undertaken; several alternative future patterns of use to be selected consistent with these studies; for each use pattern the biotic community necessary to perpetuate those uses would be described. At one end of the spectrum the uses may require the enhancement of a salmon fishery or, at the other end, simply sustaining a basic community of micro-organisms. Wherever possible, the concomitant physical parameter of quality would also be stated.

The costs associated with attaining these various levels of quality, with a margin of safety, will be determined. In addition, the imposed disbenefits will be assessed primarily in terms of present and future depreciated values, costs of restoration of quality for particular uses, costs of opportunities foregone such as lost recreational opportunities, reduced tourism, the impairment of health and other social costs.

The Act specifies that any plan for maintaining a stated level of water quality in a specific body of water must be published in the Canada Gazette for four weeks and the government cannot approve for another week thus giving the public and those affected who had not been part of the planning process five weeks to familiarize themselves with the plan and to make their views known. After governmental approval, standards, fees and charges specified in the plan would be promulgated as regulations under the Canada Water Act for that designated area and the agency would be empowered to implement the plan. Inspectors would be provided to aid in enforcement.

One of the options set forth in the Act is the employment of effluent discharge fees along with effluent standards as an incentive for the waste disposer to reduce his waste loadings and to upgrade his treatment. These fees also assist the agency in financing the construction of treatment works. By providing for the payment of such effluent discharge fees, a cost is attached to waste disposal which will force the disposer to consider four options:

1. To change the waste producing process to reduce the amount of waste being disposed or their nature.
2. To pay effluent fees and meet effluent standards.
3. To pay waste treatment charges at communal facilities.
4. To treat the wastes himself and avoid charges and reduce the fees.

The basic principle here is that it is the polluter who must initially pay, recognizing, of course, that there are mechanisms for transferring these costs to the consumer of goods and services. Some have argued that this system of effluent fees is a licence to pollute. On the contrary the present system of permitting waste disposal, provided the receiving water quality is above a certain minimum, is a licence to deprive the downstream user of his benefits without any compensation whatsoever. The payment of an effluent discharge fee, even though effluent standards have been met, is at least partial, indirect payment to the affected downstream users for the use of the environment to absorb treated wastes.

Nutrient control

In addition to the new mechanism introduced by effluent fees, the Canada Water Act has attempted to achieve changes at the manufacturing level so that the wastes themselves do not appear anywhere in the

351

system. Under Part III of the Act a scheme is presented whereby the nutrient content of cleaning agents and water conditioners can be controlled throughout Canada. Regulations issued from time to time would specify nutrients, the cleaning agents and water conditioners to be regulated, and the percentage content of nutrients permitted at the manufacturing stage in Canada or in imports to Canada. Inspectors and analysts will be employed to monitor these aspects.

The first regulations came into force August 1, 1970. It specifies that no more than 20 percent phosphates, measured as phosphorus pentoxide, is permitted in laundry detergents. The Act does not specify the substitutes that are to be used but it is worth noting that the federal government has a major research effort underway now examining the ecological and other effects of NTA and that it is carefully monitoring the research results of other countries on substitutes.

It is expected that additional regulations will be issued reducing the percentage of phosphates permitted. In keeping with the recommendation of the International Joint Commission, the Minister of Energy, Mines and Resources has announced his intention to seek a near total ban of phosphates in laundry detergents in 1972. The option, of course, still remains of including in subsequent regulations other nutrients and other cleaning agents and water conditioners as the need arises.

Inspection and enforcement

Part IV of the Act sets forth provisions for inspectors and analysts, the conditions under which they may enter premises, other than private dwellings, and seize goods. It also sets forth a maximum fine of $5,000 per day for the unlawful deposit of wastes and the unlawful manufacture or import of cleaning agents and water conditioners. It also makes interference with an inspection an offence and provides that the courts may order a person to refrain from further violation or cease activity, that is to close his offending plant.

Public information and participation programs

Part VI provides for a public information program to inform the public respecting any aspect of the conservation, development or utilization of the water resources of Canada. This short provision is of much greater significance than is indicated by the little attention it has received. All experts and consultants in this field have stressed the great effect for a relatively small financial input of a large educational program that would acquaint the public in some depth with the issues in water resource development and utilization, particularly the steps necessary to preserve environmental quality. The normal governmental communications network is obviously insufficient. The public campaigns being conducted by university students in a number of groups across Canada is a useful help in this regard, but it only points to the real need for a broader educational program. No statute or governmental initiative would have such a profound impact as the creation of a new ethic in Canada for the protection and preservation of our environment. And no other hallmark would establish the Canadian identity so clearly throughout the world as an abiding respect for a clean and attractive environment.

While public awareness is essential, so also is public participation, particularly during the planning process. To this end, it would be useful to employ private citizens on joint planning boards and committees, a step that New Brunswick has taken for the Saint John Basin. We also see the need for far more in-depth analysis of local perceptions and attitudes by behavioural scientists and we will be employing such measures in the Qu'Appelle and Saint John Basins. There is a need too for advisory bodies to the Ministers and officials and the provision for such bodies is included in the Act.

Unilateral action

There are special provisions in the Act under which the federal government can move directly to plan and develop or to manage the quality of water bodies within a province. If the Governor-in-Council is satisfied that all reasonable attempts by the Minister in charge have been made to reach an agreement with the province concerned and those efforts have failed then the Minister may undertake the following directly:

(a) Water quality management on interjurisdictional waters for which water quality management is a matter of urgent national concern.

(b) Comprehensive planning, but not implementation, on other interjurisdictional waters for which there is a significant national interest in the water resource management thereof.

(c) Planning and implementation on international and boundary waters where there is significant and national interest in the water resource management thereof.

This is a new use of the general powers of the British North America Act in recognition of the importance of national water resources development and environmental quality to the country nationally.

It should be noted that the question of "significant national interest" or "urgent national concern" is testable in the courts. Clearly unilateral action would rarely be considered. It is more likely the judgment of the people collectively rather than the uncertainty of a court decision which would be the critical factor facing a federal cabinet in deciding to take unilateral action on a major water body in the face of provincial opposition.

The process of institutional interaction

In setting up new intragovernmental and intergovernmental groups the question arises as to the nature of the interaction process. Each government normally has numerous subdivisions with different responsibilities in water matters requiring a specific process of co-ordination and decision making.

Under the Canada Water Act five separate but sequential phases are envisaged for this process:

1. Internal policy formulation within each government
2. Negotiations between governments

3. Joint planning
4. Joint implementation of plans
5. Operation, maintenance and inspection of works jointly or by single agencies.

Given complete harmony and understanding the process flows sequentially along these five stages.

The cabinets and departments within each government formulate their respective policies which are co-ordinated internally either within cabinet itself or by an interdepartmental structure. Within this policy one or more departments initiate certain proposals for joint action with another government. Consultation between governments takes place leading to draft agreements which are submitted and hopefully approved. Plans are prepared jointly under this agreement and offered for public comment. If approved by the government, an implementation agreement is subsequently negotiated and joint implementation proceeds. Upon completion of the works one or more agencies is assigned responsibility for operation and maintenance.

But it would not be realistic to expect all matters in this complex field to flow in this ideal fashion. One must contemplate honest disagreement and provide ways of avoiding stalemates. Alternative loops must be built into the decision-making chain.

In drafting the Canada Water Act the difficulties were considered when there was a failure during negotiations between governments to reach an agreement and unilateral action was necessary. The Act also covers the situation where governments having reached agreement for joint planning, could then not agree on the plans themselves, and required the termination of the planning agreement and subsequent new consulation to begin the process again. There are, of course, other possibilities for the termination of negotiations, if unsatisfactory progress is being made during the implementation phase.

Relation to other legislation

The foregoing are the key elements of the Canada Water Act, but the Act should not be viewed by itself alone. The proposed Government Reorganization Act of 1971 combines a number of programs, agencies and pieces of legislation into a new Department of the Environment, indicating a united approach covering air, water, fish, forest and wildlife.

This important step will allow the federal government to focus on the main transportation vehicles of waste, water and air, and on their biological receptors in an integrated manner. It will give the industrialists a single point of contact on most environmental problems. Certainly the current operations in these areas will be streamlined and their importance significantly elevated.

In summary, the Canada Water Act is broad, permissive, flexible legislation containing many of the more modern approaches to resource management. It is also a clear assertion of federal responsibility, but above all the Canada Water Act is a recognition that co-operation is the key element to achieve the optimum development and utilization of Canadian waters.

PART III: A SUMMARY OF FEDERAL LEGISLATION DEALING WITH RESOURCES MANAGEMENT

Although the Canada Water Act is perhaps one of the most significant federal pieces of legislation to affect resources management, it is not the only vehicle for federal action. In Part III other direct or indirect ways and means of federal interaction are briefly outlined.

1. The Canada Shipping Act

This Act, revised in 1952 with amendments to 1968, provides for Oil Pollution Prevention Regulations. The Regulations are divided into those covering: Canadian waters; non-Canadian waters; General and Enforcement.

The Act applies to ships of every nationality while they are in Canadian waters, except ships of war. Every ship must carry an oil record book, and any person who commits an infraction or contravention of the regulations is liable for a $5,000 fine or six months in prison or both. Prohibited zones

are established under the Act within which no oil may be dumped. These zones include all sea areas within 50 miles from the nearest land. Inspection and enforcement is carried out under the Minister of Transport who may designate any member of the Public Service of Canada, the R.C.M.P. or provincial, municipal or harbour police as departmental agents. Air pollution regulations in the form of density levels for smoke emissions from ships are also included.

Recently, proposed anti-pollution amendments to the Shipping Act have been introduced into Parliament and have received second reading.[2] These amendments would, if passed, raise the maximum fine to $100,000 for pollution of the sea by oil in Canadian waters; enable the government to prosecute the owners of a vessel should a spill occur; provide for bonds to be posted by ship owners up to 14 million dollars; provide for a clean-up fund by charging 15¢ per ton of oil landed or shipped from any Canadian port; and provide for inspection procedures and penalties up to $25,000 for each infraction.

2. The Fisheries Act

This Act, revised in 1952 and amended in 1970, contains pertinent sections on water pollution. Under the Act, no person may deposit or permit the deposit of wastes of any type in any water frequented by fish or in any place under any conditions where such waste may enter such waters. Exceptions to this clause may occur when these waters form part of a Water Quality Management Area pursuant to the Canada Water Act. In such a case, the standards for disposal would be the concern of the agency established by the latter Act. Any person who violates the prohibition under the Fisheries Act is liable to a fine of $5,000 for each offence. The courts also may order any person committing such an offence to cease any activity which is likely to result in committing a further offence. "Waste" under the Act is defined as any substance that, if added to any waters would degrade or alter the quality of these waters to an extent that is detrimental to their use by man or by any animal, fish or plant that is useful to man, much the same as in the Canada Water Act.

In addition, the Minister of Fisheries and Forestry may require any person, who proposes to construct any new work which might result in the deposit of waste in waters frequented by fish, to submit plans or specifications for inspection. If the Minister decides that such construction would violate the Act he may, with the approval of the Governor-in-Council, require modifications to be made or prohibit the construction entirely. The Act provides for inspectors to be appointed to carry out these provisions.

This Act employs the mechanisms of research through the Fisheries Research Board and other ecological studies as well as prohibitions and standards which designate $5,000 fines. Data collection is also carried out.

3. The Navigable Waters Protection Act

This Act contains a number of provisions concerning disposal of solid wastes or other materials in navigable waters which might hinder navigation. The Act is administered by the Ministry of Transport.

4. The National Harbours Board Act

This Act, passed in 1952, establishes the National Harbours Board which has jurisdiction in Canada's major harbours: Halifax, Saint John, Chicoutimi, Quebec, Three Rivers, Montreal and Vancouver. Certain provisions restrict anyone within the harbour confines draining or discharging into the water anything which causes a nuisance, endangers life or health or damages property. Also, no ballast or rubbish may be disposed of unless decided by the Board.

Under this Act, the mechanism of prohibitions and standards is utilized to regulate oil pollution in harbours.

5. Regulations Under the Animal Contagious Diseases Act.

These regulations prevent garbage, old vehicles, manure or other refuse to be landed or discharged in any port in Canada, except under strict conditions.

6. The National Parks Act

This Act, passed in 1952, provides for the Federal government to make regulations for the preservation, management and control of Canada's National Parks, including the prevention of and remedying of any obstruction or pollution of waterways. Regulations are included regarding sanitary facilities and the prevention of nuisances such as rubbish.

7. The Migratory Birds Convention Act

This Act, passed in 1952, provides for the federal government to make regulations protecting migratory, migratory insectivorous and migratory non-game birds that inhabit Canada during the whole or any part of the year. Under these regulations, no one may knowingly place oil or oil waste substances upon waters frequented by these birds. This Act is also likely to be transferred from the Department of Indian Affairs and Northern Development to a new environmental department.

Under this Act, the instrument of prohibitions and standards is used to provide hunting regulations.

8. The Criminal Code

Under the Criminal Code, any person who commits a common nuisance and endangers the lives, safety or health of the public, or causes physical injury to any person, is guilty of an indictable offence and is liable for imprisonment for two years. Contravention of any prohibition established in a Federal statute is a criminal offence. Pollution of water, soil and air have thus far not been prohibited under this Code, but the potentiality for doing so is generally accepted.

9. The Income Tax Act

Under regulations passed under this Act, a special depreciation allowance on water or air pollution control equipment may be deducted from the yearly income of the purchaser of such equipment, and as well includes a special and accelerated rate of capital cost allowance. In addition, the purchaser may claim the lesser of (a) 50 percent of the capital cost of such equipment such as acquired for the purpose of preventing, reducing or eliminating pollution, or (b) the amount by which the capital cost exceeds the aggregate of the amounts deducted in computing his income for previous taxation years.

Under this Act, the mechanism of tax incentives in the form of rebates is provided for companies installing this pollution abatement equipment.

10. The Boundary Waters Treaty of 1909

As this Treaty is an Empire Act, the federal government may legislate upon matters which normally would be within provincial jurisdiction. The power to implement treaties by internal legislation was given to the federal government under the B.N.A. Act, but the operation of that section is limited to treaties entered into between the "British Empire" and foreign countries. Under this Treaty, the International Joint Commission was established to investigate and make recommendations to resolve disputes regarding the use of international boundary waters. The Commission's report on all its water pollution investigations has recommended water quality objectives to be met in maintaining the waters of those streams under reference of the 1909 Treaty. Industrial waste standards, and water quality objectives for various border areas have been developed by the IJC. A major study was completed this year on the quality of the Lower Great Lakes and connecting channels which now forms the guide for international action on these important boundary waters.

11. The International Rivers Improvements Act

This Act applies to river improvement activities having trans-boundary ramifications. The provinces may legislate regarding the purely internal aspects of international waters.

12. The Railway Act

The Board of Transport Commissioners made the so-called "Air Pollution and Smoke Control Regulations." These regulations apply to all railway companies subject to the Railway Act, but only in municipalities that have passed by-laws for the regulation, control or prohibition of smoke or other air pollutants and that have a municipal officer of a municipal smoke abatement bureau duly appointed.

Since air quality management resides with the provinces, the federal role is considerably limited, but new air legislation is anticipated.

13. The Arctic Waters Pollution Prevention Act and Northern Inland Waters Act

These Acts were passed last year and are designed to protect northern areas from pollution, and promote the management of the inland waters resources of the Yukon and Northwest Territories. The N.I.W. Act sets up a water quality management system. The A.W.P.P. Act establishes a 100-mile quarantine area in Arctic waters and strict regulations on the shipment of oil through the Arctic archipelago. The Acts are administered by the Department of Indian Affairs and Northern Development.

Under the Arctic Waters Pollution Prevention Act, no person or ship may deposit wastes of any type in the Arctic waters or in any place on the mainland or island where such wastes would enter the waters. The Act dovetails with the Canada Water Act in that the section on waste deposit does not apply when a water quality management area is set up under the Canada Water Act. Also, any person who engages in exploration, development or exploitation of any natural resource adjacent to or under Arctic waters must provide evidence of financial responsibility in the form of insurance or an indemnity bond. With regard to ship waste regulations, any person found guilty under this section is liable for a $5,000 fine or in the case of a ship a $100,000 fine per day as well as seizure of ship and cargo. Any person obstructing a pollution prevention officer is liable for a fine of $25,000.

The Northern Inland Waters Act covers approximately 40 percent of the land area of Canada including the Yukon and Northwest Territories. Under this Act the Yukon Territory and Northwest Territories Water Boards are created. Water Management areas will be designated and the Boards will provide for the conservation, development and utilization of the water resources within the Territories.

This Act also included the provisions under the Canada Water Act should a water quality management area be established under the latter. Also, under prohibitions and standards and problem shed planning, mechanisms and instruments for enhancing environmental quality are included.

14. The Pest Control Products Act

This Act, passed in 1969, bans the general use of DDT except for certain particular functions. Permits must be applied for to the province in which the use is intended. The mechanisms of prohibitions and standards and subsidies and grants are employed to enhance environmental quality.

15. The National Housing Act

This Act enables Central Mortgage and Housing Corporation to provide partially forgiveable loans for construction or expansion of sewage treatment projects.

Under the Act the instrument of loans to municipalities or provinces for constructing or expanding sewage treatment projects is employed. In 1969, this loan fund provided 50.2 million dollars to municipalities, an increase of 27 percent over the previous year.

16. The Proposed Clean Air Act

In February 1971, Parliament gave first and second reading to a proposed act relating to air quality and the control of air pollution.[3] Under the proposed Clean Air Act, air pollution is defined as a condition of the ambient air arising from the presence of air contaminants that either endangers the health of human and animal life or that causes damage to plant life or to property. Within the scope of this proposed act, the Minister of the Environment may establish a system of air pollution monitoring stations throughout the country, conduct research into air pollution, formulate comprehensive plans for its control, and publish information relating to air quality control. In addition, national ambient air quality objectives may be established. As well, national emission standards and guidelines would be specified where: (1) there is a significant danger to the health of persons, and (2) it is necessary to discharge international obligations—global air pollution as well as trans-boundary. This act also includes regulation of the composition of fuels and inspection powers. Federal-Provincial Agreements may be negotiated in order to facilitate these standards. Any person violating the national air quality regulations may be fined up to $200,000 for each offence. Any emergent situation involving an extremely hazardous air contaminant may be declared a national emergency by the government.

CONCLUSION

In conclusion, the federal Acts, upon which programs to enhance environmental quality are based, are many and varied to provide the array of effective instruments necessary to change the waste production and treatment system. Constitutional impediments to the establishment of a federal presence in resources management, generally under the domain of the provinces, have been circumvented in many cases by passing regulations in areas of clear federal jurisdiction with the intent of providing a spin-off effect. For example, tax deductions for pollution abatement equipment under the Income Tax Act will hopefully improve environmental quality. Shared programs such as those under the Agricultural Rehabilitation and Development Administration also contribute to an integrated federal-provincial resolution of environmental problems. It is hoped that with the passage of such legislation as the Canada

357

Water Act, a much firmer consolidation of efforts towards tackling water problems will result, and that this in turn could lead to similar legislation in other aspects of resources management.

In looking at these regulatory Acts, however, one must also recall that the government organization Acts which set up the departments have an overwhelming influence. The degree of co-ordination, the mandate for initiatives and programs, and the specific objectives of the government are often more meaningful to environmental matters than the specific Acts.

But clearly the most significant step to be taken and one with the greatest effect for each federal tax dollar invested would be to imbue in the minds of young Canadians an ethic for the protection and preservation of their great environmental heritage. No laws or governmental programs would be as influential, no mechanisms for enforcement as effective, as a sense of pride and stewardship for our environment.

REFERENCES

[1] See a study prepared for the government of Canada by Dale L. Gibson, "Constitutional Jurisdiction Over Environmental Management in Canada," June 1970. (mimeo)

[2] The amendments were passed as Bill C-27, 1 March 1971.

[3] The Clean Air Act received Royal Assent on 23 June 1971 but as of August 1971 it still had not been proclaimed.

REGIONAL DISPARITIES

An Introduction

Cultural, physical and economic diversity is a Canadian trademark. Over time, certain parts of the country have developed rapidly, other parts more slowly and others almost imperceptibly. The Great Lakes-St. Lawrence Lowlands, although occupying a very small area, contains one-third of the Canadian population and dominates the commercial, manufacturing and financial activities of the country. As the Lowlands have developed and diversified, other regions, owing to a variety of deep-rooted factors, have suffered a relative decline. Hence, regional disparities have persisted and the gap between the rapidly and slowly developing or stagnant regions is widening. The evidence is clearly revealed by divergent incomes, standards of living, and employment opportunities.

Both academics and politicians have long recognized regional differences in Canada's economic development. However, until recently, economists have concentrated their analysis on the sectoral rather than regional aspects of the economy. A good example of this was the *Royal Commission on Canada's Economic Prospects* (established in 1955), which analyzed many segments of the Canadian economy but carried out very little regional analysis.

In 1962, the Economic Council of Canada was established by an act of Parliament to study and advise the government upon medium-and long-range development of the Canadian economy. In fulfilling its advisory function, the Council's permanent staff engages in research, it commissions special studies by outside specialists, and publishes an *Annual Review*. Although the emphasis of the Economic Council is on aggregate and sectoral economic topics, it has conducted some excellent studies on regional economic problems.

Two important conferences held in 1965 indicated a growing concern for the regional aspects of economic development. The conference on "Areas of Economic Stress in Canada" brought together scholars from various disciplines to discuss the problems of identifying and treating slow growth areas [see Wood, W. D., and Thoman, R. S., (eds.), *Areas of Economic Stress in Canada* (Kingston: Industrial Relations Centre, 1965)]. At the "International Conference on Regional Development and Economic Change," examples of regional development problems and progress were presented from different parts of Canada and from other countries [see *International Conference on Regional Development and Economic Change* (Toronto: Ontario Department of Economics and Development, 1965)]. The reports of these conferences provide valuable reading for the geography student interested in the problems of regional development in Canada.

The dimensions of regional disparity can be portrayed by many criteria—income, productivity, education and health. In the abstract from the *Second Annual Review* of the Economic Council of Canada, some of the factors associated with regional differences in income, growth, redistribution of population, labour input, and earnings, are discussed. The Council concludes that the achievement of a more equitable regional economic balance will not only involve a substantial increase in investment in human and material resources, but that to accommodate certain objectives the strict application of economic judgements may well have to be suspended.

The magnitude of poverty in Canada is subject to variable interpretation depending upon the criteria employed to gauge it. In the first of two papers on this theme, Robert Irving discusses the nature of poverty, the problems in measuring poverty, and some of the proposed solutions to alleviate this condition, particularly in rural areas. In the second paper, G. E. Mortimore describes the human attitudes, feelings and living conditions he encountered in a tour of northeast New Brunswick.

The federal government has long been involved in development programs that in effect, if not in fact, have had strong regional overtones. Many of these programs have been on a shared basis with the provinces, others have been the sole responsibility of the federal government. Whatever role the federal government has to play, whether it be to orchestrate provincial-federal programs, or to initiate and coordinate new programs, it is clear that federal involvement is essential. One of the newly established federal agencies is the Department of Regional Economic Expansion whose Minister, the Honourable Jean Marchand, is reported as saying that "The problem of economic disparity is more important to the unity of Canada than the problem of French-English relations." *(Globe and Mail,* 17 August 1971, p. 14). In his paper, L. O. Gertler summarizes federal policy in the regional economic development field, and examines recent significant shifts in the federal attitude towards regional development with an emphasis on the recently established Department of Regional Economic Expansion.

The causes of regional lag both in economic and human terms are complex as are the strategies that have been proposed to alleviate these problems. In the regional context, the strategy of focussing economic development on growth centres or growth poles is now widely accepted. Indeed, this concept forms an important component in the regional development policy for Ontario (see Richard S. Thoman, *Design for Development in Ontario: The Initiation of a Regional Planning Program,* Allister Typesetting & Graphics, Toronto, 1971.) in the strategy of the Department of Regional Economic Expansion, and in proposed development strategies for the Atlantic region. It is imperative, therefore, that the characteristics of potential growth centres be examined carefully. This is the topic of Gerald Hodge's paper in which he discusses some of the relevant conclusions from his research on this topic in Prince Edward Island, Nova Scotia, Saskatchewan, and Eastern Ontario.

Atlantic Canada, as reflected in the preceding selections, exhibits the symptoms of economic malaise. To be sure, there are many other areas in Canada with similar symptoms, but the Atlantic region has been the subject of more intensive research and prescriptive remedies than any other part of Canada. In the concluding paper some of the highlights of a strategy for economic development, as proposed by the Atlantic Development Council, are described.

Regional Growth and Disparities

ECONOMIC COUNCIL OF CANADA

In Canada, the physical immensity of the country, the presence of distinct geographic barriers, a narrow, uneven chain of settlement, and a striking diversity of resources and economic structure among our major regions all make for a particularly high degree of regional differentiation. The problem of integration and balance, to assure appropriate participation of each region in the overall process of national economic development, has been an elusive goal and of continuing concern.

There are two interrelated considerations in moving towards a better regional balance. The first is to reduce the relative disparities in average levels of income as they now exist between regions. Although these disparities have shown a slight narrowing in recent years, our subsequent analysis will show that they remain large and substantially persistent. The second consideration is the need to assure that each region contributes to total national output, and to the sustained, long-run growth of that output, on the basis of the fullest and most efficient use of the human and material resources available to the region. Here again the evidence suggests significant deficiencies in past regional development in Canada.

These considerations focus attention upon the fundamental importance of improving the utilization of human and material resources in all regions. This involves their fullest possible use, combining them in the most efficient ways possible, and continually upgrading their productive capabilities. Moreover, the narrowing of interregional income differences depends upon a particularly rapid advance in productivity and income per employed person in the relatively lagging areas, and indeed, at rates appreciably faster than the average for the economy as a whole.

It might well be suggested that the concept of balanced development among the separate regions comprising the national economy implies taking adequate account of such factors as their location, physical area and dimensions of space, and other broad geographic characteristics. Elements of this kind necessarily figure much more prominently in the analysis of regional growth than in the consideration of growth for the economy as a whole. To provide a basic perspective for interregional comparison and assessment, however, the present analysis focusses mainly upon the aspect of regional income flows and their change over time.

INTERREGIONAL INCOME DISPARITIES

The best available statistical measure of interregional income disparities in Canada is the flow of personal income within each of the ten provinces. Personal income is the flow of income to individuals, and since by far the greatest part is earned by the productive factors of labour and capital, it also provides an approximate indication of the level of economic output produced in each region.[1] Per capita personal income is also a measure of comparative productivity and economic welfare per person among the separate regions. Thus the more closely bunched are the regional averages around the national average, the smaller is the degree of income disparity among the regions, and in our terms, the more balanced is the participation of the various regions in national economic development.

The regional levels of personal income per capita are shown for three groups of years in Table 1. The most striking feature of the comparisons is the substantial percentage difference in income levels between

● Excerpt reprinted from *Towards Sustained Economic Growth*, Second Annual Review of the Economic Council of Canada (Ottawa: Queen's Printer, 1965, pp. 98-126), with permission of the Director of the Council, and Information Canada.

the highest and lowest province. For the recent period, 1963,[2] personal income per capita in Ontario was about four-fifths larger than in Prince Edward Island and twice that of Newfoundland. As for the other provinces, income levels in British Columbia and the Prairies are considerably above the average for all provinces; Quebec's per capita income is fairly close to the average. All the Atlantic Provinces range well below personal income per capita in the country as a whole.

Table 1

LEVEL OF PERSONAL INCOME
PER CAPITA BY PROVINCE
(In current dollars)

	1927	1947	1963
Ontario	509	981	2,025
British Columbia	535	980	1,966
Alberta	509	923	1,750
Saskatchewan	449	818	1,749
Manitoba	455	875	1,721
Quebec	378	709	1,521
Nova Scotia	299	676	1,302
New Brunswick	277	609	1,167
Prince Edward Island	248	477	1,115
Newfoundland	......	......	1,009
Average for Provinces	407	783	1,532

Note: Provinces are ranked in order of level of personal income per capita in 1963, and the data are for three-year averages centred on the year shown. Data for British Columbia include the Yukon and Northwest Territories.
Source: Based on data from Dominion Bureau of Statistics.

A second feature is that the rankings of provinces in terms of income levels have hardly changed over a period of almost forty years. Ontario has changed places with British Columbia as the highest and second highest income provinces. Manitoba and Saskatchewan have also traded positions in the centre of the rankings and Quebec has maintained a consistent mid-position. The Atlantic Provinces have been at the bottom of the range throughout. The broad geographical distribution over time is also noteworthy. Income levels in the five western-most provinces have been generally higher than the provincial average since 1927 while those in eastern Canada have been lower. Moreover, while there has been some reduction in the percentage range of income differences, particularly with reference to peaks in disparity experienced during the early 1930's and early 1950's, the degree of disparity remains obviously large.

Aspects of Interregional Disparity

Despite variations in personal income per capita from region to region, the feature which emerges most strikingly from the record of the past four decades is the essential persistence of income disparity among the regions of Canada. In Figure 1, the width of the shaded area, representing the spread in average per capita incomes among the major regions, is fairly constant throughout; and the position of each region relative

Figure 1

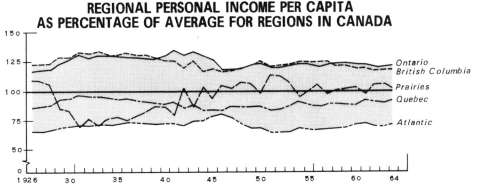

REGIONAL PERSONAL INCOME PER CAPITA
AS PERCENTAGE OF AVERAGE FOR REGIONS IN CANADA

SOURCE: BASED ON DATA FROM DOMINION BUREAU OF STATISTICS

363

to the average (represented by the horizontal line) was virtually the same in 1926 as in 1964. Furthermore, there is some evidence to suggest that the persistence in inter-regional income disparity can be traced back even further in Canadian history.

Figure 1 shows that per capita income levels in the Atlantic Region have ranged below 75 percent of the average for Canadian regions throughout the period, except for the years towards the end of World War II. After 1946, income levels in the region fell away from the average, and the inclusion of Newfoundland in the data for the Atlantic Region since 1949 has somewhat lowered the average level of income shown for that region. At the other extreme, Ontario and British Columbia, have recorded income levels roughly 25 percent above the regional average for most of the period. Per capita income in the Prairies shows an extremely wide swing away from the average during the early 1930's, reflecting the particularly adverse impact of the depression upon incomes in this region. Throughout the period, personal income per capita in Quebec has been below the regional average, but since the end of the war, the gap has been steadily narrowed.

Although Figure 1 provides an impressionistic picture of interregional income disparity, its real extent can be more accurately evaluated by means of a single index. Such an index takes account of the position of each and every region in relation to the national average and not simply the difference between the highest and lowest regions.[3] A wide scatter of regional per capita incomes around the average, such as occurred in the early 1930's in Canada, is reflected in a high value for the index. On the other hand, a concentration of regional per capita incomes around the average, as was recorded at the end of World War II, is indicated by the low value for the index. Consequently, the higher is the value of the index, the greater is the degree of imbalance in regional participation in national economic activity.

This index of income disparity among the ten provinces of Canada (including Newfoundland as of 1949) is shown in the upper curve in Figure 2. Despite short-period irregularities, the broad movements in the index over the whole period clearly emerge. Beginning in 1926, the movement in the index shows that the interregional spread of incomes widened to a maximum during the depression of the 1930's and then narrowed fairly steadily until the end of the War. In the early part of the post-war period, the degree of spread again widened until 1951,

Figure 2

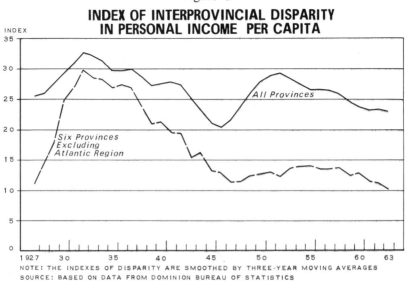

INDEX OF INTERPROVINCIAL DISPARITY IN PERSONAL INCOME PER CAPITA

NOTE: THE INDEXES OF DISPARITY ARE SMOOTHED BY THREE-YEAR MOVING AVERAGES
SOURCE: BASED ON DATA FROM DOMINION BUREAU OF STATISTICS

364

and, thereafter, drifted downward. On the historical record prior to 1946, one could say that depressed economic conditions have been accompanied by an increased interregional disparity in income while high levels of economic activity saw a narrowing in the degree of income spread. After 1946, however, it is not possible to discern a clear relationship between interregional income disparity and the level of economic activity. In view of this, it would be hazardous to assume that rapid economic advance at the national level, although providing a favourable environment, would be sufficient in itself to reduce interregional income disparity significantly.

Taking the forty-year period as a whole, the extent of relative disparity as measured by the index has shown only a slight tendency to decline, thus confirming the conclusion drawn from Figure 1. Between 1927 and 1963, the index fell by about one-tenth. It is true that from 1949 onwards the inclusion of Newfoundland, the province with the lowest per capita income of all, is responsible for a higher level in the index than would otherwise be the case. Nevertheless, calculations which exclude this province also show the same persistence of interregional income disparity, even though at somewhat lower levels, over the past fifteen years.

Figure 2 also traces the movement of the index of income disparity as calculated for only six provinces, exclusive of the Atlantic Region. The contrast between the two curves establishes that a large part of the interregional spread in incomes, and of the substantial presistence of that spread over long periods, is attributable to the relatively low income position of the entire Atlantic Region.

The foregoing analysis treats each province as a unit equal to any other province, regardless of its area or population. In technical terms, our index of disparity is not "weighted" by size of population. For certain purposes, however, it is desirable to calculate a weighted index of income disparity which takes account of differences in population among the various regions. On this basis the extent of interregional income disparity is less because the four lowest income provinces (the Atlantic Region) also have the four smallest populations and thus carry less weight in the total index. The implication of this fact is important for policy purposes, in the sense that the task of improving the quality and productivity of the resources of the lowest income provinces would involve a relatively small proportion of the total national income.

Growth of Personal Income by Region

In terms of personal income *per person,* all ten provinces of Canada have shared in national economic growth over a period of almost forty years. Moreover, the rates of increase in income per person have been closely comparable for all the provinces. Over this long period annual growth rates of this measure of income varied among the regions only from 3.5 percent in Alberta to 4.3 percent for Prince Edward Island (Table 2). Generally, the lower income areas experienced rates of increase slightly greater than the higher income regions, resulting in the modest narrowing of interregional disparity already noted.

In contrast to income per person, however, *total* personal income for each region has grown at much more variable rates. This reflects the marked differences among the regions in population growth and migration over both the longer run and the post-war period. From 1927 to 1963, the annual rates of increase in total income by region varied from a high of 6.6 percent for British Columbia to a low of 4.2 percent for Saskatchewan. In the post-war period, the high-low spread between British Columbia and Nova Scotia was almost as large (Table 2).

The differential rates of population increase (Table 2) were the key variables in these divergent patterns of per person and total income growth in various regions. In the post-war period, for example, the three provinces of Ontario, Alberta, and British Columbia recorded the highest rates of growth in total personal income. But they also experienced the most rapid population

Table 2

GROWTH OF PROVINCIAL PERSONAL INCOME AND POPULATION
(Average annual percentage change in current dollars)

Province	1927-63			1947-63		
	Personal Income Per Capita	Total Personal Income	Population	Personal Income Per Capita	Total Personal Income	Population
Ont.	3.9	5.9	2.0	4.6	7.5	2.8
B.C.	3.7	6.6	2.8	4.4	7.7	3.1
Alta.	3.5	5.8	2.2	4.1	7.6	3.4
Sask.	3.8	4.2	0.3	4.9	5.6	0.7
Man.	3.8	4.9	1.0	4.3	6.0	1.6
Que.	3.9	6.0	2.0	4.9	7.5	2.5
N.S.	4.2	5.3	1.1	4.2	5.5	1.3
N.B.	4.1	5.3	1.2	4.1	5.6	1.4
P.E.I.	4.3	4.8	0.6	5.4	6.3	0.8
Average for Provinces	3.9	5.4	1.5	4.5	6.6	2.0
(Nfld.)				(5.3)	(7.8)	(2.4)

Note: Provinces are ranked in respect of level of personal income per capita in 1963, and the period for Newfoundland is 1950-63.

Source: Based on data from Dominion Bureau of Statistics.

growth, and income per capita rose at only average or below average rates. Saskatchewan, on the other hand, experienced one of the lowest rates of expansion in total income; but with a very low rate of population increase consequent upon the rapid mechanization of agriculture and large-scale migration from the farms both to cities within the province and out of the province, its per capita income growth was one of the highest. In Nova Scotia and New Brunswick, population growth was double that of Saskatchewan, but with a lagging rate of increase in total income, the gain in per capita income in both provinces was at the bottom of the range.

It will be obvious from this that when regional economic growth is measured in terms of per capita income, it conceals a complex interaction between gains in output and population change. The latter is a highly dynamic factor in regional income growth and exerts a particularly important influence upon the degree of interregional income disparity.

REGIONAL GROWTH AND POPULATION REDISTRIBUTION

Perhaps the most striking impression of a half-century of population history in Canada is one of volatility and dynamic change. The trends in population change are summarized in Table 3. Over the forty-year period, 1921-61, population growth among the provinces ranged from only 18 percent in Prince Edward Island to 210 percent in British Columbia. Even though every province experienced some growth, the increase was entirely concentrated in the nonfarm population, which almost tripled for the country as a whole. In contrast, the farm population declined absolutely in every province, with a 30 percent decrease in the country as a whole, and the farm population fell from 37 percent to only 12 percent of the national total.

Migration between provinces, even on a net basis, has also been a prominent feature of Canada's recent development. Between 1941 and 1950 seven provinces experienced net outflows. Only Ontario and British

Table 3

TOTAL, FARM AND NONFARM POPULATION, BY PROVINCE, 1961
AND PERCENTAGE CHANGE, 1921-61

	Total Population		Farm Population		Nonfarm Population	
	1961 ('000)	% Change 1921-61	1961 ('000)	% Change 1921-61	1961 ('000)	% Change 1921-61
Newfoundland	458	—	17	—	440	—
Prince Edward Island	105	18	38	−38	67	139
Nova Scotia	737	41	83	−62	654	117
New Brunswick	598	54	100	−48	498	157
Quebec	5,259	123	651	−17	4,608	192
Ontario	6,236	112	534	−38	5,702	175
Manitoba	922	51	173	−33	749	113
Saskatchewan	925	22	307	−36	619	121
Alberta	1,332	126	290	−8	1,042	284
British Columbia	1,629	219	93	−5	1,536	261
Canada	18,201	107	2,285	−30	15,916	189

Note: Figures may not add to totals for Canada due to rounding.
Source: Based on data from Dominion Bureau of Statistics.

Columbia gained from net inflows. In the 1951-60 period, net outflows took place from six out of ten provinces, with Quebec and Alberta joining the ranks of recipient provinces. Over the twenty years, 1941-61, it is estimated that among the provinces which experienced net out-migration, this out-movement of people amounted to a total of almost 600,000 men, women and children.

The patterns of rural-urban migration and their variations confirms the drastic shift out of the farm areas of the Prairie Region, especially for Saskatchewan. The movement from the rural areas of both Quebec and all four Atlantic Provinces is also important. Although some rural migrants simply moved to rural areas of other provinces, the high rates of urban growth indicate that a significant portion of the migration took place not merely from rural to urban areas within regional boundaries, particularly within Quebec and the Prairies, but also from rural areas in one province to urban centres in another. A somewhat different picture emerges for both Ontario and British

Columbia. In both provinces, however, the apparent growth of rural population is attributable in large part to the settlement of urban workers in neighbouring rural areas, since the farm population in these provinces shared in the general nation-wide decline in farm population (Table 3).

It is also apparent that both interregional and rural-urban population movements have been reflected in widely different rates of growth in employment among the provinces. These differences are indicated in Figure 3, showing the rapid increase between 1951 and 1961 in Alberta, British Columbia and Ontario at one extreme and slower-growing provinces at the other. The importance of these variable rates of employment growth for the national economy is indicated by the regional distribution of total employment.

The broad trend of population redistribution in Canada over several past decades can be summarized as a pattern of two essentially similar movements. One has been the migration from regions of lower average income to those of higher income. The

GROWTH IN EMPLOYMENT BY PROVINCE, 1951-61

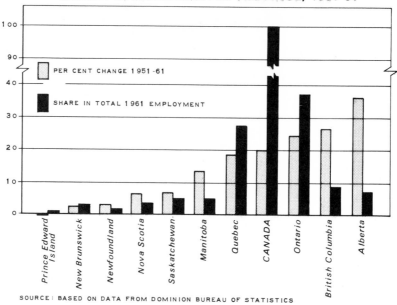

Figure 3

other has been movement from rural areas and primary industries (characterized, with some notable exceptions, by relatively lower productivity and income per worker) to urban centres and opportunities frequently offering generally higher monetary rewards. Whether or not these population movements were induced by income disparities, they have clearly contributed significantly to the process of interregional income adjustment in Canada.

REGIONAL DIFFERENCES IN LABOUR INPUT

We come now to the essential task of attempting to explain why per capita incomes differ as widely as they do among the various regions — and particularly why incomes in the Atlantic Provinces are so far below the average for the country as a whole.

In carrying out this task, it is useful to examine both the underlying factors that contribute to the flow of production and income within each region at any point in time, and the changes in the productivity of these factors over time.

On the basis of the shares in the national income received by the various factors of production, labour is by far the largest input into productive activity in the economy.

Since we are dealing with manpower, it is necessary to consider variations in the income actually *earned* by labour, rather than the broader and more inclusive concept of personal income already discussed. The patterns of interregional disparities and growth in earned income[5] per capita closely parallel those in personal income per capita, although the net effect of transfer payments is clearly to reduce the degree of interregional disparity.

The level of earned income per capita in each region obviously varies with the size of the employment base. Other things being equal, the larger the number of persons at work in relation to the total population, the larger will be the average level of income per person or per family. As Table 4 reveals, the employment base is substantially lower in the Atlantic Region than in any of the other regions. In the period 1960-64, only 27 percent of the population of the Atlantic Region was employed, on average, compared with 34 percent for the

Table 4

TOTAL POPULATION AND NUMBER OF PERSONS EMPLOYED

(1960-64 averages)

	Total Population (thousands)	Persons employed (thousands)	Persons Employed as Percent of Total Population	Index of Employment Percentage (Canada =100)
Atlantic Region	1,925	516	27	80
Quebec	5,359	1,719	32	95
Ontario	6,345	2,338	37	109
Manitoba	934	327	35	104
Saskatchewan	929	313	34	100
Alberta	1,366	479	35	104
British Columbia	1,703	554	32	97
Canada	18,561	6,246	34	100

Note: Indexes are calculated on the basis of unrounded figures.

Source: Based on data from Dominion Bureau of Statistics.

country as a whole and 37 percent in Ontario. The smaller size of the employment base accounts for roughly half of the difference in per capita earned income between the Atlantic Region and Canada as a whole. It is also an important consideration in explaining the lower average income level in Quebec.

Several factors determine the size of the employment base. The first is the age composition of the population. The Atlantic Provinces have an appreciably lower proportion of population in the working ages (Table 5) than any of the other regions. Ontario, on the other hand, has the highest proportion.

Table 5

AGE DISTRIBUTION OF THE POPULATION, BY PROVINCE, 1961

	Percent under 15	Percent 15-64	Percent 65 and over
Newfoundland	42	52	6
Prince Edward Island	36	54	10
Nova Scotia	35	57	8
New Brunswick	38	54	8
Quebec	35	59	6
Ontario	32	60	8
Manitoba	33	58	9
Saskatchewan	34	57	9
Alberta	35	58	7
British Columbia	31	59	10
Canada	34	58	8

Source: Based on data from 1961 Census of Canada.

Table 6

CIVILIAN LABOUR FORCE PARTICIPATION RATES
(1960-64 averages)

	Men	Women	Both Sexes	Index of Participation Rates (Canada=100)
Atlantic Region	73	24	47	87
Quebec	80	27	53	98
Ontario	82	33	57	106
Manitoba			55	102
Saskatchewan	80	30	53	98
Alberta			57	106
British Columbia	77	29	52	96
Canada	78	29	54	100

Note: The participation rate is the civilian labour force as a percent of the civilian population 14 years of age and over excluding inmates of institutions, Indians on reserves and residents of the Yukon and Northwest Territories.

Source: Based on data from Dominion Bureau of Statistics.

A second factor is the labour force participation rate, that is, the percentage of the adult population in the labour force, whether employed or unemployed. This factor also contributes to the low level of employment in the Atlantic Region (Table 6). Only about 47 percent of the adult civilian population was in the labour force in this region in the period 1960-64, whereas in Ontario the proportion was 57 percent.

A third factor is the unemployment rate, the percentage of the labour force out of work. The variation in these rates among the regions is relatively marked. A calculation of the average of annual rates of unemployment actually experienced in Canada and in each region over the entire postwar period shows the following:

Canada	4.4
Atlantic Provinces	7.6
Quebec	5.6
Ontario	3.2
Prairies	2.7
British Columbia	5.1

We have already noted that the labour force participation rate is lowest in the Atlantic Region where the unemployment rate is highest. Exact relationships are not easy to establish, but it is probable that the low participation rate reflects a lack of employment opportunities. It seems likely that if more jobs were available, more people would be drawn into the labour force in the Atlantic Region. Both the high unemployment rate and the low participation rate are symptoms of a substantial underutilization of manpower resources in this part of Canada.

Seasonal influences also exert disparate effects upon the level of employment among the various regions. In the period 1960-63 adverse seasonal effects upon employment in the Atlantic Region were about twice as great as in the country as a whole. They were also relatively severe in the Prairie Region, although to a lesser extent than in the Atlantic area. In Ontario, on the other hand, because of its greater concentration of activity in manufacturing and other industries that are less vulnerable to the impact of weather, the adverse seasonal impact upon employment was only half as great as the average for the whole country.

Differences in the utilization of available manpower are clearly a substantial factor in explaining differences in per capita earned income. Indeed, roughly half of the gap between the Atlantic Region and the national average can be explained in this way. But there are also substantial differences in the basic rates of earnings among the regions.

370

REGIONAL DIFFERENCES IN EARNINGS

Average earned income per employed person provides us with a rough measure of productivity per worker. An explanation of differences in this measure among the regions and in its growth over time is crucial to an understanding of regional disparities.

It will be seen (Table 7) that in the period 1960-64, earnings per employed person in the Atlantic Region were about 18 percent less than the average for all regions.

At the other extreme was British Columbia which, in spite of its shorter work week (8 percent lower than the national average), was about 19 percent above the average. Ontario was about 10 percent above the average and Quebec about 7 percent below. The Prairie Region as a whole was slightly below the regional average in the five-year period, although income per employed person in this region is subject to considerable variation from year to year because of the fluctuations in agriculture.

Table 7

AVERAGE EARNED INCOME PER EMPLOYED PERSON
(1960-64 averages)

	Average Earned Income per Employed Person	Index of Average Earned Income (Average for Regions=100)
	$	
Atlantic Region	3,080	82
Quebec	3,480	93
Ontario	4,120	110
Manitoba	3,620	97
Saskatchewan	3,660	98
Alberta	3,770	101
British Columbia	4,470	119
Average for Regions	3,740	100

Source: Based on data from Dominion Bureau of Statistics.

Table 8

HIGHEST LEVEL OF FORMAL SCHOOLING
ATTAINED BY THE LABOUR FORCE, 1961
(Cumulative percentages)

	University Degree	Some University	4-5 Years Secondary and up	1-3 Years Secondary and up	5-8 Years Elementary and up
Newfoundland	1.7	6.2	12.8	52.1	83.7
Prince Edward Island	2.3	6.5	16.9	53.9	94.5
Nova Scotia	3.7	8.0	19.2	63.0	94.7
New Brunswick	2.9	7.1	18.5	49.5	90.3
Quebec	4.5	8.6	24.7	51.6	90.8
Ontario	4.7	8.5	29.2	61.8	95.9
Manitoba	3.7	9.0	23.8	63.0	93.4
Saskatchewan	3.0	7.7	23.2	55.8	93.3
Alberta	4.3	9.8	28.8	65.4	95.5
British Columbia	4.7	11.7	37.3	72.8	96.8
Canada	4.3	8.8	27.1	59.5	93.8

Note: The labour force includes those 15 years of age and over.
Source: 1961 Census of Canada.

Differences in Educational Attainment

Level of educational attainment is an important factor explaining regional differences in income. A lower average educational level for the labour force goes hand in hand with lower levels of earnings and per capita income (Tables 7 and 8 and Figure 4). Moreover, the higher the level of education and skill of income earners, the smaller are the differences in their earnings between regions. The relatively much lower proportion of the labour force who have attended secondary school and perhaps even more particularly, the lower proportion who have attained the higher levels of secondary school and university, is particularly marked in the Atlantic Provinces.

Combined with the direct effects of educational levels on average income, there are also important indirect influences. In particular, a relatively poor structure of educational attainment in a region fails to support or attract those industries and activities which increasingly rely upon an educated and skilled work force. Together, the direct and indirect effects of educational attainment contribute significantly to interregional differences in income.

Differences in Capital Input

As in the case of labour input, differences among regions in the stock of capital, in average age and in degree of utilization help to explain interregional disparities in earnings per person. In the absence of offsetting influences, the greater the quantity of physical capital that a worker in a given industry has at his disposal, the larger his productivity tends to be and hence his own personal reward or income.

Unfortunately, comprehensive estimates of the stock of physical capital for each region—including both structures and machinery and equipment—are not available. Nevertheless an approximation of interregional differences in capital input is suggested by the comparative annual rates of new investment per person in each of the provinces, averaged in constant dollars over a lengthy period of years (Table 9).

The wide range of interregional variation is apparent. Over a period of some 15 years, average annual investment per person in Alberta, the highest province, has been about two and a half times greater than in Nova Scotia, the lowest province. For the private business sector alone, which normally accounts for about three-fifths of the

Figure 4

AVERAGE YEARS OF FORMAL SCHOOLING OF THE LABOUR FORCE BY PROVINCE, 1961

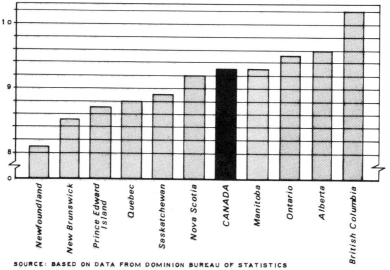

SOURCE: BASED ON DATA FROM DOMINION BUREAU OF STATISTICS

372

Table 9

INDEXES OF AVERAGE ANNUAL INVESTMENT
PER CAPITA, BY PROVINCE 1951-64
(Canada = 100)

	Total Investment	Private Investment			Public Investment
		Total	Housing	Business	
Alberta	153	154	131	161	149
British Columbia	133	133	126	135	131
Saskatchewan	110	113	75	124	99
Ontario	105	106	113	104	101
Manitoba	96	94	84	97	101
Quebec	83	85	95	81	75
Prince Edward Island	69	50	52	49	141
Newfoundland	66	59	48	62	92
New Brunswick	64	57	55	58	91
Nova Scotia	60	53	59	51	88
Canada	100	100	100	100	100

Note: Based on constant 1957 dollars. Private business investment includes trade, finance and commercial services, utilities, primary industry and construction, manufacturing and publicly owned commercial enterprises. Available public investment data cover the period 1955-63.

Source: Based on data from Department of Trade and Commerce, and from Dominion Bureau of Statistics.

total, the range in variation has been considerably greater. In this case, the three most westerly provinces have ranked considerably above the average for the country as a whole. Ontario and Manitoba are close to the national average, with Quebec distinctly behind. But all four Atlantic Provinces have achieved only 62 percent or less of the national average. In housing, as one would expect, the highest rates of construction have taken place where population growth has been most rapid. Public investment has not been marked by as extreme variation as in the private sector,

Table 10

INDEXES OF CAPITAL STOCK OF MACHINERY
AND EQUIPMENT PER CAPITA, BY PROVINCE IN 1964
(Canada=100)

	Stock of Machinery and Equipment	
	All Industries	Manufacturing
Saskatchewan	149	21
Alberta	129	59
British Columbia	114	127
Ontario	112	145
Manitoba	107	45
Quebec	77	90
Prince Edward Island	72	10
New Brunswick	68	60
Nova Scotia	66	43
Newfoundland	54	40
Canada	100	100

Note: Indexes reflect the stock of machinery and equipment in 1964, and are valued in constant 1957 dollars.

Source: Based on data from Department of Trade and Commerce, from Dominion Bureau of Statistics and estimates by Economic Council of Canada.

but even here per capita rates of additions to social capital have differed by as much as 100 percent among the regions.

This approximation of the general nature and order of interregional variation in capital input per worker is supported by preliminary estimates of regional capital stocks of machinery and equipment for all industries and for manufacturing alone. Both the ranking of the ten provinces on a per capita basis and the range of variation from the lowest to the highest (Table 10) are essentially similar to those already noted for the annual flows of new investment. Saskatchewan, with its heavily mechanized,

extensive agriculture, moves to the forefront, while the relatively undeveloped region of Newfoundland takes last place in the ranking.

The data for stock of machinery and equipment in manufacturing indicate exceptionally high levels for Ontario and British Columbia with Quebec as the only other province approaching the national average. The Ontario-Quebec contrast is of particular interest, in view of the fact that the proportion of the labour force of each province employed in manufacturing is almost identical (Table 11). It is clear that capital intensity per worker is much greater in Ontario

Table 11

REGIONAL EMPLOYMENT BY INDUSTRY, 1961
(Percentage distribution)

Industry	Canada	Nfld.	P.E.I.	N.S.	N.B.	Que.	Ont.	Man.	Sask.	Alta.	B.C.
Agriculture	10.2	1.5	27.3	5.2	7.2	7.6	7.2	17.6	37.0	21.5	4.1
Forestry	1.5	5.0	0.4	1.7	5.0	2.0	0.7	0.3	0.3	0.5	3.5
Fishing and trapping	0.5	7.5	6.2	3.2	2.1	0.2	0.1	0.3	0.3	0.2	0.8
Mines, quarries and oil wells	1.8	3.9	0.0	4.3	0.8	1.5	1.8	1.6	1.2	3.5	1.4
Total Primary	14.0	17.9	33.9	14.4	15.2	11.3	9.7	19.9	38.9	25.7	9.8
Manufacturing	21.7	10.9	8.8	14.2	16.0	26.5	26.9	13.6	4.6	8.5	19.5
Construction	6.3	7.4	6.1	6.2	5.8	6.7	6.1	5.8	5.1	7.3	5.8
Total Secondary	28.0	18.3	14.9	20.4	21.8	33.2	33.0	19.4	9.7	15.9	25.4
Transportation, communication and utilities	9.3	13.6	8.0	10.4	11.8	9.1	8.2	11.6	9.3	9.7	11.0
Trade	15.4	17.5	14.0	15.6	16.7	14.1	15.5	16.8	14.0	16.4	17.3
Finance, insurance and real estate	3.6	1.4	1.6	2.4	2.3	3.6	4.2	3.6	2.2	3.0	4.0
Community, business and personal service	19.8	16.7	16.7	19.2	20.0	20.2	19.7	18.8	18.0	19.1	21.8
Public administration and defence	7.5	11.6	8.7	15.9	10.1	5.6	7.7	7.8	5.6	7.9	8.1
Industries not stated	2.4	3.0	2.2	1.7	2.2	2.9	2.0	2.1	2.3	2.2	2.6
Total Services	58.0	63.8	51.2	65.2	63.0	55.5	57.3	60.7	51.4	58.4	64.8
TOTAL EMPLOYMENT	100.0	100.0	100.0	100.0	100.0	100.0	100.0	100.0	100.0	100.0	100.0

Note: Details may not add to totals because of rounding.

Source: Based on data from Dominion Bureau of Statistics.

than in Quebec, and influences such as these appear relevant to the differences in earnings per person employed in manufacturing (Table 12).

The limitations of the data do not permit a full evaluation of the influence of capital inputs on interregional income disparity. Nevertheless, the comparisons do indicate that the regional groupings for levels of capital input are generally the same as those defined by other factors accounting for interregional income disparity.

Differences in Regional Structures of Economic Activity

We turn now to examine whether the relative importance of different industries among the regions affects the degree of interregional income disparity. The influence of industry composition or structure arises because some branches of production yield a higher average productivity or income per worker than others. It is true that

in any particular region the average level of income will depend crucially upon the distribution of its production and employment between higher and lower income industries. Consequently, the familiar diversity in the industrial structure of economic activity among the regions of Canada has led to a belief that this is one factor which accounts for a large proportion of interregional income differences. Our analysis, however, suggests that this belief is exaggerated.

It is clear that average income per employed person varies greatly among the broad sectors of the national economy (Figure 5). Income levels in agriculture, fishing, and primary forestry are clearly much below the average for all industries, while those received in mining and financial services are much above the average.

It is also true that there is considerable diversity in the structure or composition of economic activity among the regions, as is

Figure 5

EARNED INCOME PER EMPLOYED PERSON BY INDUSTRY AS PERCENTAGE OF CANADIAN AVERAGE, 1961

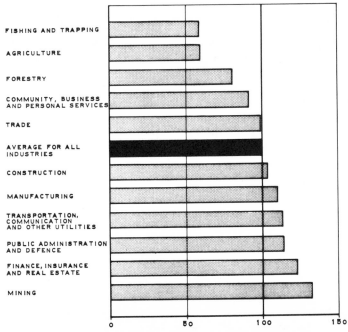

NOTE: INCOME DATA ARE BASED UPON SPECIAL 1961 CENSUS MATERIAL ON EARNED INCOME FOR THE NON-AGRICULTURAL SECTORS, AND ESTIMATES OF FARM INCOME PER EMPLOYED PERSON IN AGRICULTURE AVERAGED OVER THREE YEARS

SOURCE: BASED ON DATA FROM DOMINION BUREAU OF STATISTICS

shown by the distribution of employment in each province (Table 12). Not only are there major differences in the proportion of the labour force employed in primary, secondary and service industries among the provinces, but the distribution of employment within each of these sectors also differs widely.

The average level of earned income per employed person in each region is obviously influenced by the interaction of the two factors of industry structure and average income levels among different industries. Available data, however, indicate that because of offsetting influences, the differences in the structure of industry among the regions account for only a small part of the interregional disparity in earned income. A comparative analysis of employment and earnings per worker as between agriculture and the nonagricultural sector among the five major regions, for example, shows that this kind of structural difference is of only minor importance in explaining interregional income differences. Within the nonagricultural sector itself, and excluding fishing, trapping and certain services, a study of employment and earnings in 37 branches of industry also suggests that different regional "mixes" of industry in themselves have a similar limited effect. Moreover, while shifts of employment from industries of lower to higher productivity have been an important factor in the longer term growth of income, they have occurred more or less proportionately throughout the whole country. Consequently these structural shifts have had a fairly uniform impact on average incomes in all regions.[6]

The central reason why interregional differences in industrial structure apparently do not have a major effect on differences in average regional income levels is that there are significant disparities in productivity per worker in any given industry among the various regions. In other words, although certain industries show a high level of earned income per worker in the country as a whole, these industries do not show a consistently high level per worker in all regions of the country. Earned income per worker

(Table 12) would appear to be relatively low in all major sectors of economic activity in the Atlantic area, and relatively high in almost all industrial sectors in British Columbia and Ontario. It is, therefore, the regional differences in industrial productivity, rather than in economic structure, that exert a major influence on regional income disparity. These differences in productivity, in turn, reflect many different factors of a more basic nature concerning the availability and quality of the various economic resources among the regions, and the efficiency with which these resources are combined in the processes of production.

In summary, incomplete as the evidence may be, it appears that the varying regional distribution of economic activity by sector exerts relatively little influence upon interregional income disparity, and changes in structure have contributed only moderately to narrowing the disparity over time. This is not to say, of course, that inter-industry shifts from low to higher productivity industries within a region—particularly from the farm to the nonfarm sector—will not yield significant results in raising its average level of income. There is scope for these favourable shifts in all provinces, and particularly so in the Altantic Region and Quebec. Not only do these regions have relatively high proportions of employment in the primary industries generally (which, except for mining, provide the lowest average earnings of all the broad industrial categories throughout the country), but it is also in these same industries where the gap in average earnings as compared with other regions is the most severe. However, to the extent that this type of adjustment takes place proportionately in all parts of the country, it will be more important in raising the national average level of income than in reducing interregional income disparities.

Other Factors in Differences in Earnings

We turn now to a brief reference to certain of the obvious factors in interregional differences in output per person that although clearly important among the

376

Table 12

REGIONAL EARNED INCOME PER EMPLOYED
PERSON AS PERCENTAGE OF CANADIAN AVERAGE,
BY INDUSTRY, 1961

	Canada	Nfld.	P.E.I.	N.S.	N.B.	Que.	Ont.	Man.	Sask.	Alta.	B.C.
Agriculture	100	57	53	64	67	69	112	85	115	114	147
Forestry	100	78	48	55	65	84	118	84	61	97	152
Fishing and trapping	100	65	101	94	120	76	127	101	40	59	188
Mines, quarries and oil wells	100	90	—	71	65	91	104	101	110	124	97
Manufacturing	100	83	60	79	77	94	106	92	96	101	107
Construction	100	72	73	83	78	94	105	102	99	110	110
Transportation, communication and utilities	100	77	90	82	84	97	104	99	101	103	112
Trade	100	78	78	82	83	101	102	98	100	102	105
Finance, insurance and real estate	100	94	82	95	91	102	103	92	89	98	97
Community, business and personal service	100	69	65	77	77	96	106	97	97	102	109
Public administration and defence	100	91	88	99	91	94	105	99	98	97	104

Note: Earned income per employed person in agriculture is obtained by averaging earned income for the three years 1961, 1962 and 1963 and dividing by employment in 1961. The data for other industries are based upon a special compilation from the Population Sample Survey and covers the nonfarm labour force as of June, 1961. For the smaller provinces, therefore, the figures shown may be subject to significant error.

Source: Based on data from Dominion Bureau of Statistics.

sources of growth, are difficult to isolate and measure. Among these is the question of differences among the regions in the extent and quality of natural resources. Economists have long recognized "land" as one of the factors of production, along with labour and capital. Other things being equal, a region with an abundance of land and resources of good quality is likely to enjoy an income advantage over one less favourably endowed. We have not been able to evaluate the influence of these differences among the regions, although it seems certain that they underlie, at least in part, the marked variations in income per worker in the resource industries (Table 12).

Because of fundamental shifts in the pattern of consumption and advances in technology, however, the relative impact of resources in income differences has clearly declined. With generally rising incomes, other factors emerge more strongly in the growth process. Of major importance is the concentration of population in fairly small geographic areas, in which the most efficient production and distribution is more easily achieved. Moreover, once the process of concentration gets under way, similar powerful forces make it of cumulative importance in growth—production can be scaled still more efficiently to meet enlarging markets; business services and a versatile labour force are close at hand; new technology is more easily developed and exploited; and advanced management skills and enterprise are more readily attracted. It is on this basis that the concept of the "growth centre" as a necessary focus for regional growth has been widely advanced and accepted. The concept involves a geographic concentration of appropriate industries whose growth in the region can be effectively fostered. Once set in train, the growth process will stimulate advances in related activites that supply inputs, such as component parts, services, and fuels, to the "core" industries and make use of their output.

There is widespread evidence and experience to suggest that the process of economic "agglomeration" is an increasingly important factor in economic development. The regions of Canada, however, vary greatly in concentration of population and in locational advantages. The most striking feature is the heavy concentration of Canada's population —and hence employment and economic activity — along a narrow band which extends from Windsor, in southwestern Ontario, to Quebec City on the St. Lawrence River. The metropolitan areas within this narrow band alone account for 31 percent of Canada's total population. Moreover, much of this area is contiguous to the industrial "heartland" and major population centres of the United States, thus adding to its locational advantage.

Outside of this central industrial area, the population is widely dispersed, except for a number of metropolitan centres in widely separated parts of the country. Although these centres vary greatly in size and in effectiveness as agglomerations for self-generating growth, they clearly constitute the most important focal points for the development of their respective regional areas and for improved regional balance across the country.

FOOTNOTES

[1] Personal income is obtained from the National Accounts, published by the Dominion Bureau of Statistics. Its major components are: wages, salaries, and supplementary labour income; military pay and allowances; net income of farm operators from farm production; net income of non-farm unincorporated business; interest, dividends and net rental income of persons; and transfer payments. Thus, certain major income flows associated with corporate enterprise and government are excluded.

[2] Three-year averages are indicated by a bar over the centre year. For example, 1963 is the average of 1962, 1963 and 1964.

[3] In statistical terms it is an index of relative dispersion and is obtained by dividing the standard deviation of the distribution by the unweighted mean.

[4] *1961 Census of Canada, Vol. IV, Population Sample.*

[5] Earned income is a term used to cover all types of income from employment: wages, salaries, and supplementary labour income; military pay and allowances; and the net income of unincorporated business proprietors, including farmers. It excludes government transfer payments (such as family allowances and old age pensions), interest, dividends and net rental income.

[6] The technical analysis summarized here will be detailed in staff studies, which will also discuss possible qualifications based upon more detailed breakdowns of industrial structure.

The Problem of Poverty in Canada

ROBERT M. IRVING

In our society the benefits of economic growth and resource development are generally bestowed upon those who are best equipped to participate. Those people who are ill-endowed, and thereby participate less, if at all, receive few of the benefits. We have generally referred to this latter group as the "poor," the "poverty class," or more recently as the "economically and socially disadvantaged."[1]

In this paper I propose to briefly examine three aspects of poverty in Canada: (1) the nature of poverty; (2) some measures of rural poverty; and, (3) solutions and programs to alleviate poverty.

THE NATURE OF POVERTY

The characteristics of poverty are far removed from the traditional interpretation—that is, an inability of people to secure sufficient food to survive. In Canada this concept of poverty is no longer valid—except, perhaps, to some of the indigenous people. But, paradoxically, the emphasis has never been greater on the problems of the disadvantaged than it is nowadays.

In his study of the changing meaning of poverty, Donald Whyte concluded that in advanced industrial societies, such as Canada, "individuals and families whose resources, over time, fall seriously short of the resources commanded by the average individual or family in the community in which they live . . . are in poverty."[2] Robert Riendeau defines poverty as "the state of families and groups who . . . lack basic necessities of life, . . . and cannot satisfy legitimate aspirations considered as normal by our milieu."[3] He points out that goods which were formerly considered luxuries are now essentials, and that access to these goods (such as food, clothing, and shelter) cannot be dissociated from the attributes of health, higher education and vocational training. Poverty, then, is relative, because while most Canadians do not lack food, many nevertheless are ill-fed; likewise all Canadians have shelter, but it is often poor, crowded, and insanitary; medical care exists, but high costs preclude many from receiving adequate attention. Similarly, the opportunities for higher education exist, but more often it is those with good health and financial resources who participate in it.

Who are the Disadvantaged Canadians?

According to Menzies, poverty is unemployment or under-employment.[4] The causes of this condition are many, but the major ones are economic and technological change, poor mental or physical health, and inadequate education. While these factors account for a large percentage of the present poverty group, they can be applied equally to future conditions. Clearly, members of the middle class today, and those near the margin of poverty, may become members of the poverty class several years hence, owing to technological change or lack of training to take advantage of alternate employment. But poverty need not be causally linked with human frailties—bad luck can be a factor, or a poor business decision can induce poverty. Full-time employment at low wages, can likewise induce poverty. Similarly, an adequate retirement pension established 20 years ago may prove entirely inadequate, owing to inflation.

Few analysts agree as to who belongs to the disadvantaged group for, as they emphasize, poverty is relative.[5] Hence, in a broad sense, all Canadians, to some degree, are disadvantaged. To further confuse the problem of identification, others claim that definition is unnecessary. The Federal Minister of Health and Welfare, Mr. Mac-

● ROBERT M. IRVING *is Chairman and Professor of Geography, Department of Geography, University of Waterloo.*

Eachen, believes that "only the most insensitive can fail to recognize poverty when he sees it—poverty in material needs, poverty in needs of the spirit."[6] John K. Galbraith contends that a firm definition is unnecessary "save as a tactic for countering the intellectual obstructionism."[7] Notwithstanding these comments, it is impossible to portray the extent of Canadian poverty if an operational definition is withheld.

Social Criteria——Certain social criteria have been employed to describe the disadvantaged, and from these, some lower limits of tolerance emerge. Among these criteria are housing conditions, education, and health.

Benjamin Schlesinger lists the following as being symptomatic of social disadvantage.[8] It is imputed from these "qualities" that when the characteristic under consideration falls below an arbitrary standard, some degree of disadvantage exists. If one adopts an adequate housing standard of one room per family member, then 35 percent of Canadian families with an income of less than $3,000 per year are disadvantaged, whereas only 18 percent of the families earning more than $6,000 per year are in this category. The aged, broken families, and Indians suffer most from inadequate housing. In terms of amenities, only one-third of Indian dwellings are equipped with indoor running water, and 9 percent have indoor toilet facilities.[9]

If one adopts a minimum education standard of elementary school completion, then 47 percent of Canadians 15 years and over in 1961 were disadvantaged. About half of the unemployed in 1961 had not finished primary school, and more than 90 percent had not completed high school. The inadequately educated individual is handicapped for two reasons: first, the low level of formal education is a barrier to getting a job, or upgrading employment; second, it acts as a barrier to increasing his income so that his wife and children cannot join the affluent society around them.[10]

Minimum physical health standards are difficult to establish and measure since deviation from perfect health does not necessarily render an individual socially or economically incapacitated or handicapped. It is even more difficult to establish a normal state of mental and emotional behaviour.[11] Despite the lack of accurate criteria, estimates indicate that 1.3 million Canadians suffer from a permanent physical handicap. Of these, 570,000 (3 percent of the total population in 1961) had a severe or total disability. Half of this group is of working age and only one in four has income from employment. In the overall picture, the widest discrepancy in health conditions is between the indigenous population and the rest of the Canadian population. For example, the Eskimo infant mortality rate in 1963 was 193/1,000 live births compared with an all-Canada average of 27/1,000. In 1963, the Indian death rate was 70/1,000.

Poor people suffer other types of social deprivation as well. They participate little in community life and have a small if any role to play in making the decisions which affect them.[12] Moreover, the poor are often alienated from the rest of society because of society's apathy, prejudice or outright ignorance.

Economic Poverty——The second most common measure of poverty is financial status. This is the most easily measured standard and unquestionably is central to the concept. However, an acceptable minimum standard of income is not easily derived.

A minimum annual income of $3,000 per family is one figure that is widely employed.[13] On this basis, 23 percent of the 3.7 million urban and rural nonfarm families in Canada in 1961 had incomes below $3,000 and one family in eight received less than $2,000 annually. Half of the individual wage earners earned less than $1,500, the minimum annual income for that group. One indicator of farm poverty, based on an annual gross income of $2,500 in 1961, places 46 percent of Canadian farmers in the poverty class. Of the approximately 200,000 registered Indians living in reserves in Canada in

1965, three-quarters of the families had incomes of less than $2,000. Among the aged in our society, the median annual income for men in their late sixties is less than $2,000; this declines to an average of $900 for those in their late eighties.

John Madden, in a provocative analysis of poverty in Canada, suggests that to apply the $3,000 income level for urban and rural nonfarm families and $1,500 for individuals, as in the United States, is unrealistic in Canadian terms.[14] He states that since Canadian income is about 30 percent less than American income, the poverty limits should be accordingly reduced to $2,000 for families and $1,000 for individuals. By adopting this standard, only 13 percent of the families and 38 percent of the individuals are in the poverty category. Moreover, Madden believes that many of these poor are, in fact, better off than statistics indicate owing to cash transfers from children to parents, payment of bills by other family members, unreported part-time income from baby sitting, odd jobs, rental of rooms, and other types of casual employment.

Spiritual Poverty——Social and economic poverty are the two facets of poverty that have the greatest impact and are synonymous with anti-poverty programs. A more evasive dimension is poverty of spirit, because even in gross terms it cannot be measured, nor can it be seen. Lappin describes the origin of spiritual poverty as being rooted in the professionalism of almost all aspects of our society, and that the ultimate survival of man may not lie in solutions to economic poverty but in how man exercises control over spiritual poverty.

Increasingly, life processes nurtured by prophetic vision, artistic imagination, by ordinary intuition and folk wisdom are . . . being reconstituted as the content of professional callings. . . . The popular notion that our spiritual poverty is due to material abundance . . . is a facile judgment derived from an oversimplified view of the problem. More likely, the fact that the life force of society is progressively finding expression in diversified service commodities is the cause of our creative poverty—a poverty which, ironically, is increasing with the very proliferation of the specialists by their expertise to enhance our life style. . . . As crucial as is the war on economic poverty, the ultimate threat to the privileged and the dispossessed alike, with their ever-increasing, indeed imposed, leisure is rooted in spiritual poverty. . . . What is needed is a perspective in which the professions rather than exerting mastery over our civilization assume their proper role as its supporting institutions. . . . What this points to is an integrated way of life which must be achieved at a higher level of sophistication. The higher integration now bids man to master the immense complexities of his own making. . . . The man-made threats evoke his despair; they are internal, shadowy, evasive, and they challenge the very core of man's self.[15]

ASPECTS OF RURAL POVERTY

Knowledge of the extent of rural poverty, until recently, has been on *ad hoc* basis. A major impetus to an organized approach to its definition and identification occurred with the passage of the Agricultural Rehabilitation and Development Act of 1961. One of the first tasks of ARDA was to establish the extent and nature of rural poverty. As a result of their own studies and by commissioning other studies, the characteristics and extent of rural poverty are emerging.[16] Three dimensions of rural poverty in Canada as portrayed by ARDA are presented below.

Low Income Full-Time Farm Families

A low income full-time farm is defined on the basis of three criteria: (a) the total capital value of the farm on the open market is less than $24,950 which is considered the average minimum capitalization for the development of an economic enterprise; (b) the average gross sales of farm products in the previous year were less than $2,500— or approximately half the national average in 1961; (c) the farm operator in the preceding year worked fewer than 30 days off the farm. On this basis there are 95,410 farm families who are disadvantaged[18] (Figure 1A).

The distribution of low income farms in Canada is widespread, but several concentrations are apparent (Figure 2). Areas

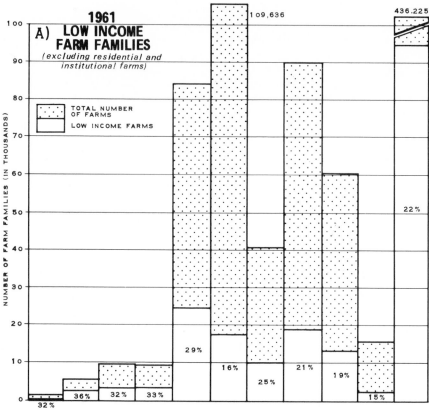

1961
A) LOW INCOME FARM FAMILIES
(excluding residential and institutional farms)

TOTAL NUMBER OF FARMS
LOW INCOME FARMS

NUMBER OF FARM FAMILIES (IN THOUSANDS)

109,636

436.225

22%

32%   36%   32%   33%   29%   16%   25%   21%   19%   15%

Figure 1

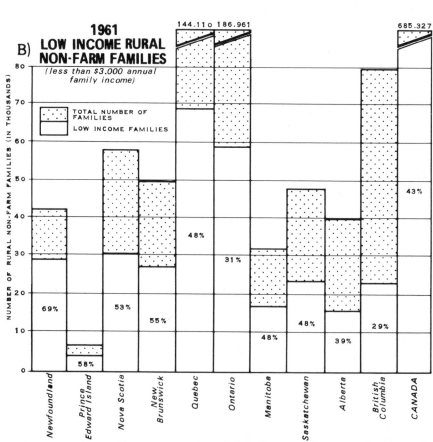

1961
B) LOW INCOME RURAL NON-FARM FAMILIES
(less than $3.000 annual family income)

TOTAL NUMBER OF FAMILIES
LOW INCOME FAMILIES

NUMBER OF RURAL NON-FARM FAMILIES (IN THOUSANDS)

144.11o   186.961

685.327

43%

69%   58%   53%   55%   48%   31%   48%   48%   39%   29%

Newfoundland
Prince Edward Island
Nova Scotia
New Brunswick
Quebec
Ontario
Manitoba
Saskatchewan
Alberta
British Columbia
CANADA

382

SOURCE: CAN. DEPT. OF FORESTRY, <u>ECONOMIC AND SOCIAL DISADVANTAGE IN CANADA</u>, OTTAWA, 1964

in which over 40 percent of the farm families are classed as low income are found: (1) along the south coast of Newfoundland; (2) the eastern county of Prince Edward Island; (3) five of the six counties fronting on the St. Lawrence River between Quebec City and Tadoussac in Quebec; (4) three counties in New Brunswick; (5) the Parry Sound District in Ontario; and, (6) in Manitoba, the area southeast of Lake Winnipeg, the Interlake District, and the area adjacent to Lake Manitoba. Within the 30-40 percent category are included many of the counties in the Atlantic Provinces, twenty counties in Quebec, and three in Ontario. West of Onatrio only thirteen counties are in this category.

For the most part, all of the areas in the over 30 percent category have physical attributes that are detrimental to agricultural development—poor soils or lack of soils, broken topography, marginal climate, and distance from markets.[19] The characteristics of low income farms in Eastern Canada (Ontario, Quebec, and the Maritime Provinces) have been described by Menzies in his study *Report of the Eastern Canada Farm Survey, 1963.*[20] The major characteristics are: (1) that about 50 percent of the farms studied are redundant, that is, their operators should leave the agricultural industry and seek employment elsewhere to the benefit of the industry and to themselves; (2) the level of education is very low; (3) most of the low income farms are too small to achieve an economic farm operation; and, (4) almost half of the cash income of low income farmers is derived from off-farm earnings, old-age pensions and welfare payments. Hence, the farm acts as a supplement to off-farm income and furnishes a relatively cheap source of housing and food.[21]

Low Income Rural Non-Farm Families

This group comprises those who reside in rural areas and in urban communities of less than 10,000 population. These families fall within the poverty bracket when annual family income is less than $3,000 or about half the national average ($5,449) in 1961.

Forty-three percent, or 294,349 rural non-farm families, are in this category (Figure 1B).

The Atlantic Provinces and Quebec have a large percentage of counties and census divisions with 50 percent or more of the rural non-farm families in this group (Figure 3). West of the Ottawa River the intensity is somewhat less, although the areas in the over 40 percent category are still numerous in Manitoba and Saskatchewan.

Education Levels

The third measure of disadvantage is reflected by the length of formal education. Here, it is considered that a person of working age with no more than four years of formal schooling is likely to experience drastically curtailed earning power and employment possibilities, and that such a person is said to be seriously disadvantaged. In Canada, 1,017,869 people, who represent 9 percent of the working age population, are in this category. In the rural areas, there are 656,566 people of school age (five years) and over not attending school, who have a grade four education or less. They represent 20 percent of all farm and rural non-farm Canadians in this category (Figure 4).

Geographically, the areas of lowest educational attainment (for rural and urban people) are found throughout Newfoundland, northern New Brunswick, in all of Quebec outside the larger cities, northwestern Ontario, northern and eastern Manitoba, and northern Saskatchewan (Figure 5). The severity of educational disadvantage, on a county or census division basis, is less west of Saskatchewan.

Factors Associated with Rural Poverty

A rigorous examination of the factors associated with rural poverty has been conducted by Brian J. L. Berry for 555 municipalities in Ontario.[22] In a factor analysis, he employed forty-seven variables, including levels of income and education, value of farm equipment and livestock, and amenities—flush toilet, hot and cold running water, and furnace heat. He found that the

384

Figure 2

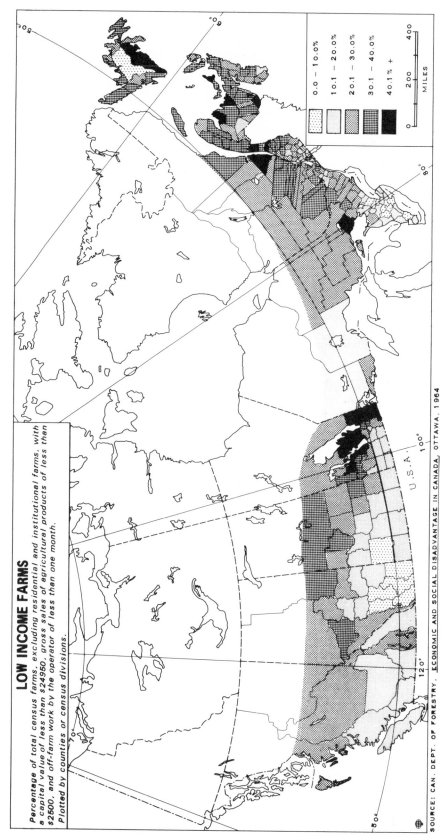

LOW INCOME FARMS

Percentage of total census farms, excluding residential and institutional farms, with a capital value of less than $24950, gross sales of agricultural products of less than $2500, and off-farm work by the operator of less than one month.

Plotted by counties or census divisions.

0.0 — 10.0%
10.1 — 20.0%
20.1 — 30.0%
30.1 — 40.0%
40.1% +

MILES

0   200   400

U.S.A.

SOURCE: CAN. DEPT. OF FORESTRY, <u>ECONOMIC AND SOCIAL DISADVANTAGE IN CANADA</u>, OTTAWA, 1964

Figure 3

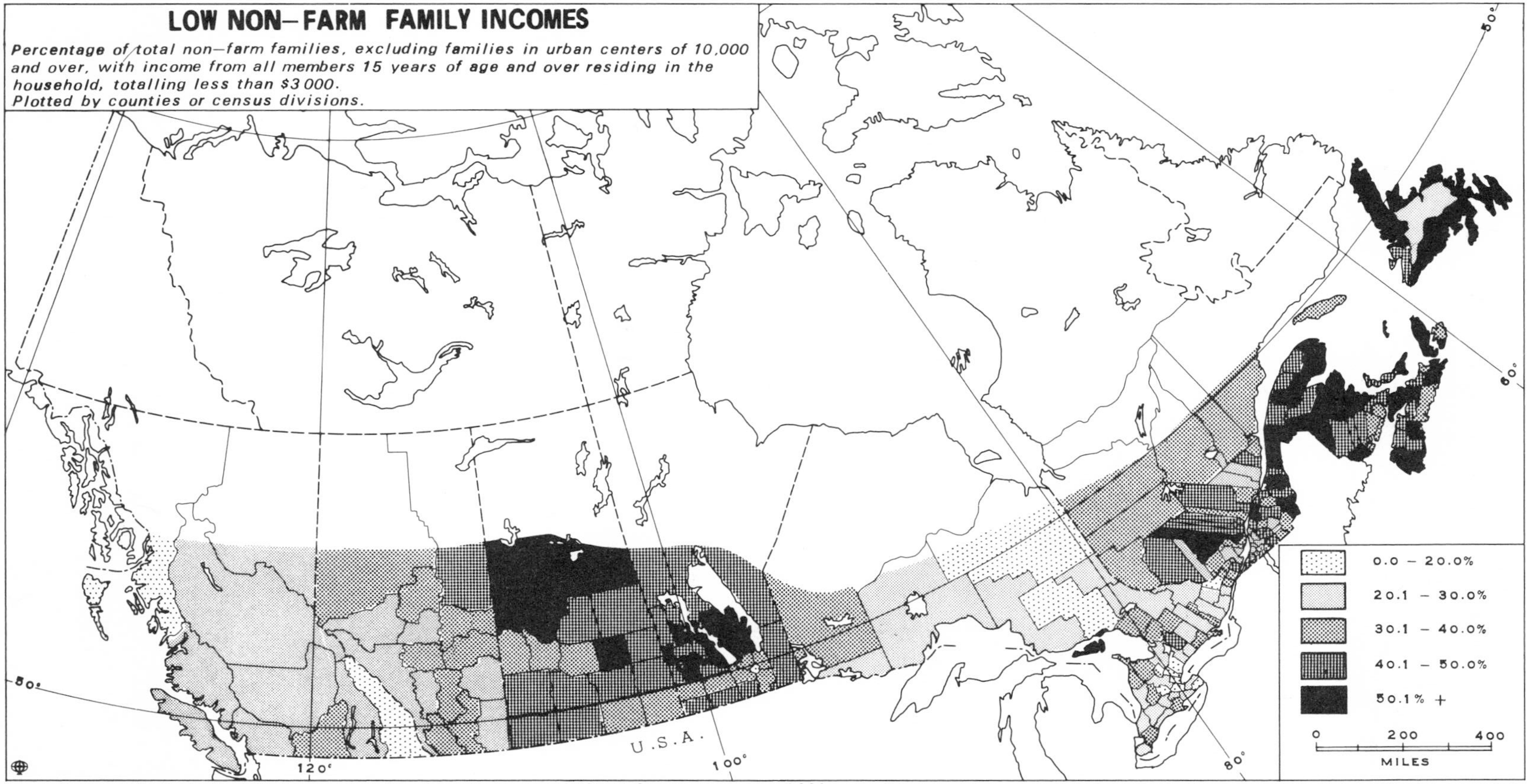

LOW NON—FARM FAMILY INCOMES

*Percentage of total non—farm families, excluding families in urban centers of 10,000
and over, with income from all members 15 years of age and over residing in the
household, totalling less than $3,000.
Plotted by counties or census divisions.*

0.0 — 20.0%
20.1 — 30.0%
30.1 — 40.0%
40.1 — 50.0%
50.1% +

MILES
0   200   400

U.S.A.

100°

120°

80°

SOURCE: CAN. DEPT. OF FORESTRY, ECONOMIC AND SOCIAL DISADVANTAGE IN CANADA, OTTAWA, 1964

385

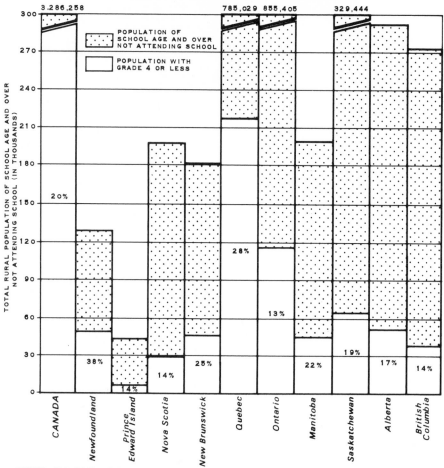

1961 LOW RURAL EDUCATIONAL LEVELS
(total rural population of school age and over not attending school, and total with grade 4 or less)

SOURCE: CAN. DEPT. OF FORESTRY, <u>ECONOMIC AND SOCIAL DISADVANTAGE IN CANADA</u>, OTTAWA, 1964

Figure 4

areas of greatest relative poverty were associated with different factors. Relative rural farm poverty appeared where the proportion of farm operators with low incomes was greatest, average incomes of farm operators were lowest and the proportion of farms with a low level of household services was high. Spatially, these areas appeared to be associated with poor physical resources. Relative rural non-farm poverty was associated with low levels of education attainment and low levels of household services. Areally, rural non-farm poverty appears to be related to the pattern of urban alignments. That is, the wealthier rural non-farm residents are found around urban areas of some

size or where several urban centres merge. The zones of high rural non-farm poverty appear in the areas between each urban agglomeration. Berry suggests that much of the rural non-farm prosperity in Ontario is related to commuting to urban employment. Conversely, the areas least accessible to urban employment opportunities are those suffering from rural non-farm poverty. In addition, the zones of rural non-farm prosperity are the zones of recent population increase, whereas the inter-urban peripheries are zones of greatest population decline and rural non-farm poverty.

In areas of relative social disadvantage, a strong correlation appears between low

Figure 5

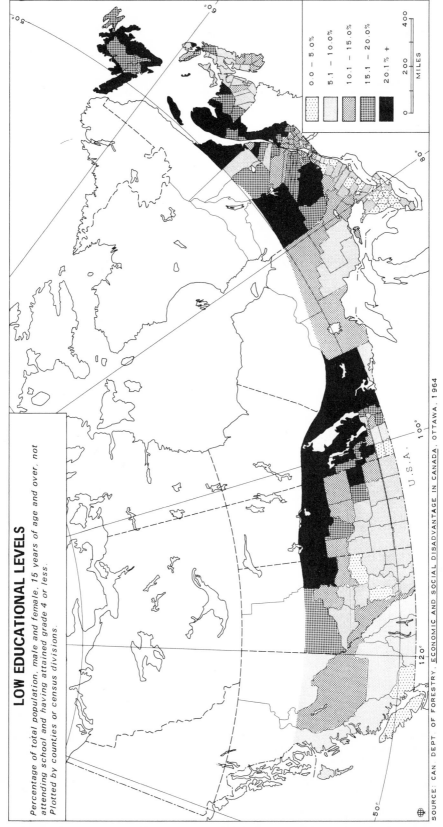

LOW EDUCATIONAL LEVELS

Percentage of total population, male and female, 15 years of age and over, not attending school and having attained grade 4 or less. Plotted by counties or census divisions.

0.0 — 5.0%
5.1 — 10.0%
10.1 — 15.0%
15.1 — 20.0%
20.1% +

MILES

U.S.A.

SOURCE: CAN. DEPT. OF FORESTRY, ECONOMIC AND SOCIAL DISADVANTAGE IN CANADA, OTTAWA, 1964

education levels and high infant mortality rates. In Ontario, the zones of French-Canadian occupancy, and those areas in which the migrant and permanent farm labour forces are high (as in Simcoe and Norfolk Counties) appear to be related to social disadvantage.

Of significance in Berry's study is the fact that the three types of relative rural poverty and the variables associated with each type are independent of one another, that is, "Location of a particular municipality on one of the scales [rural farm poverty, rural non-farm poverty, and social disadvantage] does not affect the probabilities associated with location of that municipality on either of the other two scales."[23] Berry does not yet have remedies for economic and social disadvantage; these must await the development of models to assess the effects of alternative strategies. "What is clear . . . is that different public policies will be needed to attack several distinct poverty syndromes."[24]

THE SOLUTIONS

So far in this discussion some of the characteristics of poverty in Canada have been outlined. These are that (1) about one-third of the families with income below $3,000 live in crowded housing conditions; (2) almost half of the population of working age possess less than an elementary education; (3) approximately half a million Canadians have a severe physical disability; (4) about one-quarter of the urban and rural non-farm families of Canada have an annual income below $3,000; and about half of the farmers receive less than $2,500 from the sale of agricultural products; (5) in all facets, the indigenous population appears to be in a chronic state of poverty.

There is no single solution to reduce poverty. Nor, under prevailing economic conditions and social values, can poverty be eliminated. At best, poverty can be reduced through the provision of better education opportunities, health facilities, and job opportunities. But no matter how extensive and thorough such improvements are, there will still be a group that "didn't quite make it." Moreover, as long as the philosophy prevails that wages are tied to the type of job one has, an income gap will persist. On the other hand, it is incumbent from a humanitarian point of view, if for no other consideration, to create a social and economic environment in which people can achieve a decent life. The question is how?

Federal, provincial, and municipal governments and private agencies have long been aware of poverty. But, the organization of a so-called "war" or "attack on poverty" has been realized only since 1960. The "war," as announced by the Government of Canada, is aimed at (a) achieving full utilization of human resources, and (b) the elimination of poverty. The "attack" consists of the co-ordination of six programs: Manpower Mobility Program; Area Development Program, Technical and Vocational Training Program and the Vocational Rehabilitation of the Disabled Program, the Agricultural Rehabilitation and Development Act, the Canada Assistance Plan, and the Company of Young Canadians.[25]

In this section the goals of the ARDA program will be presented, as well as some of the other solutions that have been tendered to reduce or to eliminate poverty in Canada.

ARDA

The most widespread program to focus on rural poverty in Canada is ARDA.[26] The basic objective of ARDA is to increase rural income and employment opportunities, "by improving the management and use of natural resources, and by assisting low-income people either to utilize the resources more profitably or to seek alternative opportunities in the many other fields that exist in our dynamic society."[27] Federal-provincial research and action programs are conducted in (1) developing alternative uses of land that are marginal or of low productivity for agriculture; (2) development of employment and income opportunities in agriculture, forestry, recreation, and secondary industry; and, (3) the development and conservation of soil and water resources. Within each of these categories a wide range of research and action programs is possible.[28]

The largest projects, in terms of financial investment and breadth of development, are those undertaken in the Gaspé region of Quebec, the Manitoba Interlake, and the New Brunswick North Shore. For example, in the New Brunswick region the program includes the analysis of development of educational facilities, educational television, technical and vocational training and adult education; changes in population distribution; land acquisition and land use; resettlement and rehabilitation; housing; transportation; agriculture; fisheries; industry; recreation, and employment opportunities.[29] Development of this type is clearly long range and immediate results cannot be anticipated.

The Menzies Plan

To reduce poverty levels in rural and urban Canada, M. W. Menzies contends that an attack must be based on the recognition of the low-level of productivity of a large proportion of the human and physical resources.[30] He believes that to place agriculture on a sound commercial basis will require a 50 percent reduction in the number of farms by 1975, that is, those farms with a gross sales of less than $3,750 per year (208,573 or 43.4 percent of all farms). The urban and rural non-farm poor are those workers with an income of less than $3,000 per year (572,000 rural non-farm and 546,000 urban workers).[31] To achieve his goals of farm reduction and to retrain or relocate the rural and urban poor, Menzies proposes the following voluntary program.

Farmers—Low-income farmers under the age of 55 would receive (1) an *adjustment salary* of $200 a month for one year; (2) a *resource release bonus* of $1,800, as an incentive to dispose of his land to the government; (3) if he chooses to move to an approved "growth centre", he should be offered a *relocation bonus* of $1,800 (to cover moving costs and down payment requirements for housing provided by the

Central Mortgage and Housing Corporation); (4) he should be offered a *training incentive payment* of $1,200 for one year. Menzies believes that, given a fair price for his land, plus relocation and training, most farmers under 55 should be able to achieve reasonably effective adjustment.

Farmers 55 to 64 years of age should be offered a choice of either (a) early retirement pensions equal to that of old-age pensioners, or, (b) a public works job at $2,400 minimum, plus an $1,800 relocation bonus. Those between 65 and 75 years of age should receive a relocation bonus (if they wish to leave the farm) and a special resource release bonus to accelerate farm rationalization. Any farmer over 55 years of age should, if he so desires, be permitted to remain on his land even though he has sold it to the government.

Rural Non-Farm Workers — The incentives for this group are similar to the low-income farm group, except a resource release bonus is not involved. Hence, the total outlay is $5,400. For those between 55 and 74 years of age, there should be a relocation bonus of $1,800, and those under pensionable age, a public works job at $2,400 a year if other equivalent employment is not available.

Low Income Urban Workers—The under 55 year old group should be offered an adjustment salary of $2,400 for one year, provided that retraining is accepted, plus a relocation bonus of $1,200 if a move to a new area is necessary. For the 55 to 74 age group a relocation bonus should be offered, and for those under pensionable age, a public works job at $2,400 a year, if alternate employment is unavailable.

The costs of implementing this voluntary program over a 10-year period are estimated to be $7.2 billion, or an average of $700 million annually. This cost represents slightly over one-third of the annual Canadian defence budget. Menzies does not calculate into his total budget the costs of buying land from low-income farmers, costs in administering the program, costs for constructing decent low down-payment homes by the

Central Mortgage and Housing Corporation, and training costs. He concludes, however, that "all these costs are equally investments in increased productivity having a highly favourable benefit-cost ratio."[32]

The Guaranteed Income Approach

The guaranteed income approach is yet another technique devised to lift the yoke of poverty. Few proponents of this scheme believe that it is a *permanent solution* to the low income problem, but urge its adoption as a stopgap measure. On the other hand, John Fryer has suggested that poverty in Canada would be eliminated at an annual cost of 1.6 billion dollars a year if the income levels of all families and individuals now receiving less than $3,000 and $1,500 respectively were raised to these respective levels.[33] This amount, he points out, is less than the annual Canadian defence budget.

John Morgan supports this notion too, believing that each human being has a right to a decent living, irrespective of his work or the "price tag" on his job. This he believes, could be achieved by guaranteeing a minimum annual income that "could be determined by filling out an income return form. . . . Those who have less than the guaranteed income get something, those who have more, lose something. It is called the negative income tax."[34]

The adoption of the guaranteed income is also urged by John K. Galbraith.

> I venture to think that the time has come . . . for guaranteeing minimum levels of income for those who, for whatever reason, do not earn enough for decent survival. . . . Canada pioneered a system of family allowances on this continent, and related these to her income tax. It would be an excellent thing . . . to experiment with the guaranteed income or negative income tax and show us [the United States] how it can best be done.[35]

To alleviate poverty occasioned by technological unemployment, some American writers have suggested either greatly increased unemployment benefits or even payment of full wages without time limit to those thrown out of work. Critics of this type of proposal suggest that its adoption is not only politically impracticable, but is exceedingly unhealthy to the development of human incentive. "Work is not only . . . a 'disutility' as conceived by the classical economist. It is, if not always a pleasure, the basis of self-respect and a dignified life. There is no cure for unemployment or underemployment except employment."[36]

A Dissenting View on the War on Poverty

To this point, an unchallenged assumption has been that poverty in Canada is of such magnitude and severity that nothing short of a nationwide mobilization of resources will ameliorate the condition. John Madden challenges this view and argues that in order to justify a nationwide war on poverty, the advocates who wish other citizens to foot the bill must "prove that an extension of government power over individual incomes and property is warranted both by the significance of the problem and a degree of certainty that the projected expenditures will actually do the job they are expected to do."[37]

Where, Madden asks, is the dividing line between poverty and non-poverty? What is the threshold beyond which government (public policy) should take over? We adopt this technique in dealing with unemployment, why not poverty? "It is utopian to believe that public policy can ever ensure that nobody falls below some acceptable minimum since in a dynamic society that minimum will continually shift up. . . . What is the minimum level of income and what proportion of the population must be below this particular level before a public policy is necessary?"[38]

Madden does not deny that poverty exists, nor does he contemplate leaving those in such a condition to wallow in their poverty culture. But, he argues that before any public policy with respect to low income families and individuals can be instituted, it must be selective, not all-inclusive, and it must be based on a far more substantial body of knowledge than is now available. He believes that poverty today is associated with low levels of education thirty to fifty years ago, lack of jobs in the depression of the 1930's, inflation in the early fifties, past lack of medical and hospital insurance, and with the failure of society to enable women to make their contribution to family income. Do these require public policy to solve?

> I conclude that the poor would be no better off if they had the higher income needed to enjoy the tastes of the well-to-do and they are deluded by advertisers into thinking so. What we require is a public program, a vast advertising campaign to convince those with lower incomes that their way of life is best.[39]

CONCLUSION

In this paper, some of the facts and characteristics of poverty in Canada have been exposed, but many problems and questions remain. As yet, there is no clear-cut agreement on how to measure poverty, but the basic barometer is income. Even so, the minimal poverty-line figures vary from $2,000 to $3,000 in the case of urban families; from $1,000 to $1,500 in the case of individuals; and from $2,500 (gross and net) to $3,750 (uneconomic, rather than poverty) for farmers.

The causes of poverty are complex, but they appear to be related to income, education, health, and employment opportunities. Each of these factors is important and each can have a "domino effect" on the others. But it is difficult to select one factor as always paramount and the others as always secondary. For example, a high level of education need not lead to good health and higher than average income, but the opportunities are clearly better than with a low level of education. Similarly, good health does not imply a high level of education or income, but with poor health the opportunities of receiving the latter are less. On the other hand, a combination of good health and high education levels creates much greater opportunities for an above average income than inadequate levels.

In all of the programs and solutions outlined in this paper, the emphasis is on money; education and health act only as the foundation on which greater incomes are built. And, when people are armed with the minimum amount ($3,000 for an urban family) they can march ahead to achieve the decent life. Pending agreement on the qualities of a decent life, it appears that the income-education-health triumvirate is only a partial solution to the overall problem. True, the effective upgrading of these factors will no doubt raise, but not necessarily remove, the yoke of poverty, but one cannot help but speculate whether a decent life or the "good society" can be realized by these alone. What of the environmental medium in which the decent life is to be achieved? Can this life be achieved without the simultaneous planning of urban and rural environments that will guarantee the opportunity of participating in a wide variety of activities to fulfill human needs, to enjoy and be part of the urban or rural community, and to live in housing designed for people. These qualities, ever so more difficult to measure and evaluate, are foundations of a decent life, which money alone cannot guarantee.

Many other fundamental questions can be raised about the war on poverty. If the chief goal is to enable people to enjoy a decent life, then what constitutes a decent life? Does a decent life include a vacation, the opportunity of concerts or theatres, clothing, a new automobile every five years, life insurance, and many of the other diversions enjoyed by part of our society. What level of income is required for *modest* participation in these activities or the acquisition of some of these goods? Or is a decent life simply a healthy existence at the poverty level? Moreover, is a decent life the same for urban and rural people, for French-speaking and English-speaking Canadians, for Ukrainians and Italians? Clearly, the selection of the ingredients of a decent life is an individual choice, but the minimum enabling income should be closely and realistically evaluated.

For the geographer, the analysis of poverty raises many research questions. What are the relative regional variations in poverty? For instance, is $3,000 family income adequate for the decent life in Montreal; is it more, or less, in Moose Jaw, St. John's, and Owen Sound? Is it the same in the small towns and middle-sized cities of the Prairies as in Southern Ontario? Can a $3,000 annual income buy as decent a life in the downtown core of a large city as on the periphery of the city? Where, in terms of location, should the marginal families live to utilize their financial resources most effectively?

Likewise, what are the regional variations in rural poverty? Is it accurate to suggest that all farmers earning less than $2,500 gross income a year are disadvantaged? Or is this average misleading? Could it be that a geographic analysis would reveal areal variations that would make this crude yardstick meaningless? Are the causes of farm poverty the same all over? Is the poverty-class fruit farmer of the Niagara Fruit Belt faced with the same problems as the mixed farmer in the Manitoba Interlake, or the wheat farmer in the Western Interior? Is he

poor for the same reasons? Should the government "buy them out" because they are uneconomic? Or can many of these farms be made productive through better management and marketing arrangements, thereby maintaining the people on the land? The suggestion has been made that the farm poor should be retrained for alternate employment in the cities. Why not upgrade their agricultural training and keep them on the farm?

These are some of the questions that have emerged from this brief analysis of poverty in Canada. They are raised here to indicate the depth and diversity of the problem, not as nagging, petty, diversions. But no matter what the state of poverty, the solutions will vary from one part of the country to another in accordance with regional problems, resources and opportunities.

REFERENCES

[1] The terms poverty, poor, and disadvantaged are used interchangeably to indicate those people in need. Disadvantage is used more widely today because it connotes a relative condition and is removed from poverty in terms of income only.

[2] Whyte, Donald R., "Sociological Aspects of Poverty: A Conceptual Analysis," *Can. Rev. of Sociol, and Anthrop.,* Vol. II, November, 1965, p. 178.

[3] Riendeau, Robert, "Introduction," to "Poverty in Canada in 1964," Canadian Conference on Social Welfare, Hamilton, Ontario, 1964, p. 287, (Mimeo.).

[4] Menzies, M. W., *Poverty in Canada: Its Nature, Significance and Implication for Public Policy* (Winnipeg: Manitoba Pool Elevators, 1965), p. 23.

[5] Poverty can be divided into two types, objective and subjective. We are here concerned with objective poverty. Subjective poverty does not depend on the absolute level of survival as defined by society as a whole, but from a comparison of what an individual has and what he wants to have.

[6] Quoted in Wood, K. Scott, "Profile of Poverty in Nova Scotia," Institute of Public Affairs, Dalhousie University, November, 1965, p. 1, (Mimeo.).

[7] *Ibid.,* p. 1.

[8] Schlesinger, Benjamin, *Poverty in Canada and the United States: Overview and Annotated Bibliography* (Toronto: University of Toronto Press, 1966), p. 29. A fuller summary statement is found in, "A Profile of Poverty in Canada," *Labour Gazette,* Vol. LXVI, (4), 1966, pp. 220-221. In addition to Schlesinger's bibliography on poverty, see Paltiel, Freda L., *Poverty: An Annotated Bibliography and References* (Ottawa: The Canadian Welfare Council, 1966), 136 pp.

[9] For a review of the economic, social and education status of Canadian Indians, see Hawthorn, H. B., (ed.), *A Survey of the Contemporary Indians of Canada, Part I* (Ottawa: Queen's Printer, 1966).

[10] Reid, Timothy, "Canada's Educationally Handicapped: Who are They?" *Labour Gazette,* Vol. LXVI (5), 1966, p. 293.

[11] Hood, Angus M., "Canada's Largest Health Problem," in Laskin, Richard, (ed.) *Social Problems: A Canadian Profile* (New York: McGraw-Hill, 1964), pp. 396-401, states that the reduction and treatment of mental ill health and emotional instability is Canada's largest health program and that the provision of reasonable security and wholesome living conditions contribute to the mental well-being of the individual.

[12] For example, who consults the poor about poverty? The decisions on poverty emerge from the government and middle-class professionals, not from the poor.

[13] Schlesinger, p. 31. One assumes that these minimal incomes have been adequately tested to ensure decent standards of health, education, shelter, etc. It appears to this middle-class observer that these are appallingly inadequate from any point of view.

[14] Madden, John J., "Some Aspects of Poverty in Canada," Canadian Conference on Social Welfare, Hamilton, Ontario, 1964 (Mimeo.).

[15] Lappin, B. W., "Community Development and Canada's Anti-Poverty Programs," in Schlesinger, Benjamin, *Poverty in Canada and the United States: Overview with Annotated Bibliography.* (Toronto: University of Toronto Press, 1966), pp. 54-56.

[16] See Canadian Welfare Council, *A Preliminary Report on Rural Poverty in Four Selected Areas,* 1965.

[17] Canada Department of Forestry, *Economic and Social Disadvantage in Canada: Some Graphic Indicators of Location and Degree,* October, 1964. Seven maps of social and economic disadvantage are included in this series. I have not included maps showing the distribution of: incomes of male rural non-farm wage earners; incomes of male, urban wage earners; registrations for employment (urban and rural combined); and infant mortality rates (urban and rural combined).

[18] After this study was published federal and provincial ARDA officials recognized that the $2,500 figure was too low and have adopted $3,750 cash income as the poverty line. (Menzies, *Poverty in Canada,* p. 5, Footnote 11).

[19] More detailed information on the interrelationships between land use and physical and economic factors will be available through The Canada Land Inventory—a cooperative federal-provincial program administered under ARDA. The inventory includes assessments of land capability for agriculture (see Canada Department of Forestry, *Soil Capability Classification for Agriculture,* Report No. 2, The Canada Land Inventory), forestry, recreation and wildlife; information on present land use; assessments of social and economic factors relative to land use; and an agri-climatic classification (see *The Climates of Canada for Agriculture,* Report No. 3, The Canada Land Inventory). In addition to the studies cited above, land use capability maps for agriculture have been published for several areas (for example, Winnipeg Map Sheet—62H, and Sydney (N.S.) Map Sheet—11K and J.), and many land use maps compiled by the Geographical Branch, Department of Mines, Energy and Resources, are now available.

[20] Menzies, M. W., *Report of the Eastern Canada Farm Survey,* 1963, (Winnipeg: Hedlin-Menzies and Associates, Ltd., 1963). The highlights of this study are reported in Menzies, *Poverty in Canada, Its Nature, Significance and Implications for Public Policy.* In this study Menzies defined farms with a cash income of less than $2,500 annually as "non-viable" (non-commercial or uneconomic enterprises). However, he believes that this figure should be raised to include all farms with less than $3,750 cash income per year.

[21] The situation of a low-income farm in Saskatchewan is described by Pat J. Fogarty, "Poverty in Rural Areas with Particular Reference to Saskatchewan," Canadian Conference on Social Welfare, Hamilton, Ontario, 1964 (Mimeo).

[22] Berry, B. J. L., "Identification of Declining Regions: An Empirical Study of the Dimensions of Rural Poverty," in W. D. Wood and R. S. Thoman (eds.), *Areas of Economic Stress in Canada* (Kingston: Industrial Relations Centre, 1965), pp. 22-26.

[23] *Ibid.,* p. 36.

[24] *Ibid.,* p. 49.

[25] For a brief description of each of these programs, see "The War on Poverty," *Labour Gazette,* XLV (9), 1964, pp. 794-798.

[26] The Prairie Farm Rehabilitation Act of 1935 was not designed to deal with rural poverty as such but in some respects it achieved the same results, only on a smaller scale.

[27] Canada Department of Forestry and Rural Development, *Partners in Progress* (Ottawa: Queen's Printer, 1966), p. iv. Some of the highlights of ARDA programs are outlined in this book.

[28] See Canada Department of Forestry, *The ARDA Catalogue,* 1 April, 1965 to 31 March, 1966, (Ottawa: Queen's Printer, 1967), for a listing of the projects under ARDA.

[29] For a description of the types of development in the New Brunswick program, see Canada Department of Forestry and Rural Development *Northeast New Brunswick: Federal-Provincial*

Rural Development Agreements (Ottawa Queen's Printer, 1966). A summary of the problems in the Manitoba Interlake is found in, *Guidelines for Development: The Interlake Region of Manitoba* (Winnipeg: Queen's Printer for Province of Manitoba, 1966). The program is outlined in *Interlake Area of Manitoba: Federal-Provincial Rural Development Agreement* (Ottawa: Queen's Printer, 1967).

[30] Menzies, *Poverty in Canada.*

[31] *Ibid.,* pp. 31, 32.

[32] *Ibid,* p. 39.

[33] Fryer, John, "Poverty: Can We Afford It?", *Information,* Vol. 14 (2), 1966, p. 25.

[34] Morgan, John S., "The Real Issues in a War on Poverty," *Information,* Vol. 14 (2), p. 35.

[35] Galbraith, John K., "Soaring GNP Isn't Worth Ugly Sprawl, Littered Land and Sky," *The Financial Post,* 11 June, 1966, p. 7.

[36] Myrdal G., as quoted in Menzies, *Poverty in Canada,* p. 26.

[37] Madden. "Some Aspects of Poverty in Canada." p. 308.

[38] *Ibid.,* p. 319.

[39] Ibid., p. 326.

Poverty in the Maritimes

G. E. MORTIMORE

Middle-class Maritimers often fly into a rage when writers draw attention to poverty in the Atlantic region. Most are annoyed by what they consider crass invasion of private lives by professional poverty-hunters. Some have conditioned themselves to the poor persons down the road; they look without seeing. When an outsider focuses a spotlight on things they have come to take for granted, the comfortable citizens feel threatened.

Other Maritimers are eager to tell Canada about substandard living conditions existing in great stretches of their home region. They believe Atlantic poverty is a national problem that demands national action. But they are embarrassed by a widely-believed stereotype portraying New Brunswick, Nova Scotia, Prince Edward Island and Newfoundland as a large-scale Dogpatch-on-sea.

"In places like Ottawa or Toronto," a $10,000-a-year New Brunswick Government official said, "it seems as though a chill comes over the hotel clerk's manner when I sign the register, and he sees I'm from New Brunswick. I get the feeling he is mentally checking my hair for lice."

Nevertheless, the poverty is real. The first cabins where poor people live are only a few minutes away from the elm-shaded mansions of Fredericton, this miniature capital city on the Saint John River. The worst squalor begins 80 miles to the northeast and stretches along and inland from the Gulf of St. Lawrence and the Bay of Chaleur, where Jacques Cartier sailed in to anchorage 433 years ago. Here the French-speaking descendants of pioneer Acadians live in tar-paper cabins and small frame houses lining highways and dusty back roads for mile after mile.

Motorists speeding along Highway 8 or Highway 11 toward Tracadie, Shippegan or Bathurst see them sitting in rocking chairs outside their open doorways, among hens, small children and wrecks of old cars; or

● G. E. MORTIMORE *was formerly a reporter for* The Globe and Mail *(Toronto).*

This article is slightly condensed from "The Poverty is Real in the Maritimes," *The Globe and Mail,* 21 June, 1967, p. 31, and is reproduced by permission.

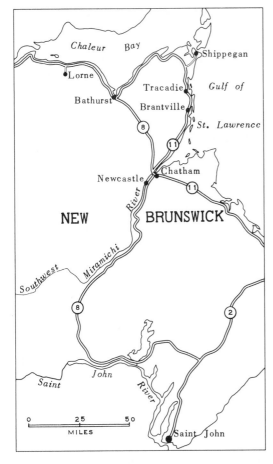

Figure 1

filling buckets of water from hand pumps, or piling pulpwood logs by the roadside, or cultivating the thin soil with horse-drawn harrows.

In St. Irenee, a village of 600 persons strung out along several miles of rough country road, nine miles inland from Tracadie on the Gulf of St. Lawrence, Mrs. Lena Benoit put down the broom with which she had been sweeping the floor and said in French: "I think he would sell if he could get a new place with a job near." She was talking about her husband Medric 47, and his possible attitude toward the federal-provincial anti-poverty plan for northeastern New Brunswick.

The plan will involve spending $100,000,-000 in 10 years—some of it for purchase of lands and houses in remote communities

and for resettlement of inhabitants in larger centres where there are jobs. St. Irenee and many other villages probably will disappear if the people listen to advice of counsellors hired to persuade them of the usefulness of resettlement and job training.

Medric Benoit works in the woods more than 100 miles away and comes home weekends. There isn't much space when the whole family is at home.

"I have had 15 children," said Mrs. Benoit, a pale, thin woman with glasses. "Eleven of them are alive. The youngest is 11 months; the oldest is 14. We have four beds."

The small living-room was clean-swept; it contained a table and a rocking chair, in which a little girl sat with a baby on her lap. A cupboard-sized bedroom adjoined; another bedroom was at the top of a ladder-like staircase directly overhead.

Mrs. Benoit went as far as Grade 2; her husband cannot read or write. "I've had 15 children; I never stop working," she burst out suddenly. "I got up at five o'clock this morning, I swept the house and I made my own bread. I don't stop."

A neighbour, Yvon Brideau, 27, said he might be interested in selling his house and small piece of land. He stays away from his wife and child all week at a pulpwood-cutting job to earn $60, out of which he pays $1.65 a day board and the balance owing on his $247 power saw.

He is married with one child. His education is above average; he went to Grade 9, as did his brother Alexis, 20, now his partner in pulpwood cutting. Alexis said he went to Montreal for a time and worked in a factory, but he hated the crowds and noise; he came home.

Many persons give up the struggle to scratch a living from submarginal farms and woodlots and uncertain small-boat fisheries. They load a few belongings in an old car or buy a train ticket and head for Toronto, Kitchener or some other city in Canada or in the New England states, where they often do poorly on the job market because they lack skills. The employment outlook is gloomy for a person who has four to eight years' education from a series of untrained and underpaid teachers in a one-room schoolhouse. Other persons stay home and live on a mixture of unemployment insurance, welfare assistance and seasonal job earnings.

New Brunswick's mining and pulp and paper industries are expanding; this is the reason why people like Mr. Benoit and the Brideau brothers are able to get jobs. But the jobs are often a long way from where the poor people live; the potential workers are barred from a large part of the employment market by geographical distance, by lack of job skills and by lack of incentive. They have been betrayed in the past by plans and promises that turned out to be illusory.

Also, wages in many of the jobs are extremely low. After counting board and power saw payments, Yvon Brideau's $60 a week is not significantly higher than the $33 a week unemployment insurance he used to draw.

Brantville, another village nearby, is better known than St. Irenee because it lies on Highway 11 linking Chatham-Newcastle with Tracadie. In Brantville, which has also been written off by economic planners as not economically viable, patches of new shingles and planks were seen on several small houses. Some houses were leaning, garbage-strewn wrecks; others were substantial and well-painted.

"A little money in these villages is like drops of water in the desert," Leo Doucet of the provincial Community Improvement Corporation had said earlier. "They start blossoming right away with better houses."

He had been talking about coastal communities a few miles north, where improved incomes for some of the fishermen had quickly been translated into brighter surroundings. The same seemed to be true of Brantville.

Joseph Savoie, 27, who keeps a small home-made-looking store and restaurant in Brantville and also commutes daily 30 miles each way to work on bridge construction in Chatham, was skeptical about the federal-provincial development program for the northeast.

"You got to be a millionaire before you get anything out of it," he said, rubbing the sleep from his eyes. His wife had roused him from bed at 9:30 p.m. to meet the visitors. He retired early so as to rise at 5 a.m. and drive to work.

"There's not a thing in it to help the poor people," Mr. Savoie said. "Say we want a factory down the shore or anything at all—we have to make the first investment. In Brantville, maybe we'd have to make a deposit of $6,000 in order to get a factory here. Why not get some industry to come in here? Surely there is one somewhere that would come. It would be work for 100 or 200 men. It would wipe out welfare.

"I left school when I was 16 years old, in Grade 10, because my mother didn't have $20 to buy books. My first job was in Berwick, Nova Scotia, 600 miles from here. Then I went to Toronto, to the DEW line, to Vancouver. But I couldn't get a job there driving a school bus.

"I think a man born in Brantville is entitled to live in his own place. How many people from Vancouver do you see living in Brantville? Why should people have to go somewhere else? It wouldn't take much— just one small factory?"

He said most fellow-residents of Brantville had the same sentiment. His store might be said to give him a vested interest in keeping Brantville alive. However, once the federal-provincial program for the northeast gets under way, the government will offer to buy stores as well as houses and farms, in order to induce people to move to larger centres where the jobs are.

In Lorne Settlement, six miles south of the Bay of Chaleur and Highway 11, reside people who have been isolated so long they have developed their own speech style, a dialect reminiscent of the West Indies. Most farms here have grown back to bush; the men travel long distances in old cars to jobs in mills and woods. The Lorners seemed quiet, dignified and self-assured as they talked in front of their small houses.

Elric Carrier, 40, keeps a tiny store, and passes bottles of pop to customers through the kitchen window. Lorne Settlement may be headed to extinction. But Mr. Carrier, his wife Gloria and their children are happy there. They don't want to leave.

Mr. Carrier cannot read or write. There was an interlude of a whole generation when Lorne had no school. Now it has a bilingual school: French in the morning, English in the afternoon. These people of English ancestry are perfectly at ease in French and in their own lilting, terminal-slurred variety of English.

Some of the Carrier children opened a small hut to introduce the two family pigs. Donald, 6, clutched a stuffed panda nearly as big as himself, which he had won on a punch board.

"My beah," he announced in Lorne style, dropping the terminal consonant, "won on punch-boa." As the Carriers waved good-bye, young Don's voice was the last heard, saying "My beah." As the car gathered speed past outlying houses and down the bush road, it occurred to me that Lorne would be a place to see again before it vanishes into history.

Regional Development in Canada

Leonard O. Gertler

Generically, regional development encompasses both theory and practice, has highly complex intellectual antecedents, extending from Isard to Goodman, and is divisible into a number of component concepts.[1] As an *economic concept* it is concerned with the problem of disparities in income, employment, welfare and rates of growth among regions.[2] As a *concept in geography* it deals with the spatial structure of a country as expressed in the distribution of people, economic activities and communities, and with the flows within and between regions.[3] As an *environmental concept*, regional development is concerned with releasing the potentials of the natural and manmade environment for the enhancement of the quality of life.[4] Viewed as a *political concept*, it has two related preoccupations:

(1) the easing of tensions between have and have-not regions within a country; and (2) the forecasting of local participation in the process of decision-making related to both the development and environmental aspects of each region.[5]

These components of regional development are closely related, but the emphasis we choose to give one or the other will vary with the scale of our concern, from the international to the local. In this paper the emphasis is on the *economic* aspects of regional development, which leads to the corollary emphasis, in Canadian terms, on the provincial or national context. It is only at these levels of government that the economic issues can be perceived in all their dimensions, and where there is capacity based on resources and jurisdiction, remedial action can be taken.

● Leonard O. Gertler *is Professor and Director of the School of Urban and Regional Planning, University of Waterloo.*

This essay is reprinted with permission of the author and the publisher from L. O. Gertler, *Prospects for Regional Planning in Canada,* (Montreal: Harvest House Ltd., 1971).

It is scarcely possible in the highly federalized Canadian state to consider the national aspects of regional development without getting involved in the relationship with provincial policies, and vice versa. But as a point of departure this paper will be concerned with the problem of interregional economic disparity as viewed at the federal level; and it will be concerned, much in the manner defined by Friedmann, with regional development policy: "a national policy for the economic development of regions, guided by objectives for the organization of the national space economy."[6]

This is a particularly auspicious time for a top-down view of regional development in Canada, because we appear to have reached an important cross-roads in federal policy. Coincident with the accession of the Trudeau Government, the past seven years of intensive effort have been appraised and apparently found wanting and a new strategy is in the process of crystallizing within the new Department of Regional Economic Expansion. The large fact that dominates this entire issue is that the interregional gap in living standards has been closed only slightly. The Economic Council of Canada has found that the index of disparity in personal income per capita for the Canadian provinces has declined by a little more than 1% (23.3% in 1961 to 21.2% in 1965.) The unemployment rate in all provinces has mercifully declined from its high level in 1961 (7.1% to 3.6% in 1966), but the regional incidence of unemployment is even more unequal; the Atlantic provinces which had 13.7% of unemployment in 1961 had 15% in 1966, and in Quebec the comparative figures are 36.1% and 37.8%[7] Structural unemployment persists. The Council report concludes that "the growth of the economy at the national level . . . is not in itself sufficient to secure major improvements in regionally balanced economic development."[8] The new federal policy, which involves a radical reorganization of its administrative and professional resources, takes up once more the task of slaying the dragon of regional distress.

398

II

The spirit of the federal policy was expressed by the Hon. Jean Marchand, Minister of Regional Economic Expansion. Speaking to "a cross-section of leaders of Central Canada's business establishment," he is reported to have made an emotional appeal to work with the government in solving the problem of regional disparities for the sake of Canadian unity.[9] In a paper presented to the Montreal Economics Association, Mr. Tom Kent, Deputy Minister of the new department, was no less explicit. He identified as the major motivations for current policy: "nationalism and social justice." "We cannot expect to have a united country, with a growing sense of national identity, if we complacently tolerate substantial region disparities." The goal of social justice he expressed in terms of closing the interregional gap in job opportunities and incomes.

The twenty-eighth Parliament has given consideration to two bills (C-173, *Government Organization Act,* 1969; Part IV. Department of Regional Economic Expansion, and C-202 [*Regional Development Incentives Act*], which together constitute the framework for the next federal thrust in regional development, i.e. "regional economic expansion."[10] These are on the record and so a detailed account of the new provisions will serve little purpose. It is of utmost importance, however, to obtain a synoptic view of this legislation as it does represent a *decidedly new departure in Canadian regional policy.* The two bills together with recent administrative changes, represent an approach which has six main elements:
(1) the bringing together in a single administration of all the regionally-oriented development programs of the federal government—Agricultural and Rural Development Act (ARDA), Prairie Farm Rehabilitation Act (PFRA), Maritime Marshland Rehabilitation Act (MMRA), Fund for Rural Economic Development (FRED), Area Development Agency (ADA), and the Atlantic Development Board (ADB). Legislation for ARDA and the farm rehabilitation programs will remain in effect, but the other legislation will be repealed, with a six-month overlap,

Figure 1

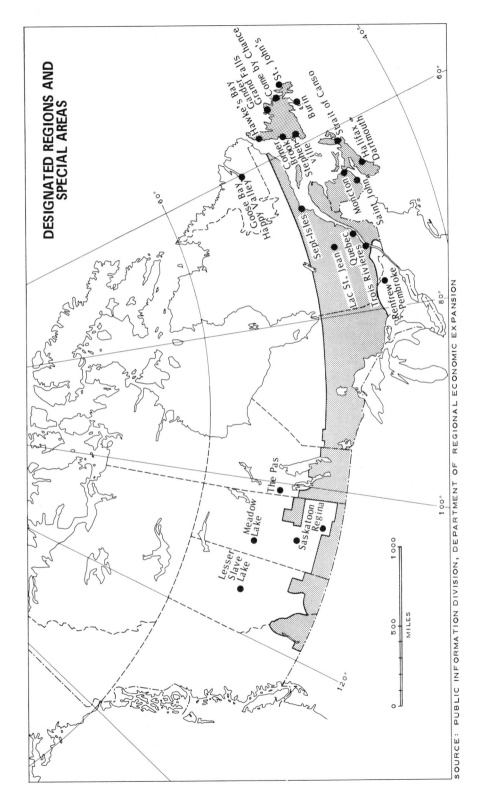

DESIGNATED REGIONS AND SPECIAL AREAS

SOURCE: PUBLIC INFORMATION DIVISION, DEPARTMENT OF REGIONAL ECONOMIC EXPANSION

July 1, 1969 to December 31, 1969, for the preceding legislation, the *Area Development Incentives Act.*[11]

(2) the formulation and implementation by the Department of Regional Economic Expansion, in cooperation with other departments, branches and agencies of the Government of Canada, of *"plans* for the economic expansion and social adjustment of special areas." The designation of special areas, which will be done in consultation with the provinces, will be based on the criterion of employment—they are areas of "exceptional inadequacy of opportunities for productive employment."[12] (see Figure 1)

(3) entering into agreements with the provinces (who may cooperate in the preparation of the economic and social plans), for the joint carrying out of such plans.[13]

(4) the provision of assistance through such agreements, for the implementation of the social and economic plans, in the form of (i) direct services and facilities of the federal government; (ii) financial contributions for programs and projects; (iii) grants and loans towards the capital cost of establishing, expanding, or modernizing facilities that will support economic expansion; (iv) the formation of joint federal-provincial corporations as development agencies; and (v) the giving of grants, loans or guarantees to commercial undertakings, whose establishment, expansion or modernization is essential to the successful implementation of an economic and social plan.[14]

(5) the provision of "regional development incentives," for the capital cost of establishing, expanding or modernizing secondary industries (which is a special case of 4 (v) above), ranging from $12,000 to $30,000 per created job, depending on degree of capitalization. The incentives are biased towards "growth poles" and urban-centred regions, by two features: the requirement that a designated region be capable of responding to the stimulus of capital incentives; and the provision that a designated region in all areas (except Prince Edward Island) must be at least 5,000 square miles—the size of the prototypical urban-centred region containing

a principal city, interacting with an area within a commuting radius of fifty miles.[15]

(6) the participation in the planning process of the provinces, persons, voluntary groups, agencies and bodies associated with the special areas. The Atlantic Development Council, set up by the Act to advise the Minister on all aspects of planning and development, constitutes a special application of this principle. To facilitate participation from the local level, through the province and the broad region to Ottawa, a devolution of administration has been provided to three regions—Canada East, consisting of the Atlantic provinces; Canada Centre made up of Ontario and Quebec; and Canada West, all of the remaining provinces.[16]

The "colour and tone" of the federal program is provided by policy addresses of the Minister and Deputy Minister of Regional Economic Expansion and by the administrative organization of the Department. There are three points of overriding importance. First, there is emphasis on a planning process—planning has been lifted to the high policy level with an Assistant Deputy Minister for Planning, coordinating the work of three offices: the Director of Economic Analysis, the Director of Social and Human Analysis, and the Director of Plan Formulation (see Figure 2). Second is the emphasis on urban growth poles—"finding and building on the points of strength in a regional economy."[17] This is a *reversal* of the regional development policy of the sixties, which through ARDA and ADA, led to investment in marginal and mainly rural areas.[18] The policy of regional economic expansion by starting with the cities will hopefully be able to work through the functional units in which solutions can be found—labour surplus in the country will be absorbed by new economic opportunities fostered in the city; and new development in the city will create demands for the resources and service industries of its tributary region. This urban emphasis, however, does not mean that the centre-periphery model[19] of the Canadian economy will be accentuated. On the contrary, the objective is to foster change *within* regions and "to ensure that the points

Figure 2

DEPARTMENT OF REGIONAL ECONOMIC EXPANSION

of strong industrial and urban growth are more widely dispersed across the country."[20] And the third policy emphasis is on the mechanism of growth in urban growth poles —on urbanization economies, industrial linkages, and manufacturing and service complexes. Incentives will not be automatic but used selectively to induce the location of what are called in Tom Kent's colourful phrase, "the herd leaders," the strategic activities that will set in motion the necessary growth processes. The federal-provincial agreements will be used to strengthen the urban location matrices by investing in infrastructure, including utilities, housing, health services, and the intellectual services of schools, research, technical and training facilities. Community and regional plans will assess the problems and potentials of slow-growth areas, and pin-point fields for investment that will produce the greatest development pay-offs.

In all of the foregoing the federal regional development policy as it is emerging is in the theoretical tradition of Perroux, Myrdal, Friedmann and Higgins. The future is staked on the premise that *poles de croissance* in slow-growth regions can become more formidable engines of growth, that "spread effects" will occur outwards from such centres to their surrounding regions, and that economic growth will occur in "the matrix of urban regions," spreading from higher to lower order centres in the urban system and ultimately extending across the entire space economy.[21]

III

To gain some perspective on our present status in regional development policy a brief historical digression may be helpful. This will be an impressionistic view, at best an historical hypothesis. The purpose of this pursuit is to explore the emergence of the overt policy concern with regional disparities at the federal level, and to throw some light on the main issues involved in the present struggle to narrow the disparities.

From the Olympian heights of an overview, there is little doubt that the dominant fact in the evolution of regional development policy in Canada was the Depression. Before that time the development of the country was under the spell of the attitudes reflected in Macdonald's "National Policy." The instruments of that policy—tariffs, land settlement, immigration, railways, freight rates, etc., had a decided regional incidence; the characteristic "centre-periphery" pattern of Canadian development was being forged. But the grievances of the periphery were generally muted at the centre. The major federal-provincial joint enterprise in the first half-century after Confederation, the Commission of Conservation (1909-1921), got involved in many basic Canadian issues, ranging from resource conservation and development to public health and town and country planning, but there is little reflection in its annual reports, or its other publications, of concern with economic disparities among the major regions of the country. Later, the boom of the twenties, with its advances in most industrial sectors, served to paper over the cracks in the Canadian economy.[22]

The trauma of the Depression produced a vast change in the orientation of public policy. The unequal regional incidence of the economic collapse led (i) to extraordinary efforts to shore up the regions that were hardest hit; and (ii) to a reappraisal of relationships, particularly fiscal, between the federal government and the provinces. The Prairie Farm Rehabilitation Act (1935) was the most typical expression of the first; and the Royal Commission on Dominion-Provincial Relations, and the system of equalization grants it inaugurated, was the vehicle for the second. Since that time, the "Canadian mind" has never been very far from the contemplation, at least in its subconscious, of the fact that "a distinguishing feature of the Canadian economy [is] that a very large proportion of the surplus—and taxable—income of the country is concentrated in a few specially favoured areas."[23]

At the policy level, however, the lag in responding fully to the regional fact has been prolonged. First the War intervened, with its inescapable preoccupations, and then the period of post-war reconstruction, understandably dominated by a Keynsian concern to avoid the hazards of the business cycle. This was the time of the White Paper on income and employment, emphasizing the manipulation of strategic national aggregates and the "shelf" of public works. During this period there were two major conferences which in retrospect, were important in the evolution of regional development policy: the National Conference on Post-War Reconstruction in 1945 and the National Conference on Renewable Resources in 1954. Through these national reappraisals the preoccupation of a staple-producing society with natural resources as such, was transformed, in a manner befitting an emergent industrial state, to a concern with resource development as an instrument of economic policy.[24]

The next period in the evolution of regional development policy were the few years 1959-61 associated with the Resources for Tomorrow Conference. It was a period of philosophic search. The vehicle was a complex structure of interacting groups consisting of the National Steering Committee of Resource Ministers, representing all provincial governments and the federal government; the Committee's Secretariat; research coordinators representing each of the renewable resource fields and "regional development"; advisory groups corresponding to each of these fields; the contributors of background research papers; and the Conference itself, which took place in October, 1961. The objectives of this intricate enterprise were, in the words of the Secretariat head, B. H. Kristjanson: "to define goals in their respective and joint resource development and management programs; to enunciate acceptable principles; and to define guidelines to action necessary to sustain an adequate rate of growth in the national, provincial and regional economies."[25]

The point of importance about the Conference is that it set in motion a number of new directions in development policy that were to exert an influence over the next decade. Three results are of particular interest. The first was the assertion of the interdependency of the use, development and conservation of renewable resources—water, agricultural land, forestry, etc.; and the corresponding need to achieve *integration at the administrative level* within both the federal and provincial governments.[26] The second was the appeal to a regional approach to economic development, both to facilitate "comprehensive development and use of resources," and "to reduce the range of disparity between regional rates of growth and limit imbalance between regions." Various administrative arrangements for this were suggested.[27] The third result, which proved to be of continuing interest and concern, was the emphasis on a participatory planning process, which would include: "(a) administrative structures that provide full opportunity for consultation with interests and regions concerned; (b) programs of public information and education that interpret resource policies; and (c) local committees to participate in planning resource developments in their own areas."[28] The work of the Resources for Tomorrow Conference together with the Senate Committee on Land Use in Canada, which conducted public hearings in the same period, prepared the way for the next major step in federal development policy, the Agricultural and Rural Development Act (ARDA).[29]

It is not the purpose of this paper to review and evaluate the regionally-oriented federal economic policies that emerged and were operational in the eight years between the passing of the ARDA legislation in May, 1961, to the enactment of the Government Organization Act, in March, 1969. Three recent appraisals, by Brewis, Buckley and Tihanyi, and the Economic Council of Canada, are helpful in this regard.[30] What is relevant to this narrative is the identification of certain features of the experience that led up to the present policy of regional economic expansion.

The first feature was the endeavour to make a direct assault (always through federal-provincial agreements, both formal and informal) on problems of resource use, economic development or living standards in the regions where these problems are found. For ARDA this was in designated "rural development areas," and in "special rural development areas," areas eligible for financial assistance from the Fund for Rural Economic Development (FRED) to carry out comprehensive development plans;[31] for the Area Development Agency (ADA) action was focussed on "designated areas," eligible for location incentives for secondary manufacturing and processing industries; and the Atlantic Development Board (ADB) was empowered to prepare a development plan for the Atlantic region. The paradox inherent in these arrangements was that for the country as a whole regional development policy was split into rural and urban components. The logic of ARDA's work led away from a concentration on land use and management to a consideration of the changes required in the entire social, cultural and economic environment that were necessary to overcome economic stress and poverty.[32] Solutions of rural poverty often pointed towards the cities, but ARDA's mandate did not extend that far. On the other hand, ADA, which was empowered to act in urban areas, gave first priority to relieving unemployment and its program of location incentives did not necessarily relate to the building up of regional growth points as a means of drawing off surplus labour in the rural areas.[33] The Atlantic Development Board with its very broad mandate could presumably overcome this dichotomy in the Atlantic area, but in practice its development strategy in some critical respects was constrained by the on-going regional policies, for example location incentives for slow growth areas of high unemployment would work against an alternative strategy giving greater impetus to growth poles.

Preparing and implementing a comprehensive regional development plan of the type initiated by ARDA in the Mactaquac and Northeastern regions of New Brunswick and in Eastern Quebec, involves some very special administrative challenges. For example, the Mactaquac Plan, with which the author was associated, required the co-ordinated action of federal and/or provincial functions representing forestry and rural development, finance, fisheries, historic sites, Indian affairs, transport, parks and recrea-

tion, water pollution control, agriculture, municipal affairs, education, highways and public power.[34] The orchestration of these diverse functions is, to say the least, not easily achieved; failure to achieve it in any respect could spell disaster for a development program. The delegates of the Resources for Tomorrow Conference wrestled with this problem, and came up with the suggestion of an interdepartmental council, at both federal and provincial levels, to co-ordinate policies and programs.[35] A planning secretariat attached to the Prime Minister's office was attempted by the Ottawa government, but its efforts did not appear to be marked by conspicuous success.[36] The problem remained unsolved during the 1961-69 period.

When the ARDA legislation was introduced into the House of Commons, the Minister responsible stressed the participatory nature of the concept of rural development, which would involve governments, groups and individuals.[37] This was not simply a matter of ideology, but a pragmatic recognition of what it takes to carry through fundamental economic and social changes within Canadian society and within a federal state.[38] The Canadian effort of rural development was to be more in the decentralizing tradition of Lilienthal than the bureaucratic centralism of Ickes.[39] Federal-provincial agreements were to be the keystone of the program, and there was much talk about "grass roots" activity, upwards from the county level. The greatest achievement along these lines appears to be the placing of federal-provincial planning and development on an operational basis, taking different forms in accordance with varying regional conditions, e.g., a federal-provincial task force in Northeastern New Brunswick, the Economic Improvement Corporation in Prince Edward Island, a regional planning board in Eastern Quebec, and so on. The follow through of the planning process from the provincial to the local level has been less certain and less consistent, with the degree of participation extending from intense involvement in Eastern Quebec, where 200 local committees and eight zonal committees played a direct role in the formulation of the Plan, to the Mactaquac region, where proposals to provide for local participation on the regional agency (The Community Improvement Corporation) were not implemented.[40] This remains a soft spot in the federally-initiated development program.

Looking at the development program of the sixties from the point-of-view of identification of the persisting issues, one additional feature needs to be recalled: the focus on elimination of rural poverty, and as a corollary of this the narrowing of the interregional income gap.[41] It is clearly too early, in relation to programs, which in many cases are long range in conception and impact, to make anything like a final judgement. But some tentative evaluations have been made. Buckley and Tihanyi have expressed doubt concerning the effectiveness of ARDA projects "as vehicles of income transfer in favour of the needy," stressing the reluctance of provincial departments, concerned with economic success, to confine projects to low-income groups.[42] A similar view has been recently asserted by the Economic Council of Canada with respect to regional economic policies generally: "The formulation of federal economic policies with significant regional implications has a long and varied history running through a number of stages and influenced by a wide variety of considerations. But the stark fact remains that the historical mix of market forces and public policy has not resulted in any significant narrowing of regional income disparities."[43]

In retrospect this last period, characterized by the first clear emergence of a regional theme in federal economic policies, was a period of experimentation. The general direction was towards comprehensive development planning, but this remained split into rural and urban components. The necessary integration of federal development functions was incomplete. Some important new ground was broken in evolving the style and method of joint planning by two levels of government; although participation at the local level was generally not sustained. At the end of the period no dramatic changes in interregional disparities were recorded.

The recently inaugurated program of regional economic expansion has united the rural and urban aspects of development policy, but the other issues here delineated, remain as a heritage and a challenge that must be faced.

IV

The issues involved in the next phase of regional development in Canada can be discussed in terms of (i) what to do about the persistence of disparities in living standards and resource utilization, and (ii) the problems of administration, i.e., of coordinating government functions in comprehensive development planning, and of involving in the planning process, decision-makers at all levels as well as the people in the affected regions.

The regional development measures that have, up to the present time, been fostered at the federal level may be summarized as investment in resources, investment in people, investment in infrastructure, and incentives to secondary and processing industry. These activities may be further characterized, in the terminology of a recent O.E.C.D. study as providing the *conditions* for development. The conditions for development might, in turn, provide a basis for a *development policy*, i.e., actions which foster a response from the outside world. "It means trying to interest businessmen from outside in locating their industries within the areas, it means demonstrating to them the economic and other advantages of doing so."[44]

Viewing the Canadian regional development experience in this light, one cannot escape the observation that we have only skirted around the fringe of the economic problem to which regional development addresses itself. There has been so far a conspicuous missing link in both the theory and practice of regional development. For if we approach the issue of disparities from the posture of a development policy that is trying to influence private decision-makers, then certain characteristics of the Canadian economy have to be understood and reckoned with. It is doubtful, for example, that regional development policy has ever seriously faced up to the highly centralized structure of Canadian industry. The Task Force on the Structure of Canadian Industry has recently commented on the trend to giantism in Canadian industry: 570 companies with assets of $25 million or more and representing about 2% of all firms had, in 1964, 46% of the assets of all companies, 35% of the sales, 53% of the taxable income, paid 67% of the dividends and accounted for 51% of the net internal cash flow.[45] In considering the implications of this economic concentration for public policy, the Task Force makes an observation that would seem to be of utmost significance for those concerned with shaping the national space economy. The oligopolistic position of the large firm, and its relative freedom from the rigours of the market, leads to the paradoxical consequence that such firms are both highly autonomous in decision-making and "more susceptible to government policies."[46] Porter draws this bow even further. He identifies an economic elite that presides "over all major segments of the corporate world in an extensive interlocking network," and this elite wields substantial power:

> They are the ultimate decision-makers and coordinators within the private sector of the economy. It is they who at the frontiers of the economic and political systems represent the interests of corporate power. They are the real planners of the economy, and they resent bitterly the thought that anyone else should do the planning. Planning, coordinating, developing, taking up options, giving the shape to the economy and setting its pace, and creating the general climate within which economic decisions are made constitute economic power in the broad sense. Nowhere is this power exercised more than in the small world of the economic elite.[47]

The strong presumption from the foregoing is that if the Canadian space economy in the period of mature industrialization still exhibits pronounced "centre-periphery" characteristics, and tenacious disparities among its regions, then it is so because the economic elite prefer it so. There is some

support for this in economic theory—Friedmann stresses the displacement of the production factors from the centre to the periphery, and the favourable terms of interregional trade exacted by the centre.[48] There is some empirical evidence in private economic organization—for example the failure of lending institutions to build an effective national network for administering housing mortgages.[49] There is a continuing thread of felt injustice in Canadian folk lore—a recent conference on northern development attended by "a cross-section of leaders and senior executives of Central Canada's industrial and business establishment" heard the financial institutions accused of having a "banana belt" complex and discriminating against the north;[50] the Atlantic Provinces Economic Council recently described the status of the Atlantic region as "colonial."[51] Could it be that the indifference to local areas often attributed to central government bureaucrats might in Canada be more appropriately ascribed to a centralized economic elite?[52] Is the imbalance in the development of our regions likely due to an indifference bred within, not the Chateau pool, but the capacious comfort of the National Club?

The conclusion that begins to emerge— the critical importance of the relationship between government and industry in a program of regional economic development— does not rest solely on such Stygian observations as the foregoing. There is a more positive side to the relationship. In Part II it was indicated that the emerging federal development strategy appears to be placing considerable reliance on the stimulation of regional growth poles by fostering the development of industrial complexes. This approach arises out of a certain school of economic thought, and out of the development experience of the Mezzogiorno in southern Italy. Since 1961 there has been an ambitious application there of the industrial growth pole concept, based on a machine industry complex, and guided by the "minimum quantum" principle—the attainment of the threshold of major, auxiliary and subsidiary industries, which as a complex provides the elements of scale, specialization and integration essential to effective competition on an international basis. In the Italian case the "minimum quantum" consisted of 25 different types of industrial establishment.[53] We may in the near future have an opportunity to see this approach worked out in Canadian terms; the research group, Italconsult, which prepared the Mezzogiorno report, is participating in a federally supported, provincially endorsed, study of a multiple industry complex in St. John, New Brunswick.[54]

The particular point that it seems necessary to stress is that the emerging policy of regional economic expansion will require unprecedented commitments from private finance and industry. The installation of a viable industrial complex is a major undertaking—and the location decisions associated with it will be of quite a different order from the occasional response of an individual firm to location incentives. The question that is inevitably begged is how, by what mechanism, will this most exacting task be accomplished? Three possibilities suggest themselves. One is that the "quantum" principle, lubricated by capital incentives, will actually work—linked industries as a group will find an economic inducement that as separate enterprises would elude them. This is the "market-carrot" approach. A second is that the government program will build up an effective countervailing power to private economic power and find ways of eliciting "cooperation." This is the "big stick" method. And the third approach would rely upon the rational response of private economic interests to competent planning in a state headed towards a post-industrial society in which there will be a suf-ficient accumulation of wealth to permit a greater investment from the rich regions in the poorer regions without a rigorous application of "the harsh criterion of the national growth rate;"[55] and a state in which a more equitable regional sharing in the national product may become the price of national survival itself. The first of the alternatives has the difficulty that quite a high cost would have to be paid to find out if the theory works. The second approach is unpleasant. The third is preferred by the author.

Whichever of the three approaches becomes operative, one thing is certain: the new policy will demand a very high level of technical and administrative competence. The bringing together of the urban and rural elements of regional development and the administrative reorganization, with its clear articulation of functions and emphasis on professional skills, are encouraging steps. There are two critical aspects of the administrative edifice, however, that appear to be shaky: the means by which the Department of Regional Economic Expansion will obtain the cooperation of "other departments, branches and agencies" in the formulation of economic and social plans; and the means by which the Department "shall make provision for appropriate cooperation with the provinces in which special areas are located and for the participation of persons, voluntary groups, agencies and bodies in those special areas."[56] On these vital relationships, which represent the two ends of the planning process in a federal state, may we not, after almost a decade of experimentation, be forgiven the expectation of a little more than another decade of pragmatic groping?

REFERENCES

[1] Reference here is to the spectrum from economic issues to environmental and social concerns, as represented by Isard's work on regional analysis; Walter Isard, *Methods of Regional Analysis: An Introduction to Regional Science.* New York: John Wiley & Sons, 1960, and Paul Goodman's 1966 Massey Lectures on CBC, entitled *The Empty Society,* in which he calls for a policy of regional reconstruction as an alternative to the ills of urban America.

[2] The Second Annual Review of the Economic Council of Canada focussed on two aspects of this concept: (i) "the importance of reducing the relative disparities in average levels of income as they presently exist among the regions," and (ii) "the need to assure that each region contributes to total national output, . . . on the basis of the fullest and most efficient use of the human and material resources available to the region." Economic Council of Canada, *Second Annual Review: Towards Sustained and Balanced Economic Growth.* Queen's Printer, 1965, p. 99.

[3] Richard S. Thorman and Gerald McGrath. "Regional Statistics and Their Uses: A Geographical Viewpoint," *Conference on Statistics, 1964*. (Editors, Sylvia Ostry and T. K. Rymes), University of Toronto Press, 1964.

[4] Leonard O. Gertler, "Regional Planning and Development," *Resource for Tomorrow Conference, Background Papers*, Volume II. Ottawa: Queen's Printer, 1961.

[5] John Friedmann, *Regional Development Policy*. Cambridge: The M. I. T. Press, 1966, pp. 18, 19, 24, 25 and 49.

[6] *Ibid.*, p. xvi.

[7] Economic Council of Canada, *Third Annual Review: Prices, Productivity and Employment*. Ottawa: Queen's Printer, 1966, pp. 244-248.

[8] *Ibid.*, p. 245.

[9] *Globe and Mail*, June 9, 1969.

[10] *Bill C-173*, An Act respecting the organization of the Government of Canada and matters related or incidental thereto. As Passed by the House of Commons, 14th March, 1969. Ottawa: Queen's Printer, 1969, and *Bill 202*, An Act to provide incentives for the development of productive employment opportunities in regions of Canada determined to require special measures to facilitate economic expansion and social adjustment. First reading. May 20, 1969. Ottawa: Queen's Printer, 1969.

[11] *Ibid.*, Bill C-202. p. 12.

[12] *Op. cit.*, Bill C-173. Section 25(1), 24.

[13] *Ibid.*, Section 26.

[14] *Ibid.*, Section 28.

[15] *Op. cit.*, Bill C-202, Section 5, Section 3 (2) (6), and Section 3(1). The minimum size of a designated region was reduced from 10,000 to 5,000 square miles in 1970-71 (Office Consolidation of the Regional Development Incentives Act, 1970-71, c. 10, 1971).
Jean Marchand, "Incentives for Regional Economic Expansion," Remarks by the Minister for Regional Economic Expansion, the Honourable Jean Marchand, relating to the introduction of the Regional Development Incentives Bill, May 27, 1969. Department of Regional Economic Expansion, May 1969.

[16] *Op. cit.*, Bill C-173. Section 25(2).

[17] Notes For Remarks by Mr. Tom Kent, Deputy Minister of Regional Economic Expansion, to the Annual Meeting of the Montreal Economics Association, the Windsor Hotel, Montreal, Quebec, Tuesday, May 20, 1969.

[18] This is a point that has been stressed by Brewis in his recent appraisal of regional policies. For example, he states: "In its efforts to achieve a change in the economic structure of impoverished areas, ARDA is handicapped by the lack of any comparable agency operating on the industrial front. The problem of rural areas cannot be solved in isolation; rural and industrial areas have to be seen in conjunction with each other." T. N. Brewis, *Regional Economic Policies in Canada*. Toronto: The Macmillan Company of Canada Limited, 1969.

[19] Friedmann has a succinct definition of the centre-periphery concept as it is used in economic theory: "a conceptual model that divides the space economy into a dynamic, rapidly growing central region and its periphery. The growth of the centre is viewed as being subsidized in part by the periphery." (Friedmann, *op. cit., p.* xvi)

[20] *Op. cit.*, Kent. pp. 10 and 13.

[21] Gunnar Myrdal, *Economic Theory and Underdeveloped Regions*. London: University Paperbacks, Methuen, 1965. Myrdal gives an account of the theory of "spread effects" that is a close approximation of the present aspirations of federal policy: "It is natural that the whole region around a nodal centre of expansion should gain from the increasing outlets of agricultural products and be stimulated to technical advance all along the line.
There is also another line of centrifugal spread effects to localities farther away, where favourable conditions exist for producing raw materials for the growing industries in the centres; if a sufficient number of workers become employed in these other localities even consumer goods will be given a spur there. These, and all other localities where new starts are being made and happen to succeed, *become in their turn, if the expansionary momentum is strong enough to overcome the backwash effects from the older centres, new centres of self-sustained economic expansion.*" (author's italics). p. 31.
Op. cit., Friedmann's observations on the structure of agricultural activity in urban-centred regions is also very germane to the present posture of federal policy. He stresses the forces that make agricultural production more efficient and farm incomes higher in the vicinity of cities. p. 30.

22 F. J. Thorpe, "Historical Perspective on the 'Resources for Tomorrow' Conference," *Resources for Tomorrow Conference, Background Papers*, Volume 1. Ottawa: Queen's Printer, 1961. pp. 1 to 6. See also Commission of Conservation, Canada, *Report*. Ottawa: King's Printer, 1910-1919.

23 Royal Commission on Dominion-Provincial Relations, *Book II, Recommendations*. Ottawa, 1940. p. 75.

24 *Op. cit.*, Thorpe, pp. 11, 12.

25 Introduction, *Resources for Tomorrow Conference Proceedings*. Ottawa: Queen's Printer, 1962.

26 *Ibid.*, Research Coordinators' Joint Statement, p. 479.

27 *Ibid.*, Research Coordinators' Individual Statements, p. 480.

28 *Ibid.*, p. 479.

29 *Op. cit.*, T. N. Brewis. See section on The Orgins of ARDA, pp. 95-107.

30 *Ibid.*

H. Buckley and E. Tihanyi, *Canadian Policies for Rural Adjustment* Prepared for the Economic Council of Canada. Ottawa: Queen's Printer, 1967; and Economic Council of Canada, *Fifth Annual Review: The Challenge of Growth and Change*. Ottawa: Queen's Printer, 1968.

31 *Op. cit.*, Brewis, pp. 108, 114, 115.

32 On this point the author tends to support the interpretation of Brewis rather than Buckley and Tihanyi. The first emphasizes the *direction* of federal policy, from an emphasis on "the quality of land" to comprehensive development planning with an emphasis on "education, training, and the provision of employment in non-primary occupations"; while the second tends to underplay the direction of policy and to demonstrate through the analysis of ARDA expenditures that the "more comprehensive attack . . . proceeds but slowly."

33 *Op. cit.*, Brewis, p. 249, for the difficulties of the rural context of ARDA. Dr. Andre Saumier, Assistant Deputy Minister, Department of Regional Economic Expansion, recently commented on the interdependence of urban and rural aspects of the regional economy. He is reported as saying that "sweeping plans for redevelopment of poverty-stricken rural areas should not be launched unless the people displaced by the programs can be absorbed elsewhere in the economy." (Comments to the Senate Committee on Poverty reported in *The Globe and Mail*, May 21, 1969.)

34 See *The Mactaquac Regional Development Plan Summary Report*. H. G. Acres and Co. Ltd. for the New Brunswick ARDA Committee, 1965.

35 *Op. cit.*, Proceedings, Resources for Tomorrow Conference, Vol. 3, p. 479.

36 The author was privileged to hear an appraisal of this experience at a closed seminar in Ottawa, Fall, 1968.

37 House of Commons Debates, Ottawa, January 25, 1961. pp. 1403-1407.

38 A recent comment on the Development Plan of Prince Edward Island is illustrative of this point: "The only simple, and perhaps the only certain way of consolidating the island farms into holdings of economic size would be through direct state action. . . . There is no possibility that Canadian public opinion would accept methods of that sort. . . . The democratic method which the plan involves . . . as the way to achieve the results you want, no doubt is less efficient but also less likely to have unwanted side effects." (W. A. Wilson, "Watchful Informality is the Word Among P.E.I. Legislators," *The Montreal Star*, June 12, 1969, p. 45).

39 Lilienthal's journals reveal his clash with Secretary of Interior, Harold Ickes, as one of the classic encounters in Western society on the centralization-decentralization issue in government. The flavour of this encounter is indicated by Ickes exasperated outburst in a meeting with Lilienthal, on October 23, 1939. "Some people believe in decentrealization . . . break things down to the constables; others believe in good organization. You can have your choice; I prefer good organization." (*The Journals of David E. Lilienthal*, Volume One, The TVA Years, 1939-1945. Harpers and Row Publishers, 1964, p. 138).

40 *Op. cit.*, Buckley and Tihanyi, p. 158 for Quebec reference. The Board of the Corporation consists entirely of public servants.

41 The attack on rural poverty was made more explicit in the second general agreement. *Federal-Provincial Rural Development Agreement*, 1965-70, Canada, Department of Forestry and Rural Development, 1966.

42 *Op. cit.*, Buckley and Tihanyi, p. 99.

43 *Op. cit.*, Economic Council, *Fifth Annual Review*, p. 177.

[44] L. H. Klassen, *Area Economic and Social Redevelopment.* Organization for Economic Cooperation and Development, Paris, 1965, p. 92.

[45] Task Force on the Structure of Canadian Industry, *Foreign Ownership and the Structure of Canadian Industry,* Privy Council Office, Ottawa, 1968, pp. 124-127.

[46] *Ibid.,* pp. 308, 309.

[47] John Porter, *The Vertical Mosaic,* University of Toronto Press, 1967, p. 255.

[48] *Op. cit.,* Friedmann, pp. 12, 13.

[49] The Hellyer Task Force had some sharp things to say about this problem:
"Particular word should be made, too, of the need to ensure an adequate regional distribution of mortgage funds within the overall national allotment. In its travels across the country, the Task Force encountered several instances where communities and even entire regions seemed almost totally dependent on the direct lending activities of CMHC for mortgage financing . . . The people of the Lakehead, for example, do not deposit their savings in their institutions with the thought only of helping to finance high-rise apartments in Toronto." (*Report of the Task Force on Housing and Urban Development.* Ottawa: Queen's Printer, 1969, p. 26.)

[50] Rudy Platiel, "For the Sake of Unity, Work to Solve Regional Disparities, Businessmen Told," *The Globe and Mail,* June 9, 1969, p. 11.

[51] Lyndon Watkins, "APEC Proposes Full Union of Maritimes," *The Globe and Mail,* April 8, 1969, p. B1.

[52] *Op. cit.,* Friedmann, p. 25.

[53] Nino Novacco, "Une Nouvelle Approche en Matiere d'Industrialisation Regionale: Le 'Pole' Industriel Dans Les Pouilles," *Ponencias del VII eme Colloque de L'Association de Science Regionale de Langue Francais,* 1968.

[54] John Carroll, "New Brunswick Regionalism Gains Strength as Ottawa Prepares Incentives," *The Globe and Mail,* May 20, 1969.

[55] *Op cit.,* Friedmann, pp. 7, 8.

[56] *Op cit.,* Bill C-173, Section 25 (1) and (2).

Urban Structure and Regional Development

GERALD HODGE

It causes hardly a ripple of interest today when one suggests that modern economic development takes place chiefly within an "urban-industrial matrix."[1] And "growth poles" have become an accepted part of the lingo of those involved in making regional development policy.[2] To state that we need to pay special attention to the spatial dimensions of economic development has the aura of stating the obvious, at least to regional scientists. However, if we examine our expertise in this realm, we find little that could help a policy-maker formulate a regional policy using the notion of an "urban industrial matrix" or "growth poles."

The theoretical knowledge we have about the role and response of urban centres in regional economic development tends to be of two types: either that "urban systems" exist in some form within regions (e.g., Christaller, Lösch, Vining, Berry) or that the process of growth of centres within a regional system is a complex function (e.g., Hassinger, Northam, Borchert, Friedmann). If we are to apply our knowledge to make more effective policy about cities and re-

● GERALD HODGE *is Associate Professor, Department of Urban and Regional Planning, University of Toronto.*

Reprinted from *Papers,* Regional Science Association, Vol. 21, 1968, pp. 101-123 with permission of the author and the editor.

gions, we shall need a convergence of these two types of knowledge. That is, to know more about how city systems perform in particular regional contexts and vice versa.[3] It is toward such a convergence that the studies reported on here have been striving.

To date, studies of the urban systems of four different regions in Canada—Saskatchewan, Eastern Ontario, Prince Edward Island, and Nova Scotia—have been completed. Two basic premises of these studies must be made clear at the outset. First, the studies all had a normative orientation; we were looking for statistical bases upon which to make judgements about the prospects of "growth centres" for policy purposes. This, of course, affected the choice of variables studied and the analytical questions posed. Second, no limiting definition of an urban system was posited. It was assumed that all the centres of a region must be considered when the prospects for growth of any single centre were being examined. In other words, all centres in a region exist as part of a system such that the effects of social and economic change redound among them. This then is a much broader perspective than that assumed in most of our current area-development programs, which tend to see the space economy either as single growth centres or as stagnant undifferentiated areas. This has allowed us to pursue more closely relationships between "growth poles" and depressed areas.[4]

All studies sought the structural features common to centres in the individual regions through factor analyses. The structural matrices were then examined through multiple regression analyses with regard to their association with growth performance among a region's centres. What seems to be of general importance emerging from these studies, and thus the aims of this report, are the comparisons that can be made among urban structures in the several regions and their relation to economic development. Succeeding sections of this paper describe the analytical technique, compare the data arrays, the structural features, and the performance analyses and indicate relations between urban and rural structures. But first, as a general orientation, the regions and their social and economic characteristics are described briefly.

THE REGIONAL SETTINGS

Saskatchewan: that part of the province roughly south of the 54th parallel to the United States border within the Great Plains, covering about 122,000 square miles with most of the 476 incorporated places functioning as rural trade centres; an area of very prosperous grain farming, extensively mechanized, and with large farms; recent oil, gas, and potash developments have enhanced the economic base; personal income $2,238 per capita in 1966.

Eastern Ontario: an 18-county area commencing east of Toronto to the Quebec border covering about 20,000 square miles in which the economy is a mix of agriculture, natural resource exploitation, manufacturing, and tourism; there are 80 incorporated centres performing various roles; agriculture in much of the region is depressed; average annual income $1,400-1,600 per capita; a *downward transitional* region.[5]

Prince Edward Island: the entire province comprising about 2,500 square miles with 25 incorporated centres most of which are farm trade centres; the economy is predominantly agriculture; per capita personal income $1,376 in 1966; a *downward transitional* region.

Nova Scotia: the entire province with its 40 incorporated centres in an area of about 21,000 square miles; fishing, coal mining, farming, manufacturing industry, and transportation comprise the main economic base; personal income $1,576 per capita in 1966; an *upward transitional* region.

THE ANALYTICAL PROBLEM AND THE APPROACH

If, as noted above, urban centres are vital to a region's development, then, for public-policy purposes, the problem in an already settled region consists in distinguishing the prospects for growth of all centres in the region. In approaching this problem, one must, of course, quickly acknowledge the

411

complexity of individual centres and of the regional system. The problem can be restated in terms of an inductive analysis; that is, to examine the many ways in which centres differ from one another and to try to determine the approach that might be "strategic and relevant" to their growth. Such an approach, though somewhat unstructured, has the advantage of not requiring a priori specifications of relationships in a realm where many variables are involved and no tenable theory exists.

Two basic postulates can be used to guide the analysis: first, urban places in relatively homogeneous regions do differ in strategic ways that affect their ability to survive; and, second, urban places do not constitute an illogical pattern of an infinite number of elements without function or meaning.[6] Urban centres may be thought of as groupings of persons interacting within a social, economic, and physical environment. The result is particular combinations of people, housing, jobs, etc., reflecting the structure of individual centres, occurring with considerable regularity and uniformity. These structural features can be used to characterize the centres under study.

Two objectives guided the analysis and thereby made it into a two-stage process: (1) to identify the underlying structural features that distinguish the development of the different urban centres in the region; and (2) to determine whether these structural features of urban development are associated with growth of urban centres in the region.

The first stage consisted of arraying all of the variables thought to have a bearing on the problem of growth and decline of urban centres and for which data was available. As many as 60 variables relating to the popula-tion characteristics, physical development, location, economic base, local government, and accessibility were examined in one study (Table 1). The intercorrelations of these variables were the basis of a principal components analysis to define the structural relations. The output of this kind of analysis is a series of clusters of the original variables, which reduces the original array to more manageable size while articulating the complexity present among the original variables. Between six and ten clusters within statistically significant limits were derived in the four regional analyses. In other words, the differences among centres in any region may be described in terms of the dimensions of urban structure that emerged.

The second stage consisted of testing the apparently valid dimensions of urban structure by means of a multiple regression model to determine whether they were linked to urban performance. The factor scores were treated as independent variables (since principal components are statistically independent), and dependent variables such as population change, change in number of retail firms, and level of family earnings were regressed against them.

This analytical approach is, thus, not a model with a set of interlocking equations so much as it is a set of interlocking statistical analyses. It develops a sequence of statistical outputs each of which has one or more uses in policy formulation as well as in regional science. Indeed, one of its merits would seem to be that while striving, ultimately, to describe the dynamics of urban systems in a regional context, the "model" also provides important information on underlying structural features of the region, and these, in turn, can be used to generate

Table 1

INPUT AND OUTPUT OF URBAN STRUCTURE STUDIES IN FOUR CANADIAN REGIONS

Region	No. Centres	No. Variables	No. Factors Used	% Variance Explained
Saskatchewan	476	40	10	66
Eastern Ontario	25	23	6	90
Prince Edward Island	80	32	7	76
Nova Scotia	40	60	10	83

statistics to describe geographical differences in structure. The importance of these analytical by-products will be highlighted in ensuing sections of this report.

THE DATA ARRAYS

In determining the "strategic and relevant" ways that urban centres differ from one another, the first issue is to choose from among the many variables that describe community differences. Factor analysis imposes no special constraints in this regard (except that the variables be numerical) and, indeed, encourages one to explore all possibilities. The sheer availability of data was often more of a constraint, especially of local data, and the problem was more pronounced in depressed areas such as Prince Edward Island. In addition to arraying data covering the broad social, economic, and physical setting of a region, particular attention was given to including variables that heightened unique features of the development situation and/or were of direct policy concern.

Data were gathered generally under five categories: demographic characteristics, social characteristics, physical development, economic base, and geographic situation. The array of Prince Edward Island, reproduced in Table 2, indicates the main features of all the arrays, the others all being more elaborate. Examples of variables specially selected for this region are those describing the distance to a "major tourist destination" and distance to the "Port Borden Ferry." The former is a reflection of the recent development and dramatic increase in use of the region's tourist attractions. The latter reflects

Table 2

VARIABLES STUDIED IN THE ANALYSIS OF URBAN STRUCTURE
IN PRINCE EDWARD ISLAND

Demographic Characteristics
  1. Population Size, 1961
  2. Population Change, 1951-61
  3. Population under 15 Years, 1961
  4. Population over 65 Years, 1961
  5. Population Density, 1961

Social Characteristics
  6. Adult Education Level, Grade 11 or Better, 1961
  7. Adult Education Level, University, 1961
  8. Quality of Elementary School Service, 1963
  9. Financial Support of Education, 1963
  10. Locally Provided Services, 1964

Physical Development
  11. Value of Private Investment per Capita, 1964
  12. Value of Private Buildings per Acre, 1964
  13. Capital Assets per Capita, 1964
  14. Dwelling Units Built before 1945, 1961
  15. Dwelling Units with Flush Toilets, 1961
  16. Median Value of Dwelling Units, 1961
  17. Housing Need Index, 1961

Economic Base
  18. Level of Retail Service, 1961
  19. Employment in Services, 1961

Geographic Situation
  20. Distance to Major Tourist Destination, 1965
  21. Distance to Port Borden Ferry, 1965
  22. Distance to Nearest Neighboring Center, 1965
  23. Highway Accessibility Index, 1965

Figure 1

COMPARATIVE FACTOR STRUCTURE FOR FOUR CANADIAN REGIONS

	SASKATCHEWAN	EASTERN ONTARIO	PRINCE EDWARD ISLAND	NOVA SCOTIA
COMMON FACTORS	Population Size V_1	Population Size V_5	Population Size V_1	Population Size V_1
	Physical Development V_3	Physical Development V_1	Physical Development V_6	——
	Population Age V_5	Population Age V_3	Population Age V_3	Population Age V_2
	——	Education Level V_6	Education Level V_4	Educational Level V_4
	Compact Development V_7	Compact Development V_9	——	N. A.
	Social Welfare V_{10}	N. A.	N. A.	Social Welfare V_7
COMMON FACTORS	Agricultural Base V_4	French/English V_2	Tourism Potential V_2	Migrant Destination V_3
	British/Protestant V_2	Industrial/ Commercial V_4	Highway Access V_5	Employment Level V_5
	Religious Affiliation V_8			Income Level V_{10}
	Community Age V_6			Transport Centre V_9
	Tertiary Employment V_9			

the importance of the major point of access to the island region.

For the Eastern Ontario region not only were more data available and, therefore, included but special characteristics were also emphasized. French-English language differences are prominent among communities in this region, for example. Also, manufacturing industry distinguishes the economic base of many centres (which was not the case in Prince Edward Island), and the importance of this element was sought through the inclusion of two variables. For Saskachewan, the array is extended mainly to include a recognition of both the polyglot ethnic orientation apparent in the settlement pattern and the role of the large-scale agriculture in urban development. The Nova Scotia study had the advantage of ready access to more data than the other studies and thus had the most ambitious data array, sixty variables. Two areas in particular were elaborated: demographic data on age structure, natural increase and migration, and employment and labour force data.

Lastly, it must be noted that the data in each study were for the incorporated cities, towns, and villages for the region. Only for these centres could the wide range of data be obtained. Moreover, few centres of importance are not incorporated in the regions.

COMPARATIVE URBAN STRUCTURE

The factor analysis for the four regions have yielded "structural matrices," as Berry calls them, which can give us insights into both internal urban structure of a region and interregional differences in structure.[7] This kind of analysis provides a concise statement of the dominant sets of relationships that exist among the variables. And, hence, it is these relationships that must be affected if it is desired to change the present course of urban development in the region.

Figure 1 arrays the structural features for each region, according to the names given them, so as to allow comparisons among the regions.

First, let us look at the broad similarities in urban structure among the regions, simi-

larities which suggest the presence of fundamental structural dimensions no matter what the region. Then we should look at the underlying relationships within and among dimensions. Lastly, the structures yield dimensions that are unique as among the four regions and that are due, largely, to social and economic differences, some of which were deliberately emphasized in the original data.

SIMILARITIES IN URBAN STRUCTURE

Among the four regions studied, two features of urban structure emerged distinctly in each—one having to do with the size of population of a centre and the other with the age structure of the population (Figure 1). The first has been called a *population size* dimension and, in all cases, portrays a syndrome of characteristics that appears with increasing population size of a centre: numbers of retail outlets increase, level of capital investment in community facilities rises, and the level of local services goes up. The *population age* dimension clearly shows communities in each region being distinguished by their populations being dominantly either old or young. Some supplementary characteristics are appended to each syndrome but do not change the orientation of the basic dimensions.

Two more urban dimensions appeared in three of the four studies. One has been called *physical development* because it reflects the average quality of housing in a centre as well as the level of investment in all physical facilities. In two of the regions, this syndrome, which showed high levels of physical development, was also correlated with good schools and higher family incomes. The *education level* of the adult population emerged as a separate dimension in three regions, with attainment of high school level and university level education being highly correlated with one another almost in isolation of other variables. Some intercorrelation did prevail, however, in two regions with either income levels or occupation.

One dimension of urban development reflecting *social welfare* case loads and social welfare expenditures emerged from the analysis in the two regions for which such data was available. In Saskatchewan, the social welfare variables were associated with the level of other local services provided by the community—the higher the level of all local services the higher the social aid case load. In Nova Scotia, a high level of social welfare provision was correlated with high unemployment and low income levels. The degree of *compactness* in physical development appeared as a separate structural dimension of centres in Eastern Ontario and Saskatchewan. Increasing population density was found closely associated with increasing intensity of capital investment per acre. In both regions, measures of accessibility were also bound in the syndrome (*highway access* in Ontario and *grain shipment volume* in Saskatchewan).

Most of the urban dimensions noted thus far are, no doubt, reasonable, but what is not expected is the independence of various traits. For example, all the variables reflecting physical development, size, and density tend to be found associated not in one dimension but in several. The orthogonal nature of principal components testifies to this independence. It means that the presence of one condition of development, as indicated by a high score of a centre on a structural dimension, does not require the presence of any other condition to a significant degree. Thus, an urban centre with a high level of physical development need not exhibit large population size, or compact development, or a young population, or a well-educated one. There is, at least, an equal chance the opposite may be true.

Another point is that these common features emerged regardless of the number of variables arrayed in the description of centres. The few times the dimensions did not appear in all regions seem more a reflection of the uniqueness of the urban system than of the amount of data one could examine about the centres. The more data that was arrayed the more different dimensions seemed to emerge, and this was more true

when the number of centres being studied grew. The latter effect is probably due to decreasing homogeneity in the regional system of centres when viewed from positions of more data or more places. That is, the amount of variance explained by the common factors decreases as the number of variables and the size of urban system decrease. Despite this qualification, their persistent appearance in the structure matrices of different regions mark these common factors as likely indicants of basic urban structure.

UNDERLYING RELATIONSHIPS IN URBAN STRUCTURE

As noted above, the principal components analysis sorts variables into independent groups according to their degree of intercorrelation. In other words, if all variables describing urban centres were all highly associated with one another, there would emerge only one significant factor. The fact that this doesn't happen when analyzing urban systems often reveals some surprising relationships existing or not existing. For example, in three of the regions the variables which reflected attributes associated with population size of centres were not significantly intercorrelated with those reflecting attributes of a centre's physical development. Or, using another example, the dichotomy into dominantly young or old populations within a system of centres was found not to be correlated with education, religious, ethnic, or employment attributes of centres. And, one final example, in all regions the economic base variables did not exhibit a close association with other attributes: in Saskatchewan, agricultural variables and other employment characteristics were independent clusters as were the employment variables in Eastern Ontario and Nova Scotia. In other words, favorable or unfavorable attributes of economic base are not regularly associated with the size of a centre, its physical development, its religious or ethnic orientation, or the educational level of its population.

The importance of viewing the internal relations in the structural matrices is that it illuminates the complex relations of variables (people, jobs, housing, etc.) which interact to produce any centre's position on each structural dimension. The problem of centres in Eastern Ontario possessing a low level of *physical development,* for instance, is that they have conditions of poor housing quality, low value dwellings, poor high school, and few public services initiated by the local community. The improvement of "physical quality" is thus a multifaceted one involving housers, planners, educators, etc. Analogies can be drawn for each region on this same dimension and on other dimensions that indicate different levels of achievement among centres. The interrelated set of variables that must be affected in programs for improvement is thereby defined.

UNIQUENESS IN REGIONAL STRUCTURE

In Figure 1 are indicated several factors in each region that do not recur in the other regions. In part, these emerge because of variables that were included to test relations with special environmental characteristics, such as the presence of a large segment of the population of Eastern Ontario who speak French. This is the essence of an independent factor showing the segregation of English- and French-speaking people into separate centres in the latter region. Ethnic and religious segregation also appear as structural features among Saskatchewan centres. The agricultural base variables— wheat yields, farm size, and soil quality—in Saskatchewan comprise an independent dimension with the clear implication that areas of well-developed agriculture need not exhibit well-developed urban centres. The tourism access variables, included for Prince Edward Island to test relations with this new thrust in the region's development, showed a distinct new dimension.

The emergence of unique factors is also, in part, due to the elaboration of data inputs in the Saskatchewan and Nova Scotia studies. The latter study, in particular, was able to include many additional variables on migration, employment, and income each of which formed their own clusters. An employment syndrome also appears in the Saskatchewan and Eastern Ontario structural matrices.

Viewed another way, the "unique" factors display several common traits. When ethnic or religious attributes of centres are included, for example, they tend to form the basis of separate structural features and not to be closely associated with other social or demographic attributes. And when employment or other economic base attributes are included, they tend to form independent clusters. Lastly, when the general geographical situation is entered into such an analysis, it also tends to be rendered in a separate dimension that can be used to describe centres. These similarities occur despite differences in the specific variables used to measure ethnic orientation or employment or geographical situation.

URBAN STRUCTURE AND URBAN PERFORMANCE

The structural matrices we have just examined provide descriptions of the complex relationships comprising the urban development situation. Beyond this, it would be useful to know if the structural features are associated with performance of centres. A multiple regression model was chosen for this test. The orthogonal principal components were considered the independent variables, and three performance measures were analyzed as dependent variables: (1) Average Family Earnings, 1961; (2) Population Change, 1951-61 and 1961-66; and (3) Change in Number of Retail Firms, 1951-61.

A score was generated for each centre in a region on each dimension on the basis of its new data scores and the factor loadings. These canonical variates were normalized scores. The performance variables dealing with change were transformed in percentage terms to a basis where 100 equalled no change in the period, and scores above and below 100 indicated growth and decline, respectively.

417

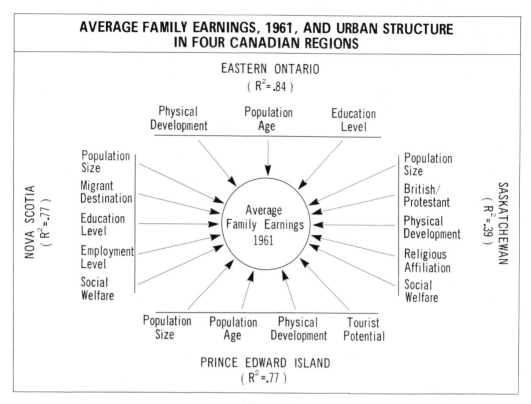

AVERAGE FAMILY EARNINGS, 1961, AND URBAN STRUCTURE IN FOUR CANADIAN REGIONS

EASTERN ONTARIO
(R^2 = .84)

Physical Development   Population Age   Education Level

Population Size
Migrant Destination
Education Level
Employment Level
Social Welfare

NOVA SCOTIA
(R^2 = .77)

Average Family Earnings 1961

Population Size
British/Protestant
Physical Development
Religious Affiliation
Social Welfare

SASKATCHEWAN
(R^2 = .39)

Population Size   Population Age   Physical Development   Tourist Potential

PRINCE EDWARD ISLAND
(R^2 = .77)

Figure 2

AVERAGE FAMILY EARNINGS, 1961

The regression model of urban structural dimensions provided good estimates of the 1961 income levels of urban families in three regions. Seventy percent of the places in Nova Scotia, Eastern Ontario, and Prince Edward Island were estimated within 10 percent. For the Saskatchewan region, although the regression was not as strong, several significant relations were apparent.

The structural features contributing most to the explanations in each region are shown in Figure 2. Several of these recur. A high level of physical development of a centre is closely associated with incomes in three regions. In Nova Scotia, the components of such a dimension are in other clusters, which contribute strongly to the explanation of income differences. Young populations and high education levels of a centre's adult population also have a strong association. In Nova Scotia and Saskatchewan, high family income is linked with large population aggregations. But in Prince Edward Island, it is linked with small population centres. The latter is explained by high incomes of town-dwelling farmers in rich agricultural areas where small centres abound.

This analysis suggests that a structural profile could be drawn of the features which characterize urban places with high average family income. It would be comprised of scores for above average *physical development,* predominantly young *population age,* and above average adult *education level.* *Population size* scores would be included in the profile, above or below average, depending upon the region. This, then, would be an approximation of the kind of urban development structure that is conducive to generating high average family incomes.

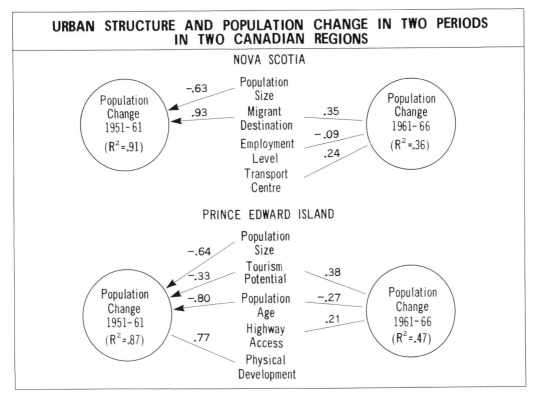

URBAN STRUCTURE AND POPULATION CHANGE IN TWO PERIODS IN TWO CANADIAN REGIONS

NOVA SCOTIA

Population Change 1951-61 (R^2=.91)

Population Size −.63
Migrant Destination .93

Population Change 1961-66 (R^2=.36)

Population Size .35
Migrant Destination
Employment Level −.09
Transport Centre .24

PRINCE EDWARD ISLAND

Population Change 1951-61 (R^2=.87)

Population Size −.64
Tourism Potential −.33
Population Age −.80
Highway Access
Physical Development .77

Population Change 1961-66 (R^2=.47)

Tourism Potential .38
Population Age −.27
Highway Access .21

Figure 3

POPULATION CHANGE

Population growth rates for the 1951-61 decade were well estimated for the centres of all regions. For three regions, the coefficient of determination was over 80 percent. The least efficient estimate was for the large sample of centres in Saskatchewan, but it was still 63 percent.

The dominant structural feature associated with the urban growth rates in Prince Edward Island, Eastern Ontario, and Saskatchewan was age structure of the population. Young populations contributed to positive growth rates in these three regions. At the same time, none of the other contributing factors were the same within these three regions. Physical development and population size were important in Prince Edward Island; education level and commercial development were important in Ontario; and community age, social welfare level, and British orientation were important in Saskatchewan.

The strong regression results for the Nova Scotia centres were attributable to explanations of two structural features, migrant destination and population size. A centre was more likely to experience population growth if it was a destination for migrants, but this was not as likely for centres of large populations. Or, stated another way, given two centres of equal population size, the one attracting the most migrants would experience faster over-all population growth.

The above findings warrant the following assumption: *an extant urban structure contains the accumulation of factors that affected past population growth rates.* The urban structures we have derived are essentially for 1961 and they have allowed a close approximation of 1951-61 population growth rates. Whether this assumption could be extended to include a close approximation of future growth rates is the next logical question. That is, if the structural features contain the ingredients that made for past growth, do they also constitute the set of elements that will determine future growth? Fortunately, the Canadian quincennial census of 1966 affords the opportunity for testing this assumption, if only for population changes.

The centres of all four regions were tested again, using the population growth rate for the 1961-66 period as the dependent variable in a regression model. Two interesting findings emerged: first, the estimates for population growth rates were well below the level achieved for the earlier period, and, second, the structural features involved in contributing most to the explanations differed from the set explaining past growth rates. About the only plausible reason for the failure of the structures to provide good estimates of future population growth is that either new structural arrangements now prevail to affect population growth, or that external forces have more effect than local structure, or both.

The changes that occurred within the sets of structural features explaining growth in the two periods are significant. Figure 3 portrays the different sets according to a path analysis for Prince Edward Island and Nova Scotia regions.[8] Brief verbal descriptions of the characteristics correlated with urban population growth in all four regions for the two periods are presented in Table 3. In Prince Edward Island and Ontario, while the fertility measure and population age remained important, the economic base characteristics changed. Other changes can be determined from this chart.

RETAIL CHANGE, 1951-1961

This performance variable was well estimated for only two regions. The structural features involved in each case are shown in Figure 4. The relationship can be stated as follows for Eastern Ontario: small, predominantly commercial centres with young and well-educated populations but not necessarily well-developed physically can be expected to experience higher rates of growth of retail firms; and for Prince Edward Island: centres with young populations and good highway access but not necessarily connected with tourist development can be expected to increase their retail firms faster.

Some of the above relations may be unexpected, i.e., retail growth is not high be-

Table 3

CHARACTERISTICS OF CENTRES EXPERIENCING POPULATION GROWTH IN TWO PERIODS FOR FOUR CANADIAN REGIONS

High Population Growth Rates Occurred in	
1951-61	*1961-66*
Saskatchewan	
Newer, compact, generally British centres with young populations and low social welfare rolls.	Smaller, newer, well-developed centres with older populations.
Eastern Ontario	
Commercial centres with young populations and high adult education levels.	Well-developed industrial centres with young populations.
Prince Edward Island	
Smaller, well-developed centres with young populations and high tourism potential.	Nontourist centres with good highway access and young populations.
Nova Scotia	
Smaller centres with ability to hold their own population and/or attract migrants.	Centres which attract migrants, have low unemployment, and tend to specialize in transport and exchange activities.

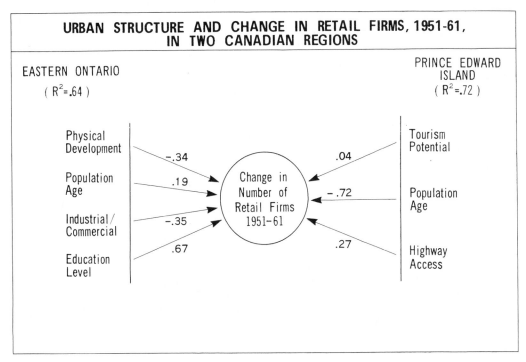

URBAN STRUCTURE AND CHANGE IN RETAIL FIRMS, 1951-61, IN TWO CANADIAN REGIONS

EASTERN ONTARIO
(R^2=.64)

PRINCE EDWARD ISLAND
(R^2=.72)

Physical Development — -.34
Population Age — .19
Industrial/ Commercial — -.35
Education Level — .67

Change in Number of Retail Firms 1951-61

Tourism Potential — .04
Population Age — -.72
Highway Access — .27

Figure 4

cause physical development or population is high (in Ontario) nor because it is associated with tourist development (in Prince Edward Island). In all cases, it must be remembered that we are dealing with growth *rate,* not growth *increments.* Thus, larger centres may add many new firms on an already large base for a small percentage change.

From the analyses of past performance, there is evidence that differences in structural features between centres of a region effect performance substantially. That is, the possession of certain structural attributes as defined by the factor analyses reflect capabilities for certain kinds of urban growth within a regional system of centres. The importance for regional policy seems clear. If efforts are made to improve or otherwise change the urban development situation, these efforts will have to start with structural conditions that presently exist. And it may turn out that some conditions are constraints on an improvement program.

INTERRELATIONS OF URBAN AND RURAL STRUCTURE

For two of the regions studied, Eastern Ontario and Prince Edward Island, it is possible to compare the structural features of both urban and rural sectors of the region. Each rural sector was also subjected to a factor analysis and a juxtaposition of spatially coincident urban structures can thus be made.[9] The importance of this in regions where agriculture is a major economic sector is that, ultimately, the future of both towns and rural areas is entwined.

Earlier, cruder analyses of urban and rural relations by this writer led to the general proposition that *poor rural areas tend to have poor urban centres.*[10] This proposition has been further elaborated in a recent study by Paris.[11] By taking factor scores for urban centres and for the townships surrounding them and subjecting them to both chi-square and correlation tests, he has concluded: (a) what happens on farms seen as units of production has very little to do with what happens in the urban centres in the vicinity; but (b) what happens to rural non-farm families is influenced in important ways by what happens in the towns and villages around.[12]

These findings confirm the inference made above on the basis of the structural relations for Saskatchewan, viz., the fact of an independent *agricultural base* dimension shows a lack of significant correlation between conditions affecting farming performance and urban performance. By distinguishing the difference between farm and nonfarm development and the relation of the latter to urban development, Paris articulates an important issue for regional policy in rural regions. Namely, that poorly developed urban centres, although not necessarily the cause of nonfarm poverty, do little to improve the problem. The added analytical power afforded by bringing together the rural and urban structural matrices shows still further uses for such analyses.

SOME PROPOSITIONS AND PROSPECTS

Having now completed four regional studies of urban systems and observed some interesting threads running through the fabric of each, it seems both possible and useful to state some propositions about urban structure in regional contexts. The regions have not encompassed any sample designed to reflect universal differences in regions, but they do seem to provide relevant comparisons. In any event, the propositions espoused here are testable, and it is hoped that this would be done in order to add to our knowledge of urban structure in regional development.

1. Common structural features underlie the development of all centres within a region.

Numerous factor analyses attest to the presence and significance of structural matrices in describing differences among centres in a region.

2. Structural features of centres tend to be the same from region to region regardless of the stage or character of regional development.

These studies showed little variation in structural matrices where the same descriptor variables were employed even though regional development differed as much as that of Prince Edward Island and Saskatchewan in terms of agricultural prosperity or between Nova Scotia and Eastern Ontario in terms of type of economic base.

3. Urban structure may be defined in terms of a set of "independent" dimensions covering at least (a) size of population, (b) quality of physical development, (c) age-structure of population, (d) education level of population, (e) economic base, (f) ethnic and/or religious orientation, (g) welfare, and (h) geographical situation.

The presence of this range of dimensions is not unusual when considering social, economic, and physical characteristics of spatial units. Analogies for the names used here for the dimensions can be found in other studies.[13] But beyond the names of dimensions it is vital to point out their statistical independence. That is, although each centre possesses each structural feature to a greater or lesser degree, there is no requirement that any two conditions will be present in the same degree. An important subdivision of this proposition might be stated:

3a. Economic base of urban centres tends to act independently of other urban structural features.

There are, of course, several other "independence relationships" that could prove just as interesting.

4. The structural matrix contains the accumulation of ingredients constituting past population growth.

It appears from our studies that a good reproduction of population growth rates can be obtained for an immediate past period when the structure is known for the end of the period. That same structure is not as useful for prediction purposes in the subsequent period.

5. As growth occurs over considerable time, the structural features conducive to such growth change.

This is a reflection of changed environmental conditions as a result of growth and of new forces acting upon the community. This much can be inferred from our studies of population growth. The lack of strong associations with the structural matrix in our attempts at "prediction" is probably indicative of how rapidly changes can occur to affect the structure (and possibly render it obsolete).

6. The structural conditions conducive to high average family income in urban centres in a region are high quality of the physical environment, the youth of the population, high educational level of the adult population, and relative population size.

The coincidence of these conditions, exclusive of economic base considerations, is associated with good income levels. This finding, alone, points to the complexity of regional developmental planning. If income is to be the welfare criterion, one must face up to a multifaceted problem encompassing the knowledge and talents of many disciplines.

REFERENCES

1 No little credit for this enlightened state of affairs is due to such as Eric Lampard, John Friedmann, Walter Isard, and William Nicholls.

2 Their case is well stated in J. R. Boudeville, *Problems of Regional Economic Planning* (Edinburgh: Edinburgh University Press, 1966).

3 The relatively early study by Brian J. L. Berry, "City Size Distribution and Economic Development," *Economic Development and Cultural Change*, Part I (July, 1961), pp. 573-88; and Gerald Hodge, "The Prediction of Trade Center Viability in the Great Plains," *Papers of the Regional Science Association*, XIII (1965), pp. 87-115, have this thrust.

4 The Saskatchewan study is reported in Gerald Hodge, "Branch Line Abandonment: Death Knell for Prairie Communities?" *Canadian Journal of Agricultural Economics*, XVI, No. 1 (1968), pp. 54-70; Eastern Ontario :————, *The Identification of Growth Poles in Eastern Ontario*, A Report to the Ontario Department of Economics and Development (Toronto, 1966); Prince Edward Island:————, *Rural and Urban Development Capability in Prince Edward Island* (Toronto: Acres Research and Planning, 1967); The Nova Scotia study will be described in a forthcoming report of the Department of Municipal Affairs, Community Planning Division.

5 This terminology of regional types is according to John Friedmann, *Regional Development Policy* (Cambridge: The M.I.T. Press, 1966).

6 A broad review of pertinent literature is found in Gerald Hodge, "Do Villages Grow?" *Rural Sociology*, 31 (June, 1966), pp. 183-96.

7 Brian J. L. Berry, *et al.*, *Essays on Commodity Flows and the Spatial Structure of the Indian Economy* (Chicago: University of Chicago, Department of Geography, 1966), Research Paper No. 111.

8 Otis Dudley Duncan, "Path Analysis: Sociological Examples," *American Journal of Sociology*, 72 (July, 1966), pp. 1-16.

9 The Eastern Ontario rural study is reported in Brian J. L. Berry, "The Identification of Areas of Rural Poverty," in E. Wood and R. Thoman, eds., *Areas of Economic Stress in Canada* (Kingston: Queen's University, 1965). The Prince Edward Island rural study is in Gerald Hodge, *Rural and Urban Development Capability in Prince Edward Island* (Toronto: Acres Research and Planning, 1967).

10 Gerald Hodge, *The Identification of Growth Poles in Eastern Ontario*, A Report to the Ontario Department of Economics and Development (Toronto, 1966) and Hodge, *op. cit.*, footnote 9.

11 J. D. Paris, "Policy Criteria for Redevelopment in Depressed Rural Areas," unpublished thesis in the Department of Urban and Regional Planning, University of Toronto (October, 1967) pp. 181-223.

12 *Ibid.*, p. 220.

13 See, for example, C. T. Jonassen and S. H. Peres, *Interrelationship of Dimensions of Community Systems* (Columbus: Ohio State University Press, 1960); and Brian J. L. Berry and Robert Murdie, *Socio-Economic Correlates of Housing Conditions* (Toronto: Metropolitan Toronto Planning Board, Urban Renewal Study, 1965).

A Strategy for the Economic Development of the Atlantic Region, 1971-1981

ATLANTIC DEVELOPMENT COUNCIL

I THE REGIONAL PROBLEM IN HISTORICAL PERSPECTIVE

The Maritime Provinces enjoyed a period of great prosperity during the first half of the nineteenth century, the so-called "Golden Age" of wood, wind and water. Under the mantle of a protective British mercantile policy, the region's cod was sold in the British West Indies, and Maritime timber found ready markets in Britain. The goods involved in this "triangular trade" were carried in wooden bottoms, and with accessible timber resources, it was not long before an extensive shipbuilding industry developed. By the middle of the nineteenth century, the region had become a major commercial maritime power, standing fourth in the world in registered tonnage.

A feature of the regional economy during the "Golden Age" was its integration in a pattern of international trade conducted under British sponsorship and with British protection, which assured the opportunity for exploitation of the region's available resources. This external integration was complemented at the same time by a considerable degree of internal integration of the region's economic structure. As the Royal Commission on Dominion-Provincial Relations noted: "All their minor and subsidiary occupations were nicely geared to support and develop these leading industries. This balance in their economy and the perfection of their skills signified a confident maturity which enabled them to compete on even terms in the deep-sea carrying trade."[1]

The institutional arrangements behind this Maritime prosperity began to be undermined even before Confederation, when changes in international conditions led to the abolition of colonial trade preferences.

Technical change was a more telling factor, however, in removing the basis of the region's prosperity. The wood, wind and water basis of ocean transportation was replaced by steel and steam. There were then fewer natural advantages for the Maritimes, and the remote, thinly populated region was not able to maintain its importance in shipbuilding and the carrying trade.

While the Maritime Provinces were losing their advantage in ocean transport, they were attempting to adapt the new technology of steam and steel to land transportation. In particular, a number of schemes were advanced to construct a railway linking Canada and the Maritimes. Although some 380 miles of railways were built in Nova Scotia and New Brunswick by the time of Confederation, it was not until 1876 that the Intercolonial Railway was completed. With it the Maritimes hoped to be able to penetrate markets in central Canada, and to capture for the ports of Halifax and Saint John an important share of the shipments from central Canada to export markets.

The great hope that the pro-Confederation spokesmen held out was access for Maritimes manufacturers to the markets of Ontario and Quebec. But the optimistic outlook was to prove unjustified by events, as a new framework of policy emerged.

In 1878 Canada adopted the National Policy, consisting of three means of providing the new nation with an economic identity. First, a transcontinental railway would be built to connect the West to the East. Secondly, the West would be settled by immigration not only to strengthen the Dominion's claim to the West, but also to provide freight for the railways. Finally, protective tariffs would be imposed on manufactured

● This paper has been abstracted from a report by the Atlantic Development Council, *A Strategy for the Economic Development of the Atlantic Region*, January, 1971, and is reproduced with permission of the Council's chairman, William Y. Smith.

goods to encourage the development of secondary manufacturing and strengthen the east-west flow of trade and, in the process, utilize the excess capacity of the railroads.

The major employment and income effects of the National Policy were conferred on the central provinces. Although the Maritimes participated in the railway boom and enjoyed the attendant benefits of the development of Nova Scotia's coal and iron and steel industries, these effects were much less pronounced and much less cumulative than those in central Canada. Soon many of the small manufacturing industries of the Maritime Provinces began to disappear in the face of competition from the mass-produced goods of central Canada.

The 1920's were a period of reasserted growth for the Canadian economy following a brief recession at the end of the First World War. New developments in hydro-electric power production, in minerals, in pulp and paper and the increasing use of the internal combustion engine, all contributed to progress, which was felt mainly in the central provinces and British Columbia. The Maritime Provinces remained merely on the fringe of this national growth.

An attempt was made to improve the competitive position of regional industry in central Canadian markets through reduced railway rates under the Maritime Freight Rates Act of 1927. Nevertheless, the polarization of manufacturing industry in central Canada continued until the onset of the Second World War. More small-scale establishments in the Maritimes found it necessary to close down. The only regional industry that experienced real growth in this period was pulp and paper.

For the most part, however, the economic activities of the Maritime Provinces proved competitive with, rather than complimentary to, the industries of central Canada. Apart from fish and coal, the goods produced in the Maritimes were available more cheaply from sources in Ontario and Quebec. While Maritime industry was unable to compete in central Canada, the hope of drawing substantial Canadian traffic to Maritime ports

did not materialize. Although railways had been built linking Halifax and Saint John with the interior of the country, the higher cost of railway over water transport resulted in Canadian traffic being drawn to St. Lawrence and American ports.

Thus the Maritime Provinces, their old economic base removed, failed to respond to the new policies of transcontinentalism and became a peripheral, lagging region of the Canadian economy. With Newfoundland's entry into Canadian Confederation in 1949, the region became the Atlantic Provinces. Prior to Confederation, Newfoundland had been cut off, politically, geographically and economically, from the growth taking place in North America. As Dr. Copes has noted:

When Newfoundland joined with Canada in 1949, it ranked behind every Mainland province by almost every criterion of economic comparison.[2]

During the first eight years of Confederation (1949-1957) Newfoundland experienced rapid economic development. However, the initial gap was so great that by 1957 Newfoundland still ranked last in terms of per capita economic performance.

During the decade following the Second World War, the maturing Canadian economy enjoyed one of the greatest expansions in its history. But the Maritimes and Newfoundland shared only to a limited extent in this prosperity. With the nation enjoying "full employment" by Canadian standards, with unemployment ranging between three and four and one-half percent during the mid-fifties, the Atlantic Provinces' rates were notably higher.

During the decade of the 1950's, as employment declined in the primary industries, there was practically no expansion in employment in the region's manufacturing industries. The required structural changes had not developed through the working of market forces. The weakness of the region's economy became most marked when the postwar period of rapid national growth terminated in 1957. Increasingly informed opinion both within and without the region, accepted the argument that national monetary and fiscal policies should be supple-

425

mented with finer instruments that could deal more directly with the problems of the Atlantic Provinces and, indeed, other lagging areas within the national economy.

In the 1960's, a number of new organizations, policies and programs relevant to regional development came into being. This was the decade of the Atlantic Development Board (ADB), Agricultural Rehabilitation Development Act (ARDA), the Fund for Rural Economic Development (FRED), and the Area Development Agency (ADA). Each of these agencies had responsibilities and functions relevant to the problem of regional disparity, but none of them had a comprehensive mandate in this respect.

It was to provide such an integrated, co-ordinated approach to regional development that the Department of Regional Economic Expansion was established in April 1969. New areas, which are sufficiently large to encompass a development potential, were designated for industrial assistance. The entire Atlantic region, except Labrador, was designated as eligible for industrial incentives. A new regional development incentives program was inaugurated there and special growth centre areas have been manned in areas where additional types of assistance are available because of the impact they could have on the regional economy. Where financial assistance is available for the provision of utilities and services it is possible to make the areas more attractive for industrial growth.

II   THE ECONOMY IN THE 1960's

In order to develop a strategy to guide the region's development in the 1970's, it is necessary to examine the trends in the economy of the Atlantic Provinces during the 1960's.

Growth Trends

Average Personal Income — Personal income per person in the Atlantic Provinces rose steadily from a level of $1,099 in 1960 to $2,033 in 1969 (Table 1). This represented an increase of approximately 85 percent, as compared with an increase of 80 percent in Canada as a whole. Personal income per person in the Atlantic Provinces amounted to 67.9 percent of personal income per person in Canada in 1960 and rose to 69.8 percent by 1969. It should be noted, however, that in absolute terms the difference between regional and national income per person increased from $519 in 1960 to $880 in 1969, a rise of $361 or 70 percent during the period.

Components of Personal Income — The main components of personal income are wages, salaries, and supplementary labour income; net income received by farm operators from farm production; net income of non-farm unincorporated business; interest, dividends, and net rental income of persons; and government transfer payments, excluding interest.

In 1960, earnings from wages, salaries and supplementary labour income accounted for

Table 1

PERSONAL INCOME PER PERSON, ATLANTIC PROVINCES AND CANADA, 1960-69

Year	Nfld.	P.E.I.	N.S.	N.B.	A.P.	Canada
			(dollars)			
1960	895	942	1,242	1,104	1,099	1,618
1961	932	943	1,256	1,099	1,111	1,613
1962	951	1,047	1,307	1,147	1,156	1,720
1963	998	1,056	1,370	1,217	1,213	1,802
1964	1,070	1,165	1,452	1,311	1,298	1,898
1965	1,154	1,248	1,562	1,416	1,398	2,066
1966	1,274	1,367	1,713	1,571	1,538	2,283
1967	1,398	1,514	1,905	1,739	1,703	2,461
1968	1,487	1,682	2,072	1,907	1,851	2,660
1969	1,613	1,818	2,307	2,083	2,033	2,913

Source: DBS, National Accounts.

426

60.8 percent of personal income in the Atlantic region and 66.7 percent for Canada. By 1968 earnings accounted for 64.6 percent of personal income in the Atlantic Provinces but had risen to 69.8 percent for Canada. The second most important source of personal income, government transfer payments (excluding interest), contributed between 13.4 and 15.8 percent of total personal income in the Atlantic region during the period 1960 to 1968, and between 8.1 and 10.7 percent in Canada. Therefore the Atlantic Provinces are more reliant on transfer payments, and derive a smaller percentage from earning in personal income, than does Canada as a whole.

Factors in Growth

Population — The population of the Atlantic Provinces increased from 1,867,000 in 1960 to 2,012,000 in 1969 for an increase of 7.8 percent. Canada's increase over the same period was 17.9 percent. Therefore, the Atlantic Provinces' share of Canadian population fell from 10.5 percent in 1960 to 9.6 percent in 1969 (Table 2).

Estimated Net Migration — The slow rate of growth in population and labour force population in the Atlantic Provinces is directly related to the net movement of people out of the region. Between 1961 and 1969, approximately 150,000 people (on a net basis) left the region—36,000 from Newfoundland, 7,300 from Prince Edward Island, 58,000 from Nova Scotia, and 48,900 from New Brunswick (Table 3). Furthermore, it is estimated that almost two-thirds of the 106,100 people who left the region on a net basis between 1961 and 1966 were between the ages of 15 and 34.[3]

Labour Force — During the period 1960 to 1969 the Atlantic Provinces labour force increased by 18.7 percent (from 550,000 to 653,000) while the Canadian labour force increased by 27.3 percent. This slower growth is due to the lower labour force participation rate and the slow growth of the labour force population. In the former case the participation rate (the labour force as a percentage of the population over 14 years) in the Atlantic region fluctuated during the 1960-69 period from 46.8 to 48.6 percent. In 1969, the participation rate in the region was 48.1 percent against 55.8 percent in Canada. The labour force population (14 years of age and over), grew by 16.5 percent between 1960-69 whereas the Canadian labour force grew by 23.7 percent. In the Atlantic Provinces only Newfoundland's labour force grew more rapidly than the Canadian labour force during the period 1960-69. The region's share of national labour force declined from 8.6 percent in 1960 to 8 percent in 1969. The Atlantic region's share of total Canadian employment declined from 7.8 percent in 1969 from a high of 8.4 percent in 1961. The employed force in the region in 1969 was 605,000—Nova Scotia, 244,000; New Brunswick, 194,000; New-

Table 2

POPULATION, ATLANTIC PROVINCES AND CANADA, 1960-69

Year	Total A.P.	Canada	A.P. as Percent of Canada
		(thousands)	
1960	1,867.0	17,870.0	10.5
1961	1,897.4	18,238.3	10.4
1962	1,926.0	18,583.0	10.4
1963	1,944.0	18,931.0	10.3
1964	1,958.0	19,291.0	10.2
1965	1,968.0	19,644.0	10.0
1966	1,974.7	20,014.9	9.9
1967	1,986.0	20,405.0	9.7
1968	2,001.0	20,744.0	9.7
1969	2,012.0	21,061.0	9.6

Source: DBS, *Population 1921-1966*, and *Population Estimates by Marital Status, Age and Sex, for Canada and Provinces*.

foundland, 131,000; and Prince Edward Island, 36,000.

Unemployment Rates — The unemployment rate is the percentage of those persons in the labour force who are not able to obtain gainful employment. The unemployment rate for the Atlantic Provinces during the period 1960-69 ranged between a high of 11.2 percent in 1961 and a low of 6.4 percent in 1966. This compares with a high of 7.1 percent and a low of 3.6 percent in Canada as a whole. The unemployment rate in the region throughout the period remained about half again as large as the Canadian rate.

Investment — Total investment in the Atlantic Provinces rose from $819 million in 1960 to $1,725 million in 1969. This represents an increase of 111 percent compared with a Canadian total of 94 percent.

Hence the region's share of total Canadian investment increased from 7.3 percent in 1960 to 7.9 percent in 1969. Although investment figures vary from year to year, certain changes in the investment pattern are apparent. In the primary industries and construction, capital investment ranged from 13 to 24 percent from 1960-69. For Canada during this period investment ranged from 14.8 to 16.7 percent. In 1969 investment in utilities was 24.9 percent compared with 20.6 percent in Canada. Capital investment in housing, institutional services, and government departments ranged from 37.5 percent to almost 50 percent during the period 1960-69 as compared with 33.9 to 39.2 percent in Canada. In 1969 the share of investment in manufacturing accounted for 16.4 percent in the Atlantic Provinces and 17.5 percent in Canada.

Table 3

NET MIGRATION, ATLANTIC PROVINCES AND REGION, 1961 TO 1969
(YEARS BEGINNING JUNE 1)

	1961-1966	*1966-1967*	*1967-1968*	*1968-1969*	*Total 1961-1969*
	(*thousands of persons*)				
Nfld.	—24.4	—4.6	—3.6	—3.4	—36.0
P.E.I.	—4.8	—1.2	—0.2	—1.1	—7.3
N.S.	—41.7	—7.3	—4.8	—4.2	—58.0
N.B.	—35.1	—5.0	—3.2	—5.6	—48.9
A.P.	—106.1	—18.0	—11.6	—14.4	—150.1

Note: May not total due to rounding.

Source: Atlantic Provinces Economic Council, *Atlantic Report*, Vol. IV, No. 4, October, 1969.

Table 4

UNEMPLOYMENT RATES, ATLANTIC PROVINCES, REGION AND CANADA, 1960-69

Year	*Nfld.*	*P.E.I.*	*N.S.*	*N.B.*	*A.P.*	*Canada*
	(*percentage*)					
1960	18.0	—	—	—	10.7	7.0
1961	19.5	—	—	—	11.2	7.1
1962	17.1	—	—	—	10.7	5.9
1963	15.1	—	—	—	9.5	5.5
1964	10.3	—	—	—	7.8	4.7
1965	11.3	5.6	5.4	7.5	7.4	3.9
1966	7.9	5.6	5.2	6.9	6.4	3.6
1967	8.4	5.6	5.6	6.9	6.6	4.1
1968	9.7	5.6	5.9	7.2	7.3	4.8
1969	10.3	5.3	5.4	8.5	7.5	4.7

— not available.

Source: DBS, *The Labour Force,* Special Surveys Division.

Industrial Structure — Of the four major industrial categories — primary, electric power, construction, and manufacturing, the percentage contribution of manufacturing to the net value of commodity production in the Atlantic region has been increasing but it is still well below the Canadian average. In 1960, manufacturing provided 37.7 percent of the net value of commodity production in the Atlantic Provinces and 54.7 percent for Canada. By 1967 the Canadian percentage had increased to 56.9 whereas the Atlantic Provinces production had fallen to 36.9 percent.

In the remaining categories—primary (agriculture, forestry, fishing, trapping and mining), electric power, and construction — the percentage contribution from 1960-1967 generally exceeded the Canadian average.

Summary

In the 1960's many of the gaps between the Atlantic Provinces and Canada as a whole did not change significantly. While there was some increase in regional income per person in relation to Canadian income per person, the absolute difference widened from $519 in 1960 to $880 in 1969. Both total population and labour force population advanced slowly in the region, compared with national increases. In 1969 the regional labour force participation rate was still considerably below the national rate, and the regional unemployment rate was 7.5 percent compared with the national rate of 4.7 percent.

In terms of the number of persons employed there was a movement out of the primary industries, with the exception of mining, which held roughly to its relative position. In spite of the absolute increases that occurred in the number of manufacturing employees, the manufacturing industry as an employer of non-agricultural labour held roughly to its relative position (Table 5). While the number of employees in manufacturing accounted for between 13 and 14 percent of non-agricultural employment throughout the 1961-68 period in the region, in Canada the number of employees in manufacturing accounted for upward to one-quarter of non-agricultural employment.

The growth of the service industries has brought about a major structural change in both Canada and the Atlantic Provinces. The number of employees in trade, finance, insurance and real estate, community, business and personal services, and in public administration, accounted for 43.4 percent of non-agricultural employment in Canada in 1961 and for 47.9 percent in 1968. In the Atlantic Provinces, employees in this group of service industries represented 44.7 percent of non-agricultural employment in 1961 and 46 percent in 1968. Thus the shift to these four service industry groups was not as pronounced in the region as in the country as a whole. However, the major difference in structure between the economy of the Atlantic Provinces and that of Canada as a whole is the significantly smaller role played by manufacturing in the region.

III THE FRAMEWORK FOR REGIONAL ECONOMIC DEVELOPMENT IN THE 1970's

The Broad Economic Trends[4]

A strategy of regional economic development must operate within the broad economic trends that are changing the economy over time. In all the highly industrialized countries, one phenomenon, in particular, embodies the major forces at work affecting economic activity and its location. This is the process of urbanization. The same process is at work in the economy of the Atlantic Provinces.

Figure 1

PERCENTAGE CONTRIBUTION BY INDUSTRY TO THE NET VALUE OF COMMODITY PRODUCTION, ATLANTIC PROVINCES AND CANADA, 1967

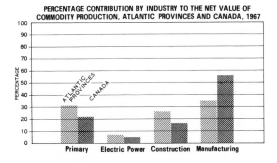

429

Table 5

DISTRIBUTION OF EMPLOYEES BY INDUSTRY AS A PERCENTAGE
OF NON-AGRICULTURAL EMPLOYMENT, ATLANTIC PROVINCES AND CANADA
1961 AND 1968

Category	1961 A.P.	1961 Canada	1968 A.P.	1968 Canada
Forestry	3.3	1.3	2.0	0.9
Mines, Quarries & Oil Wells	3.4	2.0	3.3	1.7
Manufacturing	13.2	24.2	13.4	23.4
Construction	5.4	5.4	5.6	5.2
Community, Business & Personal Service	20.9	19.8	22.8	24.0
Transport & Communications & other utilities	13.1	10.6	10.7	9.2
Trade	13.8	13.8	13.5	14.1
Finance, Insurance & Real Estate	2.1	3.6	2.4	4.0
Public Administration	7.9	6.2	7.3	5.8

Source: DBS, *Estimates of Employees by Province and Industry.*

Note: This information includes only hired employees and does not include employment in agriculture or fisheries. The figures account for approximately 75 percent of the total employed labour force in the Atlantic Provinces.

In this regard, it is useful to consider three things: first, the manifestations of the urbanization process; second, the forces, economic and social, which underlie this process and propel it along; and finally, some of the implications of the forces and of the urbanization process for economic development in the Atlantic region.

One of the major factors behind the rapid and accelerating process of urbanization is changes in the patterns of production. Over the past century, there have been dramatic shifts in the kinds of goods that society wants and the inputs that are required to produce them efficiently. In brief, there have been major shifts in consumption patterns as more and more Canadians achieve both the income with which to enjoy the good things of life and the leisure in which to enjoy them.

One of the most obvious effects of industrialization and economic growth has been a steady decline in the extent to which people live directly off the resource industries. Fifty years ago, about 50 percent of the Canadian labour force was employed in these industries. Today, it is less than 10 percent. This is, in part, because the demand for the products of nearly all the primary industries has not kept pace with the

long-term growth of income and consumption in Canada. As people become richer, they tend to spend their increased incomes on consumption that cannot be satisfied by the products of farm or forest, sea or mine. There is another very important force at work. Not only is the demand for the products of the sea and land-bound primary industries steadily decreasing in proportion to the total demands for goods and services in Canada, but the number of man-hours required to produce any given quantity of these primary products has been dropping sharply and persistently.

If the decline in the importance of primary industries has provided the push toward urban areas, the pull has come from the increasing importance of other industries. The fastest-growing segment of the economy as regards employment has been that vast complex of functions generally lumped together as the "service" or "tertiary" sector, i.e., trade and distribution activities, finance, advertising, research, and central office administration, education, medical care, repair services, and entertainment, etc.

Of importance also is the fact that the kinds of labour or skills required for the production of services are very different from those required in farming or mining.

The need for physical strength and manual dexterity has been largely replaced by the need for a high level of literacy, for professional and technical knowledge and for clerical skills. People with the kind of background, education and inclinations to encourage the development of these attributes are generally found in the large cities that form the core of metropolitan areas.

What are the implications for metropolitan development of the manufacturing sector, the third major sector of the Canadian economy besides the resource and service sectors?

One most important factor is the steadily decreasing dependence of manufacturing on specific raw material inputs. Manufactured goods are subjected to a constantly increasing amount of processing. The effect of all this processing is to reduce the role played by raw material costs in the total cost of the finished product.

A second factor is that as technology has reduced the dependence of manufacturing on raw materials, it has increased the dependence of one manufacturing operation on another. As the chain of fabrication and assembly has grown longer and more complex, the outputs of a number of manufacturing industries have increasingly become the inputs of others. Manufacturers of intermediate or producer goods must be close to industrial markets of sufficient size to support their large-scale operations—close, that is, to a large cluster of other manufacturing firms.

All in all, then, the forces pulling manufacturing employment into larger urban areas seem to be increasing rather than diminishing in strength.

It is within the context of these kinds of social and economic forces at work regionally, nationally, and world-wide that we must survey the economy of the Atlantic Provinces and determine the appropriate development policies to deal with the resulting opportunities and challenges.

The Planning of Urbanization in the Atlantic Provinces

During the 1970's, the main thrust of regional development policies must be to expand output and employment in manufacturing. We have seen that industrialization and urbanization are part of the same process. The present population distribution in the region is largely the result of the past dominance of extractive operations in the regional economy. The more rapid growth of manufacturing must lead to a more urbanized region. The success of the development program proposed by the Atlantic Development Council should lead to over 70 percent of the population living in urban centres by 1981. In 1961, only 50 percent of the population lived in urban centres. The Atlantic Development Council recommends that the urbanization of the region should be carefully planned around three types of population concentrations: the growth centre, the resource centre, and the service centre.

The Growth Centre — The growth centre approach to regional development is based on the proposition that growth, once started in certain selected urban centres, will spread to other parts of the region and thus promote the growth of the region as a whole. In other words, the objective of stimulating growth in selected centres is a means to wider regional purpose — the raising of incomes throughout the region.

There is no firm consensus on the minimum population size for a growth centre. Centres of less than 50,000 population would normally find it .difficult to create the necessary conditions for large-scale industrial development. A population of 100,000 is probably a more realistic minimum.

To be effective, growth centres must contain "propulsive" industries—industries that can give a major stimulus to income and employment in the region. These "propulsive" industries must be geared to rapidly expanding markets outside the region. Thus they play a key role in the process of regional growth, serving as a means by which growth impulses are transmitted from other regions and, in turn, transmitting growth impulses to other industries and locations within the region.

Essentials of a Growth Centre Policy — A growth centre policy is needed to meet certain key objectives of Atlantic regional development. Since the expansion of manufacturing must provide the central thrust to the region's development in the 1970's, growth centre policy should be oriented, primarily, to the expansion of the region's manufacturing base.

Emphasis should be given to those projects specifically designed to serve industrial needs. For example, high priority should be assigned to providing an adequate supply of serviced industrial land.

It will not be enough merely to expand the infrastructure and improve the services in the region's large urban centres and to offer financial incentives to industry locating in them. A more positive approach is needed. Effective industrial intelligence must delineate the industries that would appear to be viable in the long run in the major growth centres of the region. At the present time, a comprehensive growth centre policy has been developed for Saint John, New Brunswick. The future development of the Saint John area is being planned in relation to a carefully planned group of "propulsive" metal-working industries.

To a considerable degree, the stimulating effects emanating from a growth centre will operate more or less automatically. But steps should also be taken to assist this process, to ensure that the benefits of faster growth are felt throughout the regional economy. The task, in this regard, will be to increase the extent of the complementarity between the growth centres and the rest of the region to the greatest degree possible, and, in particular, to develop economic functions in other towns and cities that can support the growth centres and become the means whereby growth impulses can be transmitted from the growth centres to other centres and areas within the region.

The Resource Centre — The resource centre is an economically efficient agglomeration of processing industries and a range of supporting services, all largely dependent upon similar inputs. They may be designated by the type of resource they process—

forest product centres, fish-processing centres, agricultural-processing centres, mineral products centres.

The concentration of similar and interlinked processing units in the same area should increase, substantially, their efficiency through the provision of infrastructure and supporting services of a scale and variety not feasible in a number of scattered locations.

The Service Centre — A service centre's primary function is to supply a range of services to its area. Growth centres, resource centres, and service centres are not mutually exclusive categories. Growth centres and resource centres will also provide modern services to their areas. Indeed, the vast majority of services required by the people of the region will be furnished by these two main types of population concentration. There will be, however, some areas relatively remote from growth centres and resource centres. In these instances, and they will be mainly in Newfoundland, special service centres should be developed. They must be strategically located within the area they are designed to assist.

Conclusion

During the 1970's, the accelerated growth of the economy of the Atlantic Provinces must take place within the following framework:

(1) The national economy must be operating at or near its potential.

(2) The manufacturing sector must provide over the period about 50,000 net new jobs. A substantial portion of this growth must come through the development of groups of related industries. The growth of manufacturing is required to open up jobs to those displaced by the rationalization of the primary industries and to provide stimulus to the growth of service industries.

(3) The urbanization of the region must be carefully planned so as to preserve the amenities of the region and to provide impetus to the growth process.

The present manufacturing base of the Atlantic Provinces is relatively narrow. This is a major, if not the main, factor in the region's relatively weaker economic performance. In 1967 the Atlantic region, with 9.7 percent of the Canadian population, accounted for 4.4 percent of the total number of employees in Canadian manufacturing and for only 3.1 percent of the total value added in manufacturing. Ontario, on the other hand, with 35.0 percent of the Canadian population, accounted for 49.5 percent of manufacturing employees and 53.9 percent of value added.

During the period 1961-67, the period for which comparable data is available, there has been little change in the position of the Atlantic Provinces in Canadian manufacturing activity as far as employment is concerned. Manufacturing employment in the Atlantic region, as a percentage of the Canadian total, declined slightly from 4.6 percent in 1962 to 4.4 percent in 1967. This occurred in spite of the fact that between 1961 and 1967 employment in manufacturing rose by 17.9 percent in Newfoundland, 34.0 percent in Prince Edward Island, 18.7 percent in Nova Scotia, 16.7 percent in New Brunswick, and by 18.3 percent in the region.

Between 1961 and 1967, value added in total manufacturing activity increased by 43.1 percent in the Atlantic region. However, during the same period, Canada as a whole experienced an increase in manufacturing value added of 61.4 percent. Throughout these seven years manufacturing value added was approximately 15 percent of Gross Regional Product for the Atlantic Provinces, while manufacturing value added accounted for about 28 percent of Gross National Product for Canada.

The economy of the Atlantic region is heavily dependent on natural resources. Manufacturing industry is heavily weighted by the processing of these natural resources. In 1967 value added in total activity of manufacturing industries was $580 million in the region. Of this amount $333 million, or 57 percent, was accounted for by fish products, other food processing, sawmills and planing mills, other wood industries, and pulp and paper.

IV MANUFACTURING

Role of Manufacturing in Development

The main requirement for a faster rate of manufacturing development in the Atlantic region is the broadening of the manufacturing base by the establishment of growth and high-technology industries. This is not meant to imply that there is not an opportunity for further growth of the existing manufacturing base which, as previously noted, consists largely of resource-processing activities. In fact, the processing of natural resources will in all probability form the major activity in a number of areas of the region. However, in the growth centres the main development potential will likely come from new manufacturing activities, which in turn will generate a more rapid rate of general development.

Factors Influencing Manufacturing Location

With this leading role assigned to manufacturing it becomes necessary to examine some of the observations, of both a theoretical and an empirical nature, that have been made on industrial location.[5] There are several significant location factors. Among them are raw material supply, labour availability and cost, market accessibility, the availability of suitable factory buildings, as well as the presence of industrial centres. Other determinants of industrial location include personal considerations, proximity to rivals, availability of capital, climate, and the cost and availability of fuel, power and water. The locational choice by the industrialist is not likely to be an exact science for a number of reasons, including the difficulty of quantifying the variables concerned and the significance of non-economic location factors.

Moreover it would appear that the relative weights of the forces that determine industrial location are changing, with the location of raw materials and close proximity to markets becoming of declining significance. In modern manufacturing, industrial interrelationships have become of dominant importance in industrial location. As Mr. Tom

Kent, the Deputy Minister of Regional Economic Expansion, stated:

> Modern technology has reduced the dependence of most manufacturing industries on proximity to their sources of raw materials and markets. What has been substituted is the dependence of industries on one another. They need each other's products and each other's know-how.[6]

Industry now is much more mobile, and the variables that ultimately determine location are more susceptible to policy influence. These factors are favourable to the potential of the Atlantic region to attract secondary manufacturing activities.

Manufacturing Growth

In considering the question of what policy is most likely to promote effectively the development of secondary manufacturing in the region, it is relevant to examine the effect of past and present industrial incentive programs. Although it is difficult to evaluate precisely the effects of incentive programs, it would appear that the Area Development Agency program did not succeed in significantly changing the economic base in the Atlantic Provinces. In a study of 49 firms in New Brunswick assisted under the program in the period 1963-68, it was noted that the majority were resource-based (31) or linked to the regional consumer markets (14), and that the ADA program failed to attract a number of activities of the "growth industry" type.[7] A similar study for Nova Scotia noted the failure of the program to attract much investment in industries that displayed a high growth record elsewhere in the country.[8] Both studies noted that to some extent the program encouraged the dispersal of industry to areas where it would not otherwise have gone and where it was not always well located.

To October 31, 1970, a total of 90 locations or expansions in the Atlantic region were aided under the Department of Regional Economic Expansion's Regional Development Incentives Act, representing an estimated capital cost of $43.8 million and an estimated additional employment of 3,906. Of these 90 new locations or plant expansions, approximately 15 were in the fish-processing industry, 20 in the field of other food products and beverages, 16 in wood and paper industries and 39 in other areas of manufacturing activity. Among the latter 39 are included such operations as the rebuilding of engines, a machine shop, furniture manufacturing, manufacture of containers, bookbinding, clothing, the manufacture of ventilators, the manufacture of electronic equipment, the manufacture of railway equipment, the manufacture of brushes, and the manufacture of construction materials. The addition of these new jobs is very welcome. It is, however, too early to form definitive judgements on the effectiveness of the new incentive program; nevertheless, the heavy representation of existing, rather than new, types of activities should be noted.

In July 1970, the Honourable Jean Marchand, Minister of Regional Economic Expansion, announced incentive grants totalling $15.5 million to the Sydney steel mill; these grants are in respect of a proposed $94 million modernization and expansion program for the Sydney mill. The statement announcing these grants made reference to the achievements at the plant since its transfer to provincial ownership, but emphasized the need for extensive modernization if operations and employment were to be made more secure.

The strengthened performance of the Sydney steel plant, coupled with the improved outlook for coal, are very encouraging for both the Cape Breton area and the region and emphasize the importance of concern for the strength of existing industries, measures to aid them in remaining competitive, and attempts to maximize their contribution to the local and regional economies. The major program of expansion and modernization proposed at the Sydney mill would appear to provide an excellent opportunity to explore fully the scope for new and expanded operations in the light not only of the external markets but also of a potentially much more vigorous and diversified regional economy in the next decade.

The Environment for Regional Manufacturing in the 1970's

The decade of the 1960's saw many countries including Canada attempt to broaden their secondary manufacturing base in the hopes of accelerating economic development. It is reasonable to assume that this process will continue into the 1970's with the prospect of even greater competition among the various areas. The question then arises as to the Atlantic region's relative strength in this competitive effort.

In a recent book, J. J. Servan-Schreiber outlined the actual and potential penetration of Western Europe by American investment.[9] He stressed that European firms must take the initiative if they want to remain viable. One method for these large European firms to remain competitive is to establish plants in North America to compete more effectively with American firms. The Atlantic region would appear to be well suited for such operations due to its proximity to the large U.S. markets and its geographic location at the periphery of the American continent. The establishment of Volvo (Canada) Ltd., and Michelin Tire Mfg. Co. of Canada Ltd., in the region may be an indication of a future trend in this direction.

The possession of deep-water ports is another advantage of the Atlantic region, and one which should be of growing significance in the decade ahead. Increasingly they can be expected to generate increased activity in both transshipment and processing operations.

Strategy for Manufacturing Development

Despite the growth of manufacturing output and employment over the last decade or so, the manufacturing base of the Atlantic Provinces has only been marginally expanded. To a large extent, additions to the base have been in the form of activity catering to regional needs or concerned with the industrial activities already represented in the region. It is clear that efforts will have to become much more ambitious and much more effective in shifting the pattern of investment if manufacturing is to play the role in the regional economy that this strategy has assigned to it. It is also clear that an approach that relies upon the haphazard attraction of a number of isolated and unrelated industries will not significantly change the structure of the Atlantic economy.

The Council believes that the region's accelerated development must be based on growth centres and that in such centres, strategy should be based primarily on promoting inter-industry linkages and, where possible, industrial complexes. By "industrial complex" is meant the establishment of dynamic industries, linked together through market and supply relationships.[10] This provides an environment in which individual industries can grow together and can establish additional linkages to both manufacturing and service enterprises and where the economies of agglomeration can be achieved to the maximum extent possible. The metal-working complex being planned for the Saint John area could be a major factor in promoting the industrialization of the Atlantic region in the 1970's.

The industrial complex approach may have a special role to play in Newfoundland, where the internal market is relatively small and scattered and the regional market less accessible than it is to the other Atlantic Provinces.

Policies for Manufacturing Development

Growth Centres——The Council believes that the necessary expansion and structural change of the region's manufacturing base requires a growth centre policy. The Department of Regional Economic Expansion has designated 12 "special areas" in the Atlantic region.[11] They include the following:

In New Brunswick,
> the city of Saint John, and
> the city of Moncton and neighbouring areas;

In Nova Scotia,
> the metropolitan area of Halifax-Dartmouth, and
> the area around the Strait of Canso;

In Newfoundland,
> St. John's and a surrounding area in-

cluding the Conception Bay coast from Portugal Cove to Carbonear, and Corner Brook and its environs.

The designation of additional growth centres should be examined and considered as a second phase, but the concept of a concentrated propulsive effect emanating from the main growth centres should act as a check on any tendency to designate more than a limited number of such centres at this time.

Infrastructure Programs——Only an appropriate mix of industrial and social investment can initiate and sustain economic development. Infrastructure investment is nevertheless important, particularly in the growth centres, as it not only improves the industrial environment but the social environment as well. Short-term agreements have already been concluded between the federal and provincial governments respecting certain designated growth centres in the Atlantic Provinces with the announced intention of these being followed by longer-term, more comprehensive agreements. These agreements provide for the build-up and improvement of infrastructure and services in the growth areas.

Industrial Intelligence and Promotion—— To conduct a meaningful and effective industrial promotional program, it is necessary to know, as specifically as possible, the kind of industries that will achieve the purpose sought. This kind of knowledge has been inadequate in the past. As a consequence, industrial search and promotion has been to some extent handicapped and not always as productive as it might have been.

One approach is to try to determine a ranking of desirable industries based upon certain agreed criteria, such as expected growth, productivity rating, labour intensity, and the strength of forward and backward linkages, etc. On the basis of weights assigned to these and other factors, it is possible to arrive at a list of the most desirable industries which can then be tested for suitability and feasibility by examining markets, rate of return, comparative costs, and other relevant factors.

The Council is of the opinion that there must be a regional approach to industrial location, and recommends the establishment of a regional industrial intelligence and promotion agency. This agency would conduct a systematic and scientific program of studies, research, and investigation to assess the locational requirements of industry in relation to the industrial needs of the region and its potential for specific industrial projects.

Financing Industrial Development

Existing Federal Program of Industrial Incentives——Direct federal incentive assistance to industry is now provided mainly under the provisions of the Regional Development Incentives Act, which came into force July 1, 1969. The Act provides for development incentives in designated regions. The designated regions include all of the Atlantic region with the exception of Labrador, as well as portions of each of the other provinces. The regions are designated for a period of three years, to July 1, 1972.

Under this program, incentives are available for manufacturing or processing operations other than an initial processing operation in a resource-based industry. Excluded for example are oil refining, the production of pulp, newsprint, and mineral concentrates. Primary activities such as farming, fishing, logging and mining are not eligible, but such operations as the processing of fish and farm products, petrochemical processes, the production of paper and paperboard from pulp and the processing of mineral concentrates to produce metals are eligible.

Under this legislation a primary incentive of up to 20 percent of approved capital cost, subject to a maximum of $6 million, can be provided for the establishment of a new facility or the expansion or modernization of an existing facility in a designated region. In addition, a secondary development incentive may be authorized for the establishment of a new facility or the expansion of an existing facility to enable the manufacture or processing of a new product, i.e. one not previously manufactured or processed in the operation. The maximum secondary incentive is 5 percent of approved capital cost plus $5,000 for each job created directly in the operation.

A number of changes in the Regional Development Incentives Act were made in December 1970. These changes included provision for special development incentives. The maximum special incentive is 10 percent of capital costs for expansion and modernization; for a new plant or expansion of an existing facility to produce a new product it is 10 percent of capital costs plus $2,000 per job. The maximum total incentive assistance under the incentives program is the lesser of $30,000 for each job created directly in the operation or one half of the capital to be employed in the operation. The original ceiling of $12 million on the size of any development incentive was removed in the recent changes. Special incentives have a time limitation of three years before the time limitation in respect of primary and secondary development incentives. Projects to be assisted by a special development incentive must be brought into commercial production by December 31, 1973; the corresponding date for primary and secondary incentives is December 31, 1976. A press release on the Minister's explanation of the proposals indicated that, in the Atlantic Provinces, the special incentive would be available in addition to existing incentives. The maximum incentive in the Atlantic Provinces is thus 30 percent of capital cost for expansions and modernizations and, for new plants, 35 percent of capital costs plus $7,000 per job.[12]

Specified incentives are maximums. The incentives are discretionary and a number of factors are taken into consideration in determining the actual amount of incentive, if any, to be provided in any one instance. No assistance is to be authorized where it appears probable that the development, whether it be a new operation, an expansion or modernization, would proceed without such an incentive. The legislation indicates that the extent of assistance from other government sources is to be taken into consideration in the amount of any federal incentive.

The changes in the Regional Development Incentives Act in December 1970 also made provision for a system of loan guarantees. The loan guarantees are available for operations eligible for industrial incentives as well as for commercial operations. The loan guarantees are applicable to facilities brought into commercial production by December 31, 1976. They can be provided for loans up to 80 percent of the capital cost of the undertaking net of any federal, provincial or municipal grants, if sufficient financing would not otherwise be available on reasonable terms. The regulations define the classes of commercial undertakings to which loan guarantees are applicable throughout the designated regions as convention facilities, hotel accommodation and recreational facilities. In addition, in large centres of population, business offices, warehousing and freight-handling facilities, and shopping centres are eligible, provided that they are on a major scale in relation to the size of the community. A minimum of $1,000,000 total capital costs is specified for an eligible project in a large centre of population and $500,000 elsewhere.

The foregoing deals with incentive grants and with loan guarantees available under the Regional Development Incentives Act. In addition, there is provision under the Government Organization Act, 1969, for special assistance to certain undertakings in special areas. "Special areas" are chosen by the federal government in consultation with the province concerned, and are those that are deemed to require special measures to facilitate economic expansion and social adjustment. For such a special area, development plans are drawn up by the federal government in consultation with the province concerned, and assistance through loans and grants is available for works and facilities considered essential to the plan for the special area.

In the special areas, there is also provision for assistance to certain undertakings. Assistance can be provided for the establishment, expansion or modernization of a commercial undertaking deemed essential to the successful implementation of a plan for a special area. Such assistance may be provided by:

(1) guarantee of payment of principal or interest of loan;

(2) payment of a grant or a loan in respect of a part of capital cost;

(3) payment of a grant in respect of such part of cost of bringing into commercial production and operating, incurred in the first three-year operating period, as is attributable to factors associated with location of the undertaking in the special area.

There are, however, restrictions on these types of assistance. For example, where an undertaking could be given a development incentive, it cannot be given the capital loan or grant or the operating grant; it may be given assistance by way of guarantee of principal or interest only if the capital cost of the undertaking would exceed $75,000 for each job created directly in the undertaking, or $30 million.

It is considered too early at this stage to reach a conclusive judgement on the adequacy of the present industrial incentives program. The original program has been operative for just over one year, which may be said to represent the "gearing-in" phase. The special incentives and the new system of loan guarantees have only just been introduced. The Council is of the opinion, nevertheless, that certain changes must be made for the program to achieve maximum effectiveness. The areas where modifications and changes should be introduced relate to: depreciation regulations on incentive grants, "settling-in" grants, and loan guarantee powers.

The Role of Provincial Governments In Industrial Financing——It is the opinion of the Council that the main public burden of financing industrial development should be federal, and that the role of the provincial governments should be complementary. Much needs to be done—aside from direct financial assistance—to aid existing firms improve their competitive position and expand. Many of the programs required to

assist small business are best suited to sponsorship under provincial auspices. Moreover, federal incentives are not available below certain minimum limits of capital expenditure—$60,000 for a new firm and $30,000 for an expansion. Nevertheless, business developments that fall in this category may be very significant individually and in some localities, and should merit the attention of provincial governments.

Finally, there are various types of activity that may need and warrant assistance but which are not eligible under the federal incentives program, which is concerned primarily with secondary manufacturing.

The Needs of Existing Business

Existing business in the Atlantic region should play a significant role in the development effort of the coming decade. Some firms are well equipped to play such a role and are already undertaking expansion or venturing into new product lines and new markets. Others need special help in overcoming markets. Others need special help in overcoming problems that hinder their efficiency and potential growth. Some of these problems stem basically from the small size of the firm; others stem from deficiencies in supporting institutions and services. Specifically, there is a widespread need for more and better long-term corporate planning.

The Council also recommends a Regional Management Institute, which should be of general benefit to business in the region in ensuring that its management training requirements are recognized and made available.[13] However, such changes will not solve all the problems faced by business in the region. Indeed there is evidence that in many cases small businesses are unable to make full and effective use of existing services.

In this connection, mention should be made of the Management Engineering Services Program of the Nova Scotia government. This program relies on the voluntary participation and co-operation of businesses in a program of problem diagnosis and therapy. The approach adopted by Nova Scotia is promising and recommends itself

438

as a very practical approach toward helping small firms meet and solve their varied problems.

Employment Targets in Manufacturing

In the period 1971-81 the rate of economic growth and the level of employment in the Atlantic Provinces will be determined primarily by the growth of manufacturing. This is implicit in the policies of the Department of Regional Economic Expansion that relate most directly to the manufacturing sector.

During the decade of the 1960's, the annual average increase in the region's labour force in manufacturing was about 2,300. This was during a period when the development effort was almost completely unco-ordinated, when there was the converse of a growth centre policy, little inter-provincial co-operation, and practically no systematic industrial promotion and intelligence. Moreover, the financial incentives presently available to industries expanding or locating in the Atlantic Provinces are far stronger than those prevailing in the 1960's under the Area Development Incentives Act. Accordingly a target of 50,000 jobs in manufacturing for the decade of the 1970's, roughly double the number actually achieved in the 1960's, appears to the Council to be capable of attainment. Indeed, the success of the Multiple Industry Complex in Saint John could provide, in itself, over a quarter of this total.

With an increase in manufacturing employment in the order of 50,000, the region will have no difficulty generating an accompanying increase in employment and in income in the other sectors—particularly services—to carry it to a high level of overall performance.

Economic and Physical Planning

The proposed strategy for economic development entails a greater degree of urbanization, a growth centre approach that seeks to build on strength, and special emphasis on secondary manufacturing. A well co-ordinated approach to economic and physical planning will be necessary to ensure that the problems of pollution, congestion and haphazard urban sprawl, which beset many major industrial areas on the North American continent and elsewhere, are anticipated and prevented.

The Provinces should give high priority to efforts to bring present problems of pollution, both municipal and industrial, under control. For the future, they must exercise rigorous control to ensure that future developments incorporate all necessary safeguards for the protection of the air, water and land environment.

Conclusion

An industrial research and promotion program will only be successful if it changes significantly the structural pattern of investment in the Atlantic region. Inevitably this will also entail changes in the distribution of population and activity and in the institutions of the region. Programs of man-power training and mobility must be co-ordinated in both timing and kind with the industrial development program. Finally, the industrial development program must be kept under constant review to determine whether it is providing the number of new jobs necessary to achieve the overall development objectives. The program has not yet been adequately tested under conditions that can determine conclusively its effectiveness. It may be that changes will be necessary. For example, it may prove necessary to provide special incentives over and above those now available for key high-technology industry in the main centres.

The Council realizes that some residents in the region see no need for further economic development, particularly if it involves greater industrialization. However, there are far too many people in the region for whom circumstances and conditions do not offer any reasonable opportunity for a satisfying and rewarding life in which they can make a worthwhile contribution to their society. While it is true that the process of industrialization poses hazards, the alternative is the continued out-migration of people who would not leave if given the opportunity to work in their own region, and the wastage of lives unfulfilled in jobs where they are underemployed.

REFERENCES

[1] *Report of the Royal Commission on Dominion-Provincial Relations*. Ottawa: Queen's Printer, 1954, p. 23.

[2] Copes, P., *St. John's and Newfoundland — An Economic Survey*. Sponsored by the Newfoundland Board of Trade, St. John's, 1961, p. 125.

[3] Atlantic Provinces Economic Council, *Second Annual Review, The Atlantic Economy*. Halifax, 1968.

[4] This section is based largely on an address "Atlantic Canada — the 1970's Opportunities and Challenges," by George E. McClure to the Annual Conference of the Atlantic Provinces Economic Council, Halifax, Nova Scotia, November 9, 1970.

[5] Such an examination was conducted by Joseph C. Mills in 1965 and is contained in an unpublished report to the Atlantic Development Board, *Industrial Location with Special Reference to the Atlantic Provinces*.

[6] Tom Kent, Deputy Minister of Regional Economic Expansion, "Notes for an Address to the Annual Meeting of the Atlantic Provinces Economic Council," Halifax, Nova Scotia, October 27, 1969.

[7] Larsen, H. K., *A Study of the Economic Impact Generated by ADA-Assisted Manufacturing Plants Located in the Province of New Brunswick*. University of New Brunswick, Fredericton, March, 1969.

[8] Comeau, Robert L., *A Study of the Impact of the Area Development Agency Program in Nova Scotia, a report submitted to the Area Development Agency*. Dalhousie University, Halifax, August, 1969.

[9] Servan-Schreiber, J. J., *The American Challenge*. New York: Athenium, 1968.

[10] Isard, W., Schooler, E. W., and Vietorisz, T., *Industrial Complex Analysis and Regional Development*. New York: Wiley, 1959; W. Isard and Associates, *Methods of Regional Analysis: An Introduction to Regional Science*. New York: Wiley, 1960, Chapter 9, page 375.

[11] In addition, the following areas in the Province of Newfoundland and Labrador, which were of a special nature in relation to the resettlement program, were designated Special Areas: the part of the Burin Peninsula that includes Fortune, Grand Bank, Garnish, St. Lawrence, Burin and Marystown; the area of Grand Falls, Botwood, Lewisporte and Gander; the area of Stephenville and St. George's; the area of Hawke's Bay, Port Saunders and Port-au-Choix; the area of Come by Chance, Arnold's Cove and Goobies and the Happy Valley area in Labrador.
The operations of the Cape Breton Development Corporation are, of course, of special relevance to the Sydney, Nova Scotia, area.
In Prince Edward Island the main special instrument of development is the Development Plan for Prince Edward Island under the agreement of March 1969 between Canada and the Province.

[12] News Release, Department of Regional Economic Expansion, December 1970.

[13] On November 23, 1970, the Council sponsored an Atlantic Management Development Conference at Halifax, N.S. The Conference was attended by representatives of business, the universities and government. A resolution for the establishment of a Management Institute for the Atlantic Provinces was passed and a Founding Committee to establish terms of reference named.